Lecture Notes in Artificial Intelligence   8686

Subseries of Lecture Notes in Computer Science

LNAI Series Editors

Randy Goebel
University of Alberta, Edmonton, Canada
Yuzuru Tanaka
Hokkaido University, Sapporo, Japan
Wolfgang Wahlster
DFKI and Saarland University, Saarbrücken, Germany

LNAI Founding Series Editor

Joerg Siekmann
DFKI and Saarland University, Saarbrücken, Germany

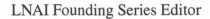

Adam Przepiórkowski
Maciej Ogrodniczuk (Eds.)

Advances in Natural Language Processing

9th International Conference on NLP, PolTAL 2014
Warsaw, Poland, September 17-19, 2014
Proceedings

 Springer

Volume Editors

Adam Przepiórkowski
Institute of Computer Science
Polish Academy of Sciences
ul. Jana Kazimierza 5
01-248 Warsaw, Poland
E-mail: adamp@ipipan.waw.pl

Maciej Ogrodniczuk
Institute of Computer Science
Polish Academy of Sciences
ul. Jana Kazimierza 5
01-248 Warsaw, Poland
E-mail: maciej.ogrodniczuk@gmail.com

ISSN 0302-9743   e-ISSN 1611-3349
ISBN 978-3-319-10887-2   e-ISBN 978-3-319-10888-9
DOI 10.1007/978-3-319-10888-9
Springer Cham Heidelberg New York Dordrecht London

Library of Congress Control Number: 2014946924

LNCS Sublibrary: SL 7 – Artificial Intelligence

Typesetting: Camera-ready by author, data conversion by Scientific Publishing Services, Chennai, India

Printed on acid-free paper

Springer is part of Springer Science+Business Media (www.springer.com)

Preface

PolTAL 2014 was the 9th International Conference on Natural Language Processing (NLP) in the *TAL series, following JapTAL 2012 (Kanazawa, Japan), IceTAL 2010 (Reykjavík, Iceland), GoTAL 2008 (Gothenburg, Sweden), FinTAL 2006 (Turku, Finland), EsTAL 2004 (Alicante, Spain), PorTAL 2002 (Faro, Portugal), VexTAL 1999 (Venice, Italy), and FracTAL 1997 (Besançon, France). The main purpose of the *TAL conference series is to bring together scientists representing linguistics, computer science, and related fields, who share a common interest in the advancement of computational linguistics and natural language processing. This purpose was amply fulfilled at PolTAL 2014: The topics of the 47 articles in these proceedings range from theoretical linguistic considerations in computational linguistics to hardcore machine learning approaches to NLP, with some papers combining the two. 27 of these papers were accepted as long papers (talks, including three alternates), with additional 20 accepted as short papers (posters). In total, we received 83 submissions.

Additionally, three distinguished keynote speakers accepted invitations to PolTAL 2014: Johan Bos (University of Groningen, The Netherlands), Ann Copestake (University of Cambridge, UK), and Mark Steedman (University of Edinburgh, UK). This is not reflected in these proceedings, as their presentations are based on material already published or submitted elsewhere.

The papers in these proceedings are grouped into six general areas: (1) morphology, named entity recognition, term extraction; (2) lexical semantics; (3) sentence-level syntax, semantics, and machine translation; (4) discourse, coreference resolution, automatic summarization, and question answering; (5) text classification, information extraction and information retrieval; and (6) speech processing, language modelling, and spell- and grammar-checking. Like many such classifications, this one also overtly simplifies a much more complex reality: Many papers touch on a number of these areas, and some could be equally well assigned to two classes. For example, the paper on "Semantic Clustering of Relations Between Named Entities" in area 5 also fits area 1, and "Cross-Lingual Semantic Similarity Measure for Comparable Articles" in area 2 could have been classified into area 5.

When editing the proceedings, we insisted on reasonable typographic and linguistic quality of all papers, which often required intensive interaction with authors. The eight papers submitted as Word documents were edited – by Maciej Ogrodniczuk, and the other LATEX submissions by Adam Przepiórkowski. We made some effort to normalize various conventions within each format and – to the extent possible – across formats, but at the point of sending the files to the publisher some differences between the two formats remained.

We would like to use this occasion to thank: Sylviane Cardey and the rest of the *TAL Steering Committee for selecting us as the organizers of this *TAL event; the Institute of Computer Science of the Polish Academy of Sciences and the PolTAL 2014 Organizing Committee for their help in making this conference a success; and, last but not least, the Program Committee – which had to be substantially extended as more submissions were received than anticipated – for their hard work that resulted in the high quality of this volume.

July 2014   Adam Przepiórkowski
                                                       Maciej Ogrodniczuk

Organization

PolTAL 2014 was organized by the Institute of Computer Science, Polish Academy of Sciences in Warsaw, Poland. The website of the conference may be found at: http://poltal.ipipan.waw.pl/.

Program Committee

Željko Agić	University of Potsdam, Germany
Tilman Becker	DFKI, Germany
Chris Biemann	Technische Universität Darmstadt, Germany
Igor Boguslavsky	Universidad Politécnica de Madrid, Spain, and Institute for Information Transmission Problems, Russian Academy of Sciences, Russia
Lars Borin	University of Gothenburg, Sweden
Johan Bos	University of Groningen, The Netherlands
António Branco	University of Lisbon, Portugal
Caroline Brun	Xerox Corporation, France
Miriam Butt	University of Konstanz, Germany
Nick Campbell	Trinity College Dublin, Ireland
Sylviane Cardey	University of Franche-Comté, France
Özlem Çetinoğlu	Universität Stuttgart, Germany
Robin Cooper	University of Gothenburg, Sweden
Ann Copestake	University of Cambridge, UK
Dan Cristea	University of Iasi, Romania
Walter Daelemans	University of Antwerp, Belgium
Markus Dickinson	Indiana University, USA
Christiane Fellbaum	Princeton University, USA
Dan Flickinger	Stanford University, USA
Mikel L. Forcada	University of Alicante, Spain
Radovan Garabík	Ľ. Štúr Institute of Linguistics, Slovak Academy of Sciences, Slovakia
Filip Ginter	University of Turku, Finland
Jonathan Ginzburg	Université Paris Diderot, France
Elżbieta Hajnicz	Institute of Computer Science PAS, Poland
Thomas Hain	University of Sheffield, UK
Aleš Horák	Masaryk University, Czech Republic
Krzysztof Jassem	Adam Mickiewicz University, Poland
Janne Bondi Johannessen	University of Oslo, Norway

Heiki-Jaan Kaalep	University of Tartu, Estonia
Laura Kallmeyer	Heinrich-Heine-Universität Düsseldorf, Germany
Ron Kaplan	Nuance Communications, USA
Łukasz Kobyliński	Institute of Computer Science PAS, Poland
Valia Kordoni	Humboldt University, Germany
Kimmo Koskenniemi	University of Helsinki, Finland
Sandra Kübler	Indiana University, USA
Krister Lindén	University of Helsinki, Finland
Markéta Lopatková	Charles University in Prague, Czech Republic
Krzysztof Marasek	Polish-Japanese Institute of Information Technology, Poland
Małgorzata Marciniak	Institute of Computer Science PAS, Poland
Agnieszka Mykowiecka	Institute of Computer Science PAS, Poland
Géza Németh	Budapest University of Technology and Economics, Hungary
Joakim Nivre	Uppsala University, Sweden
Pierre Nugues	University of Lund, Sweden
Tomasz Obrębski	Adam Mickiewicz University, Poland
Maciej Ogrodniczuk	Institute of Computer Science PAS, Poland
Guy Perrier	Inria Lorraine, France
Piotr Pęzik	University of Łódź, Poland
Maciej Piasecki	Wrocław University of Technology, Poland
Adam Przepiórkowski (Chair)	Institute of Computer Science PAS, Poland
Aarne Ranta	Chalmers University and University of Gothenburg, Sweden
Benoît Sagot	Inria, France
Agata Savary	Université François Rabelais Tours, France
Sabine Schulte im Walde	University of Stuttgart, Germany
Koenraad de Smedt	University of Bergen, Norway
Jan Šnajder	University of Zagreb, Croatia
Mark Steedman	University of Edinburgh, UK
Sebastian Stüker	Karlsruhe Institute of Technology, Germany
Stan Szpakowicz	University of Ottawa, Canada
Izabella Thomas	University of Franche-Comté, France
Reut Tsarfaty	Weizmann Institute of Science, Israel
Veronika Vincze	University of Szeged, Hungary
Martin Volk	University of Zurich, Switzerland
Aleksander Wawer	Institute of Computer Science PAS, Poland

Additional Reviewers

Peter Exner	Lund University, Sweden
Annette Rios Gonzales	University of Zurich, Switzerland

Filip Graliński	Adam Mickiewicz University, Poland
Rafał Jaworski	Adam Mickiewicz University, Poland
Marcus Klang	Lund University, Sweden
Prasanth Kolachina	University of Gothenburg, Sweden
Vojtech Kovar	Masaryk University, Czech Republic
Magdalena Plamada	University of Zurich, Switzerland
Jan Rygl	Masaryk University, Czech Republic
Vit Suchomel	Masaryk University, Czech Republic
Jakub Waszczuk	François Rabelais University of Tours, France

Organizing Committee

Michał Ciesiołka	Institute of Computer Science PAS, Poland
Konrad Gołuchowski	Institute of Computer Science PAS, Poland
Katarzyna Krasnowska	Institute of Computer Science PAS, Poland
Mateusz Kopeć	Institute of Computer Science PAS, Poland
Maciej Ogrodniczuk (Co-chair)	Institute of Computer Science PAS, Poland
Agnieszka Patejuk	Institute of Computer Science PAS, Poland
Adam Przepiórkowski (Co-chair)	Institute of Computer Science PAS, Poland
Piotr Przybyła	Institute of Computer Science PAS, Poland
Alina Wróblewska	Institute of Computer Science PAS, Poland

Table of Contents

Morphology, Named Entity Recognition, Term Extraction

Lexical Semantics

Sentence-Level Syntax, Semantics and Machine Translation

Discourse, Coreference Resolution, Summarisation, Question Answering

Text Classification, Information Extraction, Information Retrieval

Speech Processing, Language Modelling, Spell- and Grammar-Checking

Development of Amharic Morphological Analyzer Using Memory-Based Learning

Mesfin Abate and Yaregal Assabie

Department of Computer Science, Addis Ababa University, Ethiopia
mesfin.abatey@gmail.com, yaregal.assabie@aau.edu.et

Abstract. Morphological analysis of highly inflected languages like Amharic is a non-trivial task because of the complexity of the morphology. In this paper, we propose a supervised data-driven experimental approach to develop Amharic morphological analyzer. We use a memory-based supervised machine learning method which extrapolates new unseen classes based on previous examples in memory. We treat morphological analysis as a classification task which retrieves the grammatical functions and properties of morphologically inflected words. As the task is geared towards analyzing the vowelled inflected Amharic words with their grammatical functions of morphemes, the morphological structure of words and the way how they are represented in memory-based learning is exhaustively investigated. The performance of the model is evaluated using 10-fold cross-validation with IB1 and IGtree algorithms resulting in the over all accuracy of 93.6% and 82.3%, respectively.

Keywords: Amharic morphology, memory-based learning, morphological analysis.

1 Introduction

Morphological analysis helps to find the minimal units of a word which holds linguistic information for further processing. Morphological analysis plays a critical role in the development of natural language processing (NLP) applications. In most practical language technology applications, morphological analysis is used to perform lemmatization in which words can be segmented into its minimal meaning [11]. In morphologically complex languages, morphological analysis is also a core component in information retrieval, text summarization, question answering, machine translation, etc. There are two broad categories of approaches in computational morphology: rule-based and corpus-based. Currently, the most widely applied rule-based approach to computational morphology uses the two-level formalism. In rule-based approach, the formulation of rules for languages makes the development of morphological analysis system costly and time consuming [4,11]. Because of a need of hand-crafted rules for the morphology of languages and intensive requirements of linguistic experts in rule-based approaches, there is considerable interest in robust machine learning approaches to morphology which extracts linguistic knowledge automatically from an annotated or

A. Przepiórkowski and M. Ogrodniczuk (Eds.): PolTAL 2014, LNAI 8686, pp. 1–13, 2014.

unannotated corpus. Machine learning approaches have two learning paradigm: unsupervised and supervised learning. Supervised approach learns by example whereas unsupervised approach is learning by patterns. Machine learning approaches that use supervised learning paradigm include inductive logic programming (ILP), support vector machine (SVM), hidden Markov model (HMM) and memory-based learning (MBL). These paradigms have been used to implement low-level linguistic analysis such as morphological analysis [2,3,7]. Among various alternatives, the choice of the approach depends on the problem at hand. In this work, we employed MBL to develop morphological analyzer for Amharic, partly motivated by the limitations of previous attempts using rule-based [7] and ILP [10] approaches. Memory-based learning has a promising feature in analyzing NLP tasks like part-of-speech tagging, text translation, chunking and morphophonology due to its capabilities of in-cremental learning from examples. Among the MBL algorithms, IB1 and IGtree are known to be popular. Both algorithms rely on the k nearest neighbor classifier which uses some distance metric to measure the distance between each neighbor of features [4,5,9].

The remaining part of the paper is organized as follows. Section 2 presents the characteristics of Amharic language with special emphasis on its morphology. In Sect. 3, we present the proposed system for morphological analysis. Section 4 presents experimental results, and conclusion and future works are highlighted in Sect. 5. References are provided at the end.

2   Characteristics of Amharic Language

2.1   The Amharic Language

Amharic is an official working language of Ethiopia and it is widely spoken throughout the country as a first and a second language. It is a Semitic language related to Hebrew, Arabic and Aramaic. Amharic is the second most widely spoken Semitic language, next to Arabic. It uses a unique script called 'fidel' which is conveniently written in a tabular format of seven columns. The first column represents the basic form and the other orders are derived from it by more or less regular modifications indicating the different vowels. Amharic has 34 base characters and this leads to have a total of 238 (=34*7) Amharic characters. In addition, there are about two scores of characters representing labialized sounds.

2.2   Amharic Morphology

Like other Semitic languages, Amharic is one of the most morphologically complex languages. It exhibits a root-pattern morphological phenomenon [1]. Root is a set of consonants (also called radicals) which has a basic lexical meaning [12]. A pattern consists of a set of vowels which are inserted among the consonants of a root to form a stem. Semitic languages, particularly Amharic verbal stems, consist of a '*root + vowels + template*' merger. For instance, the root verb *sbr* + *ee* + *CVCVC* leads to form the stem *seber* ('broke'). In addition to such non-concatenative morphological features, Amharic uses different affixes to create inflectional and derivational morpheme. Affixation can be prefix, infix, suffix and circumfix. The morphological complexity of the language is better understood by looking at the ford formation process through inflection and derivation.

Amharic nouns are inflected for number, definiteness, cases (accusative/ objective, possessive/genitive) and gender. Amharic adjectives, in a similar affixation process to that of nouns, can be marked for number, definiteness, cases and gender. The affixation of morphemes to express numbers is similar with nouns except with some plural formation. On the other hand, Amharic verbs are inflected for any combinations of person, gender, number, case, tense/aspect and mood. As a result of this, tens of thousands of verbs (in surface forms) are generated from a single verbal root. As verbs are marked for various grammatical units, a single verb can form a complete sentence as shown in the example *yisebreñal* ('he will break me'). This verb (sentence) is analyzed as follows.

> verbal root: *sbr* ('to break')
> verbal stem: *sebr* ('will break')
> subject: *yi...al* (he)
> object: *eñ* (me)

Amharic nouns can be derived from adjectives, verbal roots (by inserting vowels between consonants), stems, stem-like verbs and nouns themselves. Few primary adjectives (which are not derived) exist in the language. However, many adjectives can be derived from nouns, stems, compound words and verbal roots. Adjectives can also be derived either from roots by intercalation of vocalic elements or attaching a suffix to bound stems. Amharic verbs can also be derived from different verbal stems in many ways.

3   The Proposed Amharic Morphological Analyzer

3.1   System Architecture

As memory-based learning is a machine learning approach, our morphological analyzer contains a training phase which consists of morpheme annotation to manually annotate inflected Amharic words, feature extraction to create instances in a fixed length of windows, parameter optimization and algorithm selection to tune and select some of the parameters and algorithms. On the other hand, the morphological analysis component contains the feature extraction to de-construct a given text, morpheme identification to classify and extrapolate, stem and root extraction to label segmented inflected words with their morpheme functions. The architecture of the proposed Amharic morphological analyzer is depicted in Fig. 1.

3.2   Training Phase

The training process requires sample patterns of words showing the changes in the internal structures of words. Amharic morphemes may predominantly be

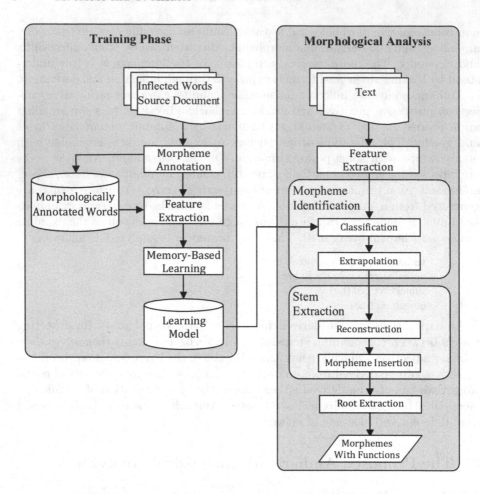

Fig. 1. Architecture of the proposed Amharic morphological analyzer

expressed by internal phonological changes in the root. These internal irregular changes of phonemes make the morphological analysis cumbersome. It is not a trivial task in finding the roots of Amharic verbs. Hence, we investigated the morphological formation of Amharic language, particularly nouns and verbs. Adjectives have similar derivation and inflection process to that of nouns. A morphological database is built after identifying the common property of all morphological formations of Amharic nouns (and adjectives) and grammatical features of all the morphemes. As to Amharic verbs, it is too difficult to find a single representation or patterns of verbs as they are different in types due to a number of morphological and phonological processes. Therefore, we consider the most significant part of the word stem which bears meaning next to the roots. In Amharic grammar, the stem of a word is the main part which remains unchanged when the ending changes. Thus, we manually annotate sample words with their patterns where the data will be used as training data.

Morpheme Annotation

Amharic nouns have more than 2 and 7 affixes in the prefix and suffix position, respectively. The affixation is not somehow arbitrary, rather they affix in ordered manner. An Amharic noun consists of a lexical part, or stem and one or more grammatical parts. This is easy to see with a noun, for example, the Amharic noun *bEtocacewn* ('their houses'). The lexical part is the stem *bEt* ('house'); this conveys most of the important content in the noun. Since the stem cannot be broken into smaller meaningful units [8], it is a morpheme (a primitive unit of meaning). The word contains three grammatical suffixes, each of which provides information that is more abstract and less crucial to the understanding of the word than the information provided by the stem: *-oc*, *-acew*, and *-n*. Each of these suffixes can be seen as providing a value for a particular grammatical feature (or dimension along which Amharic nouns can vary): *-oc* (plural marker), *-acew* (third person plural neuter), and *-n*: (accusative). Since each of these suffixes cannot be broken down further, they can be considered as a morpheme. Generally, these grammatical morphemes can have a great role in understanding the semantics of the whole word [7,12].

The following tasks were identified and performed to prepare annotated datasets used for training: *identifying inflected words*; *segmenting the word into prefix, stem, suffix*; *putting boundary marker between each segment*; and *describing the representation of each marker*. Morphemes that are attached next to the stem (as suffixes) may have seven purposes: *plurality/possessions*, *derivation*, *relativazation*, *definiteness*, *negation*, *causative* and *conjection*. The annotation is according to the prefix-stem-suffix ([P] [S] [S]) structure as shown in Table 1. The bracket ([ ]) can be filled with the appropriate grammatical features for each segmentation where *S*, *M*, *1*, *K*, *D*, and *O* indicate end of stem, plural, possession, preposition, derivative and object markers, respectively. Lexicons were prepared manually in such a way to be suitable for extraction purpose.

Amharic verbs have four slots for prefixes and four slots for suffixes [1,7,10]. The positions of the affixes are shown as follows, where *prep* is for preposition; *conj* is for conjunction; *rel* is for relativation; *neg* is for negation; *subj* is for subject; *appl* is for applicative; *obj* is for objective; *def* is for definiteness; and *acc* is for accusative.

Table 1. Example showing annotation of nouns

	Word form	Prefix	Stem	Suffix
በሰዎች	besewoc	be[K]	sew	-oc[M]
ደብተሮች	debteroc	-	debter	-oc[M]
በራዎች	berEwoc	-	berE	-woc[M]
ስለንበረታችን	slenbretacn	sle[K]	nbret	-acn[1]
ንብረታችንን	nbretacnn	-	nbret	-acn[1] -n[O]
ሰውነት	sewnet	-	sew	-net[D]
ሰውኛ	sewNa	-	sew	-eNa[D]

> *(prep|conj)(rel)(neg) subj* **STEM** *subj (appl)(obj|def)(neg|aux|acc)(conj)*

In addition to analyzing all these affixes, the root template pattern of Amharic verbs makes its morphological analysis complex. It is a challenging task representing its features into suitable memory-based learning approach. Generally, Amharic verb stems are broken into verb roots and grammatical templates. A given root can be combined with more than 40 templates [1]. The stem is the lexical part of the verb and also the source of most of its complexity. To consider all morphologically productive of the verb types, we need a morphologically annotated word list with its possible inflection forms. Then, the tokens are manually annotated in similar fashion what we did for nouns and adjectives like prefix[], stem[] and suffix[] pattern. The '[]' can be filled with the appropriate grammatical features for each segmentation. The sample annotation for verbs is shown in Table 2.

Table 2. Example showing annotation of nouns

		lemed	[V]				
		lemed	[V]	e	[6]		
		lemed	[V]	achu	[5]		
as	[A]	lemed	[V]	ec	[9]		
sle	[K]	lemed	[V]	ec	[9]		
		seber	[V]	e	[6]	n	[O]
as	[A]	seber	[V]	e	[6]	at	[O]
al	[G]	seber	[V]	ec	[9]	m	[G]

Feature Extraction

Once the annotated words are stored in a database, instances are extracted automatically from the morphological database based on the concept of windowing method [3] in a fixed length of left and right context. Each instance is associated with a class. The class represents the morphological category in which the given word posses. An instance usually consists of a vector of fixed length. The vector is built up of n feature value pairs depending on the length of the vector. Each example focuses on one letter, and includes a fixed number of left and right neighbor letters using 8-1 to 8-1 windows which yields eighteen features. The largest word length from the manually annotated data base is chosen to be the length of windows size. The input character in focus, plus the eight preceding and eight following characters are placed in the windows. Character based analysis gives concern for each character or letter to be considered. From the basic annotation, instances were automatically extracted, to be suitable to memory-based learning by sliding a window over the word in the lexicon. We used the Algorithm 1 to extract feature based on character analysis.

Input: Inflected words
Output: extracted features (instances) in a fixed-length of vector size

1. Define the length of window size.
2. Fix the middle positions of arrays as a focus letter (the focus character represents where a character is started from that position on words).
3. Read from the DB and push one step forward each character until the right context reached (filled).
4. Put 0(zero) at the class if there is no any special character like @, & and capital letters, next to the characters placed in the focus letter; if any one of those symbols exist put the value as a class(in last index)
5. Push the previous focus letter to the left and start putting each letter (as in step 3)
6. Go until it finishes that line
7. Go to the next line and repeat 3, 4, 5, 6.

Algorithm 1. Algorithm for character-based feature extraction.

For instance, the character based representation of the word *sleseberecw* is shown in Table 3. The '=' sign is used as a filter symbol which shows there is no character at that position. The construction of instances displays the 11 instances derived from the Amharic word and its associated classes. The class of the third instance is 'K' representing the preposition morpheme 'sle' ending with the prefix 'e'. Therefore, character based representation of words exhaustively transcribes their deep structure of phonological process and segments each character one at a time.

Table 3. Character-based feature extraction of the word *sleseberecw*

No	Left context								Focus	Right context								Class
1	=	=	=	=	=	=	=	=	s	l	e	s	e	b	e	r	e	0
2	=	=	=	=	=	=	=	s	l	e	s	e	b	e	r	e	c	0
3	=	=	=	=	=	=	s	l	e	s	e	b	e	r	e	c	w	K
4	=	=	=	=	=	s	l	e	s	e	b	e	r	e	c	w	=	0
5	=	=	=	=	s	l	e	s	e	b	e	r	e	c	w	=	=	0
6	=	=	=	s	l	e	s	e	b	e	r	e	c	w	=	=	=	0
7	=	=	s	l	e	s	e	b	e	r	e	c	w	=	=	=	=	0
8	=	s	l	e	s	e	b	e	r	e	c	w	=	=	=	=	=	V
9	s	l	e	s	e	b	e	r	e	c	w	=	=	=	=	=	=	0
10	l	e	s	e	b	e	r	e	c	w	=	=	=	=	=	=	=	9
11	e	s	e	b	e	r	e	c	w	=	=	=	=	=	=	=	=	F

Memory-Based Learning

Memory-based approaches borrow some of the advantages of both probabilistic and knowledge-based methods to successfully implement it in NLP tasks [5]. It performs classification by analogy. In order to learn any NLP classification problem, different algorithms and concepts are implemented by reusing data structures. We used TiMBL as a learning tool for our task [3] . There are a number of parameters to be tuned in memory-based learning using TiMBL. Therefore, to get an optimal accuracy of the model we used the default settings and also tuned some of the parameters. The optimized parameters are the MVDM (modified value difference metric) and chi-square from distance metrics, IG (information gain) from weighting metrics, ID (inverse distance) from class voting weights, and k from the nearest neighbor. These optimized parameters are used together with the different classifiers. The classifier engines we used are IGtree and IB1 which construct databases of instances in memory during the learning process. The procedure of building an IGtree is described in [6]. Instances are classified by IGTree or by IBl by matching them to all instances in the instance base. As a result of this process, we get a memory-based learning model which will be used later during the morphological analysis phase.

3.3   Morphological Analysis

The training phase is the backbone of the morphological analysis module to success-fully implement the system. The morphological analysis is implemented by using the memory-based learning model. Therefore, in this phase, the feature extraction is used to make the input words to be suitable for memory-based learning classification, the morpheme identification is applied to classify and extrapolate the class of new instances, the stem extraction process reconstructs and inserts identified morphemes, and finally the root extraction is used to get root forms and stems with their grammatical functions.

Feature Extraction

Memory-based learning learns new instances by storing previous training data into memory. When a new word is given to be analyzed by the system, it accepts and de-construct as instances to make similar representation with the one stored in memory. Feature extraction in this section is different from the one described in the training phase. The word is deconstructed in a fixed-length of instances without listing (identifying) the class labels at the last index. For example, when a new previously unseen word (which is not found in the memory) needs to be segmented, the words are similarly deconstructed and represented as instances using the same information. This instance is compared to each and every instance in the training set, recorded by the memory-based learner. In doing so, the classifier will try to find training instance in memory that most closely resembles it. For instance, the word *begoc* is segmented and its features are extracted as shown in Fig. 2.

1	2	3	4	5	6	7	8	F	1	2	3	4	5	6	7	8	Class
=,	=,	=,	=,	=,	=,	b,	e,	g,	o,	c,	=,	=,	=,	=,	=,	=,	?

Fig. 2. Feature extraction for morphological analysis

Morpheme Identification

When new or unknown inflected words are deconstructed as instances and given to the system to be analyzed, an extrapolation is performed to assign the most likely neighborhood class with its morphemes based on their boundaries. The extrapolation is based on the similarity metric applied on the training data. If there is an exact match on the memory, the classifier returns (extrapolates) the class of that instance to the new instance. Otherwise, new instance is classified by analogy in memory with a similar feature vector, and extrapolating a decision from their class. This instance is compared to each and every instance in the training set, recorded by the memory-based learner. In doing so, the classifier tries to find that training instance in memory that most closely resembles it. Taking the feature of *lenegerecw* as shown in Fig. 3, this might be instance 10 in Table 3, as they share almost all features (**L8**, **L7**, **L5**, **L3-L1**, **F**, **R1-R8**), except **L6** and **L4**. In this case, the memory-based learner then extrapolates the 9 classes of this training instance and predicts it to be the class of the new instance.

8	7	6	5	4	3	2	1	F	1	2	3	4	5	6	7	8	class
l,	e,	n,	e,	g,	e,	r,	e,	c,	w,	=,	=,	=,	=,	=,	=,	=,	?

Fig. 3. Instances for the unknown token *lenegerecw*

Stem Extraction

After appropriate morphemes are identified, the next step is the stem extraction process. In stem extraction, reconstruction of individual instances into meaningful morphemes (to their original word form) and insertions of identified morphemes in their segmentation point are performed. After stem extraction, the system searches resembling instances from previously stored patterns in memory. If there is no similar instance in memory, it uses a distance similarity matrix to find more nearest neighbor. The modified value difference metric (MVDM) which looks for the co-occurrence of the values with the target classes is used to determine the similarity of the value of features. For example, the reconstruction of the whole instances of the word *slenegerecw* is shown in Fig. 4. In the example, four non-null classes are predicted in the classification step. In the second

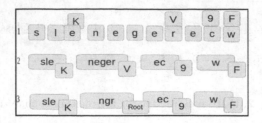

Fig. 4. Reconstruction of the word *slenegerecw*

step the letter of the morphemic segments are concatenated and morphemes are inserted. Then, root extraction can be performed in the third step.

Root Extraction

The smallest unit morpheme for nouns and adjectives is the stem. Thus, the root extraction process will not be applied on nouns and adjectives. Root extraction in verbal stems is not complex task in Amharic as roots are consonants of verbal stems. In order to extract the root from verbal stems, we simply remove the vowels from verbal stems. However, there are exceptions as vowels in some verbal stems (e.g. when the verbal stems start with vowels) serve as consonants. In addition, vowels should not be removed from mono and bi-radical verb types since they have valid meaning when they end with vowels.

4   Experiment

4.1   The Corpus

In order to evaluate the performance of the model and the capability of learnability of the dataset we conducted the experiment by combining nouns and verbs. To get unbiased estimate of the accuracy of a model learned through machine learning, it should be tested on unseen data which is not present in the training set. Therefore, we split our data set into training and testing. The total number of our corpus contains 1022 words, of which 841 are verbs and 181 are nouns (adjectives are considered as nouns as they have similar analysis). The number of instances extracted from nouns and adjectives are 1356 and from verbs are 6719 which accounts a total of 8075 instances. A total of 26 different class labels occur within these instances.

4.2   Test Results

As discussed in Sect. 3.2, we used TiMBL as a learning tool for Amharic morphological analysis. We also applied IGtree and IB1 algorithms to construct databases of instances in memory during the learning process. To get an optimal accuracy of the model we tuned some of the parameters. The optimized parameters are the modified value difference metric and chi-square from distance metrics, information gain from weighting metrics, inverse distance from class voting weights, and k from the nearest neighbor. For various combinations of parameter values, we tune the parameters until no better result is found.

Simply splitting the corpus into a single training and testing set may not give the best estimate of the system's performance. Thus, we used *10-fold* cross-validation technique to test the performance of the system with IB1 and IGtree algorithms. This means that the data is split in ten equal partitions, and each of these is used once as test set, with the other nine as corresponding train set. This way, all examples are used at least once as a test item, while keeping training and test data carefully separated, and the memory-based classifier is trained each time on 90% of the available training data. We also used *leave-one-out* cross-validation for IB1 algorithm, which uses all available data except one (n-1) example as training material. It tests the classifier on the one held-out example by repeating it for all examples. However, we found it tame consuming to use *leave-one-out* cross-validation for IGtree algorithm. Table 4 shows the performance of the system for optimized parameters.

Table 4. Test result for Amharic morphological analysis

Evaluation method	Type of algorithm	Time taken (in seconds)	Space requirement (in bytes)	Accuracy (%)
Leave-out-one	IB1	30.99000	1,327,460	96.40
10-fold	IB1	0.82077	1,213,656	93.59
	IGtree	0.03711	1,136,582	82.26

In memory-based learning the minimum size of the training set to begin with is not yet specified. However, the size of the training data matters the learning performance of the algorithm. Hence, it is crucial to draw learning curves in addition to reporting the experimental results. We perform a series of experiments by systematically increasing amounts of training data up to the currently available total dataset which is 1022. When drawing a learning curve, in most cases, the learning can be measured by fixing the number of test data against which the performance is measured. The learning curve of the system is shown in Fig. 5.

As compared to previous works, our system performed well and provided promising results. For example, in the work of Gasser [7], the system (which is rule-based) does not consider unseen or unknown words. To overcome this problem, Mulugeta and Gasser [10] developed Amharic morphological analyzer using inductive logic programming. However, our system still performs better in terms of accuracy.

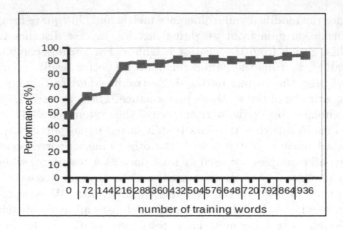

Fig. 5. Learning curve of the system

5 Conclusion and Future Work

Many high-level NLP applications heavily rely on a good morphological analyzer. Few attempts have been made so far to develop an efficient morphological analyzer for Amharic. However, due to the complexity of the inherent characteristics of the language, it was found to be difficult. This research work is also aimed at developing Amharic morphological analyzer using memory-based approach. Given the promising results, our work adds value in the overall effort to dealing with the complex problem of developing Amharic morphological analyzer. The performance of our system can be further enhanced by increasing the training data. Future work is recommended to be directed at looking into the morpheme segmentation on individual instances. Segmentation on the full words and insertions of grammatical features in each segmented morphemes is expected to boost the performance of the system.

References

1. Amsalu, S., Gibbon, D.: Finite state morphology of Amharic. In: Proc. of Inter. Conf. on Recent Advances in Natural Language Processing, Borovets, pp. 47–51 (2005)
2. Bosch, A., Busserand, B., Canisius, E., Daelemans, W.: An efficient memory-based morpho-syntactic tagger and parser for Dutch. In: Proc. of the 17th Meeting Comp. Ling. in the Netherlands, Leuven, Belgium (2007)
3. Bosch, A., Daelemans, W.: Memory-based morphological analysis. In: Proc. of the 37th Annual Meeting of the Association for Computational Linguistics, Stroudsburg (1999)
4. Clark, A.: Memory-Based Learning of Morphology with Stochastic Transducers. In: Proc. of the 40th Annual Meeting of the Assoc. for Comp. Ling., Philadelphia (2002)

5. Daelemans, W., Bosch, A.: Memory-Based Language Processing. Cambridge University Press, Cambridge (2009)
6. Daelemans, W., Bosch, A., Weijters, T.: IGTree: Using Trees for Compression and Classification in Lazy Learning Algorithms. Artificial Intelligence Review 11, 407–423 (1997)
7. Gasser, M.: HornMorpho: a system for morphological processing of Amharic, Oromo, and Tigrinya. In: Proc. of Conf. on Human Lang. Tech. for Dev., Egypt (2011)
8. Hammarstrom, H., Borin, L.: Unsupervised Learning of Morphology. Computational Linguistics 37(2), 309–350 (2011)
9. Marsi, E., Bosch, A., Soudi, A.: Memory-based morphological analysis generation and part-of-speech tagging of Arabic. In: Proc. of the ACL Workshop on Computational Approaches to Semitic Languages, pp. 1–8 (2005)
10. Mulugeta, W., Gasser, M.: Learning Morphological Rules for Amharic Verbs Using Inductive Logic Programming. In: Proc. of SALTMIL8/AfLaT (2012)
11. Pauw, G., Schryver, G.: Improving the Computational Morphological Analysis of a Swahili Corpus for Lexicographic Purposes. Lexikos 18, 303–318 (2008)
12. Yimam, B.: Ye'amarigna sewasew (Amharic Grammar). Eleni Printing Press, Addis Ababa (2000)

A Finite-State Treatment of Neoclassical Compounds in Modern Greek and English

Evanthia Petropoulou[1], Eleni Galiotou[2], and Angela Ralli[1]

[1] University of Patras, Department of Philology, Rio, Patras 26504, Greece
[2] Technological Educational Institute of Athens, Department of Informatics,
Ag. Spiridona Street, Aigaleo 122 10, Greece

Abstract. This paper presents an attempt to process neoclassical compounds (NCs) in Modern Greek (MG) and English with the use of finite-state methods. Our approach is based on a theoretical background that applies to both languages and assumes that Modern Greek compounding has affected the formation of NCs in English. The processing system is proposed to be both linguistically accurate and computationally efficient.

Keywords: Neoclassical compounds, word formation, English, Modern Greek, finite-state methods.

1 Introduction

'Neoclassical compounds' is a common (often characterized as a misleading) term for describing a large number of words in many European languages, containing morphemes of Ancient Greek and Latin origin. Examples are *biography* (Eng.), *Astronomie* (Ger.), *elettromagnetico* (It.), *télescope* (Fr.) and many others. From a morphological point of view, these words pose a challenge for word-formation theories with respect to the morphological category of their constituents and the type of processes involved in their formation. From a computational point of view, their abundance in most scientific and technical vocabularies makes their processing an immediate necessity. Moreover, the fact that many of them appear as 'internationalisms' (Wexler 1969), namely as morpho-phonologically similar versions of words formed with classical elements in different languages (e.g. *biology* (Eng.), *Biologie* (Ger.), *biologia* (It.), βιολογία (*viologia*) (MG), makes their parallel processing interesting for reasons of computational economy. In the first part of this paper, we provide the theoretical background for the comparative analysis of such words in MG and English. In the second part, we present the formalism developed for their processing on the basis of finite-state methods, namely the Lexical Compiler (LEXC) by XEROX, applying the favoured morphological analysis discussed in the first part.

2 Theoretical Background

The dubious nature of the constituent elements of words that are commonly called NCs in both MG and English has been a major issue among linguists,

A. Przepiórkowski and M. Ogrodniczuk (Eds.): PolTAL 2014, LNAI 8686, pp. 14–26, 2014.

as it has further implications for the morphological processes involved in their formation and particularly in English for the word formation system itself. The principal reason for that is their bound nature, which has led to the formation of a number of different views about their status.

In MG, the words that could be characterized as NCs in fact belong to a major category of compounds, namely, synthetic compounds. While many of these compounds have been created in ancient times, the majority of them have been recently formed and continue to be formed in order to serve specific terminological needs [32, 34], as is the case with NCs in most languages. Some examples are: υδρογόνο (*idrogòno*) "hydrogen", βιογράφος (*viogràfos*) "biographer", μικροσκόπιο (*mikroskòpio*) "microscope", ακτινοβολία (*aktinovolìa*) "radiation". Although stem constituents of compounds are usually bound and may become free words through suffixation with an inflectional suffix[1] [30], the final elements of NCs in MG, remain bound even after the addition of an inflectional suffix: –γον(ος) (*-gonos*), –γραφ(ος) (*-grafos*), -σκόπ(ος) (*-skòpos*), –βόλ(ος) (*-vòlos*) [29, 30, 32, 34].

The main contradictory views that have been expressed about the nature of these final elements are the 'confix' and the 'bound stem' view. The 'confix' view, expressed by [16], is based on the fact that elements such as –γραφος (*-grafos*) in βιογράφος (*viogràfos*) "biographer" and –γραφία (*-grafìa*) in βιογραφία (*viografìa*) "biography", have gradually acquired suffixal characteristics and should therefore be categorized as a new type of morphemes, namely 'confixes' [23, 3]. Giannoulopoulou [16] characterizes as confixes both initial (e.g. ευρω– (*evro-*), νευ–(*neu-*), οικυ–(*iko-*), πολυ–(*poli-*)) and final elements (c.g. κτόνος (*-ktònos*), –κτονία (*-ktonìa*), –ποιώ (*-piò*), –ποίηση (*-pìisi*), –λόγος (*-lògos*), –λογία (*-logìa*)) and the process they participate in, namely 'confixation', preferably as a type of derivation rather than of compounding. This view considers all these different morphemes under the same category, despite the apparent differences they have by appearing in either initial or final position.

The contradictory view, namely the 'bound stem' view, expressed by Ralli [29, 30, 32–34], recognizes a number of important properties in these bound elements that appear in final position which weaken their suffixal character. Apart from the more concrete meaning they have, in relation to suffixes, they also serve as bases to prefixed words, such as υπόλογος (*ipòlogos*) "responsible for one's actions" and υπέρμαχος (*ipèrmachos*) "firm supporter", a property that belongs definitely to stems [32]. More important is the fact that the words they appear in have a compound structure, evident by both the linking element –o–, a compound marker in MG, and the recursion they often exhibit in their structures (e.g. [[ψυχ]–o–[γλωσσ–o–λόγος]] (*psixoglosològos*) "psycholinguist"). As Ralli [32] notes, many of them serve as bases for the derivation of nouns like βιογραφία (*viografìa*) "biography" with the addition of a suffix from a small set of suffixes (–ειαN (*-iaN*), –ίαN (*-ìaN*), –είοN (*-ioN*), –ιοN (*-ioN*), –ωV (*-oV*), –ικA (*-ikA*)). The structure she proposes for such compounds is [[stem]-o-[bound stem]](infl.

---

[1] E.g. νυχτο–o–λούλουδ(ο) (*niht-o-lùludo*) "nightflower" > νύχτ(α) (*nìhta*) "night", λουλούδ(ι) (*lulùdi*) "flower" [30].

suffix) and for their derivatives [[stem$_N$–o–[bound stem]$_N$]+derivational suffix [32, 33] respectively.

A similar debate about the nature of the bound elements has also been going on in English, with three different morphological classifications at play: 'afffixes', 'combining forms' and 'bound stems'. With the exception of the 'affix' view, which has early been defeated for a number of important reasons, such as the freedom of movement within a word, characterising some of these elements (e.g. *Francophile* vs. *philanthropic* [38]), the other two are equivalent to the contradictory views in MG. Bauer [7, 8] who, along others [1, 41, 10, 21, 27, 28], supports the 'combining form' view, classifies such elements depending on their position, as Initial (e.g, *astro–*, *electro–*, *hydro–*) or Final Combining Forms (e.g. *–crat*, *–naut*, *–phile*). This term is usually adopted in order to describe disputable elements that are difficult to appoint to one or the other category [8], like forms arising from blends, clippings etc. (e.g. *Euro–*, *–(a)holic*). The reason why bound elements of classical origin should also fit in this category is that some of them regularly appear not only in combinations with each other but also in combinations with native free forms, forming 'hybrid' formations [5] (e.g. *microcomputer*, *filmography*), thus resembling other combining forms. The 'combining form' view, however, does not account, in a satisfactory way, for the freedom of position that some elements of classical origin exhibit within a word.

The 'bound stem' view, supported by [38, 40, 2, 5, 29, 30, 18], favours the parallel co-existence of two word-formation systems in English [12], a native word-based one and a non-native stem-based one, the latter being responsible for the formation of words such as the NCs [18]. The most important assumptions of this view are i) the compound structure of NCs [38] and ii) the recognition of the intermediate *–o–* (or *–i–* in some cases) appearing in many of them, as a linking vowel, thus separating it from any preceding or subsequent element (e.g. **logo–*, **–ology*) [38] and allowing the participating members to move freely within a word (e.g. *histori–o–graphy* vs. *graph–o–logy*). Within this spirit, Baeskow [5] describes the prototype of neoclassical compounding as the combination of two roots of Greek or Latin origin, one of which may be free. If such a compound, which might be either free (e.g. *microscope, astronaut, phonoelectrocardioscope*) or bound (e.g. *geograph–, biolog–, hieroglyph–*) undergoes a derivational process, the result is a neoclassical compound derivative (e.g. *biolog– +y*, *geograph– #er*, *anthropomorh– +ic*, *astrolog– #er*). This view draws a clear parallel to the words that once served as models for the formation of NCs, namely Ancient Greek compounds like θεολόγος (*theològos*) "theologist", βιογραφία (*viografia*) "biography" and many others. While this fact is not generally denied, many analyses seem to ignore it to such an extent that no parallel can be drawn anymore between the structures of the two categories. And although elements of classical origin in English have moved out of the borders of the prototype of neoclassical compounds, by appearing in new contexts and adopting new roles, the prototype of neoclassical compounding seems to serve well for the majority of what we call NCs in English ([25]).

The advantages of the 'bound stem' view over the 'combining form' or 'confix' view, on both a theoretical and a computational level are quite significant. First, it allows for a clear-cut distinction between morphemes of classical origin and clipped forms or affixes. Second, the recognition of the intermediate *-o-* (or *-i-*) as a linking element and the establishment of the processes of neoclassical compounding and neoclassical compound derivation as valid word-formation processes, give rise to a list of bound stems of classical origin in the lexicon of both languages (e.g. *-crat-*, *-log-*, *-graph-*, *-scop-*, *-phil-*, *-path-* (Eng.), *-γον* (*-gon*), *-γραφ* (*-graf*), *-σκοπ* (*-skop*), *-βολ* (*-vol*) (MG)), as well as to a set of affixes (e.g. *-y*, *-er*, *-ic*, *-ous*, *-ia*, *-ist* (Eng.), *-ειaN* (*-ia*$_N$), *-ίa*$_N$ (*-ia*$_N$), *-είο*$_N$ (*-io*$_N$), *-ιο*$_N$) (*-io*$_N$), *-ων* (*-o*$_V$), *-ικ*$_A$ (*-ik*$_A$) (MG)). This is important on theoretical grounds, since under the 'combining form' view, elements such as *-scope, -scopy* and *-scopic* in English would merely be classified as distinct elements within the same class of items, with no apparent morphological relationship among them. Likewise in MG, words like *βιολόγος* (*viològos*) "biologist" and *βιολογία* (*viologìa*) "biology" would no longer be morphologically related, as they would result from separate processes of confixation (*βιο-* + *-λόγος, βιο-* + *-λογία*) [25]. The computational advantage of this view is the greater economy resulting from the "deep-level" morphological analysis imposed, both in the storage of the data (lists of bound stems and suffixes instead of non-related combining forms) and in the implementation of the word formation processes involved.

3   Morphological Processing of Neoclassical Compounds with the LEXC

3.1   Previous Work

The morphological processing of NCs in English and other languages mainly appears in information retrieval and term recognition, especially in medicine, biology and related sciences, the terminology of which consists of NCs to a large extent. Some important works on the morphological processing of medical terminology are [6, 39, 22, 24, 11]. In MG, the morphological processing of compounds has been implemented by [13–15] and [36], through PC-KIMMO, which however has proven inadequate in certain aspects, such as the implementation of stress and the specification of certain phonological rules.

3.2   The LEXC

The Lexical Compiler (LEXC), which has been used in the present work for the morphological processing of NCs in English and MG, is based on finite-state methods, one of the most interesting developments in the area of morphological processing both on a theoretical and a computational level.[2] Formally, the LEXC

---

[2] For more information on finite-state methods in morphological processing cf. [17, 9, 37] among others.

language is a kind of right-recursive phrase-structure grammar and its syntax is similar to that accepted by Koskenniemi's Two-Level formalism [19, 20] and PC-KIMMO (version 1) [4], but with a number of features that make it more suitable for the description of a language's morphotactics [9].[3]

A LEXC description compiles into a Xerox finite-state network, either a simple automaton or a transducer. It consists of an optional 'Multichar Symbols' declaration, which contains symbols that are necessary for the morphological characterization of morphemes in their 'lexical form' (e.g. +Noun, +Verb, +Sg, +3P), followed by one or more named lexicons. All morphemes (prefixes, roots, suffixes etc.) are organized into such lexicons. Each entry in the lexicon consists of two parts, a 'Form' and a 'Continuation Class'. Continuation classes, which are inherited from Koskenniemi's Two-Level Morphology notation, are the basic mechanism for describing morphotactics in LEXC as they translate into concatenation, which is indeed the way most languages primarily build words. In this way, morphemes are properly grouped under a single lexicon because they form a coherent *target* for continuations from other morphemes [9].

However, continuation classes have proved inadequate for describing certain morphotactic phenomena, such as separated dependencies, intedigitation, infixation and reduplication. Some of these problems can be solved with the use of morphotactic filters and alternation rules, which however, can sometimes cause an explosion in the size of the resulting transducer. An alternative to those are the Flag Diacritics (FDs), which have been broadly used in the descriptions presented here. FDs are meant to enforce separated constraints on the co-occurrence of morphemes within words, as well as to mark roots for idiosyncratic morphotactic behavior and generally to handle other constraints that are feature-based rather than phonological [9].

3.3   The Corpora

The LEXC descriptions for the morphological processing of neoclassical compounds in MG and English make use of two corpora with bound stems of classical origin that appear as final elements in NCs of both languages [26]. The English corpus consists of 56 bound stems of Ancient Greek origin and 4 of Latin origin (e.g. *–graph, –scop, –naut, –phon, –meter, –log, –morph, –vor*) that have been selected on the basis of frequency in use and productivity in the formation of new words. The MG corpus contains 45 bound stems of verbal origin. Both corpora also contain an extensive listing of example NCs formed with each bound stem, collected from dictionaries and corpora, along with their derivatives and the suffixes selected by each bound stem for the formation of derivatives.

3.4   The LEXC Descriptions

The descriptions in LEXC have been developed in a similar way for the two languages, as they are fully based on the adopted analyses for NCs in MG and

---

[3] For more information on the LEXC cf. Chap. 4 in [9].

English discussed earlier, which share the following characteristics: i) the compound structure, ii) the bound stem status of the constituents, iii) the recognition of the –o– as a linking element, and iv) the nominal characteristics of the bound stems. The most significant differences between the two descriptions stem from the idiosyncratic features of the two languages. The first is the rich inflection of MG contrary to the scarce inflection of English, and the second is the nature of the first constituents, which are word stems or prefixoids in MG and native words or bound stems of classical origin in their English counterparts. Here, the two descriptions are presented in parallel, so that differences and similarities between them are easier clarified.

First Elements. The lexicon 'Root', which signifies the start state of the finite-state network, contains three entries in MG, which lead to the sublexicons 'Prefixoids', 'WordStems' and 'BoundStems'. The first two concern the first elements of NCs and the last one their final constituents. The lexicon 'Prefixoid' on the one hand, contains a limited number of entries which appear as first constituents in compounds with bound stems and are not word stems, but rather pronouns (e.g. αλληλ– *allil*– "allel–", ετερ– *eter*– "heter–"), adverbs (e.g. τηλε– *tile*– "tele–") and other morphemes, collectively characterized as prefixoids. Their continuation class leads directly to the lexicon 'Compound', where the compounding process takes place. The lexicon 'WordStems' on the other hand, contains a selection of common word stems in MG,[4] which according to the corpus of NCs appear often as first constituents (e.g. αγγλ– *angl*– "English", γαστρ– *gastr*– "gastr–", καπν– *kapn*– "smoke"). Each word stem in this lexicon is assigned two entries, one leading to inflection and the other to compounding through the appropriate continuation class. Such an example pair of entries are of the word stem βαθμ– (*vathm*–) "degree, grade" appearing in words such as βαθμός (*vathmòs*) "degree, grade" and βαθμοφόρος (*vathmofòros*) "non-commisioned officer":

```
<βαθμ"+Noun":0"@U.IC1B.ON@">          InfSuff;
βαθμ                                   Compound;
```

In this entry, '+Noun' is a multicharacter symbol signifying the grammatical category of the word stem and @U.IC1B.ON@ a FD that refers to the inflectional class of this word stem and is unified with the appropriate inflectional suffix in the relevant lexicon, named 'InfSuff'. The lexicon 'InfSuff' contains 63 entries, which together with the inflectional suffix add morphological features referring to gender (+Masc, +Fem, +Neut), number (+Sg, +Pl) and case (+Nom, +Gen, +Acc). The FDs that appear in each entry in this lexicon and specify the constraints in the suffixes' combinations with entries from both the 'WordStems' and the 'BoundStems' Lexicon have been specified according to the inflectional classes of nouns and adjectives in MG [31]. Such a FD contains the number of the inflectional class (e.g. IC2) and if more than one inflectional

---

[4] The reason why this Lexicon contains only a selection of word stems is because an exhaustive listing would be out of the scope of this description, which focuses mainly on NCs.

suffix corresponds to an inflectional class, then it also contains the letter A or B. For example, the suffixes of the 3rd inflectional class which includes nouns such as μουσικ(ή) (*musikí*) "music" and γλώσσ(α) (*glòssa*) "language, tongue" have two distinct FDs, namely @U.IC3A.ON@ and @U.IC3B.ON@, each of which corresponds to one inflectional suffix ((ή)(i)/(a)(a)). The production of wrong analyses is prevented through the addition of further FDs with the value OFF (e.g. @U.IC3A.OFF@). For instance, the inflectional suffix ((ος)(os)) could be recognized as: i) the nominative singular of Inflectional Class 1 masculine nouns, such as βιολόγ(ος) (*viològos*) "biologist", ii) the nominative singular of Inflectional Class 7 neutral nouns, such as δάσ(ος) (*dàssos*) "forest" and iii) the genitive singular of Inflectional Class 8 neutral nouns, such as σώματ(ος) (*sòmatos*) "of a body". Accordingly, the corresponding entries containing all the necessary FDs are formed as follows:

```
<"@U.IC1A.ON@""@U.IC7.OFF@""@U.IC8.OFF@""+Masc":0"+Sg":0"+Nom": ος> #;
<"@U.IC7.ON@""@U.IC1A.OFF@""@U.IC8.OFF@""+Neut":0"+Sg":0"+Nom": ος> #;
<"@U.IC8.ON@""@U.IC7.OFF@""@U.IC1A.OFF@""+Neut":0"+Sg":0"+Gen": ος> #;
```

In contrast to the MG description, the English lexicon 'Root' contains two sublexicons, namely 'Nouns' and 'BoundStems'. This is explained by the fact that the two descriptions in LEXC, as it is earlier mentioned, have been developed in accordance with the analyses adopted. Therefore, although there are also bound stems that appear only as initial elements in NCs and could thus be categorized as prefixoids (as in the case of MG), since in English they have the status of a bound stem, they are preferably also included in the Lexicon 'BoundStems', which contains the sublexicons 'General' and 'Initial'. The latter distinction is necessary in order to prevent the production and recognition of wrong sequences, such as *lithomega* or *phonotele*, which do not correspond to any of the possible structures of NCs in English. For this reason, bound stems such as *aut-*, *carni-*, *di-*, *hexa-*, *hepta-*, *hom-*, *macr-*, *mega-*, *poly-* etc., which appear only as first constituents have been included in the sublexicon 'Initial', thus being separated from the others that may appear in both positions.

The lexicon 'Nouns' on the other hand, contains entries that correspond to free words in English, of native or non-native origin, that often combine with bound stems of classical origin, forming 'hybrid' formations, as we mentioned above. The number of such entries in this description is limited and is only indicative of the process, because, as already pointed out, the English description focuses mainly on prototypical NCs. The continuation class of these entries is common for all and leads to the lexicon 'Wordfrm' (=word formation), which either assigns to these entries the morphological feature '+Noun' through an empty entry, or leads to the process of compounding, which takes place in the lexicon 'Compound'. This is another point where the two descriptions differ significantly, as the rich inflection of MG requires a more complex implementation.

Linking Vowel. The lexicons 'Compound' and 'BoundStems' are structured in a rather similar manner for both languages, justifying in this way the parallel drawn between the corresponding adopted analyses. The lexicon 'Compound',

where the process of compounding takes place, with the insertion of the linking vowel and the addition of the multi-character symbol '^CM^' (=Compound
Marker) in the morphological analysis produced, has the same number of entries
in both descriptions. The first two entries of this lexicon in both descriptions, concern the actual process of compounding, taking care of the phonological appearance or absence of the linking element via the FDs "@U.VOWEL.ON/OFF@"
and leading through the continuation classes to the lexicon 'BoundStems' in the
MG description and the lexicon 'General', in the English description, a sublexicon of the lexicon 'Boundstems'. The last two entries in both descriptions
concern the appearance of the linking vowel in recursive formations such as _μορ-
φοφωνολόγος_ (_morfofonològos_) "morphophonologist" (MG) or _astrophotography_
(Eng.) and for this reason their continuation class is 'WordStems' in the MG
description and 'Initial' in the English one. Therefore, the lexicon 'Compound'
is structured in the following way in the two descriptions (English on the left
and MG on the right column):

```
<"^CM^":o "@U.VOWEL.OFF@"> General; <"^CM^":o "@U.VOWEL.OFF@"> BoundStems;
<"^CM^":0 "@U.VOWEL.ON@">  General; <"^CM^":0 "@U.VOWEL.ON@">  BoundStems;
<"^CM^":o "@U.VOWEL.OFF@"> Initial; <"^CM^":o "@U.VOWEL.OFF@"> WordStems;
<"^CM^":0 "@U.VOWEL.OFF@"> Initial; <"^CM^":0 "@U.VOWEL.OFF@"> WordStems;
```

Bound Stems. The lexicons that contain the bound stems in the two languages
have been built exclusively from the two relevant corpora of bound stems mentioned above. In the MG description, the lexicon 'BoundStems' contains 89 entries, which relate to 47 bound stems, as it also contains entries with allomorphic
variants. These serve two purposes: first, they introduce phonological allomorphs
of bound stems, indicating the phonological alterations occurring in many bound
stems during suffixation (e.g. 'πληχ:πληγ' ('plik:plig') in the word _τετραπληγία_
(_tetrapligìa_) "tetraplegia") and second, they insert stress, as this is the only way
that stress can be implemented in the LEXC (e.g. 'λογ:λόγ' ('log:lòg'), 'μαθ:μάθ'
('math:màth'), 'χλοπ:χλόπ' ('klop:klòp')).

The choice of each allomorph is determined by the continuation class of its
entry. For example, the entries for the bound stem _–παθ_ are structured as follows:

```
<"@U.VOWEL.OFF@" παθ "+Adj":  0 "@U.IC9.ON@">                InfSuff;
<"@U.VOWEL.OFF@" παθ:πάθ "+Noun":  0 "@U.ACT1.ON@">          DerSuff;
```

The first entry of the non-stressed _–παθ_ (_–path_) leads to inflection, for the formation of words such as _καρκινοπαθής_ (_karkinopathìs_) "cancer patient", while the
second one, which contains the stressed allomorph _–πάθ_ (_–pàth_), leads to the lexicon "DerSuff", where the process of suffixation takes place for the formation of
derivatives such as _τηλεπάθεια_ (_tilepàthia_) "telepathy". Both entries contain also
the FD responsible for the appearance of the linking vowel and the symbol that
concerns the grammatical category, which is generally that of a noun and in some
cases, such as the above, that of an adjective, according to the adopted analysis.
Moreover, the first entry contains a FD for the inflectional class, while the second
contains another FD, which concerns suffixation and is discussed below. In the

English description, the entries of the bound stems that appear in both positions within a word are included in the lexicon 'General', as discussed above. The majority of the bound stems in English do not have allomorphic variants, except for a very small number, such as *−crat/−crac* (e.g. *democrat, democracy*), and *−derm/−dermat* (e.g. *pachyderm, dermatology*). In order to produce and recognize, as many as possible recursive NCs, which are found in abundance in technical and scientific dictionaries, all of the entries of the lexicon 'General' lead through their continuation class first to the lexicon 'Compound', and then to suffixation, through an empty entry in the lexicon 'Compound' with continuation class 'Suffix', leading to the corresponding lexicon.

Derivation. The lexicon 'Suffix', which mainly concerns derivational processes, in the English description contains the following: i) the entries of the nominal suffixes *−y, −ia, −er, −ist* and the adjectival *−ic, −ical, −al* and *−ous*, ii) an entry that assigns the morphological feature '+Noun' to prototypical NCs that are no derivatives, and iii) an entry that apart from the feature '+Noun' adds a final *e*, which has a phonological origin and appears in prototypical NCs containing certain bound stems in final position (e.g. *−phon (telephone), −scop (microscope), −lyt (electrolyte)*. The appearance of this *e* is regulated with the FDs "@U.PHON.ON/OFF@", which also appear in the entries of the relevant bound stems in the lexicon 'General'. All the entries of the lexicon 'Suffix' subsequently lead to the lexicon 'InfSuff' which contains only three entries, one assigning the feature '+Sg' for the singular number, another assigning the feature '+Pl' along with the *s* in the surface structure for the plural and a final entry which assigns the feature '+Pl' with no surface realization, for plural adjectives.

In the MG description, the entries for the derivational nominalizing suffixes *−εια (−ia), −ία (−ia), −εί(o) (−io), −ί(o) (−io), −ισσα (−issa), −τρια (−tria)* and the adjectival *−ικ(oς) (−ikos)* are included in the lexicon 'DerSuff', a sublexicon of the lexicon 'Suffix', together with the sublexicon 'InfSuff'. The reason for this difference between the two descriptions is computational economy, as some of the entries of the lexicon 'BoundStems' in the MG description lead to both sublexicons through the collective continuation class 'Suffix'. In both languages, some of the derivational suffixes exhibit a complementary distribution, namely, some are selected by certain bound stems and some by others. For example, the suffixes *−εια (−ia)* and *−ία (−ia)* in MG and the corresponding *−y* and *−ia* in English that both signify "an action, the result of an action, a process, a state or an attribute", are distributed complementarily in their combinations with bound stems (e.g. *βιολογία (viologìa)* "biology" vs. *γλωσσομάθεια (glossomàthia)* "language learning" (MG), *astronomy* vs. *cardialgia* (Eng.)). This is regulated through the use of the same FD with different values (ON/OFF), namely "@U.ACT.ON/OFF@", so that no wrong sequences are recognized and generated. Different, however not complementary, distribution appears also between the MG feminine nominalizing suffixes *−ισσα (−issa)* and *−τρια (−tria)*, the English nominalizing ones *−er* and *−ist* and also among the adjectival *−ic, −ous* and *−al*. These are also regulated through the use of FDs, which however must be careful, in order not to create overgeneration, as for example there are cases where the adjectival suffixes exhibit parallel distribution in their combinations with bound stems (e.g. i) *zoophilic* and *zoophilous*, ii) *allogenic, indigenous* and *protogenal*).

3.5 Evaluation and Future Work

The correct use of the FDs during the development of the description for the implementation of various constraints, as well as its overall efficiency has been examined through a special controlling function of the LEXC formalism, which maps manually introduced sequences that correspond to the surface form, with their lexical form ('apply down'), thus producing their morphological analysis. The two descriptions have been checked in a number of parameters that are significant in the formation of NCs in the two languages, some of which are the following:

i) the correct distribution of the linking vowel in NCs. The formalism recognized and correctly analyzed sequences, such as πολυλογία (*polilogìa*) "chatter" but not **πολυνολογία (*poliologìa*) (MG), and *psychopaths* but not **psychpaths* (Eng.), as follows:

```
πολυ^CM^λογ+Noun+Fem+Sg+Nom          psych^CM^path+Noun+Pl
πολυ^CM^λογ+Noun+Fem+Sg+Acc
```

ii) the right selection of inflectional and derivational suffixes, correctly analysing sequences such as βιβλιοπωλείο (*vivliopolìo*) "bookstore", but not **βιβλιοπωλιο (*vivliopolìo*) (MG), *homophobia* but not **homophoby* (Eng.), as follows:

```
βιβλι^CM^πωλ+Noun+Fem+Sg+Nom          hom^CM^phob+Noun+Sg
βιβλι^CM^πωλ+Noun+Fem+Sg+Acc
```

iii) the selection of the proper allomorph, recognizing sequences such as πατροκτονία (*patroktonìa*) "patricide" but not **πατροκτόνία (*patroktònìa*) (MG) and *homicidious* and *homicidal*, but not **homicidous* and **homicidial* (Eng.):

```
πατρ^CM^κτον+Noun+Fem+Sg+Nom          hom^CM^cid+Adj+Sg
πατρ^CM^κτον+Noun+Fem+Sg+Acc          hom^CM^cid+Adj+Pl
```

iv) the correct recognition of recursive constructions, such as βιοτεχνολογία (*viotexnologìa*) "biotechnology" in MG and *electromyelographic* in English:

```
βι^CM^τεχν^CM^λογ+Noun+Fem+Sg+Nom     electr^CM^myel^CM^graph+Adj+Sg
βι^CM^τεχν^CM^λογ+Noun+Fem+Sg+Nom     electr^CM^myel^CM^graph+Adj+Pl
```

The efficiency of the developed LEXC descriptions has not yet been tested automatically through their application to corpora containing hundreds of NCs, like the ones used for the development of the current implementation. The reason for that, as we mentioned above, is the fact that, although each description contains almost all bound stems, they both lack a great number of word stems that appear as first elements in NCs. For instance, in the MG description, as discussed earlier, the lexicon 'WordStem' contains 72 indicative entries, which correspond to word stems in MG that freaquently appear in combinations with bound stems and belong to all infelctional classes. However, this number is significantly small, in comparison to the total number of word stems that appear as first elements in NCs, the inclusion of which is beyond the scope of this implementation. Therefore, both descriptions should first be expanded in this respect and then applied into a corpus with the subsequent operation of a Tokenizer and a Lookup function [9], or they could be incorporated into a general transducer for each language, which among others contains the majority of its word stems.

The main objective of this implementation of NCs' processing in MG and English, as they appear in both technical and non-technical language, without being limited to a specific area of terminology, was the construction of a processing system that is both computationally efficient and linguistically accurate. Specifically, the proposed implementation aimed at describing the formation of prototypical NCs in the two languages in the most economical, in computational terms, way, on the basis of the conclusions drawn from the comparative analysis of their formation in the two languages. Linguistic accuracy in this respect was not undermined for the sake of computational efficiency, or the other way round. Therefore, through the use of the Lexical Compiler (LEXC), the two descriptions, one for each language, were developed on the basis of the conclusions of the theoretical analysis, according to which the phenomenon of neoclassical compounding follows similar rules in the two languages, retaining at the same time the idiosyncratic features of each. What was achieved, was the adoption of a linguistically accurate computational solution for storing linguistic data and describing the morphological processes involved. The choice of the LEXC formalism in particular and the extensive use of some of its features, like the FDs, helped to maintain this linguistic accuracy and achieve greater computational efficiency.

On the basis of the present implementation, the description of the same morphological phenomenon in other languages would definitely be an interesting endeavor, as, it is already pointed out that neoclassical compounding appears on an international level. In this respect, such an implementation could be incorporated into a multi-language information retrieval, a term recognition or an automatic translation system.

References

1. Adams, V.: An Introduction to Modern English word-formation. Longman, London (1973)
2. Adams, V.: Complex words in English. Pearson Education, Essex (2001)
3. Anastasiadi-Symeonidi, A.: Neology in Common Modern Greek, Thessaloniki (1986) (in Greek)
4. Antworth, E.: PC-KIMMO: a two-level processor for morphological analysis. In: Occasional Publications in Academic Computing, vol. 16. Summer Institute of Linguistics, Dallas (1990)
5. Baeskow, H.: Lexical Properties of Selected Non-native Morphemes of English. Gunter Narr, Tübingen (2004)

6. Baud, R.H., Rassinoux, A.-M., Puch, P., Lovis, C., Scherrer, J.-R.: The power and limits of a rule-based morpho-syntactic parser. In: Lorenzi, N.M. (ed.) AMIA 1999 Proceedings of the 1999 Annual Symposium of the American Medical Informatics Association, Washington DC, November 6-10, pp. 22–26. Hanley & Melfus, Philadelphia (1999)
7. Bauer, L.: English word-formation. Cambridge University Press, Cambridge (1983)
8. Bauer, L.: Is there a class of neoclassical compounds in english and is it productive? Linguistics 36, 403–422 (1998)
9. Beesley, R.K., Karttunen, L.: Finite State Morphology. CSLI Publications (2003)
10. Cannon, G.: Bound-morpheme items: New patterns of derivation. In: Blanck, C. (ed.) Language and Civilization: A Concerted Profusion of essays and Studies in Honour of Otto Hietch, pp. 478–494. Peter Lang, Frankfurt (1992)
11. Deleger, L., Namer, F., Zweigenbaum, P.: Morphosemantic parsing of medical compound words: Transferring a French analyzer to English. International Journal of Medical Informatics 78, 48–55 (2009)
12. Durkin, P.: The Oxford Guide to Etymology. Oxford University Press, Oxford (2009)
13. Galiotou, E., Ralli, A.: Parsing Deficiencies of the PC-KIMMO System. In: Proceedings of the 2nd Hellenic Conference on Artificial Intelligence (Companion Volume), pp. 53–64. Aristotle University, Thessaloniki (2002)
14. Galiotou, E., Ralli, A.: Morpho-Phonological modelling in Natural Language processing. In: Proceedings of the International XII. Symposium on Artificial Intelligence and Neural Networks, TAINN (2003)
15. Galiotou, E., Ralli, A.: Morpho-phonological Modelling in Natural Language Processing. International Journal of Computational Intelligence (International Journal of Information and Mathematical Sciences) 1(3), 155–158 (2005)
16. Giannoulopoulou, G.: Morphosemantic Comparison of Affixes and Confixes in Modern Greek and Italian, Thessaloniki (2000) (in Greek)
17. Jurafsky, D., Martin, J.H.: Speech and Language Processing. An introduction to natural language processing, computational linguistics, and speech recognition. Pearson Education, New Jersey (2000)
18. Kastovsky, D.: Astronaut, astrology, astrophysics: About Combining Forms, Classical Compounds and Affixoids. In: Selected Proceedings of the 2008 Symposium on New Approaches in English Historical Lexis (HEL-LEX 2). Cascadilla Proceedings Project, MA (2009)
19. Koskenniemi, K.: Two-Level Morphology: A General Computational Model for Word-form Recognition and Production. In: Proceedings COLING 1984, pp. 178–181 (1983)
20. Koskenniemi, K.: A general computational model for word-form recognition and production. In: COLING 1984, pp. 143–149 (1984)
21. Lehrer, A.: Scapes, holics and thons: The semantics of English combining forms. American Speech 73, 3–28 (1998)
22. Marko, K., Schulz, S., Hahn, U.: Morphosaurus-design and evaluation of an interlingua-based, cross-language document retrieval engine for the medical domain. Methods of Information in Medicine 44, 537–545 (2005)
23. Martinet, A.: Grammaire fonctionnelle du français. Didier, Paris (1979)
24. Namer, F.: Guessing the meaning of neoclassical compounds within LG: the case of pathology nouns. In: Proceedings of Generative Approaches to the Lexicon, Geneva, pp. 175–184 (2005)

25. Petropoulou, E.: On the parallel between neoclassical compounds in English and Modern Greek. Patras Working Papers in Linguistics, Special Issue: Morphology 1 (2009)
26. Petropoulou, E.: Bound-stem compounds in English and Modern Greek. Theoretical Analysis and Electronic Processing. PhD Dissertation. University of Patras (2011) (in Greek)
27. Prćić, T.: Prefixes vs initial combining forms in English: A lexicographic perspective. International Journal of Lexicography (2005)
28. Prćić, T.: Suffixes vs. final combining forms in English: a lexicographic perspective. International Journal of Lexicography 21 (2008) (prepublication)
29. Ralli, A.: Eléments de la Morphologie du Grec Moderne: La Structure du Verbe. Ph.D. Dissertation, University of Montreal (1988)
30. Ralli, A.: Compounds in Modern Greek. Rivista di Linguistica 4, 143–174 (1992)
31. Ralli, A.: A Feature-based Analysis of Greek Nominal Inflection. Glossologia 11-12, 201–227 (2000)
32. Ralli, A.: Greek Deverbal Compounds with bound stems. Southern Journal of Linguistics 29(1-2), 150–173 (2008)
33. Ralli, A.: Hellenic Compounding. In: Lieber, R., Stekauer, P. (eds.) The Oxford Handbook of Compounds, pp. 453–464. Oxford University Press, Oxford (2009)
34. Ralli, A.: Compounding in Modern Greek. Springer, Dordrecht (2013)
35. Ralli, A., Galiotou, E.: A morphological processor for Modern Greek. In: Proceedings of the 3rd Conference of the European Chapter for Computational Linguistics (EACL), pp. 26–31 (1987)
36. Ralli, A., Galiotou, E.: Greek Compounds: A Challenging Case for the Parsing Techniques of PC-KIMMO v.2. International Journal of Computational Intelligence (International Journal of Information and Mathematical Sciences) 1(2), 128–138 (2005)
37. Roark, B., Sproat, R.: Computational Approaches to Morphology and Syntax. Oxford University Press, Oxford (2007)
38. Scalise, S.: Generative Morphology, 2nd edn. Foris, Dordrecht (1986)
39. Schulz, S., Romacker, M., Franz, P.: Towards a multilingual morpheme thesaurus for medical free-text retrieval. In: Proceedings of MIE 1999, pp. 891–894. IOS Press, Ljubliana (1999)
40. ten Hacken, P.: Defining Morphology. A Principled Approach to Determining the Boundaries of Compounding, Derivation and Inflection. Olms, Hildesheim (1994)
41. Warren, B.: The importance of combining forms. In: Dressler, W., et al. (eds.) Contemporary Morphology. Mouton de Gruyter, Berlin (1990)
42. Wexler, P.: Towards a structural definition of 'Internationalisms'. Linguistics 7(48), 77–79 (1969)

CroDeriV 2.0.: Initial Experiments

Krešimir Šojat[1], Matea Srebačić[2], and Tin Pavelić[2]

[1] Faculty of Humanities and Social Sciences, Zagreb
ksojat@ffzg.hr
[2] University of Zagreb
msrebaci@unizg.hr, tin.pavelic@ffzg.hr

Abstract. This paper deals with the processing of derivational morphology in Croatian and focuses on the expansion of CroDeriV – a resource with data on morphological structure and derivational relations. The purpose of CroDeriV is to systematically present the morphological structure and derivational relations of Croatian lexemes, and to use this data for the enrichment and development of existing resources and tools, as well as of new ones. One of the objectives in this ongoing project is to build an analyzer for Croatian capable of analyzing both inflectional and derivational morphemes. In this paper we present the initial experiments towards the enlargement of CroDeriV to include nouns, as well as the development of a morphological analyzer for inflectional and derivational morphemes.

Keywords: CroDeriV, Croatian, derivation, nouns.

1 Introduction

This paper deals with the processing of Croatian morphology and focuses on the expansion of CroDeriV – a resource providing data on the morphological structure and derivational relations of Croatian lexemes. Although detection of the complete morphological structure and all derivational relations could improve the performance of tools used in various natural language processing tasks, the derivational relatedness of Croatian lexemes based on a thorough and detailed analysis of their morphological structure so far have not yet been processed. Derivational relations hold between words that share the same lexical morpheme – i.e., the root – and thus form derivational families. A derivational family consists of lexemes with the same root grouped around a base form. Generally, a lexeme with the simplest morphological structure serves as a base form – i.e., a stem – for various derivational processes. Derivational processes in Croatian primarily refer to affixation and compounding. Affixation mainly consists of prefixation and suffixation. The motivation for building CroDeriV is twofold: (1) to analyze and systematically present the morphological structure and derivational relations of Croatian lexemes resulting from their mutual morphological relatedness, and (2) to use this data for the enrichment of existing resources and tools, as well as for the development of new ones. One of the objectives in this ongoing

A. Przepiórkowski and M. Ogrodniczuk (Eds.): PolTAL 2014, LNAI 8686, pp. 27–33, 2014.

project is to build an analyzer for Croatian capable of full morphological analysis – i.e., the analysis of both inflectional and derivational morphemes grouped around roots.

Inflectional classes are extensively covered by the Croatian Morphological Lexicon (HML). The HML serves as a basis for lemmatization and MSD tagging of Croatian (cf. [12] for the building procedures of the HML and [1] for the evaluation of lemmatization based on the HML). Although it is an extensive inflectional lexicon, it cannot be used for word segmentation, which is necessary for deeper morphological analysis and, consequently, for the detection of derivationally related lexemes and derivational families. The detection of derivational relations is recognized as an important task for the building of resources and tools also for other languages (cf. [4] for English; [6] for Polish; [13] for German). [7] presents an elaborate account of the methods used for the clustering of derivationally related Croatian lemmas from the corpus and also provides an extensive and detailed description of the evaluation metrics. This approach is based on morphological stems and focuses on the suffixal derivation between nouns, verbs, and adjectives. However, it does not take into account the analysis of the morphological structure of lexemes nor the recognition of mutual lexical morphemes within derivational families. An accurate and linguistically justified analysis of Croatian lexemes in terms of morphemes and affixes used in their derivation is one of our main objectives in the building of CroDeriV. In the next section we briefly describe its structure and design.

2   CroDeriV

CroDeriV is a morphological database built by means of combining rule-based processing with manual checking of results. The first phase of CroDeriV's development focused on the processing of Croatian verbs. In its present form, CroDeriV contains 14,326 verbs analyzed for morphemes and grouped into derivational families via mutual lexical morphemes. The total number of lexical morphemes is 3,386. The verbs were analyzed as follows: (1) verbs in infinitive form were collected from various on-line corpora and dictionaries; (2) the collected verbs were automatically segmented into morphemes; (3) the obtained results were manually checked; and (4) verbs sharing the same lexical morpheme were mutually linked. This procedure enabled the recognition of a general morphological structure applicable to all analyzed verbs, as well as the recognition of all lexical and derivational morphemes used in verbal derivational processes. It also enabled the detection of complete verbal derivational families in Croatian, as well as the analysis of possible combinations of affixes and roots, their frequency and productivity.[1]

Most verbs in Croatian are formed from other verbs, thus expanding their morphological structure. For example, the simplex base verb *graditi* 'to build$_{ipf}$' can be prefixed and thus can form the perfective verb *nadograditi* 'to expand by

---

[1] CroDeriV is freely available for searching at http://croderiv.ffzg.hr.

building$_{pf}$'. This verb in turn can be further suffixed to form the derived imperfective: *nadogradivati* 'to expand by building, to annex$_{ipf}$'. A thorough analysis of all verbal derivational processes has yielded the maximal morphological structure of Croatian verbs as follows:

(P4) + (P3) + (P2) + (P1) + (R2) + (I) + R1 + (S3) + S2 + S1 + *ti*

where P = prefix, R = root, I = interfix, S = suffix, ti = infinitive ending, and () = non-obligatory.

Data structured in this way has already proven valuable for various linguistic studies (cf. [10]), as well as for the enrichment of other resources for Croatian – e.g., Croatian WordNet and the HML (cf. [9]; cf.[8]). The next objective in the building of CroDeriV is its extension with other parts of speech. Further on, we present the initial steps in the processing of nouns according to the principles mentioned above and their integration into CroDeriV.

3   The Derivation of Croatian Nouns

Derivational relations in Croatian extend either between words that are the same POS (verb-to-verb, noun-to-noun) or different POS (verb-to-noun, verb-to-adjective, adjective-to-noun, etc.). The most productive word-formation processes are prefixation and suffixation.[2] Like verbs, nouns in Croatian are formed through three basic derivational processes:

(a) **suffixation** (e.g., *pis(ati)* 'write' + *-ac* > *pisac* 'writer'),
(b) **compounding** (e.g., *roman* 'novel' + *-o-* + *pisac* 'writer' > *romanopisac* 'novelist'), and
(c) **prefixation** (e.g., *su-* + *radnik* 'worker' > *suradnik* 'co-worker').

On top of that, nouns are additionally formed through two combined processes:

(a) **compounding** + **suffixation** (e.g., *vatr(a)* 'fire' + *-o-* + *gas(iti)* 'extinguish' + *-ac* > *vatrogasac* 'firefighter'), and
(b) **prefixation** + **suffixation** (e.g., *po-* + *mor(e)* 'sea' + *-ac* > *pomorac* 'sailor').

Apart from these concatenative processes, nouns are also formed via:

(a) **back-formation** (e.g., *dopisati* 'to add by writing' > *dopis* 'letter'), and
(b) **conversion** (e.g., *mlada* 'young, female, adjective' > *mlada* 'bride, noun').[3]

Suffixation is by far the most productive derivational process in the derivation of Croatian nouns.[4]

As mentioned, CroDeriV has been built through the application of rules for morpheme segmentation and the manual checking of results. Generally, the main problems that we have faced in the automated processing of Croatian morphology (cf. Sect. 2) are caused by the following factors: (1) the graphical overlapping of various morphemes, (2) phonological changes at morpheme boundaries,

---

[2] Compounding is not as prominent and will not be discussed here.
[3] Two extensive accounts of Croatian derivation are [2] and [5].
[4] There are more than 90 productive suffixes used in Croatian noun derivation [2].

(3) phonological changes within lexical morphemes (which frequently result in several allomorphs), and finally, (4) numerous instances of homonymy (for a more detailed account of all instances and examples, cf. [11]). Unfortunately, we face all these problems in the processing of nouns, as well. For example, 36 different suffixal structures end in *-ac* (e.g., *bor-ac* 'fighter', *gled-a-l-ac* 'viewer', *jedanaest-erac* 'penalty kick', etc.). Frequently, suffixal structures of Croatian nouns consists of two or more suffixes (e.g., *grad* 'city' > *grad-i-(ti)* 'to build' > *grad-i-telj* 'builder' > *grad-i-telj-ic-a* 'female builder' > *grad-i-telj-ič-in* 'female builder's'. Since the list of possible suffixal combinations or the rules for the restrictions of these combinations in Croatian do not exist, the design of rules for their recognition and accurate segmentation is challenging and time-consuming work.

4   Experiment

In order to obtain a basic stock of nouns necessary for the expansion of CroDeriV with other POS and to provide a foundation for the development of tools for comprehensive morphological analysis, we have decided to use the nouns from the HML. In the experiment described below, we focused on the subset of the HML's nominal part tagged as common nouns.[5] This choice was motivated by the fact that the HML extensively covers nominal inflectional classes and therefore provides a good source for the detection of the most frequent suffixes and their combinations. The processing was divided into several steps. In the first step, we wanted to detect single nominal suffixes and obtain an initial snapshot of the suffixal productivity. In the second step of the experiment, we focused on nouns derived from verbs through non-concatenative derivational processes, namely back-formation. In the final step of the experiment, we wanted to detect possible suffixal combinations and nominal stems not recognized in the previous steps. The whole experiment is based on data already present in CroDeriV – i.e., roots and stems used primarily in the derivation of verbs from other Croatian verbs. This data was used for matching with nouns from the HML as described below.

4.1   Step 1

In the first step of the experiment, we created a set of rules for the detection and segmentation of single suffixes and applied it to the test sample. The test sample, which consisted of common nouns in the HML, comprised 20,554 nouns. The rules for the recognition and segmentation of single suffixes yielded a total of 4,933 nouns with correctly recognized stems and 22 single suffixes. The five most frequent single derivational suffixes detected in this way are: *-nje* (e.g., *pis-a-nje* 'writing'), *-ač* (e.g., *pis-ač* 'writer'), *-ica* (e.g., *s-pis-a-telj-ica* 'female writer'), *-telj* (e.g., *s-pis-a-telj* 'writer'), *-na* (e.g., *pis-ar-na* 'writing office'). Their respective frequencies are presented in Table 1.

---

[5] Lemmas in HML are tagged according to MulTextEast specifications v.4.0 ([3]).

Table 1. Five most frequent suffixal combinations

Suffix	-nje	-ač	-ica	-telj	-na
No. of occurences	2766	224	108	104	100

As the numbers indicate, the suffix *-nje* is the most frequent single suffix of common nouns in the HML. This suffix is used in Croatian almost exclusively for the derivation of gerunds from verbal stems – i.e., from their past participles. The total number of gerunds among the common nouns is 4,586. In other words, almost 25% of all common nouns in the HML are verbal nouns. Although this fact may be surprising as far as the structure of the HML is concerned, it simplified the further processing of this subset of common nouns.

A comparison of verb lists from CroDeriV and the HML revealed that all the verbs from the HML are also present in CroDeriV. Since all the verbs in CroDeriV have been analyzed for morphemes, by slightly adjusting the rules, we were able to automatically determine the full morphological structure of gerunds – i.e., their roots as well as their prefixal structures – inherited from the base verbs. After manual validation, we used these results in the next step of the experiment.

4.2   Step 2

The total remaining number of nouns to be analyzed was thus 15,968. As mentioned in Sect. 3, nouns in Croatian are formed from other POS via affixation, but also through non-concatenative processes, such as back-formation; as in:

dopisati 'to add by writing' → *dopis* 'memo'
opisati 'to describe' → *opis* 'description'
upisati 'to record' → *upis* 'record'

Consequently, all nouns derived from verbs through back-formation inherit the morphological structure of their base forms, apart from their suffixal part. For the detection of nouns derived in this manner, we again matched the data from CroDeriv and the HML. In this way we detected 3,367 common nouns tagged as candidates for the expansion of derivational families in CroDeriV. The manual checking of the results revealed that 1,200 nouns were not correctly segmented, whereas 2,167 nouns were correctly segmented and correctly assigned to the corresponding derivational families from CroDeriV.

Although the recall of 38.61% scored in this part of the processing is low, the high precision of 85.02% enabled their straightforward integration into CroDeriV. The total number of nouns used in further steps was thus reduced to 13,801.

4.3   Step 3

In the final step of the experiment, we randomly chose around 40% of this remaining set of nouns and manually analyzed them for morphemes. The primary objectives of this analysis were to detect possible suffixal combinations as input for rules capable of dealing with multiple suffixes and to detect nominal stems not recognized in previous steps due to the complex morphological structure of their lexemes. The general aim of the whole procedure was to speed up the detection of derivationally related nouns and to check whether they can be linked to derivational families from CroDeriV without further manual segmentation. For this purpose we took the following steps:

(1) we automatically removed manually obtained suffixal combinations and used only nominal stems in further processing – e.g.:

glas-ač-ic-a 'female voter' → glas 'voice',
vid-ovit-ost 'clairvoyance' → vid 'sight',

(2) we extracted all the stems and roots from CroDeriV, and

(3) we matched them with the list of obtained nominal stems from the HML.

The recall of the whole procedure was 100%. In order to measure precision, we manually evaluated 5,520 randomly selected nouns. Out of this number, 33.88% of the roots from CroDeriV were correctly assigned to nominal stems from the HML. As expected, although the recall was 100%, the precision of the automated root assignment significantly decreased. However, the final results of the conducted experiment can be considered satisfactory. A total of 1,773 roots out of 3,386 roots in CroDeriV (53.4%) has been correctly assigned to at least one noun from HML, thus enabling the automated expansion of derivational families. On the top of that, we obtained 1,753 new nominal roots through manual evaluation, which can be used in the further processing. From the initial set of 20,554 nouns, this simple automated approach assigned the correct root to more than half (12,227) of the nouns.

5 Conclusion

In this paper we have shown the initial steps towards the enlargement of CroDeriV to include another POS, namely nouns. With a combined approach using simple automatic processing and manual checking, we have obtained two noun sets:

(1) a set of nouns which are derivationally related to the verbs in CroDeriV and can be used for the enrichment of already existing derivational families via mutual root, and

(2) a set of nouns that are not derivationally related to the verbs in CroDeriV.

However, many of these nouns are mutually derivationally related and can be used for the inclusion of new derivational families in CroDeriV. Moreover, the list of newly recognized nominal roots, as well as the list of the most frequent nominal suffixes and their combinations, can be used for further improvement of our morphological processing tools.

References

1. Agić, Ž., Tadić, M., Dovedan, Z.: Improving Part-of-Speech Tagging Accuracy for Croatian by Morphological Analysis. Informatica 32(4), 445–451 (2008)
2. Babić, S.: Tvorba riječi u hrvatskome književnom jeziku. HAZU: Nakladni zavod Globus, Zagreb (2002)
3. Erjavec, T.: MULTEXT-East: Morphosyntactic Resources for Central and Eastern European Languages. Language Resources and Evaluation 46(1), 131–142 (2012)
4. Habash, N., Dorr, B.: A Categorial Variation Database for English. In: Proceedings of the Anuual Meeting of the NAACL, pp. 96–102 (2003)
5. Marković, I.: Uvod u jezičnu morfologiju. Disput, Zagreb (2012)
6. Piasecki, M., Ramocki, R., Maziarz, M.: Recognition of Polish Derivational Relations Based on Supervised Learning Scheme. In: LREC 2012 Proceedings, pp. 916–922 (2012)
7. Šnajder, J.: DerivBase.hr: A High-Coverage Derivational Morphology Resource for Croatian. In: Calzolari, N., et al. (eds.) Proceedings of the Ninth International Conference on Language Resources and Evaluation (LREC 2014). ELRA, Reykjavik (2014)
8. Šojat, K., Srebačić, M.: Morphosemantic relations between verbs in Croatian Word-Net. In: Orav, H., Fellbaum, C., Vossen, P. (eds.) Proceedings of the Seventh Global WordNet Conference, pp. 262–267. GWA, Tartu (2014)
9. Šojat, K., Srebačić, M., Pavelić, T., Tadić, M.: From Morphology to Lexical Hierarchies. In: Vetulani, Z. (ed.) Human Language Technologies as a Challenge for Computer Science and Linguistics (LTC 2013 Proceedings), pp. 474–478 (2013)
10. Šojat, K., Srebačić, M., Štefanec, V.: CroDeriV i morfološka raščlamba hrvatskoga glagola. Suvremena lingvistika 39(75), 75–96 (2013)
11. Šojat, K., Srebačić, M., Tadić, M.: Derivational and Semantic Relations of Croatian Verbs. Journal of Language Modelling (1), 111–142 (2012)
12. Tadić, M., Fulgosi, S.: Building the Croatian Morphological Lexicon. In: Proceedings of the EACL 2003 Workshop on Morphological Processing of Slavic Languages, pp. 41–46. ACL, Budapest (2003)
13. Zeller, B., Šnajder, J., Padó, S.: DERIVBASE: Inducing and Evaluating a Derivational Morphology Resource for German. In: Proceedings of the 51st Annual Meeting of the ACL (2013)

Named Entity Matching Method Based on the Context-Free Morphological Generator

Jan Kocoń and Maciej Piasecki

Institute of Informatics, Wrocław University of Technology,
Wybrzeże Wyspiańskiego 27, Wrocław, Poland
{jan.kocon,maciej.piasecki}@pwr.edu.pl

Abstract Polish named entities are mostly out-of-vocabulary words, i.e. they are not described in morphological lexicons, and their proper analysis by Polish morphological analysers is difficult.The existing approaches to guessing unknown word lemmas and descriptions do not provide results on a satisfactory level. Moreover, lemmatisation of multi-word named entities cannot be solved by word-by-word lemmatisation in Polish. Multi-word named entity lemmas (e.g. included in gazetteers) often contain word forms that differ from lemmas of their constituents. Such multi-word lemmas can be produced only by tagger- or parser-based lemmatisation. Polish is a language with rich inflection (rich variety of word forms), therefore comparing two words (even these which share the same lemma) is a difficult task. Instead of calculating the value of form-based similarity function between the text words and gazetteer entries, we propose a method which uses a context-free morphological generator, built on the top of the morphological lexicon and encoded as a set of inflection rules. The proposed solution outperforms several state-of-the-art methods that are based on word-to-word similarity functions.

Keywords: Morphological generator, similarity of proper names, word similarity metric, Named Entity Recognition, Information Extraction.

1 Introduction

Lexicons of proper names (henceforth, PNs) are valuable resources for many natural language processing tasks, especially for Named Entity Recognition (henceforth, NER). Most PN inflected forms in text cannot be straightforwardly matched in lexicon. In the case of inflectional langauges, including Polish, *basic morphological forms* (called here shortly lemmas) of PNs are used as the entry forms in PN lexicons. For instance:

1. Inflected PN: $[Lidze_{\text{lemma}=0} \ Polskich_{\text{lemma}=0} \ Rodzin_{\text{lemma}=0}]_{\text{PNlemma}=0}$
2. Lemma: $[Liga_{\text{lemma}=1} \ Polskich_{\text{lemma}=0} \ Rodzin_{\text{lemma}=0}]_{\text{PNlemma}=1}$
3. Lemmatiser: $[Liga_{\text{lemma}=1} \ Polski_{\text{lemma}=1} \ Rodzina_{\text{lemma}=1}]_{\text{PNlemma}=0}$

In the case of multi-word PNs, the PN lemma (2) of the inflected PN form (1) is not identical to the lemma sequence (3) produced by a form lemmatised (e.g.

A. Przepiórkowski and M. Ogrodniczuk (Eds.): PolTAL 2014, LNAI 8686, pp. 34–44, 2014.

based on a tagger) for (1). PN lemmatisation is challenging task and for the Polish language (and for other Slavic languages too) that is not still solved by the existing language tools. This is a specific case of the general unknown word recognition problem, i.e. the recognition of words not covered by the existing morphological analyser [5,10]. Many of such words are infrequent domain-specific PNs. Language resources providing extensive coverage of the inflected PN forms are rare. Having them, comparison of PNs occurring in text with the lexicon PN entries would be easier task, but still discontinuous PNs must be taken into account. However, due to the huge number of PNs and their forms, building a language resource of that kind is laborious and expensive, even for domain-specific lexicons. Moreover, it is hardly possible to find and collect from text enough PN forms to build an extensive lexicon for many specific domains.

PN lexicons (e.g. gazetteers) are relatively large. This increases computational complexity of searching and matching. For the sake of wide applicability, we assume that the unknown word recognition is performed out of context, i.e. we do not use additional external knowledge sources and the information from the occurrence context of an unknown word form. There is also no information whether the token being processed is a true word or a non-word symbol. Polish is a language with rich inflection and each lemma corresponds to many morphological forms on average. Identification of a proper threshold for the similarity measures allowing for proper matching unknown words against the gazetteer entries can be difficult.

Our goal is to develop a method for effective recognition and classification of unknown word forms in Polish texts, in general, with a special focus given to the recognition of the unknown inflected PN forms that are included in a large PN lexicon. The method can be also used in more sophisticated NER tasks [6] e.g. recognition of the words composing multi-word PNs.

2   Related Works

The issue of word-to-word similarity measure has been intensively studied and many solutions were proposed in the literature, including methods dedicated to inflected languages such as Polish. Such metrics take two strings as the input and return a real number from the range $[0, 1]$. Evaluation of their performance is not straightforward – the direct interpretation of the similarity value between two words by the human is difficult, if possible at all. Another option is an indirect evaluation by application – the performance of a language tool utilising the metric in some text processing task is measured.

Several metrics applied to PN matching task were analysed in [1], like *Overlap coefficient*, *Soundex* or *Levensthein*. Recommendations concerning their suitability for different PN data sets were formulated, but the experimental results in [1] on different real data sets showed that there is no single best technique. In [7,8] several known and unique proposed metrics (e.g. *Common Prefix* δ) for Polish were evaluated in a task of assigning named entities (NEs) included in a gazetteer to text words. The best results in one-word NE matching were obtained with *Common Prefix* δ, mainly due to favouring certain suffix pairs by the δ parameter [7].

In [3] it was shown that a combination of several different similarity metrics can improve the overall result. First, a combination of the selected metrics from the *SimMetrics* library[1] was considered, namely: *Cosine Similarity, Euclidean Distance, Jaro, Jaro-Winkler, Matching Coefficient, Overlap Coefficient, Q-grams Distance, Soundex*. *Jaro* and *Jaro-Winkler* metrics were also analysed in [7,8]. They also proposed a very efficient *Common Prefix* δ (CP_δ) metric, mentioned earlier, which is based on the longest common prefix of two strings, using simple rules that were derived from the analysis of similarity examples:

$$CP_\delta(s, t) = \frac{(|\operatorname{lcp}(s, t)| + \delta)^2}{|s| \cdot |t|}$$

where $\operatorname{lcp}(s, t)$ is the longest common prefix of given strings: s & t, δ – a parameter, that equals 1 if one of the two given strings ends with a, and the second ends with one of the following: o, y, q, e, else 0.

The work of [3] is based on the idea of combining individual metrics into a complex one. It was noticed that the dependency of the overall result on the constituents can be very complicated. So, a classifier-based approach was proposed: a vector of individual single metric values is classified into two classes: *similar* and *non-similar*. A decision function value is produced as an additional description. On the basis of the experiments, *Logistic Regression (LR)* classifier was chosen. It provides binary classification (similar/not similar), but it is also possible to obtain decision function value [4], which can be used to describe word pair similarity strength. As a result, the complex word similarity function called NamEnSim[2] was constructed, associated with the initial selection of candidates (performed as a simple morphological filtering applied to compared words). Details of the solution and the previous evaluation results are presented in [3].

In the work presented here we decided to apply the same evaluation process as proposed in [3] in order to show that the solution proposed here outperforms the method combining several similarity functions and the single metric approaches too.

3   Context-Free Morphological Generator

The proposed method originated from the problem of matching NEs against a gazetteer, but it has been next generalized to match any pair of words which share the same lemma. It means that it is not necessary to have a dictionary (e.g. gazetteer) in which all constituents of one and multi-words (e.g. including words comprising NEs) are lemmas, but multi-words can be stored in their proper forms. Whereas the source dictionary remains not processed by lemmatiser or

---

[1]   http://sourceforge.net/projects/simmetrics
[2]   NamEnSim – Named Entity Similarity function.

stemmer, we try to generate all possible word forms of any unknown text word and match the result against the known set of constituents from gazetteers (e.g. PNs constituents).

Contrary to previous approaches, our method does not use a similarity metric or a preliminary candidate selection for possible similar words. The idea is to generate candidates on the basis of the *ending* (even if it is empty, see Sect. 3.2) of the input word w exclusively. The *ending* of the word w is the part of the word, which is formed by removing the longest common prefix from the set of all possible inflected forms of the word w (assuming that we know that set). Such a collection is called a *group* of the word w or set of words having the same lemma and morphological description consistent with the definition of similarity (see Sect. 4).

In the following, let:

s be an unknown word (string)

w – a word

k – a word ending

$N = \{s^1, s^2, ..., s^{|N|}\}$ – a PN dictionary comprised of a set of strings

$G = \{w^1, w^2, ..., w^{|G|}\}$ – a group – a set of words from the morphological dictionary such that they have the same lemma and morphological description which is consistent with the definition of similarity (see Sect. 4)

$S = \{G^1, G^2, ..., G^{|S|}\}$ – a set of groups

$K_G = \{k_G^1, k_G^2, ..., k_G^{|G|}\}$ – a set of word endings from group G

$C_{K_G}^2 = \{\{k_G^1, k_G^2\}, \{k_G^1, k_G^3\}, ..., \{k_G^1, k_G^{|G|}\}, \{k_G^2, k_G^3\}, ..., \{k_G^{|G|-1}, k_G^{|G|}\}\}$ – – 2-combination of the set K_G

$A_S = \displaystyle\sum_{i=0}^{|S|} C_{K_{S^i}}^2$ – a set of all 2-element-combinations of word endings from all groups

$B_S = \{\{A_S^1, a^1\}, \{A_S^2, a^2\}, ..., \{A_S^{|A_S|}, a^{|A_S|}\}\}$ – a set of all 2-combinations of word endings from all groups in a dictionary with the global counter a which is a number of pairs of endings occurrences in all groups.

3.1   Inflection Rules

Consider the following *group* G, presented as a set of triples *word, lemma, tag*[3] (each triple in a separate line):

```
sprawom    sprawa    subst:pl:dat:f
sprawie    sprawa    subst:sg:loc:f
sprawą     sprawa    subst:sg:inst:f
sprawie    sprawa    subst:sg:dat:f
sprawę     sprawa    subst:sg:acc:f
```

---

[3] In the given example tags come from the National Corpus of Polish Tagset [9] and denote the following attributes (separated by the colon) – *grammatical category : number : case : gender*. For the details of the similarity definition, see Sect. 4.

```
sprawo     sprawa   subst:sg:voc:f
sprawy     sprawa   subst:pl:voc:f
sprawy     sprawa   subst:pl:nom:f
sprawy     sprawa   subst:pl:acc:f
sprawami   sprawa   subst:pl:inst:f
sprawach   sprawa   subst:pl:loc:f
spraw      sprawa   subst:pl:gen:f
sprawa     sprawa   subst:sg:nom:f
sprawy     sprawa   subst:sg:gen:f
```

The construction of the inflection rules starts from the initial identification of *groups* (see Sect.3) by aggregating words from the dictionary due to their morphological descriptions that are consistent with the definition of the similarity (see Sect. 4). As a result, the set S is built. Next, for each group G in S the longest common prefix lcp_G is determined (for all words belonging to G). For the given example lcp_G = 'spraw'. After that for each word w in G the word ending k is determined by removing lcp_G from the beginning of w. As a result, K_G set is created. For the given example, K_G = {'om','a','ie','ą','ę','o','y','ami','ach',ø}. On the basis of K_G we build a set of all 2-element-combinations of word endings from all groups, denoted as $C^2_{K_G}$. The part of this set from the given example looks as follows:

$$C^2_{K_G} = \{$$
{'om','a'},{'om','ie'}, {'om','ą'}, ...,
{'om',ø}, {'a','ie'}, {'a','ą'}, ...,
{'a',ø}, ..., {'ach',ø}
}

The last step is to create the set, which contains all possible pairs of endings from groups in S with global counter a (the number of pairs of endings occurrences in all groups), denoted as B_S. Assuming that the group given as an example is the only one:

$$B_S = \{$$
{{'om','a'}, 1},{{'om','ie'}, 2}, {{'om','ą'}, 1}, ..., {{'om','y'}, 4}, ...,
{{'om',ø}, 1}, {{'a','ie'}, 2}, {{'a','ą'}, 1}, ...,{{'a','y'}, 4}, ...,
{{'a',ø}, 1}, ..., {{'ach',ø}, 1}
}

3.2 Inflected Forms Generator

The main method responsible for the identification of word candidates that are similar to the input string s is the *context-free morphological generator*. This method can be used as a part of a NER language tool. According to the introduced definitions, the algorithm consists of the following steps:

Input:

    s – a *word* (unknown) for which a list of similar word candidates is generated,

    S – a dictionary of the inflected word forms, on the basis of which the B_S set is created

Output:

    $W = \{\{gs^1, as^1\}, \{gs^2, as^2\}, ...\}$ - a set of pairs, where gs is a word candidate similar to s, and as is the number of pairs of $\{gs, s\}$ endings occurrences in S

In the following, let:

$s = \{\text{'Warszawie'}\}$

$B_S = \{\{\{\text{'ie','a'}\}, 4\}, \{\{\text{'om','ie'}\}, 2\}, \{\{\text{'om','ą'}\}, 1\}, \{\{\text{'e',ø}\}, 1\}\}$

1. Select a subset BS_S of the set B_S of such elements, in which at least one of the endings is the ending k of word s and each of the endings has a non-zero length.
2. For each element BS_S such as $\{\{k, ks\}, a\}$ (e.g. $BS_S = \{\{\{\text{'ie','a'}\}, 4\}$, $\{\{\text{'om','ie'}\}, 2\}, \{\{\text{'e',ø}\}, 1\}\}$):
   (a) create a word gs by removing the ending k from the word s and adding the ending ks,
   (b) add the pair $\{gs, a\}$ to the set W, e.g. $W = \{\{\text{'Warszawa'}, 4\}$, $\{\text{'Warszawom'}, 2\}, \{\text{'Warszawi'}, 1\}\}$.
3. Select a subset $BS2_S$ of the set B_S of such elements, in which at least one of the endings is zero length (e.g. $BS2_S = \{\{\{\text{'e',ø}\}, 1\}\}$).
4. For each element $BS2_S$ such as $\{\{\varnothing, ks\}, a\}$:
   (a) create the word gs by adding the ending ks to the word s,
   (b) add the pair $\{gs, a\}$ to W, e.g. $W = \{\{\text{'Warszawa'}, 4\}$, $\{\text{'Warszawom'}, 2\}, \{\text{'Warszawi'}, 1\}, \{\text{'Warszawiee'}, 1\}\}$.
5. Return the set W.

3.3   Morphological Generator as a RuleSim Similarity Function

In the following, let:

$s = \{\text{'Warszawie'}\}$

$N = \{\text{'Warszawa', 'Kraków', 'Wrocław', 'Werszawa', 'Warszawom'}\}$ – a proper names dictionary as a set of strings

In order to calculate the similarity value for words $\langle s_1, s_2 \rangle$, first we have to determine the longest common prefix lcp_s for words: s_1, s_2. Next, the prefix is removed from beginnings of words of the pair $\langle s_1, s_2 \rangle$ and the pair of endings $\langle ks_1, ks_2 \rangle$ is preserved. According to definitions in Sect. 3, if an element $\{\{ks_1, ks_2\}, a\}$ is in B_S, then a is returned as the value of RuleSim(s_1, s_2) similarity function. In other case 0 is returned.

The inflection rule-based method uses the algorithm described in Sect. 3.2 to generate inflected forms for the input word w. The given set:

$$W = \{\{\text{'Warszawa'}, 4\}, \{\text{'Warszawom'}, 2\}, \{\text{'Warszawi'}, 1\}, \{\text{'Warszawiee'}, 1\}\}$$

contains the similarity function values for each candidate in W. The result is a subset of $\{gs, a\}$ from W where $gs \in N$. In the following example the result is $\{\{\text{'Warszawa'}, 4\}, \{\text{'Warszawom'}, 2\}\}$.

4   Evaluation

We adapted the evaluation process proposed in [3]. The similarity function is applied to find the lemma of an input word w_I or any of its morphological word forms in the NE lexicon if they are included in it. We define *similar* words as words which share the same lemma and agree with respect to all available attributes, except *case* and *number*. We took under consideration only words which belong to the following grammatical classes: *noun (subst), depreciative form (depr), gerund (ger), non-3rd person pronoun (ppron12), 3rd-person pronoun (ppron3), main numeral (num), collective numeral (numcol), adjective (adj), active adj. participle (pact), passive adj. participle (ppas)*[4]. These words can be described with *case* and *number* attributes. Also chosen classes cover all one-word PNs and most of multi-word PNs' components.

For the purposes of the tests, we used *NELexicon*[5] – a very large lexicon of about 1.4 million Polish PNs and NEs available on the Creative Commons licence. It includes not only lemmas, but also inflected word forms for some PNs.

In practice (and also in the prepared evaluation set), the set P of similar words returned by the similarity function for w_I should contain its proper lemma w_L. Moreover, the decision function value for all pairs $\langle w_I, w_O \rangle$ such that $w_O \in P$ and $w_O \neq w_L$ should be lower than for $\langle w_I, w_L \rangle$. This task is different than morphological guessing, e.g. [5] where authors performed generation of lemmas for unknown words on the basis of an *a tergo* index. However, generating of all possible inflected forms of the given word form is a generalization of the morphological guessing (as presented in [5]), in which, instead of *a tergo* index, we use a set of inflection rules (also with the observed rule frequency in morphological dictionary, see Sect. 3.2). The verification process of existence (in the given PN dictionary) is also performed for each generated word form for the input word.

For the sake of comparison with [7], we reproduced test sets from [7,3], i.e. analogical test sets were prepared on the basis of the description in [7,3]. During experiments with single similarity metrics (baseline tests) we obtained the same results as presented by the authors. So the reproduced test sets seem to be a good approximation of the original ones and can be used for the comparison of our own solutions with the methods of [7,3]. They concentrated on selecting a lemma (from the search space) for a PN inflected form on the input. On the basis of *NELexicon* the following test sets of pairs: lemma – inflected form, were generated:

---

[4] Morphosyntactic tags come from the National Corpus of Polish Tagset [9]

[5] http://nlp.pwr.wroc.pl/nelexicon

person_first_nam – a set of Polish first names,
country_nam – Polish country names,
city_nam – Polish city names – not used in [7],
person_full_nam – Polish person full names (multi-word).

Because the considered similarity functions are limited only to one-word PNs, the last test set was not used. Following [7,8] all experiments were performed in two variants, for two different search space sizes:

a *small search space (0 mode)* – only base forms of the test examples,
a *large search space (1 mode)* – all base forms from the named entity category.

Table 1 shows the size of test sets and search spaces for different experiment modes and categories.

Table 1. Test sets and search spaces for different experiment modes and categories

Category	Size		
	Small search space (s_space)	Large search space (l_space)	Tests
person_first_nam	480	15208	1720
country_nam	157	332	621
city_nam	8144	38256	30323

Let:

a be the number of all test examples,
s – the number of tests, in which a single result was returned,
m – the number of tests with more than one result returned,
sc (*single correct*) – the number of tests, in which a single result was returned and it was correct,
mc (*multiple and correct*) – the number of tests with more than one result, but including the correct one,
mc2 (*multiple with best one correct*) – the number of tests with more than one result and with the correct result as the top one (i.e. having the highest decision function value assigned).

We used the three measures proposed by [7]:

All answer accuracy: $AA = \frac{sc}{s+m}$
Single result accuracy: $SR = \frac{sc}{s}$
Relaxed all answer accuracy: $RAA = \frac{sc+mc}{s+m}$

In a similar way to [3] we decided to use the modified version of AA measure (by adding *mc2* parameter) in order to better analyse the cases in which more than one result was returned. Because the complex similarity function and the function utilising morphological generator always return a decision function value as a value of similarity (not only binary decision), we used mACC measure (see [3]) aimed at the comparison of different similarity functions in the domain of their values:

Modified all answer accuracy: $\text{mAA} = \frac{sc+mc2}{s+m}$

Modified global accuracy: $\text{mACC} = \frac{sc+mc2}{a}$

As a baseline we used the similarity metrics included in *SimMetrics* package (as also proposed in [3]). The experiment was performed on the person first name test set in two variants: with small (person_first_nam0) and large (person_first_nam1) search space (see Table 1 for the given sets details). The results are presented in Table 4.

Table 2. Baseline test for person_first_nam with small (s_space) and large (l_space) search space

Similarity metric	s_space variant			l_space variant		
	AA	SR	RAA	AA	SR	RAA
ChapmanLengthDeviation	0.32260	0.57387	0.53503	0.06	0.30721	0.40293
Jaro	0.83164	0.86895	0.87062	0.30501	0.64447	0.63590
JaroWinkler	0.84859	0.87275	0.87514	0.55599	0.64407	0.66517
MatchingCoefficient	0.74011	0.96608	0.97119	0.32133	0.93148	0.94260
Soundex	0.66158	0.97502	0.97401	0.63084	0.68395	0.69893
OverlapCoefficient	0.76780	0.82815	0.83164	0.61171	0.67016	0.68149
QGramsDistance	0.85198	0.86717	0.86893	0.61902	0.67568	0.68542

Single metrics expressed good accuracy in tests with small search spaces, but the results are not satisfactory in tests with large search spaces. Values for the modified evaluation measures are not presented in Table 4.

Table 4 presents the size of sets returned by NamEnSim5. Evaluation measures are based on these results. Resources with suffix *0* are variants with small search space, and with suffix *1* are variants with large search space.

Table 3. Examples of sets size returned in experiments for NamEnSim5 (description of parameteres in Sect. 4)

similarity metric	resource	a	s	m	sc	mc	mc2
NamEnSim5	person_first_nam0	1720	1296	372	1289	371	356
	person_first_nam1	1720	758	923	751	909	746
	country_nam0	621	572	27	572	27	25
	country_nam1	621	501	99	500	99	93
	city_nam0	29492	20495	8121	20258	8022	6674
	city_nam1	29492	11858	17177	11579	16701	12404

Table 4 shows values of the evaluation measures. Presented results for the method utilising morphological generator (RuleSim) are compared with the similarity function NamEnSim5 described in [3] and single similarity metric (CP_δ) described in [7]. RuleSim significantly outperforms other methods in most categories and cases, especially for two important measures: mAA (modified all answer accuracy) and mACC (modified global accuracy). NamEnSim5 has better global accuracy only for test cases with a small search space, but in practice (where the search space is large) the results of NamEnSim5 and CP_δ are not satisfactory. In most cases the results achieved by RuleSim are much better than similar results achieved on lemmatisation of Slovene words for which the method proposed in [2] (utilising a statistics-based trigram tagger) achieves the accuracy of 81%. RuleSim results are also better than one-word similarity methods described in [7,8,3].

Table 4. Evaluation results for the method utilising a morphological generator (RuleSim) in comparison with CP_δ and NamEnSim5

similarity metric	resource	mAA	SR	RAA	mACC
NamEnSim5	person_first_nam0	0.9862	0.9946	0.9952	0.9564
	person_first_nam1	0.8905	0.9908	0.9875	0.8703
	country_nam0	**0.9967**	**1.0000**	**1.0000**	0.9614
	country_nam1	0.9883	0.9980	0.9983	**0.9549**
	city_nam0	0.9412	0.9884	0.9883	0.9132
	city_nam1	0.8260	0.9765	0.9740	0.8132
CP_δ	person_first_nam0	0.9683	0.9810	0.9812	0.9593
	person_first_nam1	0.7915	0.8636	0.8662	0.7878
	country_nam0	0.9885	0.9950	0.9951	**0.9678**
	country_nam1	0.9672	0.9866	0.9869	0.9501
	city_nam0	0.9175	0.9322	0.9306	0.9168
	city_nam1	0.7734	0.8168	0.8170	0.7733
RuleSim	person_first_nam0	**0.9924**	**0.9981**	**0.9982**	**0.9826**
	person_first_nam1	**0.9366**	**1.0000**	**0.9982**	**0.9273**
	country_nam0	0.9950	**1.0000**	**1.0000**	0.9533
	country_nam1	**0.9899**	**1.0000**	**1.0000**	0.9485
	city_nam0	**0.9880**	**0.9998**	**0.9998**	**0.9328**
	city_nam1	**0.8887**	**0.9984**	**0.9986**	**0.8401**

5   Summary

Experiments have shown that it is possible to obtain reasonable results, even better than previously proposed complex similarity function [3], without using a complete morphological dictionary. The quality of the morphological base forms (lemmas) produced by the proposed generator for the unknown words (i.e. not covered by the morphological analyser) is very high.

The achieved results are better than the results of NamEnSim5 complex similarity function which is based on Logistic Regression applied to combine results produced by the selected single similarity metrics (see [3]). The proposed method does not require linguistic knowledge for the identification of endings and should achieve similar results for other languages (where the suffix plays the major role in word inflection), especially for languages with rich inflection (e.g. Slavic).

As in the previous work, we used realistic data set (also containing misspelled words) that might cause slightly worse results than expected, but still the results achieved by morphological generator are better than those of the methods proposed in [2,3,7,8].

References

1. Christen, P.: A comparison of personal name matching: Techniques and practical issues. In: International Conference on Data Mining Workshops, pp. 290–294 (2006)
2. Džeroski, S., Erjavec, T.: Learning to Lemmatise Slovene Words. In: Cussens, J., Džeroski, S. (eds.) LLL 1999. LNCS (LNAI), vol. 1925, pp. 69–88. Springer, Heidelberg (2000), http://dx.doi.org/10.1007/3-540-40030-3_5
3. Kocoń, J., Piasecki, M.: Heterogeneous Named Entity Similarity Function. In: Sojka, P., Horák, A., Kopeček, I., Pala, K. (eds.) TSD 2012. LNCS, vol. 7499, pp. 223–231. Springer, Heidelberg (2012),
   http://dx.doi.org/10.1007/978-3-642-32790-2_27
4. Lubenko, I., Ker, A.D.: Steganalysis using logistic regression. In: Proc. SPIE 7880, p. 78800K (2011)
5. Piasecki, M., Radziszewski, A.: Polish Morphological Guesser Based on a Statistical A Tergo Index. In: Proceedings of the International Multiconference on Computer Science and Information Technology — 2nd International Symposium Advances in Artificial Intelligence and Applications (AAIA 2007), pp. 247–256 (2007), http://www.proceedings2007.imcsit.org/pliks/150.pdf
6. Piskorski, J.: Named-Entity Recognition for Polish with SProUT. In: Bolc, L., Michalewicz, Z., Nishida, T. (eds.) IMTCI 2004. LNCS (LNAI), vol. 3490, pp. 122–133. Springer, Heidelberg (2005), http://dx.doi.org/10.1007/11558637_13
7. Piskorski, J., Sydow, M.: Usability of String Distance Metrics for Name Matching Tasks in Polish. In: Human Language Technologies as a Challenge for Computer Science and Linguistics, Proc. of LTC 2007, pp. 403–407. Wydawnictwo Poznańskie, Sp. z o.o (2007)
8. Piskorski, J., Sydow, M., Kupść, A.: Lemmatization of Polish person names. In: Proceedings of the Workshop on Balto-Slavonic Natural Language Processing: Information Extraction and Enabling Technologies, ACL 2007, pp. 27–34. Association for Computational Linguistics, Stroudsburg (2007),
   http://dl.acm.org/citation.cfm?id=1567545.1567551
9. Przepiórkowski, A., Bańko, M., Górski, R.L., Lewandowska-Tomaszczyk, B. (eds.): Narodowy Korpus Języka Polskiego. Wydawnictwo Naukowe PWN, Warsaw (2012)
10. Woliński, M.: Morfeusz – a Practical Tool for the Morphological Analysis of Polish. In: Kłopotek, M.A., Wierzchoń, S.T., Trojanowski, K. (eds.) Intelligent Information Processing and Web Mining. AISC, vol. 5, pp. 511–520. Springer, Berlin (2006)

NER in Tweets Using Bagging
and a Small Crowdsourced Dataset

Hege Fromreide and Anders Søgaard

Center for Language Technology, University of Copenhagen

Abstract. Named entity recognition (NER) systems for Twitter are
very sensitive to cross-sample variation, and the performance of off-the-
shelf systems vary from reasonable (F_1: 60–70%) to completely useless
(F_1: 40–50%) across available Twitter datasets. This paper introduces a
semi-supervised wrapper method for robust learning of sequential problems with many negative examples, such as NER, and shows that using a
simple conditional random fields (CRF) model and a small crowdsourced
dataset [4], leads to good NER performance across datasets.

Keywords: Twitter, semi-supervised learning, bagging, crowdsourcing,
named entity recognition, unlabeled data.

1   Introduction

Supervised named entity recognition (NER) is the task of learning to identify
and classify names of people, companies, locations, products, etc., in text from
manually annotated data. Supervised NER systems are useful in information extraction (IE), but performance is very domain-dependent [11]. Standard datasets
like CoNLL 2003,[1] MUC-7[2] and ACE 2004[3] are annotated news corpora, and
models induced from such corpora have not proven successful for NER in social media like Twitter [12]. To illustrate the drop in performance from news
to Twitter, we train a CRF model on the CoNLL 2003 training data and evaluate it on the (in-domain) CoNLL 2003 test data, as well as (out-of-domain)
manually annotated Twitter data. Named entities are detected and labeled as
either location (LOC), organization (ORG) or person (PER). While the model
has close to state-of-the-art performance on in-domain data (average F_1 across
LOC, ORG and PER: 90.1%), it performs much worse when evaluated on an
out-of-domain Twitter dataset annotated for the purpose of this paper (53.7%).
This huge drop in performance is obviously prohibitive for down-stream IE in
Twitter. The system proposed in Ritter et al. [12], which is an attempt to adapt
NER to Twitter using manually annotated tweets, does not improve over our
supervised baseline. On the same data, their system obtains a similar result (see
Table 1 below).

---

[1] http://www.cnts.ua.ac.be/conll2003/ner/
[2] LDC2001T02.
[3] LDC2005T09.

A. Przepiórkowski and M. Ogrodniczuk (Eds.): PolTAL 2014, LNAI 8686, pp. 45–51, 2014.
© Springer International Publishing Switzerland 2014

The main reason for the drop from news to Twitter is a change in topics and linguistic conventions [12]. Eisenstein [3] shows that topics and linguistic conventions on Twitter change *very* rapidly. This may explain the relatively poor performance of the system proposed by Ritter et al. [12] on our data, which is sampled differently than their training data. Language drift reduce the utility of a few months old training data from Twitter when applied to tweets sampled differently. In other words, evaluation of NER for Twitter on held-out data from the same sample of tweets may be very misleading.

Our contributions in this paper are as follows:

- We are, to the best of our knowledge, the first to consider using only crowd-sourced data, available at larger volumes, and labeled data from the newswire domain to learn named entity taggers for Twitter, but we are, nevertheless, still able to outperform state-of-the-art supervised taggers,
- we evaluate a wide range of combinations of semi-supervised wrapper methods across several datasets,
- and finally, we introduce two new sizeable evaluation datasets for Twitter NER.

2   Our Approach

The standard baseline model in NER is a linear CRF [7, 14]. This model is similar to structured perceptron [2], but linear CRF minimizes a logistic loss function and provides probability estimates, making re-ranking and semi-supervised learning with confidence thresholds possible. The linear CRF is induced from sequences of words (sentences) labeled manually with symbols indicating whether words are named entities or not. Since this manually labeled data is costly to produce, as it typically requires trained linguists (however, see [4] and Rodrigues et al. [13]), several authors have proposed using semi-supervised learning algorithms for NER model induction [15, 16, 8]. The algorithm presented here is a combination of well-known techniques, but we are the first to show that a semi-supervised approach to NER can make expert annotations of in-domain data superfluous — or at least that systems induced from crowdsourced in-domain data (and a little bit of out-of-domain labeled data) in some cases can outperform state-of-the-art supervised systems.

Our approach is sketched in Fig. 1. We begin by creating five bootstrap samples from the concatenation of the crowdsourced Twitter data and our labeled newswire data. The Twitter data (T) is resampled with replacement, and each sample has the same size (N) as the original dataset. To each sample, we add a copy of the high-quality newswire data (T') that is never altered through the semi-supervised procedure (Fig. 1, line 8). From each bootstrap sample, we learn a linear CRF model (line 9). These five models now form a product-of-experts model. In each iteration of semi-supervised learning, we use this ensemble model to label the unlabeled data (line 12). In each iteration, we add unlabeled data points with predicted labels to our labeled data. The parameter N' denotes the number of unlabeled tweets to be added in each iteration. To prevent our model from becoming too conservative, we balance the unlabeled data by removing low-confidence negative predictions. The parameter M decides how many low-confidence negative predictions to be removed from the new labeled data. After the semi-supervised procedure, we return the final product-of-experts model (line 16).

1: T, T' labeled training data, $T_i = \emptyset$
2: C crowdsourced data
3: U unlabeled data
4: S evaluation data
5: $\sigma(N, \cdot)$ bootstrap sample N datapoints with replacement
6: **for** $iter \in I$ **do**
7:     **for** $i \in [1...5]$ **do**
8:         $T_i \leftarrow \sigma(|C|, T \oplus C) \oplus T'$
9:         $\mathbf{w}_i^* = \Sigma_{i=1}^n \log p(y_i \mid \mathbf{x}_i; \mathbf{w}) - \frac{\lambda}{2}||\mathbf{w}||^2$
10:     **end for**
11:     $U_{iter} \leftarrow \sigma(N', U)$
12:     $LU_{iter} = \{\langle \arg\max_y \prod_i^5 \Sigma_j^m \mathbf{w}_i \cdot \Phi(\mathbf{x}, i, \mathbf{x}_{i-1}, \mathbf{x}_i), \mathbf{x}\rangle \mid \mathbf{x} \in LU_{iter}\}$
13:     $T \leftarrow T \oplus$ **remove_lowconf_negs**(M, LU_{iter}) with $M < N'$
14: **end for**
15: **for** $(y, \mathbf{x}) \in S$ **do**
16:     $y_s = \arg\max_y \prod_i^5 \Sigma_j^m \mathbf{w}_i \cdot \Phi(\mathbf{x}, i, \mathbf{x}_{i-1}, \mathbf{x}_i)$
17: **end for**

Fig. 1. CRF bagging and bootstrapping (parameter setting: $I = 30$, $\lambda = 1$)

3   Other Related Work

Several authors have proposed rule-based NER systems for Twitter, e.g. [10]. Off-the-shelf rule-based approaches may actually be less sensitive to drift than current state-of-the-art data-driven approaches, but we see this as motivating further research in robust data-driven approaches to NER for Twitter. [17] use distant supervision to improve NER for Twitter, but results are much worse than the ones presented here, e.g. 48% F_1 on RITTER. We do think this is an interesting direction for further research, however. The combination of distant supervision and semi-supervised learning seems like a powerful way of leveraging the information available in unlabeled data without running the risk of being led astray by this data, but here we confine ourselves to semi-supervised learning methods.

4   Data Description

The crowdsourced Twitter data provided by Finin et al. [4] were collected during 2008 and consists of 12,800 unique tweets annotated by 266 different annotators from Amazon Mechanical Turk.[4] For development, we held out and manually correct 2,900 tweets from this dataset (DEV-FININ). We used 9,715 tweets for training, containing 165,704 tokens and 8,607 named entities. Most of the tweets were annotated at least twice (95%). We call the training dataset FININ. To select the most likely labels from the redundant annotations, we used MACE [6]. MACE applies EM to detect which annotators are trustworthy, and recover the most likely answer. On our held-out data, MACE led to a small, but significant, improvement over majority voting. We used the default parameters for MACE (50 iterations, 10 restarts, no confidence threshold) to adjudicate between the turkers. The training data from CoNLL 2003 contains 12,690 sentences with 197,517 tokens and 28,039 named entities. Names are more frequent in newswire data than in Twitter, and the inclusion of the out-of-domain data more than triples the number of named entities in training. For evaluation, we use three different datasets collected at different points in time. We use the entire dataset from Ritter et al. [12] (RITTER) collected during 2010. The data were originally annotated with more fine-grained categories, but were easily mapped to our tagset. We also use the dataset from the MSM13 shared task[5] consisting of 1,450 tweets. These data were sampled in 2010 and 2011. And finally, we introduce a more recent in-house dataset, sampled in June 2013 and containing 1,545 tweets (IN-HOUSE).[6]

5   Experiments

Baselines. We compare our system to two off-the-shelf baselines, namely the Stanford NER tagger [5][7] as well as Alan Ritters system [12].[8] Moreover, we use a supervised CRF model trained on a combination of crowdsourced data (FININ) and newswire data (CONLL) as a baseline (in-house baseline). We use a fairly standard feature model, very similar to [14], but with Twitter-specific Brown clusters [9]. The concatenation of crowdsourced data and newswire data is the same we trained our system on, but in the baseline model we do not add semi-supervised learning using pools of unlabeled data. Training the baseline model only on newswire data led to much worse results, consistently lower than any of the other baseline models. Using only the in-domain Twitter data gave similar precision score as our baseline model, but the system recognized fewer entities. Thus, including gold standard out-of-domain data increased recall and F_1.

System. Our approach is a combination of bagging [1] and self-training; cf. Fig. 1. We optimized N' for F_1 and recall (to optimize robustness) leading to slightly different models, resp. BAGGING-1 ($N' = 1000, M = 0$) and BAGGING-2 ($N' = 5000, M = 1000$). Finally, we also compare our bagging models with a co-training

---

[4] https://www.mturk.com
[5] http://www2013.wwwconference.org/
[6] This dataset will be made public after the reviewing.
[7] http://nlp.stanford.edu/software/CRF-NER.shtml
[8] https://github.com/aritter/twitter_nlp

procedure with two taggers, one trained on CONLL and one trained on FININ. In each iteration, each tagger labels 1,000 unlabeled tweets and adds them to the training data of the other tagger. We also experimented with bigger pools and confidence thresholds (increasing N' and M), but did not see improvements in performance on our development data. The system was generally less confident when predicting organization names, and increasing the confidence threshold further reduced the number of new samples for this category. This resulted in lower recall without notable increase in precision.

Results. Our results are presented in Table 1 and shows the F_1 for the baselines and the semi-supervised systems evaluated on the different datasets. The last column is the macro average of the different datasets, but leaving out DEV-FININ when calcultaing the average for the semi-supervised systems and the in-house baseline. The evaluation scores are computed by the perl script conlleval.pl from the CoNLL 2000 shared task. Our three systems all perform significantly better than all baselines (p < 0.01), but we note that co-training is best on MSM13 (except for the Stanford NER system), whereas the bagging-based approaches perform best on the in-house data, as well as the RITTER dataset (except for the system from Ritter et al. 2011). BAGGING-2 gives slightly better results than BAGGING-1, mainly because removing low-confident negative predictions from the unlabeled data resulted in better recall for all categories in all datasets.

Table 1. NER results. *Ritter et al. (2011) is a supervised system, evaluated by 4-CV.

	DEV	TEST			
	FININ	IN-HOUSE	RITTER	MSM13	AV
Baselines					
Stanford NER	63.6	61.1	50.8	**80.4**	64.0
Ritter et al. (2011)	43.1	52.4	*67.1	74.0	59.2
In-house baseline	69.7	66.6	60.4	70.8	65.9
Semi-supervised systems					
CO-TRAINING	70.9	65.9	61.3	79.5	68.9
BAGGING-1	**72.0**	68.1	61.6	75.6	69.1
BAGGING-2	71.1	**70.5**	63.5	76.7	**70.2**

6   Conclusion

We showed that it is possible to learn a named entity tagger for Twitter that outperform state-of-the-art named entity taggers without adding any new gold standard data. Adding new in-domain Twitter data to training boost the performance, but due to significant language drift in Twitter, the effect of such annotations seems to diminish over time. Thus, investing in expert annotations for Twitter seems to be a poor long-term investment if the objective is to induce a robust model for identifying named entities in Twitter. Outsourcing the task to a large crowd is a cheaper and more efficient alternative, but the annotations are of worse quality.

The drop for Ritter et al.'s system when evaluated on our training data could possibly be explained by conceptual differences in the annotation scheme, but our error analysis did not reveal any evidences for such misconceptions. The performance of our in-house baseline is also reduced when applied to later datasets. This emphasize the importance of evaluating NER systems on data sampled differently than the data used in training.

Our results shows that low quality crowdsourced data from the Twitter domain together with an existing out-of-domain dataset can be used to obtain at least as good results as state-of-the-art models that relies on gold standard annotations. Further, we showed that a more robust NER system can be induced using semi-supervised wrapper methods, exploiting the vast amount of unlabeled Twitter data freely available online. All of our three methods outperformed the baselines, and bagging gave the best overall result. Removing low-confident negative predictions from training resulted in a more robust system with better recall and F_1 for all datasets, with exception of the development data.

References

1. Breiman, L.: Bagging predictors. Machine Learning 24(2), 123–140 (1996)
2. Collins, M.: Discriminative training methods for Hidden Markov Models. In: EMNLP (2002)
3. Eisenstein, J.: What to do about bad language on the internet. In: NAACL (2013)
4. Finin, T., Murnane, W., Karandikar, A., Keller, N., Martineau, J., Dredze, M.: Annotating named entities in Twitter data with crowdsourcing. In: NAACL Workshop on Creating Speech and Language Data with Amazons Mechanical Turk (2010)
5. Finkel, J., Grenager, T., Manning, C.: Incorporating non-local information into information extraction systems by Gibbs sampling. In: ACL (2005)
6. Hovy, D., Berg-Kirkpatrick, T., Vaswani, A., Hovy, E.: Learning whom to trust with MACE. In: NAACL (2013)
7. Lafferty, J., McCallum, A., Pereira, F.: Conditional random fields: probabilistic models for segmenting and labeling sequence data. In: ICML (2001)
8. Liu, X., Zhang, S., Wei, F., Zhou, M.: Recognizing named entities in tweets. In: ACL (2011)
9. Owoputi, O., O'Connor, B., Dyer, C., Gimpel, K., Schneider, N., Smith, N.A.: Improved part-of-speech tagging for online conversational text with word clusters. In: NAACL (2013)
10. Piskorski, J., Ehrmann, M.: Named entity recognition in targeted Twitter streams in Polish. In: ACL Workshop on Balto-Slavic NLP (2013)
11. Poibeau, T., Kosseim, L.: Proper name extraction from non-journalistic texts. In: CLIN (2000)
12. Ritter, A., Clark, S., Etzioni, M., Etzioni, O.: Named entity recognition in tweets: an experimental study. In: EMNLP (2011)

13. Rodrigues, F., Pereira, F., Ribeiro, B.: Sequence labeling with multiple annotators. Machine Learning, 1–17 (2013)
14. Sha, F., Pereira, F.: Shallow parsing with conditional random fields. In: HTL-NAACL (2003)
15. Suzuki, J., Isozaki, H.: Semi-supervised sequential labeling and segmentation using giga-word scale unlabeled data. In: ACL, Columbus, Ohio, pp. 665–673 (2008)
16. Turian, J., Ratinov, L., Bengio, Y.: Word representations: a simple and general method for semi-supervised learning. In: ACL (2010)
17. Wang, C.-K., Hsu, B.-J., Chang, M.-W., Kiciman, E.: Simple and knowledge-intensive generative model for named entity recognition. Technical report, Microsoft Research (2013)

Yet Another Ranking Function
for Automatic Multiword Term Extraction

Juan Antonio Lossio-Ventura[1], Clement Jonquet[1],
Mathieu Roche[1,2], and Maguelonne Teisseire[1,2]

[1] University of Montpellier 2, LIRMM, CNRS - Montpellier, France
{juan.lossio,clement.jonquet}@lirmm.fr
[2] Irstea, CIRAD, TETIS - Montpellier, France
mathieu.roche@cirad.fr, maguelonne.teisseire@teledetection.fr

Abstract. Term extraction is an essential task in domain knowledge
acquisition. We propose two new measures to extract multiword terms
from a domain-specific text. The first measure is both linguistic and statistical based. The second measure is graph-based, allowing assessment
of the importance of a multiword term of a domain. Existing measures
often solve some problems related (but not completely) to term extraction, e.g., noise, silence, low frequency, large-corpora, complexity of the
multiword term extraction process. Instead, we focus on managing the
entire set of problems, e.g., detecting rare terms and overcoming the low
frequency issue. We show that the two proposed measures outperform
precision results previously reported for automatic multiword extraction
by comparing them with the state-of-the-art reference measures.

1 Introduction

The huge amount of data available online today is often composed of plain text
fields, e.g., clinical trial descriptions, adverse event reports, electronic health
records [14], customer complaint emails or engineers' repair notes [9]. These
texts are often written with a specific language (expressions and terms) used by
the associated community. There is thus a need for formalization and cataloguing
of these technical terms or concepts. But this task is very time consuming.

Automatic Term Extraction (ATE) or Automatic Term Recognition aim to
automatically extract technical terminology from a given corpus. Technical terminology is a set of terms used in a domain. Therefore term extraction is an essential task in domain knowledge acquisition, because the technical terminology
can be used for lexicon update, domain ontology construction, summarization,
named entity recognition, information retrieval. Technical terms are useful to
gain further insight into the conceptual structure of a domain. These may be:
(i) single-word terms (simple), or (ii) multiword terms (complex). The proposed
work focuses on mutliword term extraction.

Term extraction methods usually involve two main steps. The first step extracts candidates by unithood calculation to qualify a string as a valid term.

A. Przepiórkowski and M. Ogrodniczuk (Eds.): PolTAL 2014, LNAI 8686, pp. 52–64, 2014.

The second step verifies them through termhood measures to validate their domain specificity. Formally, unithood refers to the degree of strength or stability of syntagmatic combinations and collocations, and termhood is defined as the degree to which a linguistic unit is related to domain-specific concepts [11]. ATR has been applied to several domains, e.g., biomedical [13] [14] [6] [25] [17], ecological [4], mathematical [22], social networks [15], banking [5], natural sciences [5], information technology [17], and legal.

There are some well-known ATE issues such as: (i) extraction of non-valid terms (noise) or omission of terms with low frequency (silence), (ii) extraction of multiword terms having complex and various structures, (iii) manual validation efforts of the candidate terms [4], and (iv) management of large-scale corpora.

In response to the above problems, two new measures are proposed in this paper. The first one, called *LIDF-value*, is a statistical- and linguistic-based measure and addresses issues i), ii) and iv). The second one, called *TeRGraph*, is a graph-based measure and deals with issues i), ii) and iii). The main contributions are: (1) enhanced consideration of the term unithood, by computing a degree of quality for the term unithood, and, (2) the consideration of the term dependence in the ATE process. The quality of the proposed method is underlined by comparing the results obtained with the most commonly used baseline measures. The experiments were conducted despite difficulties in comparing ATE measures, mainly because of the size of the corpora used, and the lack of available libraries associated with previous works. Our two measures improve the process of automatic extraction of domain-specific terms from text collections that do not offer reliable statistical evidence.

The paper is organized as follows. We first discuss related work in Sect. 2. Then, the two new term extraction measures are detailed in Sect. 3. Precision evaluation is presented in Sect. 4 followed by the conclusions in Sect. 5.

2 Related Work

Recent studies have focused on multiword (n-grams) and single-word (unigrams) term extraction. Term extraction techniques can be divided into four broad categories: (i) *Linguistic*, (ii) *Statistical*, (iii) *Machine Learning*, and (iv) *Hybrid*. All of these techniques are encompassed in Text Mining approaches. Graph-based approaches have not yet been applied to ATE, although they have been successively adopted in other Information Retrieval fields and they could be suitable for our purpose.

2.1 Text Mining Approaches

Linguistic Approaches. These techniques attempt to recover terms via pattern formation. This involves building rules to describe naming structures for different classes by using orthographic, lexical, or morphosyntactic characteristics, e.g., [7]. The main approach is to (typically manually) develop rules describing common naming structures for certain term classes using orthographic or lexical clues, or more complex morpho-syntactic features.

Statistical Methods. Statistical techniques chiefly rely on external evidence presented through surrounding (contextual) information. Such approaches are mainly focused on the recognition of general terms [23]. The most basic measures are based on frequency. For instance: *term frequency (tf)* counts the frequency of a term in the corpus; *document frequency (df)* counts the number of documents where a term occurs. A similar research topic, called Automatic Keyword Extraction (AKE), proposes to extract the most relevant words or phrases in a document using automatic indexation. Keywords, which we define as a sequence of one or more words, provide a compact representation of the document's content. Such measures can be adapted to extract terms from a corpus as well as ATE measures. In [14] [13], two popular AKE measures, *Okapi BM25* and *TF-IDF* (also called weighting measures), are used to automatically extract biomedical terms; *residual inverse document frequency (RIDF)* compares the document frequency to another chance model where terms with a particular term frequency are distributed randomly throughout the collection; *Chi-square* [16] assesses how selectively words and phrases co-occur within the same sentences as a particular subset of frequent terms in the document text. This is applied to determine the bias of word co-occurrences in the document text, which is then used to rank words and phrases as keywords of the document; *RAKE* [20] hypothesised that keywords usually consist of multiple words and do not contain punctuation or stop words. It uses word co-occurrence information to determine the keywords.

Machine Learning. Machine Learning (ML) systems are often designed for specific entity classes and thus integrate term extraction and term classification. Machine Learning systems use training data to learn features useful for term extraction and classification. But the avaibility of reliable training resources is one of the main problems. Some proposed ATE approaches use machine learning (ML) [4] [24] [17]. Although ML may also generate noise and silence. The main challenge is how to select a set of discriminating features that can be used for accurate recognition (and classification) of term instances.

Hybrid Methods. Most approaches combine several methods (typically linguistic and statistically based) for the term extraction task. *GlossEx* [12] considers the probability of a word in the domain corpus divided by the probability of the appearance of the same word in a general corpus. Moreover, the importance of the word is increased according to its frequency in the domain corpus. *Weirdness* [1] considers that the distribution of words in a specific domain corpus differs from that in a general corpus. *C/NC-value* [6] combines statistical and linguistic information for the extraction of multiword and nested terms. This is the most well-known measure in the literature. While most studies address specific types of entities, *C/NC-value* is a domain-independent method. It has also been used for recognizing terms in the biomedical literature [8] [14]. In [25], the authors showed that *C-value* obtains the best results compared to the other measures cited above. Another measure is *F-TFIDF-C* [13], which combines an ATE measure *(C-value)* and an AKE measure *(TF-IDF)* to extract terms, thus obtaining better results than *C-value*. Moreover, *C-value* has also been applied to different languages other than English, e.g., Japanese, Serbian, Slovenian, Polish, Chinese [10], Spanish [2], Arabic, and French [14]. That is why we have chosen *C-value* and *F-TFIDF-C* as baselines for the proposed experiments.

2.2 Graph-Based Approaches

Graph modeling is an alternative for modeling information, which clearly highlights relationships of nodes among vertices. It also groups related information in a specific way, and a centrality algorithm can be applied to enhance their efficiency. An increasingly popular recent application of graph approaches to Information Retrieval (IR) concerns social or collaborative networks and recommender systems [18]. Graph representations of text and scoring function definition are two widely explored research topics, but few studies have been focused on graph-based IR in terms of both document representation and weighting models [21]. First, text is modeled as a graph where nodes represent words and edges represent relations between words, defined on the basis of any meaningful statistical or linguistic relation [3]. In [3], the authors developed a graph-based word weighting model that represents each document as a graph. The importance of a word within a document is estimated by the number of related words and their importance, in the same way that PageRank [19] estimates the importance of a page via the pages that are linked to it. Another study, [21], introduces a different representation of document that captures relationships between words by using an unweighted directed graph of words with a novel scoring function.

In the above approaches, graphs are used to measure the influence of words in documents like automatic keyword extraction methods (AKE) while ranking documents against queries. These approaches differ from ours as they use graphs that are focused on the extraction of relevant words in a document and computing relations between words. In our proposal, a graph is built such that the vertices are multiword terms and the edges are relations between multiword terms. Moreover, we focus especially on a scoring function of relevant multiword terms in a domain rather than in a document.

3   Two Measures for Multiword Term Extraction

3.1   A New Ranking Measure Based on Linguistic and Statistical Information: *LIDF-value* (Linguisitic Patterns, IDF, and C-value Information)

Three steps are involved in computing the *LIDF-value*:

(1) **Part-of-Speech tagging:** a part-of-speech is applied to the whole corpus to obtain the lemma of words and to extract linguistic patterns. Part-of-Speech (POS) tagging is the process of assigning each word in a text to its grammatical category (e.g., noun, adjective). This process is performed based on the definition of the word or on the context in which it appears.

(2) **Candidate term extraction:** before applying any measures, we select terms having a syntactic structure appearing in the pattern list.
(3) **Ranking of candidate terms:** finally the *LIDF-value* is computed for each term.

These steps are explained in the next subsections and detailed in Algorithm 1.

From the Linguistic-Based Approach. The objective is to give greater importance to the term unithood in order to detect low frequency terms.

As in related work, we supposed that terms of a domain have a similar syntactic structure. Therefore, we build a list of the most common linguistic patterns according the syntactic structure of technical terms present in a dictionary. In our work, we chose UMLS[1] which is a biomedical dictionary. We conduct part-of-speech tagging of the domain dictionary using the Stanford CoreNLP API (POS tagging)[2], and then compute the frequency of syntactic structures. Patterns among the 200 highest frequencies are selected to build the list. From this list, we compute the weight associated with the probability that a candidate term could be a domain term if its syntactic structure appears in the linguistic pattern list. In our experiments, 2 300 000 terms were used to build the list of patterns. Table 1 illustrates the computation of the linguistic pattern probability.

Table 1. Example of pattern construction (where *NN* is a noun, *IN* a preposition or subordinating conjunction, *JJ* an adjective, and *CD* a cardinal number)

Pattern	Frequency	Probability
NN IN JJ NN IN JJ NN	3006	3006/4113 = 0.73
NN CD NN NN NN	1107	1107/4113 = 0.27
	4113	1.00

To the Statistical-Based Approach. Our method *LIDF-value* is aimed at computing the termhood for each term, using the *probability* calculated as defined above, the *idf*, and the *C-value* of each term. The inverse document frequency *(idf)* is a measure indicating the extent to which a term is common or rare across all documents. It is obtained by dividing the total number of documents by the number of documents containing the term, and then taking the logarithm of that quotient.

The *probability* and the *idf* improve the extraction of low frequency terms. The *C-value* measure is based on the term frequency. The aim of the *C-value* (see (1)) is to improve the extraction of nested terms, i.e., this criteria favors a candidate term that does not often appear in a longer term. For instance, in a specialized corpus (Ophthalmology), the authors of [6] found the irrelevant term "soft contact" while the frequent and longer term "soft contact lens" is relevant.

---

[1] http://www.nlm.nih.gov/research/umls
[2] http://nlp.stanford.edu/software/corenlp.shtml

$$C\text{-}value(A) = \begin{cases} \log_2(|A|) \times f(A) & \text{if } A \notin \text{ nested} \\[2mm] \log_2(|A|) \times \left(f(A) - \frac{1}{|S_A|} \times \sum_{b \in S_A} f(b) \right) & \text{otherwise} \end{cases} \qquad (1)$$

Where A represents multiword terms, $|A|$ the number of words in A, $f(A)$ the frequency of A in the documents, S_A the set of terms that contain A and $|S_A|$ the number of terms in S_A. In a nutshell, $C\text{-}value$ uses the frequency of the term if the term is not included in other terms (first line), or decreases this frequency if the term appears in other terms, by using the frequency of those other terms (second line). The algorithm 1 describes the applied process.

These different statistical information items (i.e., *probability* of linguisitic patterns, *C-value*, *idf*) are combined to define the global ranking measure *LIDF-value* (see (2)); where $\mathrm{P}(A_{LP})$ is the probability of a multiword term A which has the same linguistic structure pattern LP, i.e., the weight of the linguistic pattern LP computed in Sect. 3.1.

$$LIDF\text{-}value(A) = \mathrm{P}(A_{LP}) \times idf(A) \times C\text{-}value(A) \qquad (2)$$

Algorithm   1.   ComputeLIDF-value   *(Corpus,   Patterns, min_{freq}, num_{terms})*

Data: *Corpus* = set of documents of a specific-domain;
Patterns = $HT_{patterns}(pattern, probability)$ //Hashtable of linguistic patterns with its probability;
min_{freq} = frequency threshold for candidate terms;
num_{terms} = number of terms to take as output
Result: L_{terms} = List of ranked terms
begin
    Tag the *Corpus*;
    Take the *lemma* of each tagged word;
    Extract candidate terms A by filtering with *Patterns*;
    Remove candidate terms A below min_{freq};
    for *each candidate term $A \in$ Corpus* **do**
        $LIDF\text{-}value(A) = \mathrm{P}(A_{LP}) \times idf(A) \times C\text{-}value(A)$;
        add A to L_{terms};
    end
    Rank the L_{terms} by the value obtained with $LIDF\text{-}value$;
    Select the first num_{terms} terms of L_{terms} ;
end

As an improvement, we propose to take into account graph-theoretic information to highlight relevant terms, as explained in the following subsection.

3.2 A New Graph-Based Ranking Measure: *TeRGraph* (Terminology Ranking Based on Graph Information)

This approach aims to improve the precision of the top k extracted terms. As mentioned above, in contrast to the work cited before, the graph is built with a list of terms obtained according to the steps described in Sect. 3.1, where vertices denote multiword terms linked by their co-occurrence in the sentences in the corpus. Moreover, we apply the hypothesis that the term representativeness in a graph, for a specific-domain, depends on the number of neighbors that it has, and the number of neighbors of its neighbors. We assume that a term with more neighbors is less representative of the specific-domain. This means that this term is used in the general domain. Figure 1 illustrates our hypothesis. The graph-based approach is divided into two steps:

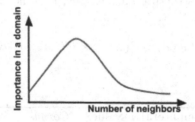

Fig. 1. Importance of a term in a domain

(1) **Graph construction:** a graph (see Fig. 2) is built where vertices denote terms, and edges denote co-occurrence relations between terms, co-occurrences between terms are measured as the weight of the relation in the initial corpus. This approach is statistical because it links all co-occurring terms without considering their meaning or function in the text. This graph is undirected as the edges imply that terms simply co-occur, without any further distinction regarding their role. We take *Dice coefficient*, a basic measure to compute the co-occurrence between two terms x and y, defined by the following formula:

$$D(x,y) = \frac{2 \times P(x,y)}{P(x) + P(y)} \tag{3}$$

(2) **Representativeness computations on the term graph:** a principled graph-based measure to compute term weights (representativeness) is defined. The aim of this new graph ranking measure, *TeRGraph*, see (4), is to derive these weights for each vertex, (i.e., multiword term weight), in order to re-rank the list of extracted terms.

$$TeRGraph(A) = \log_2 \left(1.5 + \frac{1}{|N(A)| + \sum\limits_{T_i \in N(A)} |N(T_i)|} \right) \tag{4}$$

Where A represents a vertex (multiword term), $N(A)$ the neighborhood of A, $|N(A)|$ the number of neighbors of A, T_i the neighbor i of A. The intuition for (4) is as follows: the more a term A has neighbors (directly with $N(A)$ or by transitivity with $N(T_i)$), the more the weight decreases. Indeed, a term A having a lot of neighbors is considered too general for the domain (i.e., this term is not salient), then it has to be penalized via the associated score. Figure 2 shows an example to calculate the value of *TeRGraph* for a term in different graphs. These graphs are built with different co-occurrence thresholds (i.e., Dice's value between two terms). In this example, A_1 and A_2 represent the term *chloramphenicol acetyltransferase reporter* in Graphs 1 and 2 respectively.

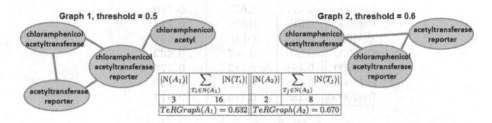

Fig. 2. *TeRGraph*'s value for ***chloramphenicol acetyltransferase reporter***

4 Experiments and Results

4.1 Data, Protocol, and Validation

In our experiments, the standard GENIA[3] corpus was used, which is made up of 2 000 titles and abstracts of journal articles derived from the Medline database, with more than 400 000 words. GENIA corpus contains linguistic expressions referring to entities of interest in molecular biology, such as proteins, genes and cells. The GENIA technical term annotation covers the identification of physical biological entities as well as other important terms.

4.2 Results

The results are evaluated in terms of *precision* obtained over the top k extracted terms (P@k) for the two proposed measures and baseline measures for multiword terms. In the following subsections, we narrow down the presented results by keepingfor the graph-based measureonly the first 8 000 extracted terms.

---

[3] http://www.nactem.ac.uk/genia/genia-corpus/term-corpus

Linguistic and Statistical Results. Table 2 presents and compares the results of multiword term extraction with the best baseline measures, such as, *C-value*, *F-TFIDF-C*, and our measure *LIDF-value*. The best results were obtained with *LIDF-value* with an improvement in precision of 11% for the first hundred extracted multiword terms. The precision of *LIDF-value* will be further improved with *TeRGraph*.

Table 2. Precision comparison of *LIDF-value* with baseline measures

	C-value	*F-TFIDF-C*	***LIDF-value***
P@100	0.690	0.715	**0.820**
P@200	0.690	0.715	**0.770**
P@300	0.697	0.710	**0.750**
P@400	0.665	0.690	**0.738**
P@500	0.642	0.678	**0.718**
P@600	0.638	0.668	**0.723**
P@700	0.627	0.669	**0.717**
P@800	0.611	0.650	**0.710**
P@900	0.612	0.629	**0.714**
P@1000	0.605	0.618	**0.697**
P@2000	0.570	0.557	**0.662**
P@5000	0.498	0.482	**0.575**
P@10000	0.428	0.412	**0.526**
P@20000	0.353	0.314	**0.377**

We evaluated *LIDF-value* and baseline measures within a sequence of n-gram terms (i.e., n-gram term is a multiword term of n words), for this we require an index term to be a n-gram terms of length $n \geq 2$. Table 3 shows the ranking of 3-gram terms with the baseline measures and *LIDF-value*. For 3-gram terms *C-value* obtains 2 irrelevant terms, *F-TFIDF-C* obtains 3 irrelevant terms while *LIDF-value* obtains only 1 irrelevant term.

Graph Results. Our graph-based approach is applied to the first 8 000 terms extracted by the Linguistic and Statistical approach. The objective is to re-rank the 8 000 terms while trying to improve the precision by intervals. One parameter is involved in the computation of graph-based term weights, namely the *threshold* of Dice value which represents the relation when building the term graph. This involves linking terms whose *Dice value* of the relation is higher than *threshold*. We vary *threshold* (δ) within $\delta = [0.25, 0.35, 0.50, 0.60, 0.70]$ and report the precision performance for each of these values. Table 4 gives the precision performance obtained by *TeRGraph* and shows that it is well adapted for ATE.

Summary. Table 5 presents a precision comparison of our two measures. In terms of overall precision, our experiments produce consistent results from the GENIA corpus. In most cases, *TeRGraph* obtains better precision with a *threshold* of 0.60 and 0.70 (i.e., better precision in most P@k intervals), which is very

Table 3. Comparison of top-10 ranked 3 gram terms (irrelevant terms are italicized and marked with *)

C-value	F-TFIDF-C
human immunodeficiency virus	*kappa b alpha**
*kappa b alpha**	nf kappa b
tumor necrosis factor	jurkat t cell
electrophoretic mobility shift	human t cell
nf-kappa b activation	mhc class ii
*virus type 1**	cd4+ t cell
protein kinase c	*c-fos and c-jun**
long terminal repeat	peripheral blood monocyte
nf kappa b	t cell proliferation
jurkat t cell	*transcription factor nf-kappa**

LIDF-value
i kappa b
human immunodeficiency virus
electrophoretic mobility shift
human t cell
mobility shift assay
*kappa b alpha**
tumor necrosis factor
nf-kappa b activation
protein kinase c
jurkat t cell

Table 4. Precision performance of *TeRGraph* when varying δ parameter

	$\delta \geq 0.25$	$\delta \geq 0.35$	$\delta \geq 0.50$	$\delta \geq 0.60$	$\delta \geq 0.70$
P@100	0.840	0.860	0.910	**0.930**	0.900
P@200	0.800	0.790	0.850	**0.855**	**0.855**
P@300	0.803	0.773	0.833	**0.830**	0.820
P@400	0.780	0.732	**0.820**	**0.820**	0.815
P@500	0.774	0.712	0.798	**0.810**	0.806
P@600	0.773	0.675	0.797	**0.807**	0.792
P@700	0.760	0.647	0.769	**0.796**	0.787
P@800	0.756	0.619	0.748	**0.784**	0.779
P@900	0.748	0.584	0.724	0.773	**0.777**
P@1000	0.751	0.578	0.720	0.766	**0.769**
P@2000	0.689	0.476	0.601	0.657	**0.694**
P@3000	0.642	0.522	0.535	0.605	**0.644**
P@4000	**0.612**	0.540	0.543	0.559	0.593
P@5000	**0.574**	0.546	0.544	0.554	0.562
P@6000	0.558	0.539	0.540	0.549	**0.561**
P@7000	**0.556**	0.540	0.540	0.545	0.552
P@8000	**0.546**	**0.546**	**0.546**	**0.546**	**0.546**

Table 5. Precision comparison of *LIDF-value* and *TeRGraph*

	LIDF-value	*TeRGraph* ($\delta \geq 0.60$)	*TeRGraph* ($\delta \geq 0.70$)
P@100	0.820	**0.930**	0.900
P@200	0.770	**0.855**	**0.855**
P@300	0.750	**0.830**	0.820
P@400	0.738	**0.820**	0.815
P@500	0.718	**0.810**	0.806
P@600	0.723	**0.807**	0.792
P@700	0.717	**0.796**	0.787
P@800	0.710	**0.784**	0.779
P@900	0.714	0.773	**0.777**
P@1000	0.697	0.766	**0.769**
P@2000	0.662	0.657	**0.694**
P@3000	0.627	0.605	**0.644**
P@4000	**0.608**	0.5585	0.593
P@5000	**0.575**	0.5538	0.562
P@6000	0.550	0.549	**0.561**
P@7000	0.547	0.545	**0.552**
P@8000	**0.546**	**0.546**	**0.546**

good because it helps alleviate the problem of manual validation of candidate terms. The performance of our graph-based measure depends somewhat on the value of the co-occurrence relation between terms. Specifically, the value of the co-occurrence relation affects how the graph is built (whose edges are taken), and hence it is critical for computation of the graph-based term weight. Another performance factor of our graph-based measure is the quality of the results obtained with *LIDF-value* due to the fact that to re-rank *TeRGraph* the list of terms extracted with *LIDF-value* is required as input, in order to construct the graph, where nodes denote terms, and edges denote co-occurrence relations.

5   Conclusions and Future Work

This paper defines and evaluates two measures for automatic multiword term extraction. The first one, *LIDF-value*, a linguistic and statistical-based measure, improves the precision of automatic term extraction in comparison with the most popular term extraction measure. This measure overcomes the lack of frequency information with the values of *linguistic pattern probability* and *idf*. We experimentally show that *LIDF-value* applied in the biomedical domain outperformed a state-of-the-art baseline for extracting terms (i.e., *C-value* and *F-TFIDF-C*), while obtaining the best precision results in all intervals (i.e., P@k).

The second one, *TeRGraph*, is a graph-based measure. It enables a reduction in the huge human effort required to validate candidate terms. The graph-based measure has never been applied for automatic term extraction. *TeRGraph* takes into account the neighborhood to compute the term representativeness in a specific domain. Our experimental evaluations reveal that *TeRGraph* has better precision than *LIDF-value* for all intervals.

As a future extension of this work, we intend to use the relation value within *TeRGraph*. Moreover, we plan to test this general approach in other domains, such as ecology and agronomy. Finally, future work includes the use of other graph ranking computations, e.g., PageRank, adapted for automatic term extraction.

Acknowledgments. This work was supported in part by the French National Research Agency under JCJC program, grant ANR-12-JS02-01001, as well as by University of Montpellier 2 and CNRS.

References

1. Ahmad, K., Gillam, L., Tostevin, L.: University of Surrey Participation in TREC8: Weirdness Indexing for Logical Document Extrapolation, Retrieval (WILDER). In: TREC (1999)
2. Barrón-Cedeño, A., Sierra, G., Drouin, P., Ananiadou, S.: An improved automatic term recognition method for Spanish. In: Gelbukh, A. (ed.) CICLing 2009. LNCS, vol. 5449, pp. 125–136. Springer, Heidelberg (2009)
3. Blanco, R., Lioma, C.: Graph-based term weighting for information retrieval. Information Retrieval 15, 54–92 (2012)
4. Conrado, M.S., Pardo, T.A.S., Rezende, S.O.: Exploration of a Rich Feature Set for Automatic Term Extraction. In: Castro, F., Gelbukh, A., González, M. (eds.) MICAI 2013, Part I. LNCS (LNAI), vol. 8265, pp. 342–354. Springer, Heidelberg (2013)
5. Dobrov, B., Loukachevitch, N.: Multiple Evidence for Term Extraction in Broad Domains. In: Proceeding of Recent Advances in Natural Language Processing (RANLP), Hissar, Bulgaria, pp. 710–715 (2011)
6. Frantzi, K., Ananiadou, S., Mima, H.: Automatic recognition of multiword terms: the C-value/NC-value Method. International Journal on Digital Libraries 3, 115–130 (2000)
7. Gaizauskas, R., Demetriou, G., Humphreys, K.: Term recognition, classification in biological science journal articles. In: Proceeding of the Computional Terminology for Medical, Biological Applications Workshop of the 2nd International Conference on NLP, pp. 37–44 (2000)
8. Hliaoutakis, A., Zervanou, K., Petrakis, E.G.M.: The AMTEx approach in the medical document indexing, retrieval application. Data & Knowl. Engineering 68, 380–392 (2009)
9. Ittoo, A., Bouma, G.: Term Extraction from Sparse, Ungrammatical Domain-specific Documents. Expert Systems with Applications 40, 2530–2540 (2013)
10. Ji, L., Sum, M., Lu, Q., Li, W., Chen, Y.: Chinese Terminology Extraction Using Window-Based Contextual Information. In: Gelbukh, A. (ed.) CICLing 2007. LNCS, vol. 4394, pp. 62–74. Springer, Heidelberg (2007)
11. Kageura, K., Umino, B.: Methods of automatic term recognition: A review. Terminology 3, 259–289 (1996)

12. Kozakov, L., Park, Y., Fin, T., Drissi, Y., Doganata, N., Confino, T.: Glossary extraction, knowledge in large organisations via semantic web technologies. In: Proceedings of the 6th International Semantic Web Conference, he 2nd Asian Semantic Web Conference (Semantic Web Challenge Track) (2004)
13. Lossio-Ventura, J.A., Jonquet, C., Roche, M., Teisseire, M.: Biomedical Terminology Extraction: A new combination of Statistical, Web Mining Approaches. In: Proceedings of Journées Internationales d'Analyse Statistique des Données Textuelles (JADT 2014), Paris, France (2014)
14. Lossio-Ventura, J.A., Jonquet, C., Roche, M., Teisseire, M.: Combining C-value, Keyword Extraction Methods for Biomedical Terms Extraction. In: Proceedings of the Fifth International Symposium on Languages in Biology, Medicine (LBM 2013), Tokyo, Japan, pp. 45–49 (2013)
15. Lossio-Ventura, J.A., Hacid, H., Ansiaux, A., Maag, M.L.: Conversations reconstruction in the social web. In: Proceedings of the 21st International Conference Companion on World Wide Web (WWW 2012), pp. 573–574. ACM, Lyon (2012)
16. Matsuo, Y., Ishizuka, M.: Keyword extraction from a single document using word co-occurrence statistical information. International Journal on Artificial Intelligence Tools 13, 157–169 (2004)
17. Newman, D., Koilada, N., Lau, J.H., Baldwin, T.: Bayesian Text Segmentation for Index Term Identification, Keyphrase Extraction. In: Proceedings of 24th International Conference on Computational Linguistics, Mumbai, India, pp. 2077–2092 (2012)
18. Noh, T., Park, S., Yoon, H., Lee, S., Park, S.: An Automatic Translation of Tags for Multimedia Contents Using Folksonomy Networks. In: Proceedings of the 32Nd International ACM SIGIR Conference on Research, Development in Information Retrieval, SIGIR 2009, pp. 492–499. ACM, Boston (2009)
19. Page, L., Brin, S., Motwani, R., Winograd, T.: The PageRank citation ranking: Bringing order to the web. Stanford InfoLab (1999)
20. Rose, S., Engel, D., Cramer, N., Cowley, W.: Automatic keyword extraction from individual documents. Text Mining: Theory, Applications, pp. 1–20. John Wiley, Sons, Ltd. (2010)
21. Rousseau, F., Vazirgiannis, M.: Graph-of-word, TW-IDF: New Approach to Ad Hoc IR. In: Proceedings of the 22nd ACM International Conference on Conference on Information, Knowledge Management, CIKM 2013, pp. 59–68. ACM, San Francisco (2013)
22. Stoykova, V., Petkova, E.: Automatic extraction of mathematical terms for precalculus. Procedia Technology Journal 1, 464–468 (2012)
23. Van Eck, N.J., Waltman, L., Noyons, E.C.M., Buter, R.K.: Automatic term identification for bibliometric mapping. Scientometrics 82, 581–596 (2010)
24. Zhang, X., Song, Y., Fang, A.C.: Term recognition using conditional random fields. In: International Conference on Natural Language Processing, Knowledge Engineering (NLP-KE), pp. 1–6. IEEE (2010)
25. Zhang, Z., Iria, J., Brewster, C., Ciravegna, F.: A Comparative Evaluation of Term Recognition Algorithms. In: Proceedings of the Sixth International Conference on Language Resources, Evaluation (LREC 2008), Marrakech, Morocco (2008)

Unsupervised Keyword Extraction
from Polish Legal Texts

Michał Jungiewicz[1,2] and Michał Łopuszyński[1]

[1] Interdisciplinary Centre for Mathematical and Computational Modelling,
University of Warsaw, Pawińskiego 5a, 02-106 Warsaw Poland
m.lopuszynski@icm.edu.pl
[2] Faculty of Electronics and Information Technology,
Warsaw University of Technology, Nowowiejska 15/19, 00-665 Warsaw, Poland

Abstract. In this work, we present an application of the recently proposed unsupervised keyword extraction algorithm RAKE to a corpus of Polish legal texts from the field of public procurement. RAKE is essentially a language and domain independent method. Its only language-specific input is a stoplist containing a set of non-content words. The performance of the method heavily depends on the choice of such a stoplist, which should be domain adopted. Therefore, we complement RAKE algorithm with an automatic approach to selecting non-content words, which is based on the statistical properties of term distribution.

Keywords: Keyword extraction, unsupervised learning, legal texts.

1 Introduction

Automatic analysis of legal texts is currently viewed as a promising research and application area [1]. On the other hand, keyword extraction is a very useful technique in organization of large collections of documents. It helps to present the available information to the user, aids browsing and searching. Moreover, extracted keywords can be useful as features in tasks, such as document similarity calculation, clustering, topic modelling, etc.

Unfortunately, the problem of automatic keyword extraction is far from solved. A recently conducted competition during the SemEval 2010 Workshop, showed that the best available algorithms do not exceed 30% of the F-measure, on the manually labeled test documents [2]. It is worth noticing that these tests were based on English texts. For highly inflected languages (e.g., Polish) it might be even more difficult and algorithms here are certainly less developed and verified.

In the presented paper, we employ recently proposed RAKE algorithm [3]. It was designed as an unsupervised, domain-independent, and language-independent method of extracting keywords from individual documents. These features make it a promising candidate tool for a highly specific task of extracting keywords from Polish legal texts. However, in the original paper authors evaluated RAKE only on English texts. Its performance on a very different Slavic language may deviate and is worth verifying.

A. Przepiórkowski and M. Ogrodniczuk (Eds.): PolTAL 2014, LNAI 8686, pp. 65–70, 2014.

The corpus used in this research consisted of 11 thousand rulings of the National Appeals Chamber from the Polish Public Procurement Office. In our opinion, this set of documents is particularly interesting and challenging. It contains very diverse vocabulary, not only related to law and public procurement issues, but also to the technicalities of discussed contracts coming form very different fields (medicine, construction, IT, etc.)

2   Automatic Stoplist Generation

The general idea behind RAKE algorithm is based on splitting a given text into word groups isolated by sentence separators or words from a provided stoplist. Each such a word group is considered to be a keyword candidate and is scored according to the word co-occurrence graph. The details of the method can be found in [3]. The stoplist constitutes the most important "free parameter" of RAKE, as it is the only way to adjust this algorithm to the specific language and domain. As recognized by the authors of RAKE, it is also a crucial ingredient on which the effectiveness of the algorithm strongly depends [3]. Our initial tests carried out with a standard information retrieval stoplist yielded poor results for the case of Polish legal texts. There were a lot of very long keywords, containing many uninformative words, even though our implementation did not include merging of the adjoining keyword candidates. Sample results are presented in Table 1A. To alleviate this type of problems, the authors of RAKE propose two methods of automatic stopwords generation from a given corpus [3]. However, none seems satisfactory for us. The first one is very crude, as it simply uses the most frequent words. The second one requires an annotated training set (supervised learning). Therefore, we develop our own unsupervised approach to the stoplist auto-generation problem. It is based on the observation that distribution of the number of occurrences per document for stopwords usually follows typical random variable model (e.g., Poisson distribution). Informative content words, on the other hand, occur in more "clustered" fashion and mostly deviate from the distribution of stopwords [4,5].

The simplest method of detecting this deviation is based on two variables — the number of documents in which a given word is present $\mathbf{df}$ and the cumulative collection word frequency $\mathbf{cf}$. For the randomly distributed stopwords the relation of $\mathbf{df}$ to $\mathbf{cf}$ in a large set of documents is defined by the probability theory [5]

$$\overline{\mathbf{df}}\,(\mathbf{cf}) = N(1 - P(0, \mu = \mathbf{cf}/N)), \tag{1}$$

where N is the total number of documents, and $P(0, \mu)$ is the probability of the word occurring 0 times, provided its average number of occurrences per document μ (by definition $\mu = \mathbf{cf}/N$). For the simplest Poisson model the equation reduces to

$$\overline{\mathbf{df}}\,(\mathbf{cf}) = N(1 - \exp(-\mathbf{cf}/N)). \tag{2}$$

The plot of $\mathbf{df}$ against $\mathbf{cf}$ for all words in the examined corpus is presented in Fig. 1a. The Poisson model is plotted in Fig. 1b. One can easily see that it

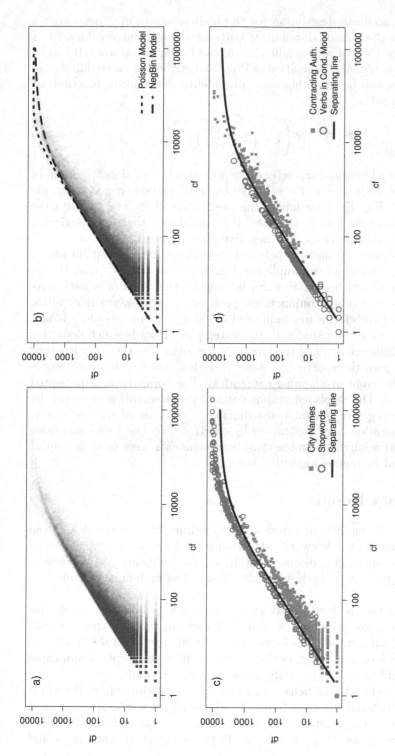

Fig. 1. Scatter plots of the number of documents with a given word (**df**) vs. its frequency in the collection (**cf**) for the whole corpus vocabulary. Panel a) shows plain scatter plot. Panel b) compares the model making use of the Poisson distribution (2) with the negative binomial approach (3). Panel c) examines the location of the city names and standard information retrieval stopwords. Panel d) contrasts the location of contracting authorities and verbs in conditional mood. Clearly, non-content words (information retrieval stopwords and verbs in conditional mood) tend to locate close to the theoretical curve given by (3). We decided to extract the non-content words for stoplist using the separating line **df/df** = 1.6, which is marked in panels c) and d)

does not give an accurate description for the high values of **cf**. Therefore, we decided to replace the Poisson distribution with the negative binomial model. It is closely related to the Poisson variable, but allows for larger variance. It can be also represented as infinite combination of Poisson distributions with different μ. After substituting the negative binomial probability distribution function for $P(0, \mu)$ in (1), we get

$$\overline{\mathbf{df}}\,(\mathbf{cf}) = N\left(1 - \left(1 + \frac{\mathbf{cf}}{Nr}\right)^{-r}\right), \tag{3}$$

where $r > 0$ is the additional parameter of the negative binomial distribution. In the case of $r \to \infty$ with fixed μ, the negative binomial variable converges to the Poisson model. In Fig. 1b, we compare the predictions of (2) and (3) with the value of $r = 0.42$, adjusted to fit the data. It is easily seen that the description of the high **cf** region improves for the negative binomial case.

To further illustrate the difference between the content and non-content words, we compared locations of a few sample word categories in (**cf**,**df**) space. We selected two groups of non-informative words, namely, the usual information retrieval stopwords (containing conjunctions, pronouns, particles, auxiliary verbs, etc.) and a class of verb forms in conditional mood, ending on -ÅĆaby -ÅĆoby. These two groups were compared with two categories of words which definitely carry important information, i.e., the names of the cities and the most frequent words extracted from the contracting authorities list (cleaned from stopwords and city names to avoid overlapping categories). The comparison is presented in Fig. 1c and 1d. The displayed graphs confirm the assumption of larger deviation from the negative binomial distribution in the case of content words. Approximate separation can be obtained by $\overline{\mathbf{df}}/\mathbf{df} < 1.6$. The terms satisfying this condition and occurring in more than ten documents were used as stoplist in RAKE keyword extraction algorithm later on.

3   Preliminary Results

After developing the method of automatically distilling the stopword list from a given corpus, we ran the keyword extraction procedure on the available documents. Since the documents did not contain any manually assigned keywords, we can do only qualitative analysis at this stage. The preliminary results are presented below.

We found that the method indeed yields useful key phrases. Its results for a sample document are presented in Table 1B and can be compared with the results obtained using standard information retrieval stoplist (Table 1A). The extracted phrases look promising, as they clearly indicate the topic of municipal waste management to which the analyzed document is related.

To get more insight into the behaviour of the algorithm throughout the whole corpus, we also analyzed the most frequently detected keywords. Top five most popular key phrases are presented in Table 1C. The result is intuitively well understood, since a considerable part of the public procurement contracts in Poland

Table 1. Summary of experiments with RAKE. Both the original keywords and their English translations are given.

A. Top 5 high-score keywords extracted from a sample document (standard stoplist)	
samej grupy kapitaÅĆowej dotyczÄĆcego wykonawcy PrzedsiÄŹbiorstwo UsÅĆug Komunalnych Empol sp.	the same capital group concerning the contractor Municipal Services Company Empol
Dzienniku UrzÄŹdowym Unii Europejskiej 23 marca 2013 r.	(in) the Official Journal of the European Union 23 March 2013
Prezesa Krajowej Izby OdwoÅĆawczej 20 czerwca 2012 r.	Chairman of the National Appeal Chamber 20 June 2012
pierwszej kolejnoÅŹci Krajowa Izba OdwoÅĆawcza winna oceniÄĞ	firstly the National Appeal Chamber should judge
Krajowa Izba OdwoÅĆawcza uwzglÄŹdniÅĆa odwoÅĆanie konsorcjum Sita MaÅĆopolska	National Appeal Chamber has upheld the appeal of the Sita MaÅĆopolska Consortium

B. All keywords extracted from a sample document (auto-generated stoplist of Sect. 2)	
PrzedsiÄŹbiorstwo UsÅĆug Komunalnych Empol	Municipal Services Company Empol
przedsiÄŹbiorstwo usÅĆug komunalnych	municipal services company
zagospodarowanie odpadÄĆw komunalnych	management of municipal waste
odbieranie odpadÄŹw komunalnych	municipal waste collection
wÅĆaÅŹcicieli nieruchomoÅŹci zamieszkaÅĆych	residential real estate owner
konsorcjum Sita MaÅĆopolska	Consortium Sita MaÅĆopolska
Sita MaÅĆopolska	Sita MaÅĆopolska
grupy kapitaÅĆowej	capital group

C. Most frequent keywords in the whole corpus (auto-generated stoplist of Sect. 2)	
roboty budowlane	construction works
robÄĆt budowlanych	construction works (different form)
konsorcjum firm	consortium of companies
ograniczonÄĞ odpowiedzialnoÅŹciÄĞ	limited liability
formularzu ofertowym	offer form

D. Most frequent keywords with four tokens (auto-generated stoplist of Sect. 2)	
PKP Polskie Linie Kolejowe	PKP Polish State Railways
Generalnej Dyrekcji DrÄĆg Krajowych	General Directorate for National Roads
samodzielny publiczny szpital kliniczny	independent public clinical hospital
wykazu wykonanych robÄĆt budowlanych	list of conducted construction works
GE Medical Systems Polska	GE Medical Systems Poland

(in the period 2007–2013, covered by the analyzed corpus) deals with large scale construction works carried out by consortia consisting of a few companies. This is clearly reflected in the obtained results.

Obviously, the most frequently occurring keywords from Table 1C are rather general. However, if we restrict ourselves to longer phrases, we can easily check that their vagueness decreases and that they still form meaningful and informative word groups. Analyzing the most popular four token key phrases (Table 1D), we found that RAKE method is capable of extracting names of large contracting authorities and companies. This also seems a very desirable behaviour of the algorithm. Of course, in order to quantify the performance of the algorithm, rigorous tests based on the human expert knowledge are necessary.

4   Summary and Outlook

In this paper, we have presented a work in progress report on the unsupervised keyword extraction from Polish legal texts. We have employed recently proposed RAKE algorithm and extended it with the automatic, corpus adopted stoplist generation procedure. Qualitative tests of the method indicate that the approach is promising. In the future, we plan quantitative tests, however, this has to involve human domain experts and hence is a lengthy process. In addition, we plan also further optimization of the method. Introducing stemming and adjusting keyword ranking scheme of RAKE algorithm seem to be the most attractive directions.

Acknowledgments. We acknowledge the use of computing facilities of the Interdisciplinary Centre for Mathematical and Computational Modelling within the grant G57-14.

References

1. Francesconi, E., Montemagni, S., Peters, W., Tiscornia, D. (eds.): Semantic Processing of Legal Texts. LNCS, vol. 6036. Springer, Heidelberg (2010)
2. Kim, S.N., Medelyan, O., Kan, M.-Y., Baldwin, T.: SemEval-2010 task 5: Automatic keyphrase extraction from scientific articles. In: Proceedings of the 5th International Workshop on Semantic Evaluation, SemEval 2010, p. 21. Association for Computational Linguistics, Stroudsburg (2010)
3. Rose, S., Engel, D., Cramer, N., Cowley, W.: Automatic keyword extraction from individual documents. In: Berry, M.W., Kogan, J. (eds.) Text Mining. Applications and Theory, p. 1. John Wiley and Sons, Ltd. (2010)
4. Church, K.W., Gale, W.A.: Poisson mixtures. Natural Language Engineering 1(02), 163 (1995)
5. Manning, C.D., Schütze, H.: Foundations of statistical natural language processing. MIT Press, Cambridge (1999)

Term Ranking Adaptation to the Domain: Genetic Algorithm-Based Optimisation of the *C-Value*

Thierry Hamon[1,2], Christopher Engström[3], and Sergei Silvestrov[3]

[1] LIMSI-CNRS, Orsay, France
thierry.hamon@limsi.fr
[2] University Paris 13, Sorbonne Paris Cité, France
[3] Division of Applied Mathematics, School of Education,
Culture and Communication, Mälardalen University, Box 883,
72123 Västerås, Sweden
{christopher.engstrom,sergei.silvestrov}@mdh.se

Abstract. Term extraction methods based on linguistic rules have been proposed to help the terminology building from corpora. As they face the difficulty of identifying the relevant terms among the noun phrases extracted, statistical measures have been proposed. However, the term selection results may depend on corpus and strong assumptions reflecting specific terminological practice. We tackle this problem by proposing a parametrised *C-Value* which optimally considers the length and the syntactic roles of the nested terms thanks to a genetic algorithm. We compare its impact on the ranking of terms extracted from three corpora. Results show average precision increased by 9% above the frequency-based ranking and by 12% above the *C-Value*-based ranking.

Keywords: Terminology, term extraction, term ranking, genetic algorithm.

1 Introduction

The scientific and technical domains are characterised by the use of terms, mainly noun phrases, referring to the specialised knowledge of the field. While this knowledge is usually recorded in terminologies, terminological resources suffer from low coverage when they are used to identify terms in corpora [2,19]. Approaches proposed to automatically extract candidate terms, i.e. potential terminological entities, from texts [5], are essential to improve the term recognition.

Traditionally, term extraction methods first take advantage of linguistic characteristics to chunk the texts and extract candidate terms [5]. However, as they face the problem of finding relevant terms among the huge amount of candidate terms, statistical information and metrics are also used to help catching the termhood of the extracted noun phrases. Also, besides the difficulty of identifying the relevant terms among the noun phrases extracted, expert users and terminologists have to examine huge lists of candidate terms [18]. We consder that ranking the terms is a crucial step before pruning the candidate term list.

A. Przepiórkowski and M. Ogrodniczuk (Eds.): PolTAL 2014, LNAI 8686, pp. 71–83, 2014.
© Springer International Publishing Switzerland 2014

Term ranking usually relies on either statistical metrics or information associated to noun phrases. The candidate term frequency is commonly used [6,14,9] but is not sufficient to fully capture the termhood [21]: it either decreases the precision [10] or the recall as many candidate terms occur only once [14,8,7]. The term identification can also be based on term length, defined as the number of words. [8] proposes to use the inverted term length (the longer is a term, the less important it is) and notes that the combination with the term frequency slightly increases precision. The *C-Value* [11] considers that the termhood is indicated by the frequency and the length of the terms as well as their nestedness[1] and their independence from other terms (see (1)).[2] Long multi-word terms, which are not components of other terms, are favoured then. Besides, when the candidate term is nested in longer candidate terms, its termhood negatively depends on the frequency of the longer candidate terms and is positively moderated by the number of candidate terms including this term. Compared to the frequency, the *C-Value* improves the results, especially with nested terms: precision increases by 31% for the nested candidate terms, but only by 1% for the whole set of the terms extracted [10]; moreover, the *C-Value* concentrates the relevant candidate terms at the top of the list [8].

$$C\text{-}Value(t) = \begin{cases} \log_2(|t|+1) \cdot f(t), & \text{if } t \text{ is not included in a term} \\ \log_2(|t|+1) \cdot (f(t) - \frac{1}{P(T_t)}\sum_{t' \in T_t} f(t')), & \text{otherwise} \end{cases} \quad (1)$$

where $f(t)$ is the frequency of the term t, $|t|$ is its length, $f(t')$ is the frequency of the term t as component of longer terms, T_t is the set of terms which include the term t, $P(T_t)$ is the number of terms including the term t.

The context of the candidate terms is also assumed to be helpful for identifying the termhood of the extracted noun phrases. Thus, the *NC-Value* combines contextual information with the *C-Value* [10]. The candidate terms ranked by the *C-Value*, are then re-ranked according to the frequency of the words occurring in the context of the top-ranked candidate terms. [10] reports improvements compared to the *C-Value*. However, the impact of the context seems depend on the corpora: [16] observe that the *C-Value* and the NC-value provide equivalent results on two biomedical corpora. Termhood can also be defined as the semantic relatedness to a domain, represented as a vector of generic words occurring in the corpus [3]. This kind of model outperforms the *NC-Value* on a biomedical corpus while the both of them are equivalent for the keyphrase extraction [3].

The work presented shows that identifying the termhood remains a difficult task and that the results depend on the corpus. Ranking metrics like *C-Value* rely on strong assumptions on the term form and reflect specific terminological practice. Besides, the *C-Value* equally considers all the candidate terms without taking into account the syntactic role of longer and shorter candidate terms. We argue that, in order to better reflect the terminological practice of a given field,

---

[1] The string inclusion of a term in another.

[2] To consider single word term as well as multi-words terms, [8] adds 1 to the length of the terms $|t|$. We use this adaptation later in the paper.

there is still room for improvement by integrating syntactic information in *C-Value* and by defining optimal weights of its intrinsic parameters. Therefore, we propose an adaptation of the *C-Value* which integrates parameters representing the syntactic role of the nested terms but also their length (Sect. 3). Because the terminological practice is usually specific to a given field, we optimise the values of these parameters with a genetic algorithm (Sect. 3.4).

After the presentation of the rule-based approach we use to extract terms (Sect. 2), we describe the new ranking measure *C-Value** (Sect. 3). We present and discuss the experiment results we perform, and evaluate the impact of the proposed measure on several corpora (Sect. 5).

2 Term Extraction

The term extraction method described in [1] and implemented in the term extractor Y$_A$T$_E$A[3] performs shallow parsing of the morpho-syntactically tagged and lemmatized texts by chunking the texts according to syntactic frontiers (pronouns, conjugated verbs, typographic marks, etc.) to identify noun phrases. Then, parsing patterns taking into account the morpho-syntactic variation, are recursively applied and provide parsed candidate terms. Syntactically ambiguous phrases are endogenously disambiguated with the already recognised candidate terms. Thus, each noun phrase, which seems to be relevant to the targeted domain, is represented by a syntactic binary tree composed of two elements describing the syntactic role of the components in the term (see for instance, the syntactic tree of *full maturation of erythrocytes* in Fig. 1): the head component (i.e. *full maturation*) is the main noun phrase of the term, and the modifier component (i.e. *erythrocytes*) modifies the head noun phrase; recursively, complex head or modifier components (i.e. *full maturation*) are decomposed in simpler head (i.e. *maturation*) and modifier (i.e. *full*). Each component can be a multi-word or a single-word term. Term parsing provides the syntactic role of terms but also information regarding their nestedness. At the end of the term extraction step, we obtain single-word terms (e.g. *maturation*) and parsed multi-word terms (e.g. *full maturation*). Statistical measures like frequency or *C-Value* are also associated to each term.

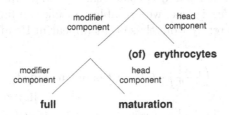

Fig. 1. Parsing tree of the term *full maturation of erythrocytes*

---

[3] http://search.cpan.org/~thhamon/Lingua-YaTeA/

3   Optimisation of the *C-Value*: *C-Value**

We present the proposed improvement of the *C-Value* by adding three types of parameters in order to better represent (i) the terminological practice concerning the length of the terms, (ii) their syntactic role when they are nested or not in longer terms, and (iii) the distribution of the nested terms.

3.1   Parametrisation of the Term Length

The basic *C-Value* gives strong influence to the term length ($|t|$): very long terms found only once are ranked higher that shorter terms. However, this choice may depend on the terminological practice of domain. Therefore, the ranking measure must be more flexible concerning the term length and their frequency. The weight of the candidate term length can be parametrised in this way. In that respect, we intend to keep the impact of the term length on the frequencies ($\log_2(|t|)$). To give more importance to shorter terms when required, we propose to change the length weight as the $\log_2\left(\frac{|t|+1}{|t|}\right)$. Besides, in order to take into account the terminological practice, we also apply the exponential parameter $\alpha \geq 0$ on the term length: $\log_2\left(\frac{|t|+1}{|t|^\alpha}\right)$. High α will give penalty to long terms, while low $\alpha < 1$ will instead penalise short terms.

3.2   Taking into Account the Syntactic Role of the Candidate Term

In the modified length weight above, the terms are considered equally whatever their syntactic role and nestedness. However, we have three different types of terms: (i) root terms which are not nested in any other terms, (ii) head terms which can be found at the head position of other terms, and (iii) modifier terms which can be found as modifiers of other terms. To reflect these three syntactic roles, we defined three distinct parameters of α: α_R for the root terms, α_H for the head terms, and α_M for the modifier terms (see (2)). We note that a given term can be both Head and Modifier term, but that Root terms cannot be combined with any other type. Moreover, for a term that is both Head and Modifier we calculate these two measures and consider their average value. One advantage of this modified *C-Value* is that we are able to rank candidate terms, that in Modifier position, differently from those that appear in Head position.

$$
\textit{C-Value}'(t) = \begin{cases} \log_2\left(\frac{|t|+1}{|t|^{\alpha_R}}\right) \cdot f(t), \text{ if } t \text{ is not included} \\ \qquad\qquad\qquad\qquad\qquad \text{in a term (Root)} \\ \log_2\left(\frac{|t|+1}{|t|^{\alpha_H}}\right) \cdot \left(f(t) - \frac{1}{P(T_t)}\sum_{t' \in T_t} f(t')\right), \\ \qquad\qquad\qquad\qquad\qquad \text{if } t \text{ is a Head term} \\ \log_2\left(\frac{|t|+1}{|t|^{\alpha_M}}\right) \cdot \left(f(t) - \frac{1}{P(T_t)}\sum_{t' \in T_t} f(t')\right), \\ \qquad\qquad\qquad\qquad\qquad \text{if } t \text{ is a Modifier term} \end{cases} \tag{2}
$$

3.3   Distribution of the Nesting Candidate Terms

Longer candidate terms including nested terms are equally considered by the
basic *C-Value*. Nested terms are penalised because of the mean frequency of the
nesting terms, without considering their distribution: the same penalty is given
to the term nested in three terms with frequencies 10, 1 and 1, and to the term
nested in three terms with the equal frequencies 4, 4, and 4. In that respect,
instead of using the average frequency, we propose to apply β-norm to give more
penalty to terms nested in several terms with unbalanced frequencies (see (3)
which defines the *C-Value*(t)*). Indeed, with $\beta > 1$, the more equally distributed
is the term among the including terms, the lower is the penalty given by the *C-
Value*(t)*; while $\beta < 1$ gives higher penalty to more equally distributed terms.
If β is close to 1, the distribution is not taken into account. If $\beta \geq 1$, the largest
possible penalty to a term is equal to $\sum_{t' \in T_t} f(t')$ and is applied when the term
is nested in only one other term ($|T_t| = 1$).

Moreover, because the length of nested and/or nesting terms can be influenced by the terminological practice, we add the parameter c which defines the
influence of nesting terms: higher c gives higher penalty if the term is nested in
other terms. Because the parameters β and c cannot be assumed to have the
same values with the Head and Modifier terms, we distinguish them according
to whether they apply to the Head terms (β_H and c_H) or to the Modifier terms
(β_M and c_M). Equation 3 summarises our improved version of the *C-Value*,
called *C-Value**.

$$
C\text{-}Value^* = \begin{cases}
\log_2\left(\frac{|t|+1}{|t|^{\alpha_R}}\right) \cdot f(t), & \text{if } t \text{ is not included in a term (Root)} \\[2ex]
\log_2\left(\frac{|t|+1}{|t|^{\alpha_H}}\right) \cdot \left(f(t) - c_H \left(\sum_{t' \in T_t} f(t')^{\beta_H}\right)^{1/\beta_H}\right), \\
\qquad\qquad\qquad\qquad \text{if } t \text{ is a Head term} \\[2ex]
\log_2\left(\frac{|t|+1}{|t|^{\alpha_M}}\right) \cdot \left(f(t) - c_M \left(\sum_{t' \in T_t} f(t')^{\beta_M}\right)^{1/\beta_M}\right), \\
\qquad\qquad\qquad\qquad \text{if } t \text{ is a Modifier term}
\end{cases} \tag{3}
$$

3.4   Genetic Algorithm Based Parameter Optimisation

We optimise the parameters α_R, α_H, α_M, β and c with a genetic algorithm
because they may depend on the corpus. They are estimated using a real-coded
genetic algorithm [22] with a fitness function. This function has to be computed
reasonably fast for the genetic algorithm to be effective. The final evaluation
measures R-precision, F-measure or average precision (see Sect. 4.2) cannot be
used since R-precision or F-measure rely on a single point and would lead to
slightly better results around the value of evaluation measure but a much worse
ranking before or after, while average precision relies on too many points to
obtain results in a reasonable time. Therefore, we choose a compromise with
a fitness function defined as the number of relevant terms at several ranks:
$f = \sum_{i \in I} r(i)$, where $I = \{N/6, 2N/6, 3N/6, 4N/6, 5N/6\}$, N is the total
number of terms, and $r(i)$ is the number of relevant terms among the i highest
ranked terms.

For the genetic algorithm, we use a population of 200 individuals. Parents are selected using a tournament selection scheme and a BLX-0.5 blend crossover scheme [13] to create new samples. We use a 20% mutation rate where we replace the old parameter with a new randomly generated one. Also, even if parameters should be estimated several times, preliminary experiments show that they converge to the same or very close values. Hence, for the experiments presented, we did not perform multiple runs in order to save computing time.

4 Experiments

We have performed several experiments to evaluate the *C-Value** and the contribution of the parameters α, β and c. We ranked the terms extracted from three corpora. In this section, we present the corpora used, the evaluation measures and the configuration of the experiments.

4.1 Corpora

The intrinsic evaluation of the term extraction is a difficult task as few corpora with annotated terms are available. We used Genia and PennBioIE corpora [16,3].[4] Each corpus has been processed through the Ogmios platform [12] configured to perform word and sentence segmentation, and to associate to each word its part-of-speech tag and its lemma with the Genia Tagger [20]. Terms have been extracted with YATEA (see Sect. 2).

Genia Corpus. The Genia corpus[5] [15] is a collection of 1,999 Medline[6] abstracts concerning the transcription factors in human blood cells (abstracts indexed by the MeSH terms *human, blood cell* and *transcription factor*). The corpus contains 18,545 sentences and 436,967 words. Each abstract is also annotated with terms (mostly noun phrases) referring to physical biological entities (organisms, proteins, cells, genes) as well as biologically meaningful terms (e.g. molecular functions). On the whole, 97,829 occurrences of terms (36,607 term types) are annotated.We consider these annotated terms as our reference for this corpus. The term extractor provides 49,249 candidate terms.

PennBioIE Corpus. The PennBioIE corpus[7] [17] is a set of Medline abstracts on two biomedical topics: drug development (sub-collection CYP450) and cancer genomics (sub-collection Oncology). Each sub-corpus has been annotated with several semantic entities related to the topic of each sub-collection. We

---

[4] We chose not to use corpora issued from the event identification challenges like I2B2 and clefeHealth, or corpora annotated with keyphrases, because this would lead to the extrinsic evaluation as the term annotation is non-exhaustive and task-oriented.

[5] Version 3.02, http://www.nactem.ac.uk/genia/genia-corpus/event-corpus

[6] http://www.ncbi.nlm.nih.gov/pubmed

[7] http://bioie.ldc.upenn.edu/

considered these two sub-corpora separately. The **CYP450 sub-corpus** contains 1,100 Medline abstracts (298,843 words) focusing on the modification of the cytochrome P450 proteins. It has been annotated with terminological entities belonging to three semantic types: CYP450 enzymes, substances related to the enzyme inhibition and quantitative measurements. The reference provides 42,337 occurrences (9,221 term types) of annotated terms. The term extractor provides a set of 47,168 candidate terms. The **Oncology sub-corpus** is a collection of 1,157 Medline abstracts (276,161 words) annotated with three types of terminological entities: gene entities, biological events identifying genomic variation, and genes malignancy description (development, behaviour, topography or morphology). The reference provides 6,704 term occurrences (2,734 term types). 39,542 candidate terms are extracted automatically.

Table 1 summarises the statistics of the three corpora used. The candidate terms are all considered during the ranking process. While comparing the number of annotated terms, we observe that the annotation process is quite different between the Genia and PennBioIE corpora. This may provide an interesting experimental context to evaluate how the optimisation algorithm can adapt the parameters of the *C-Value**.

Table 1. Corpus description with the number of abstracts, words, annotated terms and candidate terms

Corpus	Abstracts	Words	Annotated terms	Candidate terms
Genia	1,999	436,967	36,607	49,249
PennBioIE/CYP450	1,100	298,843	9,221	47,168
PennBioIE/Oncology	1,157	276,161	2,734	39,542

4.2 Evaluation

For each corpora, we compute several evaluation measures against the reference data. We study the ranking obtained through the evolution of precision, recall and F-measure at each rank of the term list. We also consider R-precision [4], i.e. precision at the rank R corresponding to the number of terms to recognise in a given corpus. This evaluation measure can also be seen as the point where precision and recall are equal, and at which the precision value should be optimal for terminology building. We also compute the average precision and its evolution because the precision evolution according to the recall is a useful information on the behaviour of the ranking models. The two baselines used are obtained with the frequency and the *C-Value*.

4.3 Experiment Configuration

In order to evaluate the behaviour of the parameters and to select the best configurations, we first perform experiments on the Genia corpus: (1) this corpus is randomly split in two sets (60% for the training set, 40% for the test set), (2) the genetic algorithm estimates the parameter values on the training set and evaluates the defined values on the test set. We also perform 10-fold cross-validation on each corpus to study the capacity of the model to adapt the parameters of the *C-Value**: the Genia corpus is randomly split in ten sets, and we consider ten times 90% for the training and 10% for the test. Finally, we rank the terms extracted from the CYP450 and Oncology corpora with the *C-Value** and the parameters estimated on the Genia corpus in order to study to which extent the parameter values are independent.

We define several model configurations for evaluating the impact of the parameters and the role of the genetic algorithm (Table 2). For instance, in the model M_1, all the parameters are set to 1 and the genetic algorithm is not used. This is kind of basic configuration to be used when no training set is available. One variation of this model is when we set only one parameter to 1 and when the genetic algorithm estimates the optimal values of the other parameters (models $M_{\beta c}$, $M_{\alpha^3 c}$, $M_{\alpha^3 \beta}$). The model $M_{\alpha \beta c}$ allows estimating three parameters α equally, and β and c separately. The model $M_{\alpha^3 \beta c}$ distinguishes the syntactic role of terms by estimating separately the three parameters α, while the model $M_{\alpha^3 \beta^2 c^2}$ also estimates the two parameters β and c.

Table 2. Parameter settings

Model	Parameters	Model	Parameters
M_1	$\alpha = \beta = c = 1$	$M_{\alpha \beta c}$	$\alpha_R = \alpha_H = \alpha_M, \beta_H = \beta_M, c_H = c_M$
$M_{\beta c}$	$\alpha = 1, \beta_H = \beta_M, c_H = c_M$	$M_{\alpha^3 \beta c}$	$\alpha_R, \alpha_H, \alpha_M, \beta_H = \beta_M, c_H = c_M$
$M_{\alpha^3 c}$	$\alpha_R, \alpha_H, \alpha_M, \beta = 1, c_H = c_M$	$M_{\alpha^3 \beta^2 c^2}$	$\alpha_R, \alpha_H, \alpha_M, \beta_H, \beta_M, c_H, c_M$
$M_{\alpha^3 \beta}$	$\alpha_R, \alpha_H, \alpha_M, \beta_H = \beta_M, c = 1$		

5   Results and Discussion

Evaluation on the Genia Corpus. We first analyse the results on the Genia corpus split in two sets (60% for training and 40% for test). The parameter estimation stops after 50 iterations using the population size 200. Although the results are likely to become better with more iterations, our preliminary experiments indicate that this improvement is very small. Table 3 presents R-precision and average precision obtained on the training and test sets. We can observe that the Genia training and test sets provide very similar results. Also, regarding the baselines, the ranking based on the original *C-Value* has lower R-precision and average precision than the ranking based on frequency. However, all the models based on *C-Value** outperform the baselines: the average precision increases by 4,5 to 9% by comparison with the frequency, and by 8.5% and 12% by comparison with the *C-Value*. Similar improvements are observed with R-precision.

Table 3. Results on the Genia corpus (60% for training, 40% for test)

Model name	R-prec$_{train}$	R-prec$_{test}$	avg Prec$_{train}$	avg Prec$_{test}$
Frequency	0.4590	0.4671	0.4338	0.4441
C-Value	0.3344	0.3594	0.3935	0.4147
M_1	0.5091	0.5090	0.5088	0.5124
$M_{\beta c}$	0.4974	0.5084	0.4910	0.5002
$M_{\alpha^3 c}$	0.5259	**0.5285**	**0.5416**	**0.5407**
$M_{\alpha^3 \beta}$	**0.5293**	0.5272	0.5387	0.5363
$M_{\alpha \beta c}$	0.5144	0.5139	0.5266	0.5269
$M_{\alpha^3 \beta c}$	0.5197	0.5207	0.5386	0.5360
$M_{\alpha^3 \beta^2 c^2}$	0.5222	0.5233	0.5330	0.5262

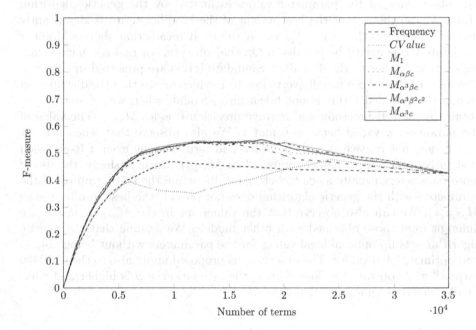

Fig. 2. Evolution of F-measure on the ranked list of candidate terms extracted from the test set of the Genia corpus

Model Analysis. We study the evolution of precision, recall, F-measure and average precision on the test set (see Fig. 2 for the evolution of the F-measure[8]): the M_1 model, the $M_{\alpha^3 c}$ model which is the best for the average precision, and the models in which all parameters are defined by the genetic algorithm. For the top ranked terms, precision and average precision are better with frequency and C-Value, while recall and F-measure are similar for all the ranking models. The evolution of the average precision shows that the $M_{\alpha^3 \beta^2 c^2}$ model is close to the

---

[8] Additional figures for other measures can be found at http://perso.limsi.fr/hamon/Files/2014PolTAL-appendix/2014PolTaL_Figures.pdf

frequency and *C-Value* rankings. The three of them provide similar rankings for the top terms. We can also observe that the models based on *C-Value** improve the ranking (precision and recall) after one hundred terms and until 70% of the whole set of candidate terms. The models based on *C-Value** rank similarly the candidate terms even if the results obtained with the model M_1 appear to be slightly lower than other models, mainly with recall and F-measure.

The comparison of models provides with several observations. The genetic algorithm has positive impact on the results, still the M_1 model can be used for ranking terms when there is no reference data usable for training. The parameters α strongly influence the results: when they are set equally it has negative effect (model $M_{\alpha\beta c}$), when they are set to 1 it leads to poor results (model $M_{\beta c}$). The observation of the parameter values estimated by the genetic algorithm (Table 4) also shows that the best weight of the Modifier terms is significantly smaller (close to 0) than the other two α values. It means that shorter Modifier candidate terms are to be penalised. On the contrary, α_R and α_H have values higher than 1: shorter Head or Root candidate terms are preferred in this way. The best parameter β is usually very close to 1, which means that the distribution among the included terms is not taken into account; when we set it to 1, we obtain the best R-precision and average precision (model $M_{\alpha^3 c}$). The value of the parameter c varies between 0 and 1. We also observe that when c is set to 1 it does not impact average precision and term ranking even if R-precision is slightly lower ($M_{\alpha^3\beta}$ model compared to $M_{\alpha^3\beta c}$). Unsurprisingly, the use of more parameters usually leads to better results, while the setting up of all the parameters with the genetic algorithm does not provide the best results (model $M_{\alpha^3\beta^2 c^2}$). We can also observe that the values set in the $M_{\alpha^3\beta^2 c^2}$ model are different from those obtained with other models. We assume that the genetic algorithm sets up optimal local values for the parameters without being able to find optimal global values. The observations proposed apply also to the CYP450 corpus,[9] while on the Oncology corpus, the estimation of β is higher and c has smaller and sometimes negative values.

Table 4. Parameter values estimated by the genetic algorithm on the Genia training part

Model	α_R	α_H	α_M	β_H	β_M	c_H	c_M
M_1	1	1	1	1	1	1	1
$M_{\beta c}$	1	1	1	0.997	0.997	0.7095	0.7095
$M_{\alpha^3 c}$	1.1014	1.0344	0	1	1	0.91	0.91
$M_{\alpha^3 \beta}$	1.1622	1.1445	0.075	1.0132	1.0132	1	1
$M_{\alpha\beta c}$	1.1604	1.1604	1.1604	1.0140	1.014	0.9953	0.9953
$M_{\alpha^3 \beta c}$	1.1067	1.0961	0.0857	0.9538	0.9538	0.8316	0.8316
$M_{\alpha^3 \beta^2 c^2}$	1.3005	0.6093	0.7381	1.5085	1.1307	1.5224	1.17

---

[9] Due to the lack of space, we do not presented the results on these corpora in this paper.

Cross-Validation Evaluation. The 10-fold cross-validation on the three corpora helps evaluating the dependence of *C-Value** on corpora. We consider here the three first models that obtain the best average precision on the Genia corpus. Table 5 presents the results of the cross-validation. As previously, the *C-Value**-based ranking on Genia and CYP450 systematically gives better R-precision and average precision than those obtained with frequency or *C-Value*. It seems to confirm that *C-Value** with optimal estimation of the parameters can be successfully applied to various text collections. However, on the Oncology corpus, the ranking of the candidate terms is similar for all measures considered. Besides, the results are disappointing: average precision is slightly better with *C-Value** than with frequency or *C-Value*, while R-precision is lower. We assume that such results are mainly due to the specificity of these reference data, in which fewer terms are annotated.

Table 5. 10-fold cross validation on the three corpora

Corpora	Frequency		*C-Value*		$M_{\alpha^3 c}$		$M_{\alpha^3 \beta}$		$M_{\alpha^3 \beta c}$	
	R-prec	avgPrec	R-prec	avgPrec	R-prec	avgPrec	R-prec	avgPrec	R-prec	avgPrec
Genia	0.3882	0.3589	0.3055	0.3509	0.4318	0.4212	0.4324	0.4098	0.4323	0.4133
CYP450	0.3079	0.2434	0.2711	0.2013	0.3540	0.3028	0.3595	0.3051	0.3621	0.3074
Oncology	0.1017	0.0615	0.1019	0.0606	0.0962	0.0643	0.0959	0.0620	0.0961	0.0643

Recycling Genia Parameters on Other Corpora. We use the parameter values estimated on Genia (see Table 4) to rank the terms extracted from the CYP450 and Oncology corpora. Table 6 presents the results. We observe on the corpus CYP450 that all the *C-Value** models outperform frequency, *C-Value* and M_1 models, which is also corpus independent. $M_{\alpha \beta c}$ proves to be the best model and indicates that the syntactic role of the nested terms has no influence on this corpus. We assume that this requires further investigation and that the impact of the syntactic roles may be due to the term length distribution or to the number of annotated terms. However, as the results obtained on the CYP450 corpus suggest, the parameters can be set on training corpus and achieve good results on other corpora. This fact must be confirmed with comparable experiments with corpora issued form other domains. Besides, some results obtained

Table 6. Evaluation with training on Genia and testing on CYP450 and Oncology

	CYP450 R-prec.	CYP450 avgPrec.	Oncology R-prec.	Oncology avgPrec.
Frequency	0.3315	0.2596	0.1498	0.0849
CValue	0.2960	0.2042	0.1450	0.0800
M_1	0.3677	0.3484	0.1441	0.0774
$M_{\alpha \beta c}$	**0.4517**	**0.3837**	0.1355	0.0771
$M_{\alpha^3 \beta c}$	0.3959	0.3515	0.1508	0.0917
$M_{\alpha^3 \beta^2 c^2}$	0.4074	0.3445	0.1450	0.0817
$M_{\alpha^3 c}$	0.3885	0.3410	0.1498	**0.0939**
$M_{\beta c}$	0.3906	0.3314	0.1412	0.0861
$M_{\alpha^3 \beta}$	0.3885	0.3552	**0.1517**	0.0879

on the Oncology corpus are difficult to interpret: the observations show that the frequency and *C-Value* models are close among them. However, the low variation in the number of terms according to the term length is a direction to investigate.

6 Conclusion

We tackle the problem of the candidate terms ranking extracted from specialised corpora. To achieve this purpose to use *C-Value**, that is an improved and parametrised *C-Value*. This value purpose is to take into account the syntactic role of the nested terms when considering their length, and to take into account the distribution of the nesting terms. Parameters are optimised with a standard genetic algorithm. Our experiments indicate that R-precision and average precision are increased respectively by 9 and 12% in two of the three corpora used. The study of precision, recall and f-measure evolution shows a notable improvement on 70% of the ranked candidate terms. The proposed measure will be included in the next release of the YᴀTᴇA term extractor.[10]

In our future work, we will analyze the behaviour of *C-Value** that may vary according to the domain studied (all the corpora currently used are related to the biomedical domain). We will also investigate how the results may depend on the term extractor.

References

1. Aubin, S., Hamon, T.: Improving term extraction with terminological resources. In: Salakoski, T., Ginter, F., Pyysalo, S., Pahikkala, T. (eds.) FinTAL 2006. LNCS (LNAI), vol. 4139, pp. 380–387. Springer, Heidelberg (2006)
2. Bodenreider, O., Rindflesch, T.C., Burgun, A.: Unsupervised, corpus-based method for extending a biomedical terminology. In: Workshop on Natural Language Processing in the Biomedical Domain, pp. 53–60 (2002)
3. Bordea, G., Buitelaar, P., Polajnar, T.: Domain-independent term extraction through domain modelling. In: Proceedings of the 10th International Conference on Terminology and Artificial Intelligence (TIA 2013), France (2013)
4. Buckley, C., Voorhees, E.: Retrieval system evaluation. In: Voorhees, E., Harman, D. (eds.) TREC: Experiment and Evaluation in Information Retrieval, ch.3. MIT Press (2005)
5. Cabré, M.T., Estopá, R., Vivaldi, J.: Automatic term detection: a review of current systems. In: Recent Advances in Computational Terminology. John Benjamins, Amsterdam (2001)
6. Daille, B.: Study and implementation of combined techniques for automatic extraction of terminology. In: Proceedings of the Balancing Act, pp. 29–36 (1994)
7. Dowdall, J., Hess, M., Kahusk, N., Kaljurand, K., Koit, M., Rinaldi, F., Vider, K.: Technical terminology as a critical resource. In: Proceedings of LREC 2002 (2002)
8. Drouin, P.: Acquisition automatique des termes: l'utilisation des pivots lexicaux spécialisés. Ph.D. thesis, Université de Montréal (2002)

---

[10] Currently freely available on CPAN
    http://search.cpan.org/~thhamon/Lingua-YaTeA/

9. Drouin, P.: Term extraction using non-technical corpora as a point of leverage. Terminology 9(1), 99–117 (2003)
10. Frantzi, K.T., Ananiadou, S., Mima, H.: Automatic recognition of multi-word terms: the C-Value/NC-Value method. International Journal on Digital Libraries 3(2), 115–130 (2000)
11. Frantzi, K.T., Ananiadou, S., Tsujii, J.: Automatic term recognition using contextual clues. In: Proceedings of MULSAIC 1997, pp. 73–79 (1997)
12. Hamon, T., Nazarenko, A., Poibeau, T., Aubin, S., Derivière, J.: A robust linguistic platform for efficient and domain specific web content analysis. In: Proceedings of RIAO 2007, Pittsburgh, USA (2007)
13. Herrera, F., Lozano, M., Sánchez, A.M.: A taxonomy for the crossover operator for real-coded genetic algorithms: An experimental study. International Journal of Intelligent Systems 18(3), 309–338 (2003)
14. Justeson, J.S., Katz, S.M.: Technical terminology: some linguistic properties and an algorithm for identification in text. Natural Language Engineering 1(1), 9–27 (1995)
15. Kim, J.D., Ohta, T., Teteisi, Y., Tsujii, J.: Genia corpus – a semantically annotated corpus for bio-textmining. Bioinformatics 19(suppl. 1), 180–182 (2003)
16. Korkontzelos, I., Klapaftis, I.P., Manandhar, S.: Reviewing and evaluating automatic term recognition techniques. In: Nordström, B., Ranta, A. (eds.) GoTAL 2008. LNCS (LNAI), vol. 5221, pp. 248–259. Springer, Heidelberg (2008)
17. Kulick, S., Bies, A., Liberman, M., Mandel, M., McDonald, R., Palmer, M., Schein, A., Ungar, L., Winters, S., White, P.: Integrated annotation for biomedical information extraction. In: Proceeding of BioLINK 2004, pp. 61–68 (2004)
18. Lame, G.: Classement automatique de documents et analyse terminologique de corpus. In: Actes de la Conférence TIA 2001, pp. 149–158 (2001)
19. McCray, A.T., Browne, A.C., Bodenreider, O.: The lexical properties of the gene ontology (GO). In: Proceedings of AMIA 2002, pp. 504–508 (2002)
20. Tsuruoka, Y., Tateishi, Y., Kim, J.D., Ohta, T., McNaught, J., Ananiadou, S., Tsujii, J.: Developing a robust part-of-speech tagger for biomedical text. In: Bozanis, P., Houstis, E.N. (eds.) PCI 2005. LNCS, vol. 3746, pp. 382–392. Springer, Heidelberg (2005)
21. Velardi, P., Missikoff, M., Basili, R.: Identification of relevant terms to support the construction of domain. In: Proceedings of the Workshop on Human Language Technologies and Knowledge Management (2001)
22. Wright, A.H.: Genetic algorithms for real parameter optimization. Foundations of Genetic Algorithms 1, 205–218 (1991)

Graph-Based, Supervised Machine Learning Approach to (Irregular) Polysemy in WordNet

Bastian Entrup

Applied and Computational Linguistics,
Justus-Liebig Universität Gießen, Germany
bastian.entrup@germanistik.uni-giessen.de

Abstract. This paper presents a supervised machine learning approach that aims at annotating those homograph word forms in WordNet that share some common meaning and can hence be thought of as belonging to a polysemous word. Using different graph-based measures, a set of features is selected, and a random forest model is trained and evaluated. The results are compared to other features used for polysemy identification in WordNet. The features proposed in this paper not only outperform the commonly used CoreLex resource, but they also work on different parts of speech and can be used to identify both regular and irregular polysemous word forms in WordNet.

1 Introduction

In [1, p. 16], regular polysemy is defined as follows:

> Polysemy of the word A with the meanings a_i and a_j is called regular if, in the given language, there exists at least one other word B with the meanings b_i and b_j, which are semantically distinguished from each other in exactly the same way as a_i and a_j and if a_i and b_i, a_j and b_j are nonsynonymous.

Often mentioned is the so-called grinding rule: the name of an animal can often also be used to refer to products gained from it. Irregular polysemy on the contrary covers those cases that do not exhibit such patterns. When for example animal names are used to denote humans, this can be done referring to different properties of animals. Calling someone a *lion* is mostly due to strength and courage, while *chicken* may refer to a lack of courage. Since the productive patterns of regular polysemy can be identified and used by computer systems while irregular cases are harder to identify, most computational approaches to polysemy based on WordNet (WN) [11] are focused on regular polysemy.

Since WN represents word senses rather than words, the classic definitions of polysemy cannot be applied to WN. Instead of looking at the binary decision of whether a *word* is polysemous or homonymous, the method proposed in this paper will look at the sense level and try to identify those (homograph) word forms that actually are related. Taking the word *bank* and all its word forms in

A. Przepiórkowski and M. Ogrodniczuk (Eds.): PolTAL 2014, LNAI 8686, pp. 84–91, 2014.

WN, the approach taken here is to connect all those forms that are related to the field of finance, while not connecting them to those forms that are related to a slope of any kind.

To achieve this goal, WN's network topology is exploited: measures such as the geodesic paths or properties of nodes (e.g., the closeness or betweenness) are used. Since it could be assumed that homonymous words – or more exactly in this context word forms – have no semantic similarity to each other, measures of semantic similarity based on paths between two nodes are taken into consideration to distinguish related and unrelated word forms.

Although the approach proposed here, especially the graph-based features, can be applied to all parts of speech (POSs) in WN, this paper is restricted to those word forms that are either nouns or potentially connected to nouns. This restriction is necessary to compare the proposed feature set to the CoreLex features.

2 Related Work

2.1 Regular Polysemy Detection in WordNet

The *CoreLex* resource [5] defines a set of 39 basic types (BTs), i.e., semantic classes of words that subsume a number of word senses in WN (e.g. *food* or *animal*). Taking advantage of the hierarchical order of nouns in WN, the BTs are assigned to anchor nodes in WN that are identified as the hypernym of the word senses belonging to the given semantic class. When looking at words, one lexeme is likely to have different word senses and hence different word forms in WN. Each word form is assigned to at least 1 of the 39 BTs, resulting in a list of BTs related to a word. This list should display patterns of regular polysemy. For example, a word like *lamb* has a meaning that belongs to the BT *animal* as well as one belonging to *food*, etc. This pattern can be found in other words as well. It therefore satisfies the definition of regular polysemy given above.

The approaches described in [2] and [16] are based on the CoreLex resource. [16] calculate a ratio of polysemy for words based on the BTs they are related to. The more words share the same pattern of BTs, the more likely those words are polysemous. Polysemy and homonymy are considered "two points on a gradient, where the words in the middle show elements of both" [16, p. 268].

[5], [16], and [2] can only be used to detect regular polysemy. The great number of irregular polysemous forms cannot be found and is, as done in [16], regarded as homonymous.

[17] takes a different approach and calculates lexical similarity of glosses of potentially related word senses sharing a common lemma. If the glosses are similar, their meaning is considered to be similar as well. This approach is applicable to other POSs than nouns since it does not rely on the hypernymy/hyponym relation – which in WN does not exist for adjectives or adverbs.

Table 1. Existing measures of semantic similarity/relatedness

Abbreviation	Source and description	POS
Resnik and Yarowsky	[14], implemented by [13]	N, V
Lin	[10], implemented by [13]	N, V
Jiang and Conrath	[8], implemented by [13]	N, V
Hirst and St. Onge	[7], implemented by [13]	N
Leacock, Miller, and Chodorow	[9], implemented by [13]	N, V
Wu and Palmer	[18], implemented by [13]	N, V
distance	geodesic path between the two nodes	all

2.2 Computing Semantic Similarity in WordNet

The most intuitive measure of semantic similarity in WN is to calculate the geodesic path (i.e., the distance) between two nodes.

A number of semantic similarity features that can be applied to WN have been proposed (see Table 1).[1] These are mainly based on the geodesic path between the two nodes in question. Most of these measures are restricted to the noun and verb subset of WN, since they rely on the hierarchical order of the noun or verb network.[2]

The features in Table 1 and others have been used in [15] to find synsets that are related and could be merged to make WN more coarse grained and thereby raise the accuracy in wore sense disambiguation based on WN. Furthermore, [15] propose calculating the distance of both word senses to their closest common hypernym.

3 Feature Set

The question of whether two word senses are related in meaning can be answered by calculating the semantic similarity of the word forms and by looking at the network toponymy (i.e., local and global features of the network). These include information on the degree of the nodes being examined, the nodes representing the synsets as well as the nodes representing the word forms connected to them. Furthermore, different centrality measures for these nodes are calculated. The centralities are thought of as giving insight to the position a node has within the network. For example, a node with a high closeness centrality [6] can be expected to show shorter geodesic paths to any other node, not only to those it is semantically related to. The betweenness [6] indicates the node's position on geodesic paths of other nodes. The eigenvector centrality [3] and the PageRank [12] are two further centrality measures that are likely to better indicate centrality than the degree of the nodes.

---

[1] Measures based on gloss overlap have been excluded.

[2] For information on the single features, see the sources given in Table 1.

Table 2. Proposed graph-based feature set

Abbreviation	Source or description	POS
isA-Rel	*is-A* relation between word forms	N,V
isDerivedFrom	is one word form derived from the other?	all
closeness	the closeness value of the node	all
betweenness	the betweenness value of a node	all
POS	the part of speech of word form	all
word sense degree	degree of the word sense nodes	all
synset degree	degree of the synset nodes	all
eigenvector centrality	the eigenvector centrality values of the nodes	all
page rank	the page rank values of the nodes [12]	all
POS	the part of speech of word form	all
sharedLemmas	number of lemmas shared by the synsets	all
minDist2SharedHypernym	proposed in [15]	N,V

Although this paper focuses on nouns, noun word forms are often homograph to word forms of other POSs. Since these homographs are also considered, the information on the POS of a node is used as a feature.[3]

The number of lemmas two synsets share is a further feature. Also the information of whether one word sense is a direct hyponym or hypernym of the other (e.g., the synset {human, man} subsumes {man} (male human being)) was considered. An overview of the graph-based features for the noun subset of WN is given in Table 2.

4   Evaluation

4.1   Evaluation of the Model and Feature Set

Each noun word sense sharing its word form with any other word sense in WN is extracted.[4] An instance consists of the features for the two word forms as reported in Tables 1 and 2. Each pair of the kind w1 : w2 or w2 : w1 is only considered once. A subset of 2,511 pairs was manually classified as either sharing a similar/common meaning, class {*yes*}, or as being just arbitrarily homograph, class {*no*}. In the set, 1,237 pairs have been classified as being related, while 1,274 have been classified as being unrelated. Using a simple prediction assigning the most common class to each instance (i.e., *no*), a model has to top a baseline of 50.74% correctly classified instances.[5]

---

[3] Especially when looking at other POSs than nouns, one can find the tendency of some POSs (e.g., adverbs and adjectives) to be more likely to share meaning with adverbs or adjectives than other POSs.

[4] These include pairs of two noun word forms as well as pairs of noun word forms that are homographic to an adjective, verb, or adverb.

[5] This number might seem high, since polysemy is expected to be more frequent than homonymy. But here word forms belonging to different word senses are considered, not words.

Table 3. Precision difference obtained by removal of the single feature

Feature	Loss	Feature	Loss
word 1 degree	0.52	word 2 pos	0.48
word 1 closeness	5.98	word 2 degree	−0.24
word 1 betweenness	1.19	word 2 closeness	0.16
word 1 eigenvector centrality	0.68	word 2 betweenness	0.24
word 1 page rank	1.23	word 2 eigenvector centrality	0.32
word 1 synset degree	3.46	word 2 page rank	0.36
distance	−1.95	word 2 synset degree	0.52
is-A rel.	0.78	Lin	0.36
Hirst and St. Onge	0.36	Resnik and Yarowsky	0.36
Leacock and Chodorow	0.48	Jiang and Conrath	0.84
Wu and Palmer	0.28	isDerivedFrom	0.52
sharedLemmas	0.92	minDis2SharedHypernym	0.20

Different algorithms have been proposed for classification tasks. Here, the best results are obtained using the random forest model [4].[6] Based on 100 random trees, each constructed while considering 17 random features and 10 seeds, using a 10-fold cross-validation, the model reaches a precision of 0.861 and a recall of 0.877 out of 1. The F-measure is thus 0.87; 86.98% of the instances were correctly classified. The model outperforms the baseline by 36.24 points.

Unfortunately, the random forest algorithm is a black box when it comes to evaluating the impact of a single feature on the overall performance. Unlike decision trees the random forest model selects the features randomly. To evaluate the contribution of a feature, an ablation study was performed. One feature is sequentially deleted and the algorithm evaluated again. The gain or loss in accuracy is shown in Table 3. A negative number indicates a gain.

An ablation study, however, does not evaluate the impact of the feature but rather its contribution to the trained model. The combination of different features has more influence on the model than the information content of the single feature. This is especially true for the random forest model, as [4] shows, and can be explained by the randomly chosen features.

Interestingly enough, removing the distance from the feature set results in a considerable gain of accuracy of nearly 2 points. The distance was thought to be a good indicator of the class {*yes*}: Almost all instances with a geodesic path shorter than 6 are of this class. It does not reliably predict the other class though. Even infinite paths are no indicator of class {*no*}.

The closeness of the first word form has the biggest impact on the model. The closeness indicates the mean length of the geodesic paths from this node to any other node in the graph. The degree of the first word form has a high

---

[6] Other classifiers (e.g., support vector machines and Bayesian models) were evaluated as well. The full results cannot be presented here. Still, the findings are very comparable to the ones that will be presented here, but the precision, recall, and F-measures are considerably lower.

impact on the performance of the model. Also the page rank and betweenness can contribute a lot to the overall accuracy. The measures of semantic similarity that were thought to indicate close relations between word senses have less impact. This is likely related to the fact that these measures only work on pairs where both word forms are nouns. Only 1,322 of the total 2,511 available instances are of this kind. The local and global network measures that were proposed in this paper as features of semantic relatedness between word forms in WN contribute the greater part to the accuracy of the model.

Leaving out all semantic similarity measures and the geodesic path shows this even more clearly. Using only the here proposed graph-based features results in an even higher accuracy of 90.12% of correctly classified instances. The model is trained using 16 random features.[7] This relatively small set of easy to use and to calculate graph measures contains enough information to suit the classification task and reach high precision, recall, and F-measure (all three 0.9). This is very well balanced: No class has a significantly higher precision or recall. In the following comparison to CoreLex BTs, the measures of semantic similarity and the geodesic path will not be considered and only the features given in Table 2 will be used.

4.2   Comparison to Using CoreLex BTs as Features

To compare the graph-based features to the CoreLex BTs, every noun word form was annotated the appropriate BT assigned by the CoreLex resource. The assumption is that those BTs show patterns of regular polysemy that the classification algorithm should be able to identify. The actual rules proposed in [5,16,2] were not used. Also, this is not a direct comparison to those results. But it should give insight into the quality of the features used.

The CoreLex approach can only be used to identify regular polysemy. The ratio of correctly classified instances can therefore be expected to drop compared to the graph-based approach. All instances of a noun word form and one of another POS will not show any significant patterns. This again can be expected to result in a drop of accuracy.

Using only the BTs to train a model and evaluating it as before results in 64.99% accuracy.[8] This is 25.13 points less accurate than the method proposed in this paper.

The BTs are not fit to identify cases involving other POSs than nouns. Using only instances containing just nouns and no other POSs, and only the BTs as features results in 66.77% accuracy. Using the graph-based measures instead results in 82.9%. Thus, they are still 16.13 points more accurate. Combining

---

[7] As the number of available features drops so does the optimal number of features used in the random forest model.

[8] Different models were trained using different algorithms. Still, the random forest model was the most accurate one. The numbers given in the following are always the highest possible rates of accuracy of a random forest model of 100 trees. The number of randomly selected features varies.

all measures further increases the accuracy up to 93.31% on the set containing instances of nouns and other POSs and 87.9% when only noun–noun pairs are used.

5 Outlook

The approach presented in this paper aims at connecting those homograph word forms in WN that share some common meaning and can hence be assumed to belong to a polysemous word. Both regular and irregular polysemy are meant to be covered. This distinguishes the proposed method from other approaches. Earlier efforts were solely focused on regular polysemy and thereby ignored the apparently quite high number of irregular cases found in WN.

The procedure described here is not limited to only nouns but can be straight-forwardly adapted to other POSs as well. By taking homographs of different POSs into account and by handling irregular polysemy as well, it outperforms models trained on the BTs proposed in [5] by far. Using the proposed features and the mentioned measures of semantic similarity including the geodesic path results in 86.86% accuracy. Using only the graph-based measures further increases the accuracy up to 90.12%. Combining the network-based features and the BTs results in 93.31% accuracy. Although it was assumed that the decision whether two homograph word forms belonging to different synsets share a common meaning was a question of semantic similarity, the geodesic path, although when < 6 a good indicator for the class *yes*, and other measures of semantic similarity did not improve the performance.

Following [16, p. 268], polysemy and homonymy are "two points on a gradient, where the words in the middle show elements of both". The method presented in this paper allows measuring the degree of homonymy a word exhibits by looking at the sense level and connecting the different senses of a word by relating the corresponding word forms. The word forms of a word like *bank* are not all connected, only those that actually share a common meaning.

The next steps will be to find fitting features for other POSs based on a deep analysis of the networks structure, to manually annotate a test and training set, and to train appropriate models on this data.

References

1. Apresjan, J.U.D.: Regular Polysemy. Linguistics 12, 5–32 (1974)
2. Boleda, G., Padó, S., Utt, J.: Regular polysemy: A distributional model. In: Proceedings of the First Joint Conference on Lexical and Computational Semantics – vol. 1: Proceedings of the Main Conference and the Shared Task, and vol. 2: Proceedings of the Sixth International Workshop on Semantic Evaluation, SemEval 2012, vol. 1, pp. 151–160. Association for Computational Linguistics, Stroudsburg (2012)
3. Bonacich, P.: Power and Centrality: A family of Measures. The American Journal of Sociology 92(5), 1170–1182 (1987)

4. Breiman, L.: Random forests. Machine Learning 45, 5–32 (2001)
5. Buitelaar, P.: Corelex: An ontology of systematic polysemous classes. In: Proceedings of the 1st International Conference on Formal Ontology in Information Systems (FOIS 1998), June 6-8. Frontiers in Artificial Intelligence and Applications, vol. 46, pp. 221–235. IOS Press, Trento (1998)
6. Freeman, L.C.: Centrality in Social Networks Conceptual Clarification. Social Networks 1(1978), 215–239 (1978)
7. Hirst, G., St-Onge, D.: Lexical chains as representations of context for the detection and correction of malapropisms. In: Fellbaum, C. (ed.) WordNet: An Electronic Lexical Database, pp. 305–332. MIT Press (1998)
8. Jiang, J.J., Conrath, D.W.: Semantic similarity based on corpus statistics and lexical taxonomy. CoRR cmp-lg/9709008 (1997)
9. Leacock, C., Miller, G.A., Chodorow, M.: Using corpus statistics and wordnet relations for sense identification. Comput. Linguist. 24(1), 147–165 (1998)
10. Lin, D.: An information-theoretic definition of similarity. In: Shavlik, J.W. (ed.) Proceedings of the 15th International Conference on Machine Learning (ICML 1998), Madison, WI, USA, July 24-27, pp. 296–304. Morgan-Kaufman Publishers, San Francisco (1998)
11. Miller, G.A.: WordNet: A lexical database for english. Commun. ACM 38(11), 39–41 (1995)
12. Page, L., Brin, S., Motwani, R., Winograd, T.: The pagerank citation ranking: Bringing order to the web. In: Proceedings of the 7th International World Wide Web Conference, Brisbane, Australia, pp. 161–172 (1998)
13. Pedersen, T., Patwardhan, S., Michelizzi, J.: Wordnet: Similarity – measuring the relatedness of concepts. In: McGuinness, D.L., Ferguson, G. (eds.) AAAI, pp. 1024–1025. AAAI Press / The MIT Press (2004)
14. Resnik, P., Yarowsky, D.: Distinguishing systems and distinguishing senses: new evaluation methods for word sense disambiguation. Natural Language Engineering 5(2), 113–133 (1999)
15. Snow, R., Prakash, S., Jurafsky, D., Ng, A.Y.: Learning to merge word senses. In: Proceedings of the 2007 Joint Conference on Empirical Methods in Natural Language Processing and Computational Natural Language Learning, June 28-30, pp. 1005–1014. ACL (2007)
16. Utt, J., Padó, S.: Ontology-based distinction between polysemy and homonymy. In: Proceedings of the Ninth International Conference on Computational Semantics, IWCS 2011, pp. 265–274. Association for Computational Linguistics, Stroudsburg (2011)
17. Veale, T.: Polysemy and category structure in wordnet: An evidential approach. In: Proceedings of the Fourth International Conference on Language Resources and Evaluation, LREC 2004. European Language Resources Association (2004)
18. Wu, Z., Palmer, M.: Verbs semantics and lexical selection. In: Proceedings of the 32nd Annual Meeting on Association for Computational Linguistics, ACL 1994, pp. 133–138. Association for Computational Linguistics, Stroudsburg (1994)

Attribute Value Acquisition
through Clustering of Adjectives

Agnieszka Mykowiecka and Małgorzata Marciniak

Institute of Computer Science, Polish Academy of Sciences,
Jana Kazimierza 5, 01-248 Warsaw, Poland
{agn,mm}@ipipan.waw.pl

Abstract. In the paper we analyse Polish descriptive adjectives which
occur in domain related texts. The experiments were done on data obtained from hospital discharge records. Prenominal adjectives selected
from these texts were filtered out of presumably relative adjectives and
clustered on the basis of a set of context related features and interword
relations derived from Wordnet. We tested if this procedure can be used
to automatically identify concept features, i.e. whether adjectives representing different values of one feature will form one cluster. The obtained
results proved to be useful as a preprocessing step in a specialized sub-
domain ontology creation procedure.

Keywords: Adjectives, clustering, property identification, domain texts.

1   Introduction

An ultimate goal of research done within an area of natural language processing is
automatic text and speech understanding. While this task is still much too hard
to solve, a lot of simpler but practically useful applications are being developed
like information extraction or semantic role labeling. To provide them, one needs
a description of searched information, e.g. list of semantic concepts together with
their features, and data on how they are expressed in texts. In the case of highly
specialized topics, the task of domain model creation has to be performed by
domain specialists who posses the required knowledge. But quite often, they are
not prepared to organize the extracts from their knowledge into a taxonomy and
they fail to enumerate "off-line" all concepts and features which are relevant to
the chosen area. As an adequate domain model is crucial to the efficiency and
reliability of any computer application built upon it, even partial automatisation
of this process could be of practical use.

Domain model creation consists of several steps which can be addressed separately. In a very rough approximation, one can distinguish the steps of concept
recognition, definition of concept attributes, and concept taxonomy creation.
The first step is usually done by terminology extraction methods which rely on
identifying (mainly nominal) phrases in the domain related texts and ranking
them according to chosen criteria approximating domain relevance. Standard approaches to automatic terminology extraction are discussed in [15], while some

A. Przepiórkowski and M. Ogrodniczuk (Eds.): PolTAL 2014, LNAI 8686, pp. 92–104, 2014.

improvements to the ranking procedure are proposed for example in [5]. At this step we can identify *left kidney* or *acute appendicitis* as domain significant concepts. Afterwards it would be useful to recognize that *left* and *acute* are the values of some concepts' attributes, which can be named for example as *location* and *type*. The next step is the recognition of relations between concepts. These relations can be of very different kinds, but for ontology creation the most important one is hypernymy-hyponymy relation. Work on hypernymy detection was inspired by research connected with Wordnet population, [12]. In [14] the idea of automating pregrouping of concepts which should be located close in the hierarchy tree was proposed.

The goal of the research presented in this paper was to elaborate a method of attribute indication on the basis of an analysis of descriptive adjectives. Most research conducted in the area of automatic ontology learning was focused on nouns and nominal phrases. Adjectives as such were less studied, but for example [2] used relational adjectives for extraction of hyponyms from medical texts; while [3] and [8] proposed using adjectives for attribute learning. In [10] a semi-supervised machine-learning approach for the classification of adjectives into property denoting (like in *deep wound*) vs. relation denoting adjectives (like in *environmental science*) is presented. In [16] corpus-based methods were used to group Polish adjectives into semantic clusters. Contrary to this latter work, in our research we focus on domain specific texts and we analyse only property denoting adjectives to address the task of property identification. We postulate that adjectives describing values of one feature are used in similar contexts and thus can be automatically identified. In domain specific texts the degree of word ambiguity is lower than in general texts, so we expect to get not very much noise while grouping adjectives themselves not their senses.

Similar work was done by [11], which was aimed at automatic identification of adjectival scales, and performed clustering adjectives as the first stage of this process. Hatzivassiloglou and McKeown utilized only the Kendall τ coefficient counted for adjective-noun pairs and the information of adjective co-occurrence within one phrase.

The paper is organized as follows. The characteristics of the collected data and the process of its linguistic analysis are briefly described in Sect. 2. Section 4 presents a set of features used in the classification process. The next section presents the results of the proposed approach, and finally an evaluation of the results is described in Sect. 6.

2   Data Characteristics

The experiments presented in the paper were performed on hospital discharge documents gathered at one of the Polish children's hospitals. Texts came from six departments (general, surgery, neurology, neonatology, infectious diseases and rehabilitation) and were written by several physicians of different specialties. Original MS Word files were converted into plain text and anonymised before further processing. Next, all texts were analysed using standard general purpose

NLP (Natural Language Processing) tools. The morphological tagger Pantera [1] based on the linguistic data from the Morfeusz analyser [21] was used to divide text into tokens and annotate them with morphosyntactic tags. The morphosyntactic description included a part of speech name (POS), a base form, as well as case, gender and number information, where they are appropriate. The end of sentence tags were also introduced. This information is then used by shallow grammars which were defined to recognize boundaries of nominal phrases.

The data set consists of 3,116 documents comprising 1,940,000 tokens. The tagset we use is quite detailed, thus words which could constitute a nominal phrase can be of one of four categories: *noun (subst), gerund (ger), brev* and *acron*. We are interested in descriptive features expressed both by *adjectives (adj)*, e.g. *duży* 'big', and *past participles (ppas)*, e.g. *powiększony* 'enlarged'. The Pantera tagger does not analyse out of dictionary forms. These forms (11,777 in total) were tagged as *ign* and were not taken into further considerations. Table 1 presents statistics for the relevant syntactical classes recognized within our data.

Table 1. Selected part of speech distribution

	POS	types	occurrences
acronim	acron	1,476	25,814
abreviation	brev	168	76,058
gerund	ger	444	13,405
substantive	subst	3,642	362,352
adjective	adj	1,743	130,611
past participle	ppas	291	23,278

3   Noun Phrases with Adjectival Modifiers

Adjective modification is one of the most typical constructions occurring inside nominal phrases. For English, the thorough studies of the role of adjectives in grammar and their types is presented in [19]. In Polish linguistic tradition, two main groups of adjectives are distinguished, basing on a difference in the type of properties they describe: qualitative (i.e. absolute, descriptive), and relative (i.e. classifying, distinguishing). A similar but a little more precise classification was introduced in [6]. This co-called BEO classification consisted of basic (i.e. qualitative, property-denoting) adjectives (like in *old box*), object-oriented (relational) adjectives (like in *environmental science*) and a new class of event-related adjectives (like *eloquent person*). This division was used for example in [7] and [10] in which a semi-supervised machine-learning approach for the classification of adjectives was presented. In [13] a very detailed classification of Polish adjectives into 56 classes made from the point of view of a machine translation application was presented. In our work, we divide adjectives into two groups without introducing a differentiation between basic and event-related adjectives and we perform a clustering experiment on all but relational adjectives.

In Polish noun phrases, adjectives can occur at the both sides of a noun and their position can help in the classification task. This work concentrates on descriptive adjectives, which in Polish occur typically before nouns, e.g. *niski poziom* 'low level', in opposition to classifying adjectives, which are located generally after a noun, e.g. *jama$_{subst}$ brzuszna$_{adj}$* 'abdominal cavity'. However, not all adjectives occurring before noun are actually descriptive. Apart from a limited number of examples of lexicalised connections of an adjective and a noun in this very order, e.g. *biały szum* 'white noise' or *ostry dyżur* 'ER', there is a systematic rule of placing at least one classifying adjective before a noun which is modified by more than one relative adjective, e.g. *wojewódzka poradnia alergologiczna* 'provincial allergology clinic' vs. *poradnia wojewódzka* 'provincial clinic' and *poradnia alergologiczna* 'allergology clinic'. In such cases an adjective which describes a more specific aspect of a concept is placed before a noun and a more general adjective is placed after a noun. As the specificity itself is subjective, the order of adjectives changes in different contexts. To be able to filter out classifying adjectives which occurred before nouns, we identified nominal phrases with adjectival modification both before and after nouns. For further processing we selected adjectives that occurred more often before nouns than after them. All phrases obeyed standard Polish gender, case and number agreement constraints. In Table 2 the numbers of types of phrases consisting of one noun and one or more adjectival modifiers which were recognized in the analysed texts, are given (subphrases included in wider phrases are also counted).

Table 2. Types of phrases in data

	nouns		length=2		length=3		length>3	
	types	occ.	types	occ.	types	occ.	types	occ.
A N	1,209	22,425	3,323	21,037	654	1,212	94	176
N A	1,070	63,857	3,067	54,510	754	6,349	375	2,998
A N A	310	4,355	—	—	906	3,426	408	929

Although in general not all adjectives which precede nouns are descriptive, an examination of the data confirmed that adjectives that occur mainly to the left of a noun being a concept are descriptive, and should be represented in the ontology as values of a feature of a concept. For further processing, from the list of adjective-noun phrases we selected adjectives which occurred at least 10 times as left modifiers and occurred relatively much more frequently (two times) as left than as right modifiers.

4   Similarity Features

Our goal was to elaborate a classification scheme which can be easily used for any new types of texts for which no specific ontological resources are available, so we had to rely on such parameters which can be derived directly from text or obtained from general linguistic resources or tools. Thus, we use standard window based features, and context derived coefficients defined specifically for this particular task. Additionally, we use information from the lexicon of closely related words and from Polish Wordnet (plWordnet [16]) – a general, very rich lexical database which contains information on different meanings of words (nouns, adjectives and verbs) and connections between them.

4.1   Lexical Context

We are looking for groups of adjectives which define similar attributes or features of concepts. Values of the same features can be identified by common lexical contexts they occur in. For example, after the expression *zwraca uwagę* 'draws attention' usually untypical aspects of a following concept are given, like *uwypuklone ciemię* 'arched crown', *spłaszczona potylica* 'flattened occiput' or *chrapliwy oddech* 'rasping breath'. Similarity of lexical contexts is a rather obvious and often used feature to define different kind of similarities of particular lexical items, e.g. [12] used patterns like "such NPx as NPy" to identify that NPx is a kind of NPy (NPx, and NPy are noun phrases). This characteristic behaviour can be described by a context consisting of a few words used either to the left or to the right of the expression. It may also be the case that in domain texts some syntactic structures are characteristic for introducing particular types of attributes, so we also decided to test a syntactic characterisation of the context by specifying grammatical classes that words surrounding an adjective belong to.

The chosen maximal length of the context is small, as the analysed documents are rather concise — their authors tend to use short informative phrases and they rapidly change topics.

The context features for an adjective are defined symmetrically for right and left text surrounding an expression consisting of the adjective and a noun (not taking into account a noun which is inside this expression but the subsequent one as the modified noun is taken into account by the $sim_{com-nouns}$ coefficient described in 4.2) and consist of:

- Sequences of base forms of 1, 2 and 3 tokens (afterwards abbreviated to: lb1...lb3 for left, and to rb1...rb3, for right contexts).
- Sequences of POS tags of 2, 3, 4 tokens (lpos2...lpos4, rpos2...rpos4).
- The base form of the nearest:
  - verb (if there are no verbs encountered within the sentence boundaries, this feature value is set to null) (lvp, rvp);
  - noun type token (e.g. nouns, gerunds, acronyms) (lnp, rnp);
  - adjective type token (e.g. adjectives, participles) (ladj, radj);
  - preposition (lpp, rpp).

While establishing contexts we do not go beyond the sentence/paragraph boundaries. In the case of the nearest base form, we ignore similarity which would arise from contexts being ends of sentences or paragraphs, punctuation marks and conjunctions (like *i* 'and'), i.e. in the formula below we do not count adjective occurrences in these contexts.

All lexical similarities were calculated according to the Jaccard coefficient scheme used on form frequencies (we tested also Jaccard coefficient on types and Dice coefficient in both versions, the results were a little lower, but did not differ much; we did not explore more existing similarity measure variants given for example in [20]). In the equation below, t is a context type, C_t is a set of all contexts of the type t, and $ctx(c, a_i)$ is the number of occurrences of an adjective a_i in a context c.

$$
\begin{aligned}
sim_t(a_i, a_j) \\
&= \frac{number\ of\ all\ occurrences\ of\ common\ C_t\ type\ contexts}{number\ of\ occurrences\ of\ all\ C_t\ contexts\ of\ both\ adjectives} \\
&= \frac{\Sigma_{c \in C_t} min(ctx(c, a_i), ctx(c, a_j))}{\Sigma_{c \in C_t} ctx(c, a_i) + \Sigma_{c \in C_t} ctx(c, a_j) - \Sigma_{c \in C_t} min(ctx(c, a_i), ctx(c, a_j))}
\end{aligned}
\tag{1}
$$

4.2 Common Modified Nouns

The classification task concerns modifying adjectives, so the most important context in this case is a context consisting of a modified noun. If more than one concept have a given attribute, e.g. size, it is likely that nouns representing these concepts will occur with the same adjectives expressing it, e. g.*mały, po-większony, niewielki* 'small, enlarged, slight'. To account for this observation we established a similarity measure whose value is equal to the division of a number of commonly modified nouns by a maximum number of nouns modified by one adjective in the analysed text:

$$
\begin{aligned}
sim_{com-nouns}(a_i, a_j) &= \frac{number\ of\ commonly\ modified\ nouns}{maximum\ number\ of\ nouns\ modified\ by\ one\ adjective} \\
&= \frac{\Sigma_{n \in N:\ frq(a_i n) > 0\ and\ frq(a_j n) > 0}\ 1}{max_{n \in N} \left(\Sigma_{a_i \in A:\ frq(a_i n) > 0}\ 1 \right)}
\end{aligned}
\tag{2}
$$

where A is a set of analysed adjectives, N is a set of nouns, $frq(a_i n)$ is a frequency of a bigram consisting of a_i and n.

4.3 Wordnet Similarity

Polish Wordnet is a large net of lexical meanings connected via different relations. In particular, it contains information on a lot of adjectives. Unfortunately, the adjective hypernymy hierarchy is very flat and most adjectives have one or at most two nodes above them so adjectives' similarity is hard to derive from there. None of the wordnet-based similarity measures provided at the plWordnet site, works for adjectives. Thus, we used information from plWordnet relations (antonymy, synsets relatedness and hypernymy) directly to implement similarity measures for adjectives. As adjectives in our data are not semantically disambiguated, we aggregate information given for all meanings.

The relation which is the most numerous in the adjectival part of plWordnet is antonymy. Although this relation, of course, does not link words with the same meaning, it may link adjectives expressing different values of one attribute, e.g. *duży, mały* 'big small', so we use it as a source of one of the similarity measures ($\text{sim}_{w_{a-ant}}$).

Two other plWordnet relations which we use are hypernymy and relatedness. They are directly connected with meaning similarity, and information coming from these two sources is represented jointly by one similarity feature ($\text{sim}_{w_{a-hyp}}$). This coefficient takes into account adjectives from the same synset.

In some cases additional information can be obtained indirectly from the antonymy relation. Words which share the same antonyms (for some of their senses) are similar so we added the coefficient $\text{sim}_{w_{a-sim}}$ basing on this information.

The three plWordnet similarity features are defined as:

$$\text{sim}_{w_{a-X}}(a_i, a_j) = \frac{number\ of\ senses\ in\ relation\ X}{smaller\ number\ of\ senses}$$
$$= \frac{\Sigma_{si \in S(a_i), sj \in S(a_j):\,((si\ X\ sj)\ or\ (sj\ X\ si))}\,1}{min\,(S(a_i), S(a_j))} \quad (3)$$

where a_i and a_j are adjectives, $S(a_i)$ is a set of senses of a_i and X is one of three possible relations: an antonymy, a sum of hypernymy and synsets relatedness, or the similarity relation defined above.

Although plWordnet contains numerous adjectives, quite a number of the words from our list are not represented there. As many adjectives are derived from nouns, for these forms we used additional information from the noun hypernymy hierarchy. Using a set of rules we transformed adjectives into possible noun forms and calculated $\text{sim}_{w_{n-hyp}}$ coefficient in the same way as above for those nominal forms which were found in plWordnet (nearly no links between adjectives and nouns are present in this data at the moment). The fourth plWordnet similarity feature defined using this information is:

$$\text{sim}_{w_{n-hyp}}(a_i, a_j) = \frac{\Sigma_{si \in S(n(a_i)), sj \in S(n(a_j)):\,((si\ hypernim\ sj)\ or\ (sj\ hypernim\ si))}\,1}{min\,(S(n(a_i)), S(n(a_j)))}$$
$$(4)$$

where $n(a_i)$ is a noun derived from a_i.

4.4   Data from the Dictionary of Polish Synonyms

As hypernymy relation among adjectives in plWordnet is not very elaborated, we additionally used data included in an open source dictionary of synonyms (http://synonimy.ux.pl). It contains 13,180 groups with 44,550 words or phrases. From this data set we obtained 85 similarity pairs for the considered list of adjectives. As, in this case, no frequency data nor number of senses described are available, the similarity is defined as:

$$sim_{dict}(a_i, a_j) = \frac{number\ of\ groups\ with\ both\ elements}{bigger\ number\ of\ groups\ for\ both\ elements} \qquad (5)$$

5   Clustering

Automatic clustering of adjectives was done using Multidendrograms tool [9] – a program which implements hierarchical clustering and solves the non-uniqueness problem found in the standard pair-group algorithm by grouping more than two clusters at the same time when ties occur. As input, it takes singular similarity values for all pairs of adjectives which were counted as a weighted sum of 26 features described in the previous section. In (6) S_{sim} is the set of all similarity coefficients described in Sect. 4.

$$sim(a_i, a_j) = \Sigma_{sim_t \in S_{sim}} weight(sim_t) \times sim_t(a_i, a_j) \qquad (6)$$

We performed several experiments with different weights assigned to different groups of features. So, we tested if any features help in obtaining better results, and which of them are the most valuable in this task. The complete linkage strategy was chosen as the clustering method.

In the first model, the weights were tuned on the basis of manual grouping of the 28 most popular adjectives that appeared at least 50 times in adjective-noun phrases in the corpus. The procedure of selecting these adjectives was the following. From the whole set of 105 adjectives that appeared at least 50 times in adjective-noun phrases we selected those that were used three times more frequently to the left of an adjective than to the right. After this step we obtained 65 adjectives. From this set we selected those that create at least two-element groups according to the information obtained from the plWordnet and the thesaurus of synonyms. This set consists of adjectives that are frequent in general language so it was relatively easy to group them manually and create a development set for tuning weights of coefficients. The grouping reflects projection of annotator's knowledge about the domain into synsets represented in plWordnet. This step resulted in 12 groups given in Table 3.

The method of constructing the starting set of adjectives was motivated by our desire to obtain a data set which includes some multi-element groups whose automatic identification can be then later tested. Due to the limited capacity of our hospital data, many properties are represented there by only one possible value so randomly chosen adjectives might create mostly one element groups.

In the process of manual tuning of the set of weights we performed several experiments described below. Their results were compared with a manually created classification using the B-cubed measure [4] which is sensitive to the presence and absence of the elements of the compared groups. The weights from the best manually tuned model are presented in Table 4. For this model we obtained F-measure of 0.799 for the division into 12 groups, while the highest F-measure

Table 3. Manual grouping of 28 adjectives

group 1: *drobny* 'small', *niewielki* 'small', *duży* 'large', *powiększony* 'enlarged';
group 2: *pozostały* 'remaining';
group 3: *stały* 'stable, constant';
group 4: *kolejny* 'next', *ponowny* 'repeated', *początkowy* 'initial',
           ostatni 'last', *pierwszy* 'first' *drugi* 'second';
group 5: *obfity* 'abundant', *liczny* 'numerous', *nieliczny* 'not numerous';
group 6: *nieznaczny* 'minor, insignificant', *znaczny* 'considerable, substantial',
           istotny 'important';
group 7: *obustronny* 'two-sided', 'mutual;
group 8: *rozluźniony* 'loose' 'relaxed';
group 9: *podwyższony* 'increased', *wzmożony* 'enhanced', *niski* 'low',
           wysoki 'high',
group 10: *różny* 'different';
group 11: *płynny* 'liquid', 'floating',
group 12: *silny* 'strong', *słaby* 'weak'

of 0.813 was obtained for 13 groups. If the information from the thesauri was neglected (the appropriate weights in the model given in Table 4 were set to a 0 value) the F-measure dropped to 0.758. The highest F-measure of 0.809 was obtained for 15 groups.

Table 4. The model

	coeff.	value	coeff.	value	coeff.	value	coeff.	value	coeff.	value
left/right POS										
	lpos2	0.12	lpos3	0.12	lpos4	0.12				
	rpos2	0.09	rpos3	0.09	rpos4	0.09				
left/right base form										
	lb1	0.12	lb2	0.12	lb3	0.12				
	rb1	0.09	rb2	0.09	br3	0.09				
left/right nearest verb, noun, adjective, preposition										
	lvp	0.12	lnp	0.30	ladj	0.25	lpp	0.06		
	rvp	0.90	rnp	0.25	radj	0.20	rpp	0.04		
thesauri										
	w_{a-sim}	0.20	w_{a-ant}	0.40	w_{a-hyp}	0.10	w_{n-hyp}	0.15	dict	0.20
common nouns	0.50									

To discover how important the particular type of coefficient is, we compared models with the only one non-zero value of the weight related to this coefficient. For all types of coefficients we checked the F-measure for the division consisting of 12 groups. We observed how many groups were created, how many adjectives were not linked and how quickly the set of adjectives was divided. The last information could be obtained by analysing thresholds for which the set of adjectives is divided into chosen number of groups. One of the questions we set out to explore was whether left and right contexts are equally important. In order to answer this, we set all left contexts based on POS and base forms to the same value. We obtained an F-measure of 0.601 in comparison with manual model, while the right contexts gave a slightly worse result (0.556). Similar differences were found comparing left and right sets of coefficients based on nearest (results for the left contexts): noun (0.609 – for threshold 0.043), verb (0.569 – 0.106), adjective (0.565 – 0.051), and preposition (0.541 – 0.0015). In this case combining left contexts give a 0.622 F-measure while the right one 0.577. So in the manual model the left contexts have slightly higher weights than the right one.

Table 5. Results for groups of coefficient types

type of similarity	F-measure
left noun/adjective/verb/preposition	0.622
right noun/adjective/verb/preposition	0.577
both nouns/adjectives/verbs/prepositions	0.699
left strings of base forms and pos	0.601
right strings of base forms and pos	0.556
both strings of base forms and pos	0.636
wordnet relations	0.655
common nouns	0.698–0.75

Table 5 presents the results for several groups of coefficient types. They show that the division closest to the manually created one is obtained for the coefficient based on common nouns modified by adjectives. Although there is no division into 12 groups, the division into 11 groups gives an F-measure of 0.698, while for 14 groups – 0.75. A somewhat surprising result was obtained for coefficients based on plWordnet and the synonym thesaurus, as it is slightly worse than the result for the combination of nearest noun/adjective/verb/preposition coefficients. The reasons for this may be twofold. First, we do not perform any word sense disambiguation, so for example, the word *stały* 'stable, constant' is connected in one group with *płynny* 'liquid', so the first meaning is preferred, while in our data it is used in another meaning *stała opieka* 'constant care', or *stały ból* 'constant pain'. The second problem is that the plWordnet hierarchy of adjectives is not very elaborated. It is being developed intensively at the moment, so we expect that the results will improve in the future.

To test to what extent the suggested method can be used in an automatic mode, without any manual parameter tuning step, we check the result for a model in which weights assigned to all similarity coefficients were given the same nonzero value. For this model we obtained the F-measure of 0.651 for 12 groups, while the best F-measure of 0.718 was for 20 groups. So, although the results are lower, they are not very much different from those obtained via manual tuning.

6   Evaluation

An evaluation of the method was done on the basis of 101 adjectives that appeared in the data at least 30 times in adjective-noun phrases and were used two times more often to the left of a noun than to the right.

A manual grouping of these 101 adjectives was done according to the subjective knowledge of an annotator familiar with the domain and data. The starting point of this task was the result of grouping the most common 28 adjectives. The grouping of 101 adjectives was checked by another annotator, who suggested only one change. This process resulted in a division consisting of 52 groups, 28 of them containing only one element.

The automatic clustering of the evaluation set using the manually tuned model described in the previous section, compared with the manual grouping for the same number of groups, gave following results: precision 0.676, recall 0.653 and F-measure: 0.664.

The evaluation data contained 28 adjectives that took part in tuning the model. If we removed these 28 adjectives from the final divisions we got 34 one element groups (6 more) and that affected the results of comparison making them better than they were. It turned out that these initially selected 28 adjectives were 'seeds' of larger groups.

The results obtained for the evaluation set using the model consisting of equal coefficients for the division into 52 groups were following: precision 0.626, recall 0.620 and F-measure 0.623. Thus, the results obtained using this non tuned model were only slighty lower for this bigger set than for the initial 28 adjectives. Again, the difference between results for the tuned model and the simple one was not very big.

Table 6 presents one group obtained by automatic clustering of 101 adjectives together with equivalents from the manual division. In the later data *dodatni* was included in the separate group together with *negatywny* 'negative'. The last line presents the respective group obtained by automatic clustering of 28 adjectives (the other 2 adjectives are not present in this smaller set).

Table 6. An examplary result

method	set	group				
automatic	101	*dodatni,*	*niski,*	*wysoki,*	*obniżony,*	*podwyższony*
		positive,	low,	high,	reduced,	increased
manual	101		*niski,*	*wysoki,*	*obniżony,*	*podwyższony*
			low,	high,	reduced,	increased
manual & automatic	28		*niski,*	*wysoki,*		*podwyższony*
			low,	high,		increased

7   Conclusion

From the performed experiments it appears that automatic clustering basing on the suggested parameters set can be used to preliminary identify adjectives describing one property in big enough and coherent data. Adjectives that are not related with other adjectives in plWordnet or other thesauri can be clustered only on the basis of the syntactic information available in the corpus and the results are still acceptable.

The most informative feature in our model was the coefficient based on common nouns modified by both compared adjectives. In further work we plan to test different ways of defining this particular similarity measure to check if these results could be even better. Moreover, the left contexts turned out to be only slightly more important than the right ones, and contexts based on the nearest nouns/adjectives/verbs/prepositions are only slightly better than contexts based on POS and base forms.

To improve the results, we plan to extend the experiments presented here by combining them with adjectives' sense disambiguation task using clustering methods allowing for the placement of one word in more than one cluster like in [18].

References

1. Acedański, S.: Tager morfosyntaktyczny PANTERA. In: [17], pp. 197–207
2. Acosta, O., Aguilar, C.A., Sierra, G.: Using relational adjectives for extracting hyponyms from medical texts. In: Lieto, A., Cruciani, M. (eds.) AIC@AI*IA. CEUR Workshop Proceedings, vol. 1100, pp. 33–44. CEUR-WS.org (2013)
3. Almuhareb, A., Poesio, M.: Attribute-based and value-based clustering. An evaluation. In: Proceedings of the 2004 Conference on Empirical Methods in Natural Language Processing, pp. 158–165 (2004)
4. Bagga, A., Baldwin, B.: Algorithms for scoring coreference chains. In: The First International Conference on Language Resources and Evaluation Workshop on Linguistics Coreference, pp. 563–566 (1998)
5. Barrón-Cedeño, A., Sierra, G., Drouin, P., Ananiadou, S.: An improved automatic term recognition method for Spanish. In: Gelbukh, A. (ed.) CICLing 2009. LNCS, vol. 5449, pp. 125–136. Springer, Heidelberg (2009)
6. Boleda, G.: Automatic acquisition of semantic classes for adjectives. PhD thesis, Pompeu Fabra University (2006)
7. Boleda, G., Badia, T.: Morphology vs. syntax in adjective class acquisition. In: Proceedings of the ACL-SIGLEX Workshop on Deep Lexical Acquisition, DeepLA 2005, pp. 77–86. Association for Computational Linguistics, Stroudsburg (2005)
8. Cimiano, P.: Ontology Learning and Population from Text. Algorithms, Evaluation and Applications. Springer (2006)
9. Fernández, A., Gómez, S.: Solving non-uniqueness in agglomerative hierarchical clustering using multidendrograms. Journal of Classification 25, 43–65 (2008)
10. Hartung, M., Frank, A.: A semi-supervised type-based classification of adjectives: Distinguishing properties and relations. In: Proceedings of the Seventh International Conference on Language Resources and Evaluation (LREC 2010). ELRA, Valletta (2010)

11. Hatzivassiloglou, V., McKeown, K.: Towards the automatic identification of adjectival scales: Clustering adjectives according to meaning. In: Schubert, L.K. (ed.) ACL, pp. 172–182. ACL (1993)
12. Hearst, M.A.: Automated discovery of WordNet relations. In: Fellbaum, C. (ed.) WordNet: An Electronic Lexical Database, pp. 131–153. MIT Press (1998)
13. Jassem, K.: Semantic classification of adjectives on the basis of their syntactic features in Polish and English. Machine Translation 17(1), 19–41 (2002)
14. Mykowiecka, A., Marciniak, M.: Combining wordnet and morphosyntactic information in terminology clustering. In: Proceedings of the 24th International Conference on Computational Linguistics (COLING 2012), Mumbai, India (2012)
15. Pazienza, M., Pennacchiotti, M., Zanzotto, F.: Terminology Extraction: An Analysis of Linguistic and Statistical Approaches. In: Sirmakessis, S. (ed.) Knowledge Mining. STUDFUZZ, vol. 185, pp. 255–279. Springer, Heidelberg (2005)
16. Piasecki, M., Szpakowicz, S., Broda, B.: A Wordnet from the Ground Up. Oficyna Wydawnicza Politechniki Wrocławskiej, Wrocław (2009)
17. Przepiórkowski, A., Bańko, M., Górski, R.L., Lewandowska-Tomaszczyk, B. (eds.): Narodowy Korpus Języka Polskiego. PWN, Warsaw (2012)
18. Tomuro, N., Lytinen, S.L., Kanzaki, K., Isahara, H.: Clustering using feature domain similarity to discover word senses for adjectives. In: Semantic Computing, ICSC 2007, pp. 370–377 (September 2007)
19. Warren, B.: Classifying Adjectives. Acta Universitatis Gothoburgensis, Göteborg (1984)
20. Weeds, J., Weir, D.J.: Co-occurrence retrieval: A flexible framework for lexical distributional similarity. Computational Linguistics 31(4), 439–475 (2005)
21. Woliński, M.: Morfeusz — a practical tool for the morphological analysis of Polish. In: Intelligent Information Processing and Web Mining. AISC, vol. 35, pp. 511–520. Springer, Heidelberg (2006)

Cross-Lingual Semantic Similarity Measure for Comparable Articles

Motaz Saad, David Langlois, and Kamel Smaïli

SMa^rT Group, LORIA,
INRIA, Villers-lès-Nancy, F-54600, France
Université de Lorraine, LORIA, UMR 7503, Villers-lès-Nancy, F-54600, France
CNRS, LORIA, UMR 7503, Villers-lès-Nancy, F-54600, France
{motaz.saad,david.langlois,kamel.smaili}@loria.fr

Abstract. A measure of similarity is required to find and compare cross-lingual articles concerning a specific topic. This measure can be based on bilingual dictionaries or based on numerical methods such as Latent Semantic Indexing (LSI). In this paper, we use LSI in two ways to retrieve Arabic-English comparable articles. The first way is monolingual: the English article is translated into Arabic and then mapped into the Arabic LSI space; the second way is cross-lingual: Arabic and English documents are mapped into Arabic-English LSI space. Then we compare LSI approaches to the dictionary-based approach on several English-Arabic parallel and comparable corpora. Results indicate that the performance of our cross-lingual LSI approach is competitive to the monolingual approach and even better for some corpora. Moreover, both LSI approaches outperform the dictionary approach.

Keywords: Cross-lingual latent semantic indexing, corpus comparability, cross-lingual information retrieval.

1 Introduction

Comparing cross-lingual articles is a challenging problem for several tasks in natural language processing and especially in machine translation and cross-lingual information retrieval. The comparison can be done in terms of topics, opinions or emotions. In this paper, we focus on how to retrieve comparable articles. A comparable corpus is a collection of articles in multiple languages which are not necessarily translations of each other, but they are related to the same topic. On the other hand, a parallel corpus can be considered as a comparable corpus in which each sentence in the source corpus is aligned to its translation in the target corpus.

There are many methods proposed in literature to compare as well as to retrieve cross-lingual articles. These methods are based on bilingual dictionaries [10,16,19], or on cross-lingual Information retrieval (CL-IR) [7,1,21] or on cross-lingual Latent Semantic Indexing (CL-LSI) [2,11,6,14].

In dictionary-based methods [10,16,19], two cross-lingual documents d_a and d_e are comparable if a maximum of words in d_a are translations of words in

A. Przepiórkowski and M. Ogrodniczuk (Eds.): PolTAL 2014, LNAI 8686, pp. 105–115, 2014.

d_e, so a bilingual dictionary can be used to look-up the translation of words in both documents. The drawbacks of this approach are the dependency on bilingual dictionaries which are not always available and the necessity to use morphological analyzers for languages that can be inflected. Moreover, word-to-word translations based on dictionaries can lead to many errors. [19] proposed binary and cosine measures based on multi-WordNet [3] dictionary to compare Wikipedia and news articles. Both binary and cosine measures proposed by [19] require the source-target texts to be represented as vectors of aligned words. Word weight for the binary measure is either 1 or 0 (presence or absence of the word), while it is term frequency for the cosine measure. The similarity of cross-lingual documents is computed as follows: the binary measure counts the words in d_a which are translation of words in d_e and then normalize it by the vector size, whereas the cosine measure computes the cosine similarity between source and target vectors which represent the frequency of the aligned words of d_a and d_e.

In Cross-Lingual Information Retrieval (CL-IR) methods, one can use Machine Translation (MT) systems in order to achieve source and target documents into the same language. Then classical IR tools can be used to identify comparable articles [7,1,21]. Query documents are usually translated into the language of indexed documents. This is because the computational cost of translating queries is far less than the cost of translating all indexed documents. The drawback of this approach is the dependency on MT systems. The performance of MT affects the performance of the IR system. Moreover, the MT system needs to be developed first if it is not available for the desired language.

In Cross-Lingual Latent Semantic Indexing (CL-LSI) methods, documents are described as numerical vectors that are mapped into a new space. Then one can compute the cosine between vectors to measure the similarity between them. The LSI method has already been used in context of CL-IR in [2,11,14]. In their approach, the source document and its translation (the target) are concatenated into one document and then LSI learns links between source and target words or documents. [2] focused their work on Greek-English document retrieval and [11] focused on French-English documents, while [14] computed the similarity of Wikipedia articles in several European languages.

In this work, we focus on CL-IR for English-Arabic document retrieval. In order to avoid using bilingual dictionaries or morphological analyzers or MT systems, we use CL-LSI to compare and retrieve English-Arabic documents. Another advantage of CL-LSI is that it overcomes the problem of vocabulary mismatch between queries and documents. We therefore use the same approach as [11], however, we apply it on Arabic-English articles and [11] used parallel corpus in their work, but we use both parallel and comparable corpus to train CL-LSI.

In this paper, we use LSI in two ways to retrieve Arabic-English comparable documents. We refer to the first way as monolingual: the English article is translated and then mapped into the LSI Arabic space; the second way as cross-lingual: Arabic and English articles are mapped into Arabic-English CL-LSI space. We also compare these methods to the dictionary-based method proposed by [19] which is described above.

Besides using CL-LSI to retrieve comparable articles, we also use it to measure the "comparability of a corpus", i.e. to inspect if a target corpus is a translation of a source one and how much they are different from each other. This enables an understanding of how much the source and target texts, in a comparable corpora, are similar to each other. This can be useful for many applications such as cross-lingual lexicon extraction, information extraction, and sentence alignment.

The rest of the paper is organized as follows: corpora and the method are described in Sect. 2, 3, and 4. Results are presented and in Sect. 5. Finally, the conclusion is stated.

2   Corpora

In this section we describe the corpora we used for our experiments. It consists of documents collected from newspapers, United Nations resolutions, talks, movie subtitles and other domains. These corpora are either parallel or comparable. A detailed description of these corpora is provided in the following subsections.

2.1   Parallel Corpora

Table 1 presents the parallel corpora. $|S|$ is the number of sentences, $|W|$ is the number of words, and $|V|$ is the vocabulary size. The table also shows the domain of each corpus. The parallel corpora that we use are: AFP[1], ANN[2], ASB[3] [12], Medar[4], NIST [15], UN [17], TED[5] [4], OST[6] [20] and Tatoeba[7] [20].

Note that OST is a collection of movie subtitles translated and uploaded by users. So the quality of the translations may vary from a user to another.

As can be noted from Table 1, in all parallel corpora, English texts have more words than Arabic. In contrast, Arabic texts have vocabulary larger than English. The reason is that certain Arabic terms can be agglutinated [13], while English terms are isolated. For instance, the Arabic term وَسَنُعطِيهِم *wasanoṭeyhm* translating to "and we will give them" in English, is an example where one Arabic term corresponds to five English words. On the other hand, Arabic has a larger vocabulary because it is morphologically rich [8,18]. For example, the English word "travellers" may correspond to three forms in Arabic: مُسَافِرون *mosā-ferwn* in masculine nominative form, مُسَافِرين *mosāferyn* in masculine accusative/genitive form or مُسَافِرَات *mosāferāt* in feminine form.

---

[1] www.afp.com

[2] www.annahar.com

[3] www.assabah.com.tn

[4] www.medar.info

[5] www.ted.com

[6] www.opensubtitles.org

[7] www.tatoeba.org

Table 1. Parallel Corpora

Corpus	$|S|$	$|W|$		$|V|$	
		English	Arabic	English	Arabic
Newspapers					
AFP	4K	140K	114K	17K	25K
ANN	10K	387K	288K	39K	63K
ASB	4K	187K	139K	21K	34K
Medar	13K	398K	382K	43K	71K
NIST	2K	85K	64K	15K	22K
United Nations Resolutions					
UN	61K	2.8M	2.4M	42K	77K
Talks					
TED	88K	1.9M	1.6M	88K	182K
Movie Subtitles					
OST	2M	31M	22.4M	504K	1.3M
Other					
Tatoeba	1K	17K	13K	4K	6K
Total	2.3M	37M	27.5M	775K	1.8M

2.2 Comparable Corpora

Table 2 shows WIKI and EuroNews comparable corpora, where $|D|$ is the number of articles, $|W|$ is the number of words and $|V|$ is the vocabulary size. Each pair of comparable articles is related to the same topic. WIKI and EuroNews corpora were collected and aligned at article level in [19]. WIKI is collected from Wikipedia website[8] and EuroNews is collected from EuroNews website.[9] WIKI articles are edited online by Wikipedia community. There is a hyperlink between articles that are related to the same topic, but each article may be written independently. Therefore, Wikipedia articles are not necessarily translations of each other.

Table 2. Comparable Corpora

	WIKI		EuroNews	
	English	Arabic	English	Arabic
$|D|$	40K	40K	34K	34K
$|W|$	91.3M	22M	6.8M	5.5M
$|V|$	2.8M	1.5M	232K	373K

3 LSI-Based Methods

The LSI method [5] decomposes a term-document matrix X using the the Singular Value Decomposition (SVD) as $X = USV^T$. The matrices U and V^T are

---

[8] www.wikipedia.org
[9] www.euronews.com

the left and right singular vectors respectively, while S is a diagonal matrix of singular values. Each column vector in matrix U maps terms in the corpus into a single concept of semantically related terms that are grouped with similar values in U. The decomposition USV^T has a rank R, where R is the reduced concept dimensionality in LSI.

For our monolingual LSI approach, X is represented as in (1). It is an $m \times n$ matrix that represents a given monolingual corpus which consists of n documents, and m terms. The entries w_{ij} are the $tfidf$ weights.

$$
X = \begin{array}{c} \\ t_1 \\ t_2 \\ \vdots \\ t_m \end{array}
\begin{array}{cccc} d_1 & d_2 & \cdots & d_n \end{array} \atop
\left(\begin{array}{cccc}
w_{11} & w_{12} & \cdots & w_{1n} \\
w_{21} & w_{22} & \cdots & w_{2n} \\
\vdots & \vdots & \ddots & \vdots \\
w_{m1} & w_{m2} & \cdots & w_{mn}
\end{array} \right)
\tag{1}
$$

$$
X = \begin{array}{c} \\ t_1^a \\ t_2^a \\ \\ \vdots \\ \\ t_l^a \\ t_1^e \\ t_2^e \\ \\ \vdots \\ \\ t_m^e \end{array}
\begin{array}{cccc} d_1^u & d_2^u & \cdots & d_n^u \end{array} \atop
\left(\begin{array}{cccc}
w_{11}^a & w_{12}^a & \cdots & w_{1n}^a \\
w_{21}^a & w_{22}^a & \cdots & w_{2n}^a \\
\vdots & \vdots & \ddots & \vdots \\
w_{l1}^a & w_{l2}^a & \cdots & w_{ln}^a \\
w_{11}^e & w_{12}^e & \cdots & w_{1n}^e \\
w_{21}^e & w_{22} & \cdots & w_{2n}^e \\
\vdots & \vdots & \ddots & \vdots \\
w_{m1}^e & w_{m2}^e & \cdots & w_{mn}^e
\end{array} \right)
\tag{2}
$$

In our cross-lingual LSI approach, X is represented as in (2). Each d_i^u is the concatenation of the Arabic document d_i^a and its corresponding English document d_i^e. Consequently, X represents a bilingual corpus consisting of n cross-lingual documents, l Arabic terms, and m English terms. So X is an $(l+m) \times n$ matrix. X, as represented in (2), can be used to represent parallel or comparable corpora. For a parallel corpus, each d_i^u represents a pair of parallel sentences, while for a comparable corpus, it represents a pair of comparable documents. Term-document matrix as formulated in (2), enables LSI to learn the relationship between terms which are semantically related in the same language or between two languages.

This method helps us to achieve our objective to retrieve comparable articles. We describe this retrieval process in the next section.

4   Experiment Procedure

As outlined in Sect. 1, for a source document in English, our objective is to retrieve the target comparable documents in Arabic. So the source document is compared with all target documents and then the most similar target documents are retrieved. This is done by describing the source and target documents as bag-of-words, then mapping them into vectors in LSI space and subsequently by comparing these vectors. If the value of cosine similarity between the two vectors is high, we consider these two documents as comparable. All English and Arabic texts are preprocessed by removing punctuation marks.

In the next sections, we describe how LSI matrices are built and how they are used to retrieve comparable articles. Then we compare the results of these two methods.

4.1   Building LSI Matrices

Steps below describe how LSI matrices are built:

1. Split English and Arabic corpora presented in Sect. 2 into training (90%) and testing (10%) subsets.
2. Use Arabic training corpus to create X as in (1). Then apply LSI to obtain USV^T, the monolingual LSI matrix (LSI-AR) as shown in left of the Fig. 1.
3. Use English-Arabic training corpus to create X as in (2). Then apply LSI to obtain USV^T, the cross-lingual LSI matrix (LSI-U) as shown in right of the Fig. 1.

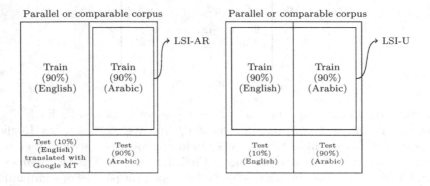

Fig. 1. LSI models

The optimal rank of USV^T in steps 2 and 3 above is chosen experimentally. According to [9], the optimal number of dimensions to perform SVD is in the range $[100\ldots500]$. We conducted several experiments in order to determine the best rank and we found that the dimension 300 optimizes the similarity for the parallel corpus. So we use the dimension 300 in all our experiments.

4.2   Retrieving Comparable Articles

The test corpus is composed of n pairs of English e_i and Arabic a_j documents (aligned at sentence level in parallel corpus and at the document level in comparable corpus). The goal is then to retrieve the a_i among all the a_j given e_i. The following steps describe the two methods:

LSI-AR:

1. For each a_j, get a'_j: $a'_j = a^t_j U S^{-1}$.
2. Translate each English document e_i into Arabic using Google MT service[10] and get a_{e_i}.
3. For each a_{e_i}, get a'_{e_i}: $a'_{e_i} = a^t_{e_i} U S^{-1}$.
4. For each a'_{e_i} and a'_j, compute $\cos(a'_{e_i}, a'_j)$.

LSI-U:

1. For each a_j, get a'_j: $a'_j = a^t_j U S^{-1}$.
2. For each e_i, get e'_i: $e'_i = e^t_i U S^{-1}$.
3. For each e'_i and a'_j, compute $\cos(e'_i, a'_j)$.

e'_i, a'_{e_i}, and a'_j in the methods above are vectors of the same nature since they have a language independent representation. After these two methods, we can use the cosine values to get the most similar Arabic document to a given English one. For each e_i, we sort a_j in descending order according to the cosine values. e_i and a_j are truly comparable if $i = j$. In other words, for each source document, we have only one relevant document. So in the sorted list of a_j, the condition ($i = j$) is checked in the top-1 (recall at 1 or $R@1$), top-5 (recall at 5 or $R@5$), and top-10 (recall at 10 or $R@10$) lists. The performance measure is defined as the percentage of a_i which are successfully retrieved in $R@1$, $R@5$, $R@10$ lists, among all e_i.

5   Results and Discussion

5.1   Retrieving Parallel Articles

The results of the LSI-AR and LSI-U approaches are presented in Table 3. Results are presented for a random sample of 100 source and target test articles because of the computational cost of doing the experiment on all the test corpus. As shown in Table 3, it is not easy to get a general conclusion about the performance of LSI since it depends on the nature of the corpus and on the desired recall ($R@1$, $R@5$ or $R@10$). For example, for AFP, ASB, TED, UN, and Medar, LSI-U is slightly better than LSI-AR. In contrast, for ANN, NIST, OST and Tatoeba, LSI-AR is better than LSI-U. The performance of LSU-U is equal to, or better than LSI-AR in 6 over 9 of corpora for $R@1$. The average value for ($R@1$) in LSI-AR and LSI-U methods are 0.71 and 0.72 respectively. Moreover, we checked the significance of these differences (McNemar's test), and we found that they are not significantly different. Therefore, both approaches obtain mostly similar performance. In addition, we recall that the LSI-U does not require a MT system. Therefore, we can affirm that the LSI-U is competitive compared to LSI-AR.

---

[10] `translate.google.com`

Table 3. LSI results for parallel corpora

Corpus	Method	$R@1$	$R@5$	$R@10$
Newspapers				
AFP	LSI-AR	0.94	0.96	0.99
	LSI-U	0.97	0.99	0.99
ANN	LSI-AR	0.80	0.91	0.94
	LSI-U	0.82	0.92	0.94
ASB	LSI-AR	0.79	0.90	0.92
	LSI-U	0.85	0.92	0.97
Medar	LSI-AR	0.56	0.76	0.81
	LSI-U	0.61	0.78	0.85
NIST	LSI-AR	0.78	0.87	0.92
	LSI-U	0.71	0.82	0.84
United Nations Resolutions				
UN	LSI-AR	0.97	1.00	1.00
	LSI-U	0.98	0.99	1.00
Talks				
TED	LSI-AR	0.52	0.73	0.82
	LSI-U	0.60	0.83	0.92
Movie Subtitles				
OST	LSI-AR	0.39	0.61	0.72
	LSI-U	0.33	0.76	0.85
Other				
Tatoeba	LSI-AR	0.70	0.85	0.94
	LSI-U	0.61	0.79	0.86

The performance of LSI-AR and LSI-U approaches on OST corpus is poor as expected because of the nature of this corpus. OST is composed of subtitles that are translated by many users as mentioned in Sect. 2.

To investigate the effect of the performance of the MT system on the performance of the LSI-AR, we run an experiment to simulate a perfect MT system. This is done by retrieving an Arabic document by providing the same document as a query. This experiment is done on all corpora and the results in terms of $R@1$ are 1.0 for all corpora. These results reveal the lack of robustness of LSI-AR according to the MT system's performance.

We compare our method with the dictionary-based method that was proposed by [19] on the union of AFP and ANN corpora. Results are presented in Table 4 where the dictionary-based method is denoted as DICT.

As shown in the table, both LSI methods achieve better results than DICT, except for $R@10$ which are slightly worse than DICT. It can be concluded that this method is better than DICT since it does not need any dictionary nor morphological analysis and it is language independent.

Table 4. Recall results the union of AFP and ANN corpora

Method	$R@1$	$R@5$	$R@10$
DICT	0.49	0.81	1.0
LSI-AR	0.87	0.95	0.96
LSI-U	0.86	0.96	0.98

5.2 Retrieving Comparable Articles

For comparable corpora, the same experimental protocol is applied. Table 5 shows the performance of recall of the LSI-U method on EuroNews and WIKI comparable corpora. As shown in the table, the performance of the LSI-U on EuroNews corpus is better than WIKI corpus.

Table 5. Testing LSI-U on comparable corpora

Corpus	$R@1$	$R@5$	$R@10$
WIKI	0.42	0.84	0.94
EuroNews	0.84	0.99	1.0

This could be due to the fact that EuroNews articles being mostly translations of each other [19], while Wikipedia articles are not necessarily translations of each other as mentioned in Sect. 2.

From Tables 5 and 3, it can be noted that LSI-U can retrieve the target information at document level and sentence level respectively with almost same performance. The evidence for that is, for parallel corpora, AFP, ANN, and ASB, 0.97, 0.83, and 0.84 $R@1$ was achieved respectively and for EuroNews comparable corpus, 0.84 $R@1$ was achieved.

5.3 Comparing Corpora

We take advantage of the used method in order to study the comparability of some supposed comparable corpora such as WIKI and EuroNews. We do that by computing the average cosine, avg(cos), for all pair articles of the test parts of these corpora. So for each corpus, the LSI-U matrix is built from the training part and used to compute the avg(cos) for the test part. This experiment is done on BEST, EuroNews and WIKI corpora. BEST is the union of AFP, ASB and UN parallel corpora. These corpora are chosen because they have the best recall performance as shown in Table 3. Statistics on comparability are presented in Table 6.

The average similarity proposes to corroborate the fact that for parallel corpus, we get better recall results than by using the other corpora. In other words, the score for BEST which is a parallel corpus aligned at sentence level is better than the one for WIKI which is considered as a real comparable corpus. For EuroNews (near parallel), which is composed of translated articles, the results are better than for WIKI, but lower than for BEST.

Table 6. Statistics on comparability

Corpus	BEST	EuroNews	WIKI
avg(cos)	0.53	0.46	0.23

6   Conclusion

In this paper we described a method which permits to measure comparability between corpora. This method is based on LSI, which we used in two ways: monolingual (LSI-AR) and cross-lingual (LSI-U). The first method needs to use a machine translation system in order to compare two vectors of the same type of data, whereas the second method merges the training data of both languages and in the test step the comparison is then done on two vectors of the same type since they contain the representation of two cross-lingual documents.

We applied this method on English-Arabic documents. The method allows us to identify comparable articles extracted from a variety of corpora. The measure we proposed has shown its feasibility since it enables distinguishing of parallel corpora from strongly comparable corpora such as Euronews and also from the weakly comparable corpora such as WIKI. The feasibility of the method has been illustrated in this paper since it has been tested on 9 different corpora. Some of them are largely used by the community and others are less popular but more difficult such as OST. The best results have been achieved for AFP corpus and the worst for OST.

In future work we will use this method in order to retrieve comparable articles from the social media to collect and build parallel corpora for languages which are under-resourced. The method developed in this paper will be expanded and adapted in order to compare the cross-lingual corpora in terms of opinions and emotions.

References

1. Aljlayl, M., Frieder, O., Grossman, D.: On Arabic-English Cross-Language Information Retrieval: Machine Translation Approach. In: Machine Readable Dictionaries and Machine Translation, ACM Tenth Conference on Information and Knowledge Managemen (CIKM), pp. 295–302. ACM Press (2002)
2. Berry, M.W., Young, P.G.: Using latent semantic indexing for multilanguage information retrieval. Computers and the Humanities 29(6), 413–429 (1995)
3. Bond, F., Paik, K.: A survey of wordnets and their licenses. In: 6th Global WordNet Conference (GWC 2012), pp. 64–71 (2012)
4. Cettolo, M., Girardi, C., Federico, M.: Wit[3]: Web inventory of transcribed and translated talks. In: Proceedings of the 16th Conference of the European Association for Machine Translation (EAMT), Trento, Italy, pp. 261–268 (May 2012)
5. Deerwester, S., Dumais, S.T., Furnas, G.W., Landauer, T.K., Harshman, R.: Indexing by latent semantic analysis. Journal of the American Society for Information Science 41(6), 391–407 (1990)
6. Dumais, S.: Lsa and information retrieval: Getting back to basics. In: Handbook of Latent Semantic Analysis, pp. 293–321 (2007)

7. Fujii, A., Ishikawa, T.: Applying machine translation to two-stage cross-language information retrieval. In: White, J.S. (ed.) AMTA 2000. LNCS (LNAI), vol. 1934, pp. 13–24. Springer, Heidelberg (2000), http://dx.doi.org/10.1007/3-540-39965-8_2

8. Habash, N.: Introduction to Arabic natural language processing. Synthesis Lectures on Human Language Technologies 3(1), 1–187 (2010)

9. Landauer, T.K., Foltz, P.W., Laham, D.: An introduction to latent semantic analysis. Discourse Processes 25(2-3), 259–284 (1998)

10. Li, B., Gaussier, E.: Improving corpus comparability for bilingual lexicon extraction from comparable corpora. In: Proceedings of the 23rd International Conference on Computational Linguistics, pp. 644–652. Association for Computational Linguistics (2010)

11. Littman, M.L., Dumais, S.T., Landauer, T.K.: Automatic cross-language information retrieval using latent semantic indexing. In: Grefenstette, G. (ed.) Cross-Language Information Retrieval. The Springer International Series on Information Retrieval, pp. 51–62. Springer, US (1998)

12. Ma, X., Zakhary, D.: Arabic newswire english translation collection. Linguistic Data Consortium, Philadelphia (2009)

13. Meftouh, K., Laskri, M.T., Smaïli, K.: Modeling Arabic Language using statistical methods. Arabian Journal for Science and Engineering 35(2C), 69–82 (2010)

14. Muhic, A., Rupnik, J., Skraba, P.: Cross-lingual document similarity. In: Proceedings of the ITI 2012 34th International Conference on Information Technology Interfaces (ITI), pp. 387–392 (June 2012)

15. NIST, M.I.G.: NIST 2008/2009 open machine translation (OpenMT) evaluation. Linguistic Data Consortium, Philadelphia (2010)

16. Otero, P., López, I., Cilenis, S., de Compostela, S.: Measuring comparability of multilingual corpora extracted from wikipedia. In: Iberian Cross-Language Natural Language Processings Tasks (ICL), p. 8 (2011)

17. Rafalovitch, A., Dale, R.: United nations general assembly resolutions: A six-language parallel corpus. In: Proceedings of the MT Summit XII, vol. 13, pp. 292–299 (2009)

18. Saad, M.: The Impact of Text Preprocessing and Term Weighting on Arabic Text Classification. Master's thesis, Computer Engineering Dept., Islamic University of Gaza, Palestine (2010)

19. Saad, M., Langlois, D., Smaïli, K.: Extracting comparable articles from wikipedia and measuring their comparabilities. Procedia - Social and Behavioral Sciences 95, 40–47 (2013), http://www.sciencedirect.com/science/article/pii/S1877042813041402, corpus Resources for Descriptive and Applied Studies. Current Challenges and Future Directions: Selected Papers from the 5th International Conference on Corpus Linguistics (CILC 2013)

20. Tiedemann, J.: Parallel data, tools and interfaces in opus. In: Chair), N.C.C., Choukri, K., Declerck, T., Dogan, M.U., Maegaard, B., Mariani, J., Odijk, J., Piperidis, S. (eds.) Proceedings of the Eight International Conference on Language Resources and Evaluation (LREC 2012). European Language Resources Association (ELRA), Istanbul (2012)

21. Ture, F.: Searching to Translate and Translating to Search: When Information Retrieval Meets Machine Translation. Ph.D. thesis, Graduate School of the University of Maryland, College Park (2013), http://hdl.handle.net/1903/14502

An Integrated Approach
to Automatic Synonym Detection
in Turkish Corpus

Tuğba Yıldız[1], Savaş Yıldırım[1], and Banu Diri[2]

[1] Department of Computer Engineering, Istanbul Bilgi University,
Eski Silahtarağa Elektrik Santrali, Kazım Karabekir Cad. No: 2/13,
34060 Eyüp, Istanbul, Turkey
{tdalyan,savasy}@bilgi.edu.tr
[2] Department of Computer Engineering, Yildiz Technical University,
Davutpasa, 34349 Istanbul, Turkey
banu@ce.yildiz.edu.tr

Abstract. In this study, we designed a model to determine synonymy. Our main assumption is that synonym pairs show similar semantic and dependency relation by the definition. They share same meronym/holonym and hypernym/hyponym relations. Contrary to synonymy, hypernymy and meronymy relations can probably be acquired by applying lexico-syntactic patterns to a big corpus. Such acquisition might be utilized and ease detection of synonymy. Likewise, we utilized some particular dependency relations such as object/subject of a verb, etc. Machine learning algorithms were applied on all these acquired features. The first aim is to find out which dependency and semantic features are the most informative and contribute most to the model. Performance of each feature is individually evaluated with cross validation. The model that combines all features shows promising results and successfully detects synonymy relation. The main contribution of the study is to integrate both semantic and dependency relation within distributional aspect. Second contribution is considered as being first major attempt for Turkish synonym identification based on corpus-driven approach.

Keywords: Synonym, near-synonym, pattern-based, dependency relations.

1 Introduction

As one of the most well-known semantic relations, synonymy has been subject to numerous studies. By the definition, synonyms are words with identical or similar meanings. The discovery of synonym relations may help to address various Natural Language Processing (NLP) applications, such as information retrieval and question answering [1–3], automatic thesaurus construction [4,5], automatic text summarization [6], language generation [7], English lexical substitution task [8], lexical entailment acquisition [9].

A. Przepiórkowski and M. Ogrodniczuk (Eds.): PolTAL 2014, LNAI 8686, pp. 116–127, 2014.
© Springer International Publishing Switzerland 2014

Various methods have been proposed for automatic synonym acquisition. Recent studies were generally based on distributional similarity and pattern-based approach. General idea behind distributional similarity is to capture the semantically related words. Distributional similarity of words sharing a large number of contexts could be informative [10]. Pattern-based approach is the most precise acquisition methodology earlier applied by Hearst [11] and relies on lexico-syntactic patterns (LSPs).

On the other hand, these methodologies themselves can be ambiguous and insufficient. Distributional similarity approach can cover other semantically related words and might not distinguish between synonyms and other relations. For example, list of top-10 distributionally similar words for orange is: yellow, lemon, peach, pink, lime, purple, tomato, onion, mango, lavender [12]. In addition, the pattern-based approach tends to capture hyponymy and meronymy relations as well, whereas it is apparently incompatible for synonyms detection. Thus, pattern-based approach or external features such as grammatical relations can be integrated into distributional similarity approach for identifying synonyms by narrowing distributional context. Although some studies have showed that classical distributional methods always have a higher recall than pattern-based techniques in this area [13], integrating two or more approaches were reported that system performance was improved [9, 13–15].

In this study, overall objective is to determine synonym nouns in a Turkish Corpus by relying on distributional similarity that is based on syntactic features (obtained by dependency relations) and semantic features obtained by syntactic patterns and LSPs respectively. The features of the proposed model consist of co-occurrence statistics, four semantic relations and ten syntactic dependency relations where a pair of words are represented with fifteen different features and a target class (SYN/NONSYN).

One of the main contributions of the study is that the system first obtains acquirable semantic relations such as hypernymy, meronymy from corpus by LSPs to extract subtle relations such as synonymy. The second contribution of the study is considered to be the first major attempt for Turkish synonym identification based on corpus-driven approach.

2   Related Works

A variety of methods have been proposed to automatically or semi-automatically detect synonyms from text source, dictionaries, wikipedia, search engines. Among them, the most popular methods are based on distributional hypothesis [10] which states that semantically similar words share similar contexts. The process of this approach was as follows: co-occurrence, syntactic information, grammatical relations of the words surrounding the target word are extracted as a first step. Afterwards target word is represented as a vector with these contextual features. At the second step, the semantic similarity of two terms is evaluated by applying a similarity measure between their vectors. The words can be ranked by their both semantic and syntactic similarity. Finally, top candidates are selected as the most similar words from ranked list.

There have been various studies [4,16,17] which used distributional similarity to the automatic extraction of semantically related words from large corpora. Distributional approaches have been applied into monolingual [4,18,19], monolingual parallel [20,21], bilingual corpora [20,22], multilingual parallel corpora [23] and monolingual dictionary [24,25], bilingual dictionaries [12]. Some of the studies [26–29] were relied on multiple-choice synonym questions such as SAT analogy questions, TOEFL synonym questions, ESL synonym-antonym questions. These studies fell into different types with respect to weighting scheme, similarity measurement, grammatical relations, etc. However most of these studies are not individually sufficient for synonyms. Because this approach also covers near-synonyms and does not distinguish between synonyms and other relations, hence, recent studies used different strategies: integrating two independent approaches such as distributional similarity and pattern-based approach, utilizing external features or ensemble method with combining the results to obtain more accuracy. Mirkin [9] integrated pattern-based and distributional similarity methods to acquire lexical entailment. Firstly, they extracted candidate entailment pairs for the input term by these methods.

Another study [31] emphasized that selection of useful contextual information was important for the performance of synonym acquisition. Therefore, they extracted three kinds of word relationships from corpora: dependency, sentence co-occurrence, and proximity. They utilized vector space model(VSM), tf-idf weighting scheme and cosine similarity. Dependency and proximity performed relatively well by themselves. The combination performance of all contextual information gave the best result. Other study of Hagiwara (2008) [14] proposed a synonym extraction method by using supervised learning based on distributional and/or pattern-based features. They constructed five synonym classifiers: Distributional Similarity (DSIM), Distributional Features (DFEAT), Pattern-based Features (PAT), Distributional Similarity and Pattern-based Features (DSIM-PAT) and Distributional and Pattern-based Features (DFEAT-PAT).

Other study [15] used three vector-based models to detect semantically related nouns in Dutch. They analyzed the impact of three linguistic properties of the nouns. They compared results from a dependency-based model with context feature with 1st and 2nd order bag-of-words model. They examined the effect of the nouns' frequency, semantic specificity and semantic class.

In one of the recent studies, [30], graded relevance ranking problem was applied to discover and rank the quality of the target term's potential synonyms. The model used supervised learning method; linear regression with three contextual features and one string similarity feature. The method was compared to two different methods [14,27]. As a result, proposed methods outperformed the existing ones.

In Turkish, recent studies on synonym relations are based on dictionary definition TDK[1] and Wiktionary[2] [32,34,35]. Within this framework, the main

---

[1] Türk Dil Kurumu (The Turkish Language Association)
[2] Vikisözlük: Özgür Sözlük

contribution of our work is its corpus-driven characteristics and it relies on both dependency and semantic relations.

3 Methodology

3.1 Data

The methodology employed here is to identify the synonym pairs from a large Turkish corpus of 500M tokens. A Turkish morphological parser, which is based on a two-level morphology [33], was used.

A good way to evaluate system performance is to compare the results to a gold standard. First, as gold standard, human judgments about the similarity of pairs of word are used. We manually and randomly selected 200 synonym pairs and 200 non-synonym pairs to build a training data set. Secondly, non-synonym pairs are especially selected from associated (relevant) pairs such as tree-leaf, student-school, computer-game, etc. Otherwise, selection of irrelevant pairs for negative examples can lead to false induction. The model is considered accurate if it can distinguish correct synonym pairs from relevant or strongly associated ones.

3.2 Similarity Measurement and Representation

Synonym pairs were gathered on the basis of co-occurrence statistics, semantic and grammatical relations. In order to compute the similarity between concepts and eliminate incorrect candidates, we used the cosine similarity measurement based on the word space model which is a representational Vector Space. In this study, words space was derived from a specialized context obtained by dependency patterns. Vector representation of words gives strong distributional indication for synonymy detection.

Similarity measurement between two vectors sometimes needs term weighting. Weighting scheme for context vectors might be normalization, pmi, dice, jaccard or raw frequency. The scheme can vary depending on the problem, therefore, it must be tested on the domain. Since we do not observe any significant improvements between the weighting formula, raw frequency is used for context vectors.

3.3 Features

Our methodology relies on the assumption that synonym pairs mostly show similar dependency and semantic characteristics in corpus. They share the same meronym/holonym relations, same particular list of governing verbs, adjective modification profile and so on, by definition. Even though it is no-use applying LSPs to extract synonymy, acquisition of other semantic relations such as meronymy could be easily done by simple string matching utilization and morphological analysis. By means of the acquisitions, the proposed model can determine if a given word pair is synonym or not. All attributes are based on relation measurements between pairs. For each synonym pair, 15 different features are extracted from different models: co-occurrence, semantic relations based on LSPs and grammatical relations based on syntactic patterns and head-modifier relation.

Feature 1: Co-occurrence. The first feature is gathered statistics about the co-occurrence of word pairs with a broad context (window size is equal to 8 from left and right) from corpora. Contrary to hypernymy and meronymy relation, it is seems impossible to directly extract synonym pairs by applying LSPs to a big corpus. Synonym pairs are not likely to co-occur together in same context and specific patterns at the same time. Therefore, first-order distributional similarity does not work for synonyms. At least, second order representation is needed. Simple co-occurrence measure might not be used for synonymy but non-synonymy. Their co-occurrence could be lower than relevant pairs. We experimentally selected dice metric to measure co-occurring feature. It is computed by roughly dividing the number of co-occurrences by summation of marginal frequencies of words.

Features 2/3: Meronym/Holonym. Detection of meronymy/holonymy is used to detect synonymy relation. After applying LSPs, some elimination assumption and measurement metrics such as chi or pmi to acquire meronym/holonym relation, we obtain a big matrix in which rows depict whole candidates, columns depict part candidates and cells represent the possibility of that corresponding whole and part are in meronymy relation. To measure the similarity of meronymy profile of two given words, cosine function is applied on two rows indexed by two given words. Applying cosine function on two columns gives the similarity of holonym profile.

For the relation, three different clusters of LSPs are analyzed in Turkish corpus; General (GP), Dictionary-based (TDK-P) and Bootstrapped patterns (BP) [37, 38]. First cluster is based on widely used general patterns. These patterns are collected from some pioneer studies and analyzed in Turkish. Second one is based on dictionary patterns that are extracted from TDK and Wiktionary. We adopted both types of patterns to extract the sentences that include part-whole relations from a Turkish corpus. Third cluster is based on bootstrapping. Some manually prepared seeds were used to induce and score LPSs. Based on reliability scores, we decided to filter out some generated patterns and finally obtained six different significant patterns. Once all three pattern clusters have been evaluated, third cluster of patterns (BP) showed significant performance. Table 1 shows six example patterns in third cluster(BP). All of the experiments in the studies, [37, 38], indicate that proposed methods have good indicative capacity.

Features 4/5: Hyponym/Hypernym. Same procedure in meronymy acquisition holds true for hypernymy and hyponymy relation. One relation matrix is built for hypernymy/hyponymy by applying LSPs and same procedure is carried out. The most important LSPs for Turkish [36] are as follows:

1. "NPs gibi CLASS" (CLASS such as NPs),
2. "NPs ve diğer CLASS" (NPs and other CLASS)
3. "CLASS lArdAn NPs" (NPs from CLASS)
4. "NPs ve benzeri CLASS" (NPs and similar CLASS)

First pattern gives strong indication of is-a hierarchy. Given the syntactic patterns above, the algorithm extracts the candidate list of hyponyms for a hypernym. The method had a good capacity to get higher precision, such as 72.5% [36].

Table 1. Bootstrapped Patterns and Examples

Patterns	Examples
NPy+gen NPx+pos	door of the house / evin kapısı
NPy+nom NPx+pos	house door / ev kapısı
NPy+Gen (N-ADJ) NPx+Pos	back garden gate of the house / evin arka bahçe kapısı
NPy of one-of NPxs	the door of one of the houses / evlerden birinin kapısı
NPx whose NPy	the house whose door is locked / kapısı kilitli olan ev
NPxs with NPy	the house with garden and pool / bahçeli ve havuzlu ev

Features 6–15. The dependency relations are obtained by syntactic patterns (or regular expression). For example, for *auto* and *car* pair, possible governing verbs bearing direct-object relations might be drive, design, produce, use, etc. The dimension of word-space model of *direct-object syntactic relation* consists of verbs and the cells indicate the number of times the selected noun is governed by corresponding verb. The more they are governed by the similar verb profile, the more likely they are synonyms. Likewise, the process is naturally applicable for other syntactic features. The more they are modified by same adjectives, the more likely they are synonym. Although 36 different patterns were extracted, eight were eliminated because of the poor results. Then we grouped them according to their syntactic structures. Representation of groups, number of patterns and examples in English/Turkish are given in Table 2.

The essential problem we face in the experiments is the lack of features of some words. Particularly, rare words cannot be represented due to lack of corpus evidence. Even in the corpus that contains about 500M words, all instances of use of Turkish language may not be present. Thus, those instances in train data that do not occur in any of dependency and semantic relations are eliminated. Especially the pairs including low frequent word cannot be represented and evaluated by means of the methodology as the number of missing values in many features increases. Out of 400 instances, about 40–50 are discarded from training data due to insufficiency.

3.4   Binary Classification for Synonym

Finally, train data turns out to contain balanced number of negative and positive examples with fifteen attributes. All the cells contain real value between 0–1. We know and accept that all features but co-occurrence feature have positive linear relationship with target class. Therefore, the data is considered to exhibit linear dependency. As a consequence of linearity, linear regression is an excellent and simple approach for such a classification. It has been widely used in statistical applications. The most suitable algorithm is *logistic regression* which can easily be used for binary classification in the domains with numeric attributes and nominal target class. Contrary to the linear regression, it builds a linear model based on a transformed target variable.

Another model would be perceptron. If the data can be separated perfectly into two groups using a threshold value or a function, it is said to be linearly separable. The perceptron learning rule is capable of finding a separating hyperplane on linearly separable data. However, our problem looks more suitable for logistic regression (transformed linear regression) than perceptron.

Table 2. Dependency Features

Features	Dependency relation	# of Patterns	Examples
G1	direct object of verb	13	I drive a car
			araba sürüyorum
G2	subject of verb	3	waiting car
			bekleyen araba
G3	direct object/subject of verb	3	-
			-
G4	modified by adjective+(with/without)	2	car with gasoline
			benzinli araba
G5	modified by inf	1	swimming pool
			yüzme havuzu
G6	modified by noun	1	toy car
			oyuncak araba
G7	modified by adjective	1	red car
			kırmızı araba
G8	modified by acronym locations	1	the cars in ABD
			ABD'deki arabalar
G9	modified by proper noun locations	1	the cars in Istanbul
			Istanbul'daki arabalar
G10	modified by locations	2	the car at parking lot
			otoparktaki araba

4   Results and Discussion

To evaluate the impact of semantic and dependency relations in finding synonyms, first, we look at their individual performances in terms of cross-validation. Picking up each feature one by one with target class, we evaluated the performance of logistic regression on the projected data. As long as the averaged f-measured score of the corresponding feature is higher than 50%, it is considered a useful feature otherwise, independent feature.

The first aim is to find out which feature is the most informative for detecting synonymy and contributes most to the overall success of the model. When evaluating the result as shown in Table 3, the semantic features are notably better than syntactic dependency models in finding true synonyms. They are called to be good indicators.

Table 3. F-Measure of Semantic Relations (SRs) Features

	co-occurrence	hyponym	hypernym	meronym	holonym
F-Measure	62.5	60.5	60	68.7	73.7

Among semantic relations, the most powerful attributes are meronymy and holonymy features with f-measure of 68.7% and 73.7%, respectively. The possible reason for the success seems to be the sufficient number of cases matched by lexico-syntactic and syntactic pattern from which semantic and syntactic features are constructed. For example, the model utilizing meronymy relations has a good production capacity and success. The Table 4 shows that meronymy-holonymy matrix has the size of 17K x 18K. The total number of instance is 1.7M. Average number of instances for each meronym is 102 and for each holonym is 96. They also show good performance. The averaged number of instances for hypernymy and hyponymy are 50 and 8, respectively. As a result of insufficient data volume, hypernymy/hyponymy semantic relation is relatively weaker than meronymy.

Table 4. Statistics for features : Mero:Meronym, Hypo: Hyponym, AVG_cpr: average case per row, AVG_cpc: average case per column

	G1	G2	G3	G4	G5	G6	G7	G8	G9	G10	Mero	Hypo
#ofrow	16K	18K	10K	13K	7K	13K	20K	6K	1.7K	13K	17K	4.3K
#ofcol	1.7K	1.7K	1.4K	5K	1.6K	13K	5.6K	1.6K	0.2K	5K	18K	29K
#ofcases	3.3M	3M	0.5M	1.6M	1M	5.3M	12M	0.1M	0.01M	1M	1.7M	0.2M
AVG_cpr	206	164	47	128	140	391	590	23	7	75	102	50
AVG_cpc	2010	1783	341	319	621	405	2106	86	51	195	96	8

Among dependency relations, G1, G4 and G7 have better performance as shown in the Table 5. Also their production capacities are sufficient as well. The poorest groups, G8 and G9, have low production capacity and their performances are worse. As a consequence of the poor results, they are called independent and useless variables. Co-occurrence feature has negative linear relation with target class and its individual performance is 62.5%. It is acceptable as a useful feature.

The successful features are linearly dependent on target class. The most suitable machine learning algorithm is the logistic regression. After aggregating all useful features which have better than the individual performances, the machine learning process was carried out and evaluated. The achievement of aggregated

Table 5. F-measure of Dependency Relations Features

	G1	G2	G3	G4	G5	G6	G7	G8	G9	G10
F-Measure	64.7	58	60.5	65	61.6	58.8	63	49.4	48.3	62.6

model was evaluated in terms of cross validation. On the aggregated data where all useful features are considered, the performance of logistic regression is f-measure of 80.3% and that of voted perceptron is 74%. The achieved score is better than the individual performance of each feature. The number of useful features is obviously the main factor to get higher scores. The proposed model utilizes only a huge corpus and morphological analyzer and it receives an acceptable score. Moreover, other useful resources might be integrated into the model to obtain better result. Dictionary definitions, WordNet, and other useful resources could be used and evaluated in future work.

5   Conclusion

In this study, synonym pairs were determined on the basis of co-occurrence statistics, semantic and dependency relations within distributional aspect. Contrary to hypernymy and meronymy relation, simply applying LSPs does not extract synonym pairs from a big corpus. Instead, we extracted other semantic relations to ease detection of synonymy. Our methodology relies on some assumptions. One is that the synonym pairs mostly show similar semantic characteristics by definition. They share the same meronym/holonym and hypernym/hyponym relations. Particular lexico-syntactic patterns can be used to initiate the acquisition process of those semantic features.

Secondly, a pair of synonym words mostly shares a particular list of governing verbs and modifying adjectives. The more a pair of words are governed by similar verb profile and modified by similar adjectives, the more likely they are synonym. We built ten groups of syntactic patterns according to their syntactic structures.

To apply machine learning algorithm, three annotators manually and randomly selected 200 synonym pairs and 200 non-synonyms. Non-synonym pairs were especially selected from associated (relevant) pairs such as tree-leaf, apple-orange, school-student. Otherwise, such negative example selection could lead to false inference. The main challenge faced in the experiments is the lack of features of some words due to their corpus evidence. Thus, such instances were eliminated. Remaining instances was classified by the most suitable algorithm which is the *logistic regression*. It can easily be used for binary classification in domains with numeric attributes and nominal target class.

As long as individual performance of any feature is higher than f-measure of 50%, it is considered as useful features or considered independent feature from target class. The aim was to find out which features are the most informative for detecting synonymy and contribute most to the overall success of the model. When comparing the results, it was clearly observed that the semantic features are notably better than syntactic dependency models in finding true synonyms. The most effective attributes are meronymy and holonymy features with weighted average f-measure of 68.7% and 73.7% respectively. The analysis indicated that the possible reason for the success is sufficiency in the number of cases from which semantic and dependency features are constructed. As a consequence of insufficient data volume, hypernymy/hyponymy relation is relatively worse than meronymy. Among dependency relations, G1, G4 and G7 outperformed the others. Likewise, it was also observed that sufficiency in the number of cases was the strong factor. After aggregating all useful features, the same learning process was carried out. The aggregated model shows promising results and performance. Regression model achieved an acceptable f-measure of 80.3%.

One of the main contributions of the study is that the system first obtains acquirable semantic relations such as hypernymy, meronymy from corpus by lexico-syntactic patterns to extract subtle relations such as synonymy. The second contribution of the study is considered to be the first major attempt for Turkish synonym identification based on corpus-driven approach.

References

1. Mandala, R., Tokunaga, T., Tanaka, H.: Combining Multiple Evidence from Different Types of Thesaurus for Query Expansion. In: 22nd Annual International ACM SIGIR Conference on Research and Development in Information Retrieval, Berkeley, CA, USA, pp. 191–197 (1999)
2. Bai, J., Song, D., Bruza, P., Nie, J., Cao, G.: Query Expansion Using Term Relationships in Language Models for Information Retrieval. In: 14th ACM International Conference on Information and Knowledge Management, Bremen, Germany, pp. 688–695 (2005)
3. Stefan, R., Liu, Y., Vasserman, A.: Translating Queries into Snippets for Improved Query Expansion. In: 22nd International Conference on Computational Linguistics, COLING 2008, Manchester, UK, pp. 737–744 (2008)
4. Lin, D.: Automatic Retrieval and Clustering of Similar Words. In: 36th Annual Meeting of the Association for Computational Linguistics and 17th International Conference on Computational Linguistics, Montreal, Quebec, Canada, pp. 768–774 (1998)
5. Inkpen, D.: A Statistical Model for Near-synonym Choice. ACM Transactions on Speech and Language Processing 4(1), 1–17 (2007)
6. Barzilay, R., Elhadad, M.: Using Lexical Chains for Text Summarization. In: Proceedings of the ACL Workshop on Intelligent Scalable Text Summarization, Madrid, Spain, pp. 10–17 (1997)
7. Inkpen, D.Z., Hirst, G.: Near-synonym Choice in Natural Language Generation. In: Recent Advances in Natural Language Processing III, Selected Papers from RANLP 2003, Borovets, Bulgaria, pp. 141–152 (2003)
8. McCarthy, D., Navigli, R.: The English Lexical Substitution Task. Language Resources and Evaluation 43(2), 139–159 (2009)
9. Mirkin, S., Dagan, I., Geffet, M.: Integrating Pattern-Based and Distributional Similarity Methods for Lexical Entailment Acquisition. In: Proceedings of the COLING/ACL 2006 on Main Conference Poster Sessions, Sydney, Austraila, pp. 579–586 (2006)
10. Harris, Z.: Distributional Structure. Word 10(23), 146–162 (1954)

11. Hearst, M.A.: Automatic Acquisition of Hyponyms from Large Text Corpora. In: 14th International Conference on Computational Linguistics, COLING 1992, Nantes, France, pp. 539–545 (1992)
12. Lin, D., Zhao, S., Qin, L., Zhou, M.: Identifying Synonyms among Distributionally Similar Words. In: IJCAI 2003, Proceedings of the Eighteenth International Joint Conference on Artificial Intelligence, Acapulco, Mexico, pp. 1492–1493 (2003)
13. Wang, W., Thomas, C., Sheth, A.P., Chan, V.: Pattern-based Synonym and Antonym Extraction. In: Proceedings of the 48th Annual Southeast Regional Conference, Oxford, MS, USA, p. 64 (2010)
14. Hagiwara, M.: A Supervised Learning Approach to Automatic Synonym Identification Based on Distributional Features. In: 46th Annual Meeting of the Association for Computational Linguistics on Human Language Technologies: Student Research Workshop, Columbus, OH, pp. 1–6 (2008)
15. Heylen, K., Peirsman, Y., Geeraerts, D., Speelman, D.: Modelling Word Similarity: an Evaluation of Automatic Synonymy Extraction Algorithms. In: International Conference on Language Resources and Evaluation, LREC (2008)
16. Hindle, D.: Noun Classification from Predicate-Argument Structures. In: 28th Annual Meeting of the Association for Computational Linguistics, Pittsburgh, Pennsylvania, USA, pp. 268–275 (1990)
17. Gasperin, C., Gamallo, P., Agustini, A., Lopes, G., Lima, V.: Using Syntactic Contexts for Measuring Word Similarity. In: Workshop on Knowledge Acquisition and Categorization, ESSLLI (2001)
18. Curran, J.R., Moens, M.: Improvements in Automatic Thesaurus Extraction. In: ACL 2002 Workshop on Unsupervised Lexical Acquisition, Philadelphia, USA, pp. 59–66 (2002)
19. van der Plas, L., Bouma, G.: Syntactic Contexts for Finding Semantically Related Words. In: Meeting of Computational Linguistics in the Netherlands (CLIN), Amsterdam, pp. 173–186 (2005)
20. Barzilay, R., McKeown, K.: Extracting Paraphrases from a Parallel Corpus. In: 39th Annual Meeting and 10th Conference of the European Chapter, Proceedings of the Conference, Toulouse, France, pp. 50–57 (2001)
21. Ibrahim, A., Katz, B., Lin, J.: Extracting Structural Paraphrases from Aligned Monolingual Corpora. In: The Second International Workshop on Paraphrasing: Paraphrase Acquisition and Applications, Sapporo, Japan, pp. 57–64 (2003)
22. Shimohata, M., Sumita, E.: Automatic Paraphrasing Based on Parallel Corpus for Normalization. In: Third International Conference on Language Resources and Evaluation, Las Palmas, Canary Islands, Spain, pp. 453–457 (2002)
23. van der Plas, L., Tiedemann, J.: Finding Synonyms Using Automatic Word Alignment and Measures of Distributional Similarity. In: 21st International Conference on Computational Linguistics and 44th Annual Meeting of the Association for Computational Linguistics, Sydney, Australia, pp. 866–873 (2006)
24. Blondel, V.D., Sennelart, P.: Automatic Extraction of Synonyms in a Dictionary. In: SIAM Workshop on Text Mining, Arlington, VA (2002)
25. Wang, T., Hirst, G.: Exploring Patterns in Dictionary Definitions for Synonym Extraction. Natural Language Engineering 18(3), 313–342 (2012)
26. Freitag, D., Blume, M., Byrnes, J., Chow, E., Kapadia, S., Rohwer, R., Wang, Z.: New Experiments in Distributional Representations of Synonymy. In: Ninth Conference on Computational Natural Language Learning (CoNLL), Ann Arbor, Michigan, pp. 25–32 (2005)
27. Terra, E., Clarke, C.: Frequency Estimates for Statistical Word Similarity Measures. In: HTL/NAACL 2003, Edmonton, Canada, pp. 165–172 (2003)

28. Turney, P.D., Littman, M.L., Bigham, J., Shnayder, V.: Combining Independent Modules in Lexical Multiple-choice Problems. In: Recent Advances in Natural Language Processing III, Selected Papers from RANLP 2003, Borovets, Bulgaria, pp. 101–110 (2003)

29. Turney, P.D.: A Uniform Approach to Analogies, Synonyms, Antonyms, and Associations. In: 22nd International Conference on Computational Linguistics, Coling 2008, Manchester, UK, pp. 905–912 (2008)

30. Yates, A., Goharian, N., Frieder, O.: Graded Relevance Ranking for Synonym Discovery. In: 22nd International World Wide Web Conference, WWW 2013, Rio de Janeiro, Brazil, pp. 139–140 (2013)

31. Hagiwara, M., Ogawa, Y., Toyama, K.: Selection of Effective Contextual Information for Automatic Synonym Acquisition. In: 21st International Conference on Computational Linguistics and 44th Annual Meeting of the Association for Computational Linguistics, Sydney, Australia, pp. 353–360 (2006)

32. Yazici, E., Amasyali, M.F.: Automatic Extraction of Semantic Relationships using Turkish Dictionary Definitions. In: EMO Bilimsel Dergi, Istanbul (2011)

33. Sak, H., Güngör, T., Saraçlar, M.: Turkish Language Resources: Morphological Parser, Morphological Disambiguator and Web Corpus. In: Nordström, B., Ranta, A. (eds.) GoTAL 2008. LNCS (LNAI), vol. 5221, pp. 417–427. Springer, Heidelberg (2008)

34. Serbetci, A., Orhan, Z., Pehlivan, I.: Extraction of Semantic Word Relations in Turkish from Dictionary Definitions. In: ACL 2011 Workshop on Relational Models of Semantics, Portland, pp. 11–18 (2011)

35. Orhan, Z., Pehlivan, I., Uslan, V., Onder, P.: Automated Extraction of Semantic Word Relations in Turkish Lexicon. Mathematical and Computational Applications (1), 13–22 (2011)

36. Yildirim, S., Yildiz, T.: Automatic Extraction of Turkish Hypernym-Hyponym Pairs From Large Corpus. In: COLING (Demos), pp. 493–500 (2012)

37. Yildiz, T., Diri, B., Yildirim, S.: Analysis of Lexico-syntactic Patterns for Meronym Extraction from a Turkish Corpus. In: 6th Language and Technology Conference: Human Language Technologies as a Challenge for Computer Science and Linguistics, pp. 126–138 (2013)

38. Yıldız, T., Yıldırım, S., Diri, B.: Extraction of Part-Whole Relations from Turkish Corpora. In: Gelbukh, A. (ed.) CICLing 2013, Part I. LNCS, vol. 7816, pp. 126–138. Springer, Heidelberg (2013)

Distributional Context Generalisation and Normalisation as a Mean to Reduce Data Sparsity: Evaluation of Medical Corpora

Amandine Périnet[1] and Thierry Hamon[2]

[1] INSERM, U1142, LIMICS, Paris, France,
UPMC Univ Paris 06, Univ Paris 13, Sorbonne Paris Cité, Villetaneuse, France
amandine.perinet@edu.univ-paris13.fr
[2] LIMSI-CNRS, Orsay, France,
Université Paris 13, Sorbonne Paris Cité Villetaneuse, France
hamon@limsi.fr

Abstract. Distributional analysis relies on the recurrence of information in the contexts of words to associate. But the vector space models implementing the approach suffer from data sparsity and from a high dimensional context matrix. If reducing data sparsity is an important aspect with general corpora, it is also a major issue with specialised corpora that are of much smaller size and with much lower context frequencies. We tackle this problem on specialised texts and propose a method to increase the matrix density by normalising and generalising distributional contexts with synonymy and hypernymy relations acquired from corpora. Experiments on a French biomedical corpus show that context generalisation and normalisation improve the results when combined with the use of relations acquired with lexico-syntactic patterns.

Keywords: Data sparsity, hypernym, synonymy, distributional analysis, specialised corpora.

1 Introduction

Assuming that words occurring in a similar context tend to be semantically close, vector space models (VSMs) easily quantify the semantic similarity between two words by measuring the distance between their corresponding vectors within an n-dimension space where each dimension is a distributional context [18,21]. Besides the high number of dimensions, VSMs mainly suffer from data sparsity within the matrix representing the vector space: many elements are equal to zero because only few contexts are associated to a target word. This is partly due to a very limited set of distributional contexts associated to each word compared to the number of words [2]. Similarity between two words is then hard to compute. Hence, distributional methods show better results when much information is available and especially with general corpora, usually of great size [23]. But if data sparsity reduction is still an important aspect with general corpora, it is also a major issue when working with specialised corpora, as the sizes are

A. Przepiórkowski and M. Ogrodniczuk (Eds.): PolTAL 2014, LNAI 8686, pp. 128–135, 2014.

smaller and the frequencies lower. We focus here on this last point and propose a method that aims at reducing distributional context diversity by normalising and generalising contexts thanks to semantic relations acquired from corpora. The context frequency is then increased and, consequently, data sparseness and the dimensions of the VSM are reduced.

Section 2 presents related work on data sparsity reduction within distributional methods. The method of generalisation and normalisation of distributional context is described in Sect. 4. Its impact on specialised corpora (Sect. 3) is evaluated and analysed through several experiments (Sect. 5).

2   Related Work

Reducing data sparsity is a main issue in distributional analysis. Methods usually aim at influencing the selection of useful contexts to modify context distribution. [3] propose to first rank contexts according to their frequency, and then take the rank into account to weight these contexts. Incorporating additional semantic information in contexts also has a positive influence on the performance of a standard distributional method [20]. [6] improves the quality of the acquired relations between low or mid frequency nouns by modifying the context weights. A set of positive and negative examples are selected with an unsupervised classifier. A supervised classifier is then applied to re-rank the semantic neighbors.

The sparseness problem may also be tackled by limiting the dimensions of the context matrix with smoothing algorithms [21] or by factorization method of matrices by Singular Value Decomposition (SVD) to counterbalance the lack of data for low frequencies [22]. Latent Semantic Analysis (LSA) [14,16] uses SVD and improves the precision of the results: the original data in the context matrix are abstracted in independent linear components, to reduce noise and highlight the main elements. However, dimension reduction requires to build an initial huge matrix. To avoid this, Random Indexing (RI) [12] incrementally builds the context matrix according to an index vector of the target words randomly generated, and reduces the matrix dimension. RI and LSA demonstrate similar performance when identifying synonyms in a similar way as the TOEFL test [13]. The selection of the best contexts combined with a normalisation of their weights improves the quality of a SVD reduced matrix [17].

As above works, we aim at incorporating semantic information within distributional contexts, but by reducing the number of contexts and increasing their frequency. Contrary to SVD based methods that remove information, we both generalise and normalise contexts through the integration of additional semantic knowledge computed from our corpora.

3   Material

Corpus. To evaluate our approach, we use the Menelas corpus [24]. It consists in a French medical text collection, on the topic of coronary diseases, and contains 84,839 words. The corpus has been analysed through the Ogmios platform [9].

The linguistic analysis includes a morphosyntactic tagging and a lemmatisation of the corpus, with TreeTagger [19], and a term extraction with Y$_A$T$_E$A [1] to identify terminological entities.

Synonymy Relations (SYN). Context normalisation is performed with 168 synonymy relations between complex terms (*infection de blessure (wound infection)* and *septicit de blessure (wound sepsis)*) proposed by the rule-based method [10].

Hypernymy Relations. For context generalisation, we use hypernymy relations acquired with several approaches:

- Lexico-syntactic patterns (LSP): we use the patterns defined by [15] to detect 98 hypernymy relations between simple or complex terms (*artère coronaire, veine (vein), artère (artery)*, and *vaisseau (vessel)*).
- Lexical Inclusion (LI): based on the hypothesis that the lexical inclusion of a term (*restriction*) in another (*restriction du débit coronaire (restriction of coronary output)*) conveys a hypernymy relation between those terms [7]. We obtain 7,187 relations (computed on the Y$_A$T$_E$A output).
- Terminological variation (TV): the method proposed by [11] to acquire terminological variants based on morphosyntactic transformation rules. The insertion rule is essentially used on our corpus and allows us to consider the 171 obtained relations as hypernymy relations (*lésion significative (significant lesion)* and (*lésion coronaire significative (significant coronary lesion)*).

4   Distributional Context Abstraction

The problem of data sparsity on specialised corpora may be addressed by increasing the density of the context matrix. Superficial variations, not statistically significant or noisy contexts may be removed or gathered. We use relations automatically acquired to generalise and normalise contexts. After a brief description of the distributional method, we present the context generalisation/normalisation.

4.1   Distributional Method

We focus on the extraction of relations between nouns and terms (simple and complex): hypernyms, co-hyponyms (*nécrose (necrosis) / ischémie (ischemia)*), equivalent terms (*ecg / électrocardiogramme*), and many domain relations *traitement (treatment) / cathéter (catheter)*. The contexts of the target words are defined within a fixed-size window: they are lemmas of adjectives, nouns, verbs and terms co-occurring with the target. Once contexts are defined, they are generalised and normalised (Sect. 4.2). Finally a similarity score is computed for each pair of targets. We use the Jaccard index, recognised as adapted to specialised corpora [8], and the cosine of the context vector. With Jaccard, we use the Relative Frequency, and no weight with Cosine. We also intend to produce a denser context matrix by applying three thresholds on distributional parameters: the target word frequency, the number and frequency of shared contexts. These thresholds help to remove relations computed from shared contexts occurring by chance and to limit the proposed relations. The threshold on the number of shared context is computed based on pair-wise shared contexts; dimensions (i.e. shared contexts) are deleted if the values associated are under the threshold, i.e. if there are not enough shared contexts. As a result, the number of words in relation is reduced, and so may be the number of contexts and target words. For each parameter, the threshold is automatically computed as the mean of the values taken by each parameter on the corpus (see Sect. 5).

4.2   Generalisation and Normalisation Rules

To reduce sparsity, we define several rules that rely on semantic relations (Sect. 3) to generalise and normalise contexts separately and in combination.

Normalisation Rule. The normalisation rule aims at reducing semantic variation with synonymy relations organised in clusters. The most frequent word is chosen as the cluster representative. Then, for each word $\text{ctxt}_i(w)$ in the context of the target word w, a synonym cluster $\mathbb{S}(R) = \{S_1, \ldots, S_n, R\}$ is built, with its representative R. We define one context normalisation rule, represented by the function $s(w)$, applied to each word $\text{ctxt}_i(w)$ in the context of a word w to substitute the context word by the representative of the cluster it belongs to: if $\exists R | \text{ctxt}_i(w) \in \mathbb{S}(R)$, then $\text{ctxt}_i(w) := R$ (the synonymy set may be empty).

Generalisation Rules. Contexts are generalised with hypernymy relations. For each word $\text{ctxt}_i(w)$ in the context of a target word w, we have a set of hypernymy relations $\mathbb{H}_s(\text{ctxt}_i(w)) = \{H_1, \ldots, H_n\}$ for each method: $\mathbb{H}_{\text{LSP}}$, $\mathbb{H}_{\text{LI}}$ and $\mathbb{H}_{\text{TV}}$, and also globally $\mathbb{H}_{\text{ALL3}}(\text{ctxt}_i(w)) = \mathbb{H}_{\text{LSP}} \cup \mathbb{H}_{\text{LI}} \cup \mathbb{H}_{\text{TV}}$. For each word $\text{ctxt}_i(w)$ in the context of a target word w, we define two substitution rules, represented by the function $h(w)$, that allow to generalise contexts:

1. if $|\mathbb{H}_S(\text{ctxt}_i(w))| = 1$, then $\text{ctxt}_i(w) := H_1$, i.e. if the word in context corresponds to only one hypernym (H_1), acquired by one or several methods S or globally, the word is replaced by this hypernym.
2. if $|\mathbb{H}_S(\text{ctxt}_i(w))| > 1, \text{ctxt}_i(w) := \text{argmax}_{|H_i|}(\mathbb{H}_S(\text{ctxt}_i(w)))$, i.e. if the context corresponds to several hypernyms acquired by one or several methods S or globally, we consider the hypernym frequency $|H_1|, \ldots, |H_n|$ in the corpus and select the most frequent hypernym.

Combining Normalisation and Generalisation Rules. The context generalisation rules may globally use the hypernymy relations, independently of the method used to acquire them (set $\mathbb{H}_{\text{ALL3}}(\text{ctxt}_i(w))$) and may also exploit separately or sequentially each set. We combine the previous rules by first applying normalisation rules and then generalisation rules on the normalised context words $s(\text{ctxt}_i(w))$ (represented by the composed function $h \circ s)(x))$. It also requires to normalise each element of the hypernymy set $\mathbb{H}_S(\text{ctxt}_i(w)) = \{H_1, \ldots, H_n\}$. Thus, the generalisation rules consider the sets $\mathbb{H}_S(s(\text{ctxt}_i(w))) = \{s(H_1), \ldots, s(H_n)\}$ to be applied to the words in context.

5   Experiments and Evaluation

5.1   Experiments

To evaluate the impact of normalisation and generalisation, we performed several experiments on the Menelas corpus. Our baseline is the VSM without context substitution. We also evaluate the impact of the thresholds on distributional parameters (Sect. 4.1). Their values are computed from the corpus and for the baseline: (i) number of shared contexts (large window: 2, small window: 1) ; (ii) number of contexts (large window: 3, small window: 2); (iii) number of targets (both windows: 3). Finally, we apply an empirically defined threshold on the similarity score: sim > 0.000999 for Jaccard and sim > 0.9699 for Cosine.

Abstraction rules are first applied individually and separately with the four relation sets $\mathbb{H}_{\text{LSP}}(\text{ctxt}_i(w))$, $\mathbb{H}_{\text{LI}}(\text{ctxt}_i(w))$, $\mathbb{H}_{\text{TV}}(\text{ctxt}_i(w))$, and $\mathbb{S}(\text{ctxt}_i(w))$. To grasp the contribution and complementarity of each source of relations, we define several experiments where the generalisation rules sequentially exploit the hypernym sets (e.g., $\mathbb{H}_{\text{LSP}}(\text{ctxt}_i(w))$ then $\mathbb{H}_{\text{LI}}(\text{ctxt}_i(w))$ and $\mathbb{H}_{\text{TV}}(\text{ctxt}_i(w))$). The union of the three hypernym sets ($\mathbb{H}_{\text{ALL3}}(\text{ctxt}_i(w))$) is also used to generalise the contexts. Finally, the previously described configuration for generalising contexts have been applied to the normalised contexts $s(\text{ctxt}_i(w))$. All the experiments have been performed with both window sizes: a 5 word ($\pm$ 2 words, centered on the target) and a 21 word window ($\pm$ 10 words, centered on the target).

5.2   Evaluation

Results are evaluated by comparing the acquired relations with the 1,735,419 relations provided by the French part of the UMLS[1] metathesaurus. These relations are hypernyms (*régime alimentaire (diet) / régime diabétique (diabetic diet)*), and many siblings (*thrombolyse (thrombolysis) / traitement anti-coagulant (antcolagulant treatment), vertige (vertigo) / douleur (pain)*). Like [5] and [6], we consider the obtained relations as neighbor sets associated to target words. In these sets, neighbors are ranked according to their similarity with the target word. The evaluation is performed with measures usually used on VSM results: R-precision and mean average precision (MAP) [4]. R-precision computes the precision at the threshold n_i defined as the number of correct neighbors expected for one target word (i.e. the potential UMLS relations between the words or terms of our corpus). Mean average precision (MAP) is based on the not interpolated precision of the semantic neighbors given their rank. It reflects the ranking quality and evaluates the relevance of the similarity measure; the method that ranks all the correct semantic neighbors on top of the list is favored while adding noisy neighbors at the end of the list has no impact on the evaluation.

---

[1] http://www.nlm.nih.gov/research/umls/

6   Results and Discussion

We report the results obtained for context generalisation based on the hypernymy relation sets taken separately, all together and some sequential combinations that inform on the usefulness and impact of the generalisation order. The analysis of the experiments reveals that the sequential exploitation of these sets has no impact on the VSM quality. We also present the results of context normalisation, performed alone and before generalisation. The results obtained for a 21 word window, with thresholds on the parameters are presented in Table 1.

Globally, Jaccard and Cosine VSMs have similar behaviors with any experimental set, with both normalisation and generalisation. However, when thresholds are applied, less UMLS relations are acquired with Jaccard than with Cosine, and in both cases, adding thresholds reduces the number of UMLS relations. The number of relations provided by the abstracted VSMs is stable with Cosine, while it consistently decreases with Jaccard. The thresholds on Jaccard similarity may be too high when generalisation and normalisation are performed, and should be computed according to the experimental sets and not only to the baseline.

Context Generalisation. Without thresholds on the parameters, generalisation with patterns obtains the best MAP and R-precision results. This generalisation then improves the ranking quality, but also decreases the number of relations acquired and found in the UMLS. When thresholds on distributional parameters are applied, the behavior of the methods used for generalisation varies according to the similarity measure. With Jaccard, generalisation using

Table 1. Results for a 21 word window and thresholds on the distributional parameters

	Acq. Rel.		Rel. in UMLS		MAP		R-precision	
	JACC	COS	JACC	COS	JACC	COS	JACC	COS
BASELINE	406	428	4	44	0.406	0.188	0.250	0.118
Generalisation								
Variants	472	424	8	40	0.280	0.188	0.143	0.118
Lexical inclusion	328	223	4	26	**0.454**	*0.110*	0.250	0.133
Patterns	398	381	6	36	0.219	**0.206**	0.000	0.000
Variants + lex. inclusion	328	223	4	26	**0.454**	0.110	0.250	0.000
Patterns + lex. inclusion	338	220	4	26	**0.454**	0.101	0.250	0.000
3 generalisation methods	336	243	4	26	0.414	0.108	0.250	0.000
Normalisation								
Synonyms (Syn)	474	424	8	40	0.280	0.188	0.143	0.118
Normal. + generalisation								
Syn/Variants	474	419	8	22	0.280	0.189	0.143	0.118
Syn/Lexical inclusion	366	279	6	14	**0.440**	0.105	0.333	0.000
Syn/Patterns	394	377	6	20	*0.219*	**0.206**	*0.000*	*0.133*
Syn/Variants + lex. incl.	324	223	4	14	**0.454**	0.110	0.250	0.000
Syn/Patterns + lex. incl.	394	373	6	20	*0.219*	**0.206**	*0.000*	**0.133**
Syn/3 general. methods	370	280	6	14	**0.454**	0.105	**0.333**	0.000

LI gives the better MAP results, even when combined with terminological variation and patterns. But as the number of relations found in the UMLS is very low, this may not be very significant. Thresholds should be tuned to each set. With Cosine, results are similar but slightly improved compared to the ones obtained without thresholds. They decrease with LI and increase with pattern generalisation. The order has no influence when hypernymy relations provided by patterns or lexical inclusion are combined with terminological variation. It means that TV generalisation is complementary of LSP or LI generalisation.

Context Normalisation. Normalisation performed without thresolds has no effect on the VSM results, but it may slightly increase the results when combined with generalisation. With thresholds, normalisation generally decreases the quality of the results. But when performed in combination with the generalisation based on the union of the three hypernym sets and with Jaccard, more relations are found in the UMLS, and both MAP and R-precision increase. With Cosine, normalisation improves the results when combined with a pattern generalisation (with or without LI). On the contrary, with Jaccard and the same generalisation, it decreases the results. But as we mentioned previously, it must be confirmed with an adaptation of the thresholds to each experimental set.

7   Conclusion

In this paper, we adress data sparsity reduction in matrices of context vector used to implement distributional analysis. We proposed to generalise and normalise distributional contexts with synonyms and hypernyms acquired from corpora. We performed some experiments on a French medical corpus combining several parameters. Although the evaluation of distributional methods is difficult, we compare our results to relations proposed by the French UMLS. The result analysis shows that a large window (21 words) and Jaccard obtain the best results. As for generalisation and normalisation, results are similar with both Jaccard and Cosine. Normalisation does not influence the VSM results when used on its own, but improves them when combined with generalisation performed with lexico-syntactic patterns. Finally, general thresholds defined according to the original VSM are not relevant when generalisation and normalisation are performed. They need to be defined specifically, according to each configuration. These results open several perspectives, among which a manual analysis of the relations and of the impact of the generalisation and normalisation process on manipulated data, as well as a comparison of our method with other dimension reduction methods, Random Indexing and LSA.

References

1. Aubin, S., Hamon, T.: Improving term extraction with terminological resources. In: Salakoski, T., Ginter, F., Pyysalo, S., Pahikkala, T. (eds.) FinTAL 2006. LNCS (LNAI), vol. 4139, pp. 380–387. Springer, Heidelberg (2006)

2. Baroni, M., Bernardini, S., Ferraresi, A., Zanchetta, E.: The wacky wide web: a collection of very large linguistically processed web-crawled corpora. Language Resources and Evaluation 43(3), 209–226 (2009)
3. Broda, B., Piasecki, M., Szpakowicz, S.: Rank-based transformation in measuring semantic relatedness. In: Gao, Y., Japkowicz, N. (eds.) Canadian AI 2009. LNCS (LNAI), vol. 5549, pp. 187–190. Springer, Heidelberg (2009)
4. Buckley, C., Voorhees, E.: Retrieval system evaluation. In: TREC: Experiment and Evaluation in Information Retrieval, ch. 3 (2005)
5. Curran, J.R.: From distributional to semantic similarity. Ph.D. thesis, Institute for Communicating and Collaborative Systems, University of Edinburgh (2004)
6. Ferret, O.: Sélection non supervisée de relations sémantiques pour améliorer un thésaurus distributionnel. In: Actes de TALN 2013, pp. 48–61 (2013)
7. Grabar, N., Zweigenbaum, P.: Lexically-based terminology structuring. Terminology 10, 23–54 (2003)
8. Grefenstette, G.: Corpus-derived first, second and third-order word affinities. In: Sixth Euralex International Congress, pp. 279–290 (1994)
9. Hamon, T., Nazarenko, A., Poibeau, T., Aubin, S., Derivière, J.: A robust linguistic platform for efficient and domain specific web content analysis. In: RIAO (2007)
10. Hamon, T., Nazarenko, A., Gros, C.: A step towards the detection of semantic variants of terms in technical documents. In: COLING-ACL 1998, pp. 498–504 (1998)
11. Jacquemin, C.: Spotting and discovering terms through natural language processing. The MIT Press (2001)
12. Kanerva, P., Kristofersson, J., Holst, A.: Random indexing of text samples for latent semantic analysis. In: Conf. of the Cognitive Science Society, vol. 1036 (2000)
13. Karlgren, J., Sahlgren, M.: From words to understanding. In: Proceedings of the ACL 2001, pp. 294–308 (2001)
14. Landauer, T., Dumais, S.: A solution to Plato's problem: The latent semantic analysis theory of acquisition, induction, and representation of knowledge. Psychological Review 104(2), 211 (1997)
15. Morin, E., Jacquemin, C.: Automatic Acquisition and Expansion of Hypernym Links. Computers and the Humanities 38(4), 363–396 (2004)
16. Padó, S., Lapata, M.: Dependency-based construction of semantic space models. Computational Linguistics 33(2), 161–199 (2007)
17. Polajnar, T., Clark, S.: Improving distributional semantic vectors through context selection and normalisation. In: Proceedings of EACL 2014 (to appear, 2014)
18. Sahlgren, M.: The Word-Space Model: Using Distributional Analysis to Represent Syntagmatic and Paradigmatic Relations between Words in High-Dimensional Vector Spaces. Ph.D. thesis, Stockholm University, Stockholm, Sweden (2006)
19. Schmid, H.: Probabilistic part-of-speech tagging using decision trees. In: New Methods in Language Processing, pp. 44–49 (1994)
20. Tsatsaronis, G., Panagiotopoulou, V.: A generalized vector space model for text retrieval based on semantic relatedness. In: EACL 2009, pp. 70–78 (2009)
21. Turney, P.D., Pantel, P.: From frequency to meaning: Vector space models of semantics. Journal of Artificial Intelligence Research 37, 141–188 (2010)
22. Vozalis, E., Margaritis, K.G.: Analysis of recommender systems' algorithms. In: Proceedings of HERCMA (2003)
23. Weeds, J., Weir, D.: Co-occurrence retrieval: A flexible framework for lexical distributional similarity. Computational Linguistics 31(4), 439–475 (2005)
24. Zweigenbaum, P.: Menelas: an access system for medical records using natural language. Computer Methods and Programs in Biomedicine 45 (1994)

A Parallel Non-negative Sparse Large Matrix Factorization

Anatoly Anisimov, Oleksandr Marchenko,
Emil Nasirov, and Stepan Palamarchuk

Faculty of Cybernetics, Taras Shevchenko National University of Kyiv, Ukraine

Abstract. This paper proposes parallel methods of non-negative large sparse matrix factorization – a very popular technique in computational linguistics. Memory usage and data transmitting necessity of factorization algorithm was analysed and optimized. The described effective GPU-based and distributed algorithms were implemented, tested and compared by means of large sparse matrices processing.

Keywords: Computational linguistics, parallel computations, non-negative matrix factorization.

1 Introduction

Non-negative matrix and tensor factorization are very popular techniques in computational linguistics. With the help of non-negative matrix and tensor factorization within the paradigm of latent semantic analysis [1] computational linguists are capable of solving practical problems such as classification, clustering of texts and terms [2, 3]), construction of semantic similarity measures [4, 5]), automatic extraction of linguistic structures and relations (Selectional Preferences) and Verb Sub-Categorization Frames), etc. [6]

This paper describes the construction of a model for parallel non-negative factorization of a large sparse matrix. Such a model can be used in large NLP systems not limited to narrow domains.

The problem of non-negative factorization for a sparse large matrix emerged in the development of a measure of semantic similarity between words with Latent Semantic Analysis usage. To cover a wide range of topics a great amount of articles from the English Wikipedia was processed to construct the similarity measure. Lexical analysis of the various Wikipedia articles was performed to calculate the frequency of using words and collocations. As a result, a large matrix Terms × Articles was constructed. It contains frequency estimation of the terms in the texts. The precise size of the matrix equals to 2,437,234 terms × 4,475,180 articles of the English Wikipedia. The frequency threshold T=3 was set to remove the noise. The resulting matrix contains 156,236,043 non-zero elements. To factorize a sparse matrix of such size it is necessary to develop a specific model for parallelizing matrix computations. The model has been implemented due to the usage of distributed and parallel computing on the GPU. Recently a

A. Przepiórkowski and M. Ogrodniczuk (Eds.): PolTAL 2014, LNAI 8686, pp. 136–143, 2014.

great number of powerful parallel models for Non-Negative Matrix Factorization (NMF) has been developed [7–10]. However, none of the developed applications for them is an acceptable solution for the defined task. Some of them do not meet the requirements of the matrix dimensions [7–9]. The model presented in work [10] performs NMF for sparse matrices of required dimensions in an acceptable time, but it requires excessively large computational resources and it is not always affordable.

2 NMF Algorithm

Non-negative matrix factorization of matrix V of size $[n; m]$ is a process of calculating two matrices W and H of size $[n; k]$ and $[k; m]$ respectively, such that $V \approx WH$.

$$F(W,H) = \sqrt{\sum_{i=1}^{n}\sum_{j=1}^{m}(V_{i,j} - (WH)_{i,j})^2} \tag{1}$$

The goal of the algorithm is to minimize the cost function that quantifies the approximation quality. There are a lot of different cost functions. In this paper the root-mean-square distance between V and WH is used (see (1)).

In [11], the authors proposed a simple iterative algorithm to approximate the matrices. It consists of two consequent updates of matrices W and H given in (2) and (3).

$$(H_t)_{ij} = (H_{t-1})_{ij}\frac{(W_{t-1}^T V)_{ij}}{(W_{t-1}^T W_{t-1}H_{t-1})_{ij}} \tag{2}$$

$$(W_t)_{ij} = (W_{t-1})_{ij}\frac{(VH_t^T)_{ij}}{(W_{t-1}H_tH_t^T)_{ij}} \tag{3}$$

In (2) and (3), H_t and W_t are the matrices obtained from iteration t. H_0 and W_0 are initialized with random values from $[0; 1)$ range.

The algorithm continues until either a stationary point is reached or a certain number of iterations is performed.

3 Model Analysis

The goal is to solve the NMF problem for different k values and compare results for all of them. Table 1 shows memory requirements for storing W and H for different k. On each iteration the algorithm described in Sect. 2 requires twice as much memory as required for matrices storage. It does not include memory required for V. Due to such excessive memory requirements of the algorithm it is difficult to execute it on a single machine, without dumping data to the hard drive. Two variants of the algorithm implementation are described below: local (with intensive hard drive usage) and distributed (with intensive network usage).

Table 1. Memory requirements for storing of W and H for different k, based on 32-bit float

k	100	200	300
W	0.98Gb	1.95Gb	2.92Gb
H	1.79Gb	3.58Gb	5.37Gb
total	2.76Gb	5.53Gb	8.29Gb

4   A GPU Version of the Algorithm

To simplify explanations the substitution $(H' = H^T)$ and transformation of (2) and (3) result in:

$$(H'_t)_{ij} = (H'_{t-1})_{ij} \frac{(V^T W_{t-1})_{ij}}{(H'_{t-1} W^T_{t-1} W_{t-1})_{ij}} \tag{4a}$$

$$(W_t)_{ij} = (W_{t-1})_{ij} \frac{(V H'_t)_{ij}}{(W_{t-1} H'^T_t H'_t)_{ij}} \tag{4b}$$

It allows us to treat both formulas in the same way, by simply substituting either H', W and V^T or W, H' and V instead of A, B and S into (5).

$$A_{ij} = A_{ij} \frac{(SB)_{ij}}{(AB^T B)_{ij}} \tag{5}$$

From this point, only evaluation of (5) with a configuration W, H' and V is discussed, since other configuration can be obtained in the same way.

Formula (5) can be calculated as a series of four steps as in (6).

$$C = SB \tag{6a}$$

$$K = B^T B \tag{6b}$$

$$D = AK \tag{6c}$$

$$A_{ij} = A_{ij} \frac{C_{ij}}{D_{ij}} \tag{6d}$$

This order of computation (5) requires a minimal number of calculations. The steps have computational complexity of $O(k * (nnz(S) + n))$, $O(k^2 m)$, $O(k^2 n)$ and $O(kn)$ correspondingly, where $nnz(S)$ is a number of non-zero cells in matrix S. The first three steps are natively supported by CUDA cuSPARSE [12] and cuBLAS [13] libraries (or other similar libraries for AMD). The fourth step requires custom GPU kernel implementation, but at the same time it is a relatively cheap operation and thus it can be performed on CPU.

Also these matrices are too large to be stored in the memory of GPU, thus operations should be performed by parts in a manner that reduces amount of excessive memory copying.

So for (6a) matrices can be written as $S = (S_1'|S_2'|...|S_t')^T$ and $B = (B_1|B_2|...|B_r)$ and each cell of C calculated as shown in (7). Since B is larger than S in terms of memory usage, the multiplications should be grouped by pieces of B (to upload them only once). Also it is rational to minimize r and keep t reasonably small, otherwise most of GPU cores will be idle. C is a matrix of size $[m; k]$ for H' and $[n; k]$ for W.

$$C = \begin{pmatrix} S_1'B_1 & ... & S_1'B_r \\ ... & & ... \\ S_t'B_1 & ... & S_t'B_r \end{pmatrix} \tag{7}$$

For (6b) it is preferable to write matrices as $B = (B_1'|...|B_t')^T$ and $K = B_1'^T B_1' + ... + B_t'^T B_t'$, because it doesn't require any redundant matrix uploads to GPU. K is a matrix of size $[k; k]$.

For (6c) matrix K should be kept in memory and A should be multiplied by blocks of rows. D is a matrix of size $[m; k]$ for H' and $[n; k]$ for W.

There is no need to store D in memory if (6d) is applied on the piece of matrix A that was used to obtain a piece of matrix D.

The complexity of operation is $O(nm)$ and straightforward to be implemented with CUDA toolkit.

5   Distributed Algorithm

The next step to improve the performance is to use a distributed grid of PCs of the same configuration. There are several distribution models. There are following three distribution models in case we have 2 nodes in the grid:

1. **W and H' are separately calculated on different nodes.** Both nodes work in one of the two modes alternatively. They either support the other node (supplying data to the other node) or lead (calculating by using the data received form supporting node). In this distribution model, on each iteration it is necessary to transmit over the network the amount of data equal to sizeof(W) + sizeof(H), where sizeof(X) is the amount of memory required to store matrix X. A lead node will also be mostly idle, because (6a) is the most resource-demanding step out of all 4.
2. **W and H' are split in chunks of columns and evenly distributed between nodes.** Where $H' = (H_1'|H_2')$ and $W = (W_1|W_2)$, the first node is responsible for H_1', W_1 and the second node – for H_2', W_2. In this model each node behaves as supporting and leading node at the same time. Nodes have to transmit the amount of data equal to $1.5 * $ (sizeof(W) + sizeof(H)) over the network.

3. **W and H' are split in chunks of rows and evenly distributed between the nodes.** Where $H' = (H_1|H_2)^T$ and $W = (W_1'|W_2')^T$, the first node is responsible for H_1, W_1' and the second – for H_2, W_2'. In this model, similarly to the previous one, each node functions in both modes at the same time. Nodes have to transmit the amount of data equal to sizeof(W) + sizeof(H) over the network because it is possible to calculate pieces of H^T*H, W^T*W on each node separately and there is no need to transmit H_2' to H_1' and W_2 to W_1 as in second model.

Nodes also have to transmit one or several matrices of size $[k; k]$ in each of the above models, but their total size is neglectable comparing to the size of W and H. Metrics calculation for both the first and the third model requires transmitting the amount of data equal to $\frac{(\text{sizeof}(W)+\text{sizeof}(H))*K}{2}$, where K is the number of nodes in the grid, the second model requires $(K - 1)$ times more transmitted data.

The last model is the most preferable, because it is better in both network and GPU utilization, thus it is used in the implementation.

Also it should be mentioned that in case of the grid expansion, the total amount of the data transmitted over the network rises polynomially, but per node it will be limited by $2 * (\text{sizeof}(W) + \text{sizeof}(H))$.

Since V is a sparse matrix, it may contain an unevenly distributed amount of non-zero cells and this may badly impact on the distributed algorithm performance. To optimize distribution of work between the nodes it is reasonable to rearrange the rows and columns of V in a way that equalizes amount of non-zero cells in each large cell of matrix V.

The third model is implemented to perform NMF of input matrix and used on a grid of four nodes, so this case will be described.

Where matrices W, H' and V are partitioned according to the selected model

3: $W = (W_1'|W_2'|W_3'|W_4')^T$, $H' = (H_1|H_2|H_3|H_4)^T$, $V = \begin{matrix} V_{11} & ... & V_{14} \\ ... & & ... \\ V_{41} & ... & V_{44} \end{matrix}$.

The algorithm consists of three main phases: initialization, iterations and metrics calculation. At initialization phase W, H and V are distributed between all the 4 nodes. Node i gets W_i', H_i and $V_{ki}, V_{ik}, k = \overline{1,..,4}$, this phase is represented by the scheme in Fig. 1.

The iteration phase consists of two similar steps, one for calculation of H' and the other for W. Each of them is subdivided into 3 smaller sub-steps as it is described further.

At the first sub-step each node calculates $k \times k$ matrix $W_i' * W_i'^T$ and sends it to the aggregator. The aggregator sums up all received pieces into one matrix K_w and sends the aggregated result to all the nodes. This sub-step is represented by the scheme in Fig. 1.

At the second sub-step each node calculates its own $(V_{1i}^T|V_{2i}^T|V_{3i}^T|V_{4i}^T)^T * W_i'$. The resulting matrix has the same size as H'. Finally each node divides its matrix according to the initial partitioning of matrix H and transmits these pieces to the corresponding nodes. This sub-step is represented by the scheme in Fig. 2.

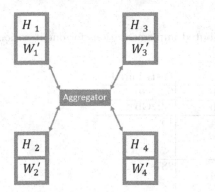

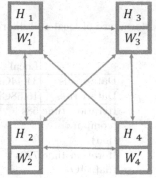

Fig. 1. Initial partitioning in the model with 4 nodes

Fig. 2. The iteration phase of the distributed model with 4 nodes

At the third sub-step the nodes calculate matrix $H_i * K_w$ and perform an in-place update of matrix H_i. This sub-step does not require any network communication.

These three sub-steps are intended for calculating matrix H'. After updating H', the same sub-steps should be made for W. Specifically next products should be calculated $H_i * H_i^T$, $(V_{i1}|V_{i2}|V_{i3}|V_{i4}) * H_i$ and $W_i' * K_h$.

At the metrics calculation phase each node transmits its piece of matrix H' to all other nodes. After receiving a piece of matrix H' each node calculates the corresponding part of the metrics. This phase is also represented in Fig. 2.

6 Results of Analysis

The previously described distributed algorithm with GPU usage has been implemented. The local GPU algorithm that dumps and reads data from the local hard drive has also been implemented to compare the performance of the models.

Both implementations are executed with the same input matrix. The local version is executed on one node with the same memory restrictions.

We used the following hardware configuration for the tests: Intel Core i7 CPU, NVIDIA GeForce GTX560 1Gb, 8Gb of RAM (available 6 Gb), 1Gbit LAN and SATA III hard drive.

Table 2 shows the time and resources required for each version of the algorithm to perform the iteration. The data for distributed model are per node, so the total data IO (read & write together) across all 4 nodes is 49.76Gb. Table 3 shows comparison of metrics calculation. The data in both tables are obtained for $k = 300$.

The experiments show that the process of matrices calculation converges after approximately 100 iterations. Therefore, the calculation of the non-negative factorization for the given sparse large matrix with the proposed model takes approximately 9.6 hours for the distributed implementation and almost 21 hours for the local.

Table 2. Performance of local and distributed implementations for one iteration

	Local	Distributed
Data reads	34.44Gb	6.22Gb
Data writes	16.58Gb	6.22Gb
Iteration time (computation)	58s	15s
Iteration time (data IO)	729s	287s

Table 3. Performance of local and distributed implementations for metrics calculation

	Local	Distributed
Data reads	13.66Gb	6.22Gb
Data writes	0	6.22Gb
Time (computation)	45865s	11371s
Time (data IO)	192s	280s

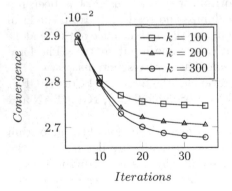
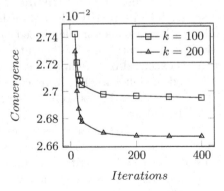

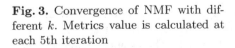

Fig. 3. Convergence of NMF with different k. Metrics value is calculated at each 5th iteration

Fig. 4. Convergence of NMF with $k = 200$ and $k = 300$. Metrics value is calculated at each 100th iteration

7   Conclusion

We have combined the GPU-based and distributed algorithms, and also paid special attention to memory usage, which allows larger input matrices to be factorized. The experiments showed the constructed model is effective. It can be used to perform the tasks of industrial scale to factorize sparse matrices of large dimension with an acceptable time using available computing resources.

Proposed distributed model can be easily modified to speed up non-negative factorization of large tensors.

References

1. Deerwester, S., Dumais, S.T., Furnas, G.W., Landauer, T.K., Harshman, R.: Indexing by latent semantic analysis. Journal of the American Society for Information Science, 391–407 (1990)
2. Xu, W., Liu, X., Gong, Y.: Document Clustering Based on Non-negative Matrix Factorization. In: Proceedings of the 26th Annual International ACM SIGIR Conference on Research and Development in Informaion Retrieval, SIGIR 2003, pp. 267–273. ACM, New York (2003)
3. Shahnaz, F., Berry, M.W., Pauca, V.P., Plemmons, R.J.: Document Clustering Using Nonnegative Matrix Factorization. Inf. Process. Manage., 373–386 (2006)
4. Landauer, T.K., Foltz, P.W., Laham, D.: An introduction to latent semantic analysis Discourse processes, pp. 259–284. Ablex Publishing Co. (1998)
5. Mihalcea, R., Corley, C., Strapparava, C.: Corpus-based and knowledge-based measures of text semantic similarity. In: AAAI 2006, pp. 775–780. AAAI Press, Menlo Park (2006)
6. Van de Cruys, T.: A non-negative tensor factorization model for selective preference induction. Natural Language Engineering, 417–437 (2010)
7. Bader, B.W., Kolda, T.G.: MATLAB Tensor Toolbox Version 2.5 (2012), http://www.sandia.gov/~tgkolda/TensorToolbox/
8. Kanjani, K.: Parallel Non Negative Matrix Factorization for Document Clustering Texas A & M University (2007)
9. Kysenko, V., Rupp, K., Marchenko, O., Selberherr, S., Anisimov, A.: GPU-Accelerated Non-negative Matrix Factorization for Text Mining. In: Bouma, G., Ittoo, A., Métais, E., Wortmann, H. (eds.) NLDB 2012. LNCS, vol. 7337, pp. 158–163. Springer, Heidelberg (2012)
10. Liu, C., Yang, H.-C., Fan, J., He, L.-W., Wang, Y.-M.: Distributed Nonnegative Matrix Factorization for Web-scale Dyadic Data Analysis on Mapreduce. In: Proceedings of the 19th International Conference on World Wide Web, WWW 2010, pp. 681–690. ACM, Raleigh (2010)
11. Lee, D.D., Seung, H.S.: Algorithms for Non-negative Matrix Factorization In NIPS, pp. 556–562. MIT Press (2000)
12. Naumov, M., Chien, L.S., Vandermersch, P., Kapasi, U.: CUDA CUSPARSE Library NVIDIA, San Jose, CA (2010)
13. NVIDIA: CUBLAS Library User Guide (2013), http://docs.nvidia.com/cublas/index.html
14. Cickocki, A., Zdunek, R., Phan, A.H., Amari, S.-I.: Non-negative matrix and tensor factorizations: applications to exploratory multiway data analysis and blind source separation Fabulous, Singapore, pp. 237–240 (2009)

Statistical Analysis of the Interaction between Word Order and Definiteness in Polish

Adrian Czardybon, Oliver Hellwig, and Wiebke Petersen

SFB 991, University of Düsseldorf, Germany

Abstract. Although (in-)definiteness is semantically relevant in Polish, the language lacks explicit linguistic features for marking it. The paper presents the first quantitative, statistical evaluation of the correlation between word order and definiteness. Our results support previous qualitative theories about the influence of the verb-relative position on definiteness in Polish.

1 Introduction

The paper presents the first quantitative assessment of linguistic strategies for expressing definiteness in Polish using statistical evaluation of an annotated corpus.[1] We define definiteness as referential uniqueness of a noun or noun phrase (NP; details in Sect. 2). In contrast to languages such as English or German, Polish lacks definite and indefinite articles. Therefore, definiteness is usually not marked explicitly at the sentence level. This contrast between Polish and English is illustrated by example (1) which represents the first sentence of the Polish translation of George Orwell's novel *Nineteen Eighty-Four* [1, p. 7]. While no explicit markers of definiteness are found with the nouns *dzień* 'day' and *zegary* 'clocks' in the Polish sentence, articles mark the definiteness of the words '*day*' and '*clocks*' in the English translation.

(1)   Był jasny, zimny dzień kwietniowy i   zegary biły   trzynastą.
       was bright cold   day April   and clocks struck thirteen
       "It was a bright day in April, and the clocks were striking thirteen."

Although Polish lacks articles, previous research leaves no doubt that definiteness is a relevant semantic feature in Polish. Szwedek states that "[a]lthough there is no article in Polish we seldom have doubts whether a noun in a text is definite or indefinite" [2, p. 203]. Researchers have discussed several linguistic structures that may be used for expressing definiteness in Polish, one of the most frequently mentioned being the position of an NP in relation to the position of the main verb [3,4,5,6].

While previous studies have dealt with definiteness in Polish mainly from a qualitative perspective, the present paper is, to our best knowledge, the first

---

[1] We would like to thank our annotators Joanna Strzępek and Helena Zamolska. The work on this topic is financed by the Deutsche Forschungsgemeinschaft (DFG) through CRC 991.

A. Przepiórkowski and M. Ogrodniczuk (Eds.): PolTAL 2014, LNAI 8686, pp. 144–150, 2014.

quantitative evaluation of definiteness strategies. Following the ideas found in previous research, the paper focusses on the verb-relative positions of NPs as an indicator of definiteness. Apart from submitting existing scientific hypotheses to a statistical assessment, the computational and statistical framework developed for this paper serves a more far-reaching purpose. If we are able to validate strategies that have an influence on the definiteness of the Polish NP, these strategies can also be used for developing machine learning algorithms that determine the definiteness of an NP in unannotated Polish corpora automatically. Such algorithms are a major building block for assessing Löbner's theory of concept types and determination [7,8] for computationally under-resourced languages such as Polish.

The following sections introduce the concept of definiteness and the linguistic features used to express it (2), describe the corpus (3) and its evaluation (4), and summarize our results (5).

2   Theoretical Background and Linguistic Features

This section formalizes the notion of definiteness for NPs. In addition, we present a short survey of previous research on linguistic factors that are said to influence definiteness in Polish, with a special focus on word order.

2.1   Definiteness of NPs

For this study, we follow Löbner [7,8] in assuming that uniqueness is the underlying concept of definiteness: If a noun is definite, there is only one referent that fits the definite NP in the given linguistic context. Löbner distinguishes between semantic and pragmatic uniqueness. Individual nouns such as *John*, *Pope* or *moon* are semantically, i.e. inherently unique, because they have only one referent in their contexts of utterance. This is also true for functional nouns such as *father*, *head* or *difference* which are two- or more-place predicates in contrast to the individual nouns. Functional nouns are inherently unique since each person can only have one father. Thus, they express a one-to-one relation between two entities (for example the father and the person who he is the father of). In contrast, sortal (*dog, book, chair*) and relational nouns (*brother, finger, uncle*) are not inherently unique. They require (extra-)linguistic context in order to achieve unique reference. Since a person can have more than one brother or none, relational nouns are not inherently unique expressing a one-to-many relation in contrast to the functional nouns.

We annotated NPs as definite if they refer uniquely. In this context, it was not relevant whether unique reference was due to the semantics of the noun (individual and functional nouns) or whether the unique reference was established from the context (pragmatic uniqueness).

2.2   Features

Word order has been mentioned frequently as one of the most important strategies for expressing definiteness in Polish [3,4,5,6]. Błaszczak claims that "in a postverbal position ... a nominal phrase not accompanied by any determiner ... is in principle ambiguous (definite or indefinite)" [5, p. 11]. Furthermore, she writes that "[i]n a preverbal position a nominal is normally interpreted as definite" [5, p. 15]. The theory that preverbal bare NPs are mainly interpreted as definite, whereas postverbal bare NPs can be definite or indefinite will be assessed in Sect. 4.

Apart from the verb-relative position of NPs, other strategies for expressing definiteness in Polish including perfective and imperfective aspect ([9], [10]) as well as case marking[2] ([10, p. 35], [14, pp. 30, 48–49, 86]). We are planning to examine the influence of these features in follow-up studies, along with the roles of pronouns such as possessive, demonstrative (*ten, tamten, ów, taki*), and indefinite pronouns (*jakiś* 'some', *jakikolwiek* 'any', *niektóry* 'some', *niejaki* 'some', *żaden* 'none', *pewien* 'certain', *inny* '(an)other', *jeden* 'one', numerals, quantifiers (*wszystek* 'all', *wiele* 'many/much', *kilka* 'a few/several', *parę* 'a few', *oba* 'both'), restrictive linguistic structures such as relative clauses or prepositional phrases, and NPs with ordinals and superlatives.

3 Data and Annotation

We based our study on the first 479 sentences of a Polish translation of George Orwell's novel *Nineteen Eighty-Four* [15], which is annotated with morphosyntactical information according to the TEI standard. Frequently, the 1-million-word subcorpus of the National Corpus of Polish ("NKJP") [16] is used for such annotation tasks. However, the fact that the NKJP does not consist of coherent text passages of more than 40 to 70 words [16, p. 54] would have been a major drawback in our case, because context plays a crucial role when it comes to deciding whether an NP is definite or not in Polish.

We used MMAX2 [17] for annotating the data. Annotation was carried out independently by two native speakers of Polish. Since we had to develop annotation guidelines while performing this initial study, guidelines were adapted during the process of annotation.

For each noun the three main categories (1) "part of an idiom/proverb", (2) "multiword lexeme", and (3) "(in-)definite noun" were assigned. Furthermore, there was always the option to choose "don't know". The nouns contained in the category "part of an idiom/proverb" (*w końcu* 'finally', *na czas* 'in time', *zdać sobie sprawę z czegoś* 'to realize sth') were excluded from the further analysis because they are normally not referential. The monolingual dictionary of Polish [18] was consulted in unclear cases of idioms/proverbs and multiword lexemes (*klatka piersiowa* 'chest', *hokej na trawie* 'field hockey'). The definiteness of the NPs assigned to categories (2) and (3) was chosen from among the subcategories (i) generic, (ii) indefinite, (iii) definite, explicitly marked by a demonstrative, (iv) definite due to other reasons, and (v) ambiguous between definite and indefinite

---

[2] It is argued that verbs such as *kupić* 'buy', *dać* 'give', and *pożyczyć* 'lend/borrow' allow for a case alternation of the direct object ([11, p. 83], [12, pp. 316–317], [13, p. 72]).

reading. Generic NPs were excluded from the further evaluation because we were only interested in referential NPs. Option (iii) was included as a preparatory step for a follow-up study in which the role of demonstratives in marking definiteness in Polish will be investigated.

The annotation produced a total of 8664 word tokens, including 2447 nouns. Out of these nouns, 2059 were annotated with definiteness information, while the remaining ones belonged to the category "part of an idiom/proverb" and "don't know". Nouns having definiteness information were derived from 1079 different lemmata, out of which 696 were hapax legomena, 306 occurred 2-5 times, and the remaining 77 more frequently. Annotation yielded a $\kappa = 0.985$ according to [19]. This high value is certainly due to the fact that the guidelines were developed along with the initial annotation, and a clear drop of κ may be expected in follow-up annotations.

4   Statistical Evaluation

For assessing the sentence-position hypothesis that is examined in this paper (refer to Sect. 2), we needed to determine the positions of nouns in relation to the positions of the main verbs in each sentence. Because syntactic substructures are not marked in [1], we split input sentences into syntactic chunks that contain exactly one main (non-auxiliary) verb. For this sake, we used a heuristic function that describes typical sentence structures in Polish in terms of regular expressions. Subsequent statistical analysis was restrained to these one-verb chunks. It should be noted that each of these chunks may consist either of a main clause or of a subordinate clause. We were able to extract 304 chunks with exactly one main verb from 46.6% of all 479 sentences in this way, while the remaining sentences had unclear chunkings. As this study focusses on bare NPs, we further excluded 101 nouns and NPs that were used with a determiner such as a demonstrative, indefinite, or possessive pronoun, because these determiners influence the definiteness of the noun at the NP level. For each resulting chunk, we recorded the number of nouns occurring before and after the main verb, and the respective definiteness annotations of the nouns. Raw counts of this procedure are given in Table 1.

To test the research hypothesis that definiteness of NPs is related to their verb-relative positions, we constructed a 2×2 contingency table, using both columns

Table 1. The positions of nouns with definiteness annotations relative to the main verb

Type	postverbal	preverbal
ambiguous	7	0
definite (demonstr. pron.)	4	9
definite (not explicit)	222	197
generic	3	1
indefinite	155	49

and the two rows "definite (not explicit)" and "indefinite" from table 1. The content of this table is displayed as a bar plot in Fig. 1, grouped by definiteness (left) and the verb-relative position (right). Because the expected frequencies for all cells of the 2×2 contingency table were higher than 5, we applied a χ^2 test as a statistical test for count data to this table. The null hypothesis of the test claims that definiteness of NPs is not related to the verb-relative position, while the alternative hypothesis postulates such a relationship. Because this paper is an exploratory study, we chose a comparatively high significance level of $\alpha = 10\%$, which produces decisions that are in favour of the alternative hypothesis. The χ^2 test yields a value of 30.367 for the 2×2 contingency table constructed from Table 1, showing highly significant differences between the factors at the given significance level α.

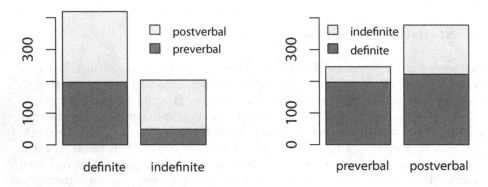

Fig. 1. Absolute frequencies of (in-)definite nouns in the pre- and postverbal position

The right plot in Fig. 1 clearly supports what Błaszczak [5, pp. 11–15] and other authors state about the influence of the verb-relative position on definiteness (refer to Sect. 2.2): Bare nouns in the preverbal position are mostly interpreted as definite, whereas the postverbal position is ambiguous in our corpus. In addition to these ideas formulated in previous research, the left subplot in Fig. 1 demonstrates that indefinite nouns show the tendency to occur in the postverbal position, while no positional preference is found with definite nouns.

5   Conclusion

The results of our study show a strong interaction between the definiteness of an NP and its position in relation to the main verb. This is in accordance with the observations made in the literature ([20, p. 235], [21, pp. 232–233], [6, p. 217]). The quantitative evaluation in Sect. 4 showed that the postverbal position is basically ambiguous in terms of definiteness, while the preverbal position is strongly associated with definite NPs. Analyzing our data in the opposite direction, the syntactic position of definite NPs cannot be predicted, whereas indefinite NPs are prominently found in the postverbal position, as can be observed in Fig. 1. The comparatively high number of 49 indefinite preverbal NPs (refer to Table 1) is unexpected and should be submitted to a closer examination.

The results of this study indicate several directions for future research. First, we will focus on sentences with more than one NP placed postverbally, and investigate whether there is a tendency of placing indefinite NPs rather in sentence-final position in contrast to the postverbal, but not the sentence-final position. This approach is motivated by Szwedek's ([22, p. 80]) observation that the postverbal, but not sentence-final unstressed NP is always interpreted as definite. For this task, we need to annotate syntactic chunks either manually or by using a shallow syntactic parser (chunker). Second, it can be observed that inherently unique nouns such as individual and functional nouns are interpreted as definite regardless of their placement within the sentence. Löbner's theory of concept types and determination could explain our observation that definite nouns do not show clear positional preferences, as stated above and shown in Fig. 1. Therefore, we are planning to annotate the concept type of the nouns in our corpus in a second step. This additional layer of information will make it possible to obtain a much more detailed picture of the connection between the syntactic position, definiteness, and the concept types. A working hypothesis for such a follow-up study would claim that sortal nouns have a tendency to be definite if placed preverbally, whereas they tend to be indefinite in the postverbal position, which is not the case with functional and individual nouns.

References

1. Orwell, G.: Rok 1984. Warszawskie Wydawnictwo Literackie MUZA SA (2008)
2. Szwedek, A.: Some aspects of definiteness and indefiniteness of nouns in Polish. Papers and Studies in Contrastive Linguistics (2), 203–211 (1974)
3. Szwedek, A.: Word Order, Sentence Stress and Reference in English and Polish. Linguistic Research, Edmonton (1976)
4. Grzegorek, M.: Thematization in English and Polish. A study in Word Order. PhD thesis, Drukarnia Uniwersytetu IM. A. Mickiewicza, Poznań (1984)
5. Błaszczak, J.: Investigation into the Interaction between the Indefinites and Negation. Akademie Verlag, Berlin (2001)
6. Mendoza, I.: Nominaldetermination im Polnischen. Die primären Ausdrucksmittel (Nominal determination in Polish. The primary means of expression). PhD thesis, Ludwig-Maximilians-Universität, München, Habilitationsschrift (2004)
7. Löbner, S.: Definites. Journal of Semantics 4(4), 279–326 (1985)
8. Löbner, S.: Concept Types and Determination, vol. 28. Oxford University Press (2011)
9. Wierzbicka, A.: On the Semantics of the Verbal Aspect in Polish. In: Jakobson, R. (ed.) To Honor Roman Jakobson. Essays on the Occasion of his Seventieth Birthday. Mouton, The Haque (1967)
10. Witwicka-Iwanowska, M.: Artikelgebrauch im Deutschen. Eine Analyse aus der Perspektive des Polnischen (The article use in German. An analysis from the Polish perspective). Narr Verlag, Tübingen (2012)
11. Topolińska, Z.: Remarks on the Slavic Noun Phrase. Wydawnictwo Polskiej Akademii Nauk, Wrocław (1981)

12. Topolińska, Z.: Składnia grupy imiennej (The syntax of the noun phrase). In: Topolinska, Z. (ed.) Gramatyka współczesnego języka polskiego. Składnia, pp. 301–386. Państwowe Wydawnictwo Naukowe, Warszawa (1984)
13. Tokarski, J.: Fleksja polska (Polish inflection). Wydawnictwo Naukowe, Warszawa (1973, 2001)
14. Sadziński, R.: Die Kategorie der Determiniertheit und Indeterminiertheit im Deutschen und im Polnischen (The category of determinedness and indeterminedness in German and Polish). WSP, Częstochowa (1995)
15. Kotsyba, N., Radziszewski, A., Derzhanski, I., Erjavec, T.: Multext- East cesAna: Nineteen Eighty-Four, Polish, Warszawa (2010)
16. Przepiórkowski, A., Bańko, M., Górski, R.L., Lewandowska-Tomaszczyk, B. (eds.): Narodowy Korpus Języka Polskiego (The National Corpus of Polish). Wydawnictwo Naukowe PWN, Warszawa (2012)
17. Müller, C., Strube, M.: Multi-Level Annotation of Linguistic Data with MMAX2, pp. 197–214. Peter Lang, Frankfurt (2006)
18. Dubisz, S., Sobol, E.: Uniwersalny Słownik Języka Polskiego (Universal dictionary of the Polish language). Wydawnictwo Naukowe PWN, Warszawa (2006)
19. Fleiss, J.L.: Measuring nominal scale agreement among many raters. Psychological Bulletin 76(5), 378–382 (1971)
20. Weiss, D.: Indefinite, definite und generische Referenz in artikellosen slavischen Sprachen (Indefinite, definite, and generic reference in article-less Slavic languages). In: Mehlig, H. (ed.) Referate des VIII. Konstanzer Slavistischen Arbeitstreffens. Kiel 28.9-1.10.1982, pp. 229–261. Verlag Otto Sagner, München (1983)
21. Lyons, C.: Definiteness. University Press, Cambridge (1999)
22. Szwedek, A.: A Linguistic Analysis of Sentence Stress. Gunter Narr Verlag, Tübingen (1986)

Stanford Typed Dependencies:
Slavic Languages Application

Katarzyna Marszałek-Kowalewska[1], Anna Zaretskaya[2], and Milan Souček[3]

[1] Adam Mickiewicz University, Poznań, Poland
k.marszalek.kowalewska@gmail.com
[2] University of Málaga, Spain
azaretskaya@gmail.com
[3] Lionbridge Technologies, Inc., Tampere, Finland
milan.soucek@lionbridge.com

Abstract. The Stanford typed dependency model [7] constitutes a universal schema of grammatical relationships for dependency parsing. However, it was based on English data and did not provide descriptions for grammatical features that are fundamental in other language types. This paper addresses the problem of applying the Stanford typed dependency model for Slavic languages. Language features specific to Slavic languages that are presented and described include ellipsis, different types of predicates, genitive constructions, direct vs. indirect objects, reflexive pronouns and determiners. In order to maintain cross-language consistency we try to avoid major changes in the original Stanford model, and rather devise new applications of the existing relation types.

Keywords: dependency parsing, treebank, Slavic languages, Stanford typed dependencies.

1   Introduction

There has been a major development in building language resources in the last couple of decades. Various annotated language corpora could be collected thanks to newly developed computational linguistics methods applied to large volumes of data available via the Internet. Also, specialized corpora like dependency treebanks are now available for many languages. Monolingual dependency treebanks have been recently built for a majority of Slavic languages as well (e.g. [1] for Russian; [13] for Czech; [12] for Slovenian; [26] for Croatian; and [30] for Polish). One of the next targets for building language resources is preparation of multilingual data with consistent annotation format that could be used for cross-language research and mainly for Natural Language Processing (NLP) tasks. In terms of dependency treebanks, similar tasks have been tackled in the last decade in the series of the Conference on Computational Natural Language Learning (CoNLL) shared tasks, where various approaches to parsing multilingual data from different treebank sources were presented in [4, 14, 17, 24]. Lately, there also have been activities related to building universal

A. Przepiórkowski and M. Ogrodniczuk (Eds.): PolTAL 2014, LNAI 8686, pp. 151–163, 2014.

multilingual treebank resources [16], using unified Part of Speech (POS) [20] and dependency relation (deprel) [7] schemas. For similar cross-language activities, such unified annotation tag sets are required in order to describe universal grammar relations in languages of different types. The Stanford typed dependencies model (SD) [7] was designed to provide a simple description of the grammatical relationships in a sentence that also could easily be used for tasks not strictly related to linguistic research. This dependency model seems to be suitable for multilingual tasks where grammatical relationships in different languages need to be transformed into a universal schema while a certain level of simplification needs to be introduced. The SD model was initially defined for English and built on English reference data, which means that its application to other languages brings challenges in any cases where new types of grammatical relations need to be handled. Some improvements for the SD schema already have been suggested in various works, both by the original authors of the model (see e.g. [8–10]) and by teams that have been using SD when building treebank data (e.g. [16]). Furthermore, SD was developed on the basis of Wall Street Journal data, i.e. newswire data, which may lack constructions that are characteristic for other data types. To be more precise, although the SD model provides a good description of basic grammatical relations such as subject, object, noun phrase relations, or subordinate clauses, it lacks coverage for questions, discourse particles, ellipsis [9], or grammatical relations which are expressed by bound morphemes in morphologically rich languages. In this paper, we present the SD application for Slavic languages as it was defined for Slavic treebanks that have been built for the Universal Dependency Treebank project.[1] Since dependency treebanks following the SD model are already available for multiple languages (see [5] for Chinese; [15] for Finnish; [22] for Persian; [28] for Hebrew; [3] for Italian; [16] for German, English, Swedish, Spanish, French and Korean, and [23] for French, Spanish, German and Brazilian Portuguese), including Slavic languages resources to this group would be beneficial in terms of extending the language coverage for the cross-language studies and multilingual NLP applications. In the following sections we present problematic examples and provide proposals for applying Stanford dependency model on grammatical relations that are not covered in the basic SD model, but need to be taken into account for Slavic languages.

1.1 Slavic Language Peculiarities

Slavic languages constitute one of the major modern language families in the world. It is the fourth largest sub-family within Indo-European languages with around 300 million speakers [25].

Word order in Slavic family is more free than in Germanic languages, i.e. the order of major constituents (subject, object) is determined more by pragmatic rather than syntactic factors. The major factors that influence Slavic word order are conventional word order, constituent structure and the marking of grammatical relations by means other than word order [25]. However, the order within individual constituents is more fixed, e.g. demonstratives and numerals usually precede nouns (but not always).

---

[1] https://code.google.com/p/uni-dep-tb/

All Slavic languages retain a rich set of morphological categories. As Comrie et al. claim [6] Slavic morphology is mainly fusional, i.e. an affix can combine a number of grammatical categories. In contrast to English, genitive case in Slavic languages is expressed in terms of declension. Another typical Slavic feature is the use of impersonal constructions which can cause difficulties in differentiating between subject and direct object.

The following subchapters provide more detailed information on a few specific features of Slavic languages with regard to applying the SD model. We will focus here on determiners, modality, reflexive pronouns, copular verb ellipsis, difficulties in differentiating between subject and direct object, genitive constructions, and indirect objects.

2   Slavic Application

2.1   Determiners

A determiner is a word that modifies a noun or a noun phrase, e.g. *a, the* in English, or *der, die, das* in German. As a POS category, determiners usually include articles and pronouns (demonstrative, interrogative or possessive pronouns) [29].

In the basic SD model the deprel *det* is used to cover the relation between the head of an NP[2] and its determiner [7]:

The man is here	*det*(man, the)
Which book do you prefer?	*det*(book, which)

For Germanic languages the most obvious determiner is the article. Although Slavic languages lack articles (apart from Macedonian and Bulgarian), their sets of determiners are quite extensive due to rich morphology, i.e. declension of determiners is sometimes known as 'special adjective declension' [25]. To illustrate this, let's compare the demonstrative pronoun *this* in English and Polish:[3]

Eng: [*this*]
Pol: [*ten, tego, temu, tym, ta, tej, tę, tą, to*]

Following the basic SD model, for the Slavic languages application, the relation between demonstratives and interrogatives and the noun that they directly modify is expressed as *det*:

Cz: *Znám toho mu e* "I know that man"	*det*(muže, toho)
Cz: *Který den je dnes?* "What is the day today?"	*det*(den, který)

---

[2] NP stands for a noun phrase. This construction has either a noun or a pronoun as its head [29].
[3] We use following language abbreviations in examples: Eng for English, Cz for Czech, Pol for Polish, Rus for Russian and Slk for Slovak.

The SD model also provides a label for the relation between the head of an NP and a word that precedes and modifies the meaning of the NP determiner (*predet* deprel), e.g.

All the boys are here   *predet*(boys, all)

Also in Slavic languages the relation *predet* can be used with words like *all*, *whole* etc., if they precede another determiner, e.g.:

Cz: *Všichni naši přátelé* "All our friends"   *predet*(přátelé, všichni)
Rus: *Все эти люди* "All these people"   *predet*(люди, все)
Pol: *Wszyscy ci ludzie* "All those people"   *predet*(ludzie, wszyscy)

The interesting point is that in some cases, the reverse order of determiners is also possible in Slavic languages. For such multiple determiners, we decided to keep *det* assigned to the demonstrative pronoun and *predet* to other pronouns.

Pol: *Ci wszyscy ludzie* "All those people"   *predet*(ludzie, wszyscy)

The Universal Dependency Treebank [16] proposes to merge *det* and *predet* into just one deprel *det*. If such application is preferable, *predet* easily can be converted into *det* for Slavic treebanks as well.

Although possessive pronouns are often considered determiners as well, the basic SD schema uses the deprel *poss* to describe the relationship between possessive pronouns and the noun that they modify, e.g.:.

Pol: *Moja siostra jest tu* "My sister is here"   *poss*(siostra, moja)

2.2   Reflexive Pronoun

A reflexive pronoun refers back to the subject of the clause in which it is used, e.g. *myself* in English, *sich* in German or *si* in Italian [29].

In Slavic languages the reflexive pronoun may take several forms. It can function as an object-like reflexive pronoun or as a purely reflexive marker of the related verb. In East Slavic, the once-present reflexive-as-clitic has disappeared and now it is expressed as a verbal suffix joined to the verb, e.g.:

Rus: *Она одевает*ся   "She dresses herself"

In other languages, the reflexive acts as a clitic that may be found at different positions, e.g.:

Pol: *Musisz się umyć*   "You need to wash yourself"
Pol: *Musisz umyć się*   "You need to wash yourself"
Pol: *Drzwi otworzy y się*   "Door opened (itself)"

The basic SD model does not provide a special deprel representing the relation between a reflexive and its head (verb). For the Slavic application, depending on its syntactic function, the reflexive pronoun is treated either as a functional pronoun with the label *dobj* or *iobj*, when it has a recognizable pronoun function, e.g.:

Pol: *Piotr umy się* "Peter washef himself"   *dobj*(umył, się)

Or as a particle with deprel *prt* in cases when it is purely reflexive, e.g.:

Pol: *uśmiechnąć się* "smile (oneself)"   *prt*(uśmiechnąć się)

In the latter example, the reflexive cannot be treated on the basis of syntactic functions; it does not behave as an object. Therefore, we decided to use the deprel *prt* which in the original SD model was used to identify phrasal verbs and described the relation between a verb and its particle, e.g.:

They broke up   *prt*(broke, up)

2.3   Copular Verbs Ellipsis in Russian

Copular verbs or copulas are verbs with one complement that serve as a link to what the referent of the subject is or becomes. The most common copula in English is *be* ("*The Earth is round*"). Other verbs used as copulas in English provide additional meaning to the mere linking, like the verbs *become, appear, seem, feel,* etc. [11].

The SD manual [7] defines the *cop* relation as 'the relation between the complement of a copular verb and the copular verb'. In this relation, the copular verb depends on its complement:

Bill is big   *cop*(big, is)

In the standard version of the Universal Dependency Treebank project [16], some function words, such as copulas and adpositions, are treated as heads of their complements. We follow this same approach for the Slavic SD application. This means that the copula complement depends on the copular verb. In addition, we discarded the *cop* relation as well as deprels *acomp* and *attr*. Instead, we introduce the dependency label *scomp* (subject complement). The new label *scomp* is defined as 'a verb complement that refers to the subject of the clause. If the complement is capable of inflection, it will agree with the subject in number and gender'. Thus, the nature of the verb (i.e. whether it is copulative or not) is not crucial. This allows the verb to occupy its natural position in the root node of the sentence, while both the subject and the complement depend on the verb, e.g.

Bill is big   *scomp*(is, big); *nsubj*(is, Bill)

There seem to be no drawbacks in this representation until it is applied to the Russian language, where the main copular verb *быть* "to be" is almost always omitted in present tense. Consider the three examples provided below, all of which contain a copulative structure. The ellipsis occurs in sentence (1), which on a semantic level, has the verb "to be" in present tense. However, the same sentence in past tense (2) does contain a verb form, as well as sentence (3), where another copular verb is used instead of the verb "to be".

(1) Rus: *Этот студент – лингвист*   "This student is a linguist"
(2) Rus: *Этот студент был лингвистом*   "This student was a linguist"
(3) Rus: *Этот студент работает лингвистом* "This student works as a linguist"

In the case of copula ellipsis, both the subject *nsubj* and subject complement *scomp* are left without a head node, therefore a decision has to be made to define their relationship. One of the two apparent solutions is to consider the subject as the head of an *scomp* relation with the complement: *scomp*(студент, лингвист). The other solution is to have the subject depend on the complement: *nsubj*(лингвист, студент). The two dependency structures are shown in Figures 1 and 2.

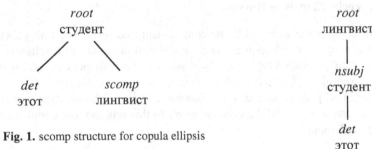

Fig. 1. scomp structure for copula ellipsis

Fig. 2. nsubj structure for copula ellipsis

We concluded that the relation between the verb and complement is closer than the one between verb and subject because – in terms of phrase constituents – they are both parts of a verb phrase and thus form one unit. Therefore it seems more natural, when the verb is omitted, to fill its *root* position by the complement. As a result, we selected the second solution for the Slavic SD application.

2.4   *Nsubj* and *iobj* in Russian

In Stanford typed dependencies, a nominal subject (*nsubj*) is a noun phrase which is the syntactic subject of a clause [7]. Some Russian verbs and constructions require a subject in dative or genitive case. Therefore, it is often difficult to decide whether a noun in oblique case is a subject or an indirect object (*iobj*). Similar examples can be found in Polish as well:

Rus: *Марии нет дома*   "Maria(*gen*) is not at home"
Rus: *Ему нельзя выходить*   "He(*dat*) is not allowed to go out"
Pol: *Jemu nie wolno wyjść*   "He(*dat*) is not allowed to go out"

For the Slavic SD application, we propose to enlarge the definition of *nsubj* so that it also covers the most frequent constructions where the subject is used in oblique case:

- A genitive NP depending on the predicative word *нет*, which in fact is the negated copular verb *to be* in present tense. This also works for all negated structures with verbs expressing existence:

Rus: *Здесь никого нет*   "There is nobody"
Rus: *Письма не пришло*   "The letter didn't come"

- Dative NPs that depend on PRED words or adverbs with predicative function:
  Rus: *Мне нужно уйти*   "I need to go"

- Dative NPs with impersonal verbs in third person singular:

Rus: *Ему пришлось подчиниться*   "He had to obey"

In all 3 cases, the non-nominative subject is still analyzed as *nsubj*. In other cases, dative and genitive NPs are treated as regular indirect objects.

2.5   Genitive Constructions

Slavic genitive constructions have been the object of syntax and semantics studies for a long time. Below, we outline the issues that arose in relation to the SD model application to Russian genitives.

For the Slavic SD application, we introduce a genitive modifier relation *gmod* to the SD model as a genitive attribute that modifies an NP. Consider the following:

Rus: *ножка стула*   "leg of the chair"
Pol: *pisk opon*   "screech of tires"

The problem occurs, however, when a noun in genitive case appears in other contexts that are different from *gmod*. Thus, NPs in genitive case are used in the position of direct object in negative constructions [2]:

Rus: *Я не заметил водки на столе*   "I didn't notice any vodka(*gen*) on the table"

It is important to note that, despite the genitive case, the noun *водки* "vodka" is clearly a direct object, since it is governed by a transitive verb. The genitive case appears only due to negation, and in an affirmative sentence the use of genitive is not possible. Genitive is often used as *dobj* when it has semantics similar to the partitive case, and means 'some amount of the whole' [19]:

Rus: Он дал мне денег   "He gave me some money(gen)"

Another use of genitive as direct object is referred to as intentional in [2]:

Rus: Он ждал сигнала   "He waited for a signal(gen)"

In other words, the noun serves as an argument of intentional verbs, such as *to wait, to expect,* and *to search,* which normally take non-referential arguments. Such arguments do not correspond to a defined entity in the real world, and their existence is not clear (waiting for a signal does not necessarily mean that it will appear). Therefore, genitive case serves here to express this non-referentiality.

Finally, genitive forms appear also in prepositional object *pobj* when the preposition requires the genitive case:

Rus: Я убежал от опасности   "I escaped from the danger(gen)"

Thus, the following rules were applied in order to handle different types of genitive objects:

1)   *dobj* is used for genitive object with negation, intensional verbs, or similar to partitive

Rus: Он искал поддержки   "He was looking for support(gen)"
dobj(искал, поддержки)

2)   *pobj* is used for genitive object with preposition:

Rus: Цитата из книги   "A quote from a book(gen)"
pobj(из, книги)

Another puzzling issue of Russian morpho-syntax is the genitive of quantification [21]. It appears that if the NP containing a numeral stands in a position which syntax assigns a direct case (nominative or accusative), the case of the noun is assigned by the numeral within the NP, regardless of the syntactic position:

Rus: Я купил пять машин   "I bought five cars(gen)"
Rus: Я купил машину   "I bought a car(acc)"

However, this does not happen when the NP is syntactically assigned and in the oblique case or in case of quantifiers:

Rus: *Я восхищаюсь пятью машинами*   "I admire five cars(*instr*)"
Rus: *Я купил много машин*   "I bought many car(*gen*)"

We treat these cases consistently, i.e. the noun is always the head of the numeral/quantifier, even if the case actually is assigned by the numeral/quantifier to the noun, e.g. *num*(машин, пять), *advmod*(машин, много).

2.6 Slavic Indirect Objects: *iobj* and Noun Head

The presence of cases in Slavic noun morphology creates some challenges for handling different types of objects within the SD model. The Russian language has 6 cases that are also common in other Slavic languages: nominative, accusative, genitive, dative, instrumental, and locative (Russian has almost lost the vocative case that still remains in other languages, e.g. Czech.)[4].

Nominative and accusative forms have corresponding dependency relations, which are covered in the initial version of the SD guidelines [7], i.e. nominal subject (*nsubj*) relation for nominative forms and direct object (*dobj*) for accusative forms. Locative case is used in Slavic languages mainly with prepositions, therefore it is analyzed using the *pobj* relation. Nouns in genitive case, when used as non-prepositional verb-objects, are mostly direct objects, or, if used as noun modifiers, they are mostly *gmod*. Thus, dative and instrumental objects are left uncovered, and it should be decided whether these objects need separate deprel tags as well, or if they can be merged under one generic indirect object (*iobj*) relation.

Since English only has dative indirect objects, the Russian *iobj* relation would be significantly different from the one defined in Stanford typed dependencies, where it is specific to the dative case: 'The indirect object of a VP[5] is the noun phrase which is the [dative] object of the verb' [7]:

She gave me a raise   iobj(gave, me)

Moreover, there are a number of sentences in Russian where dative and instrumental indirect objects are both present, like in the following sentence:

Rus: *Она пишет ему письмо ручкой*   "She writes him a letter with a pen"

Here, both *ему* "him" and *ручкой* "with a pen" will be indirect verb objects, even though they have different functions and correspond to different semantic roles, which can lead to syntactic ambiguity. Verbs that can take two indirect objects are, however, quite rare, and we decided to merge the two object relations into one in order to maintain cross-language consistency. Thus, dependency guidelines for Russian have the following definition for *iobj*: 'The indirect object of a VP is the

---

[4] Some researchers also distinguish so-called partitive genitive (*чашка чаю* "cup of tea"), and second locative cases (*в лесу* "in the forest").

[5] VP stands for a verb phrase. The VP has a verb as its head [29].

noun phrase without prepositions, which is the ablative, dative, or genitive object of the verb. This relation does not cover the genitive object with negated verbs and the genitive object denoting part of the whole that we treat as *dobj*).'

Another important issue related to Russian nominal cases arises from the fact that NPs in oblique cases can depend on other NPs. Consider the following phrase:

Rus: *письмо Лизе* "a letter to Liza(dat)"

As nouns do not normally take arguments, it is hard to say whether in these examples the nouns in oblique cases are objects or modifiers. This problem is sometimes referred to as 'argument-modifier ambiguity/distinction' [18]. Clear cases of noun argument occur in verb nominalization or in a noun with semantics of action, as in *удар кулаком* "a punch with a fist". Even though it does not fully comply with the initial definition of *iobj* and with the concept of object in general, we decided to treat these structures as indirect object relations in order to avoid creating a new deprel.

3   Slavic SD

In chapter 2, we proposed some modifications for the Slavic application of the SD schema. The main modifications are related to a new application of existing deprels, as in the case of *iobj* or *prt*. We also propose the addition of three new labels, two of them replacing some old labels. The addition of the deprel *gmod* is proposed for handling relatively frequent Slavic genitive modifier relations that do not have appropriate representation in the original SD schema. Original labels *attr*, *acomp*, and *cop* are replaced with labels *scomp* and *ocomp* in our model for improving consistency in labeling verbal complements. Also labels *abbrev* and *rel* are removed following proposals in [9]. Deprel *possessive* is removed, since possessive marker is not used in Slavic languages and this dependency relation is therefore obsolete. For deprel *ref*, we decided to analyze clause referents using deprels corresponding to the actual function of referent words in the sentence, so we analyze e.g. pronouns in relative clauses using the deprels *nsubj* or *dobj* depending on the internal head of the relative clause. Other deprels are used consistently with the original SD model [7], for the full set used in our model, see Table 1 below. The data annotated during the course of this project has become a part of the Universal Dependency Treebank (GSD[6]), and as such, some harmonization rules are applied [16]. The Universal SD (USD) presented in [10] brings some further cross-language consistency applications for multilingual treebanks. In order to compare between these related SD models and identify the differences between them, Table 1 also contains corresponding deprels used in GSD and USD. Our project-specific labels can be easily converted to match with the GSD or USD models, to achieve further consistency between different treebanks.

---

[6] We use GSD and USD as they are used in [10].

Table 1. List of deprels for Slavic SD application and corresponding GSD and USD deprels labels

Deprel	Gloss	GSD	USD
advcl	Adverbial clause modifier	advcl	advcl
advmod	Adverbial modifier	advmod	advmod
agent	Agent	adpmod	case
amod	Adjectival modifier	amod	amod
appos	Apposition	appos	appos
aux	Auxiliary verb	aux	aux
auxpass	Passive auxiliary	auxpass	auxpass
cc	Coordinating conjunction	cc	cc
ccomp	Clausal complement	ccomp	ccomp
complm	Clausal complement marker	mark	mark
conj	Conjunct	conj	conj
csubj	Clausal subject	csubj	csubj
csubjpass	Passive clausal subject	csubjpass	csubjpass
dep	Undetermined Dependent	dep	dep
det	Determiner	det	det
dobj	Direct object	dobj	dobj
emot	Emoticon	dep	dep
expl	Expletive	expl	expl
gmod	Genitive modifier	poss	poss
infmod	Infinitival modifier	infmod	nfincl
interj	Interjection	dep	dep
iobj	Indirect object	iobj	iobj
mark	Marker	mark	mark
mwe	Multi-word expression	mwe	mwe
neg	Negative particle	neg	neg
nn	Noun compound modifier	compmod	compound/name
npadvmod	NP adverbial modifier	nmod	nmod
nsubj	Nominal subject	nsubj	nsubj
nsubjpass	Passive nominal subject	nsubjpass	nsubjpass
num	Numeric modifier	num	nummod
number	Element of compound number	num	nummod
ocomp	Object complement	acomp/attr/cop	cop/xcomp
p	Punctuation	p	punct
parataxis	Parataxis	parataxis	parataxis
partmod	Participial modifier	partmod	nfincl
pcomp	Prepositional complement	adpcomp	ncmod
pobj	Prepositional object	adpobj	nmod
poss	Possession modifier	poss	poss
preconj	Preconjunct	cc	preconj
predet	Predeterminer	det	predet
prep	Prepositional modifier	adpmod	case
prt	Phrasal verb particle	prt	prt
purpcl	Purpose clause modifier	advcl	advcl
quantmod	Quantifier phrase modifier	advmod	advmod
rcmod	Relative clause modifier	rcmod	relcl
root	Root	root	root
scomp	Subject complement	acomp/attr/cop	cop/xcomp
tmod	Temporal modifier	advmod	advmod/tmod
xcomp	Open clausal complement	xcomp	xcomp

4   Conclusion

In this paper, we have presented an updated version of the SD schema for Slavic languages application. In order to keep cross-language consistency, we avoid introducing major changes to the schema that was already used for producing data resources for several languages. Proposed changes are applied using existing dependency labels for new language features that were not present in languages based on which the original SD schema was developed. Several new labels are introduced to improve consistency in handling verbal complements and in the case of *gmod*, the new label handles a language-specific feature that was not taken into account in the original SD model. In terms of consistency with existing SD treebanks, these new labels can be easily converted to labels used by other SD resources and vice versa. The updated SD schema expands coverage potential for SD treebanks and can be used both when building new SD based treebanks as well as for converting existing dependency treebanks into the SD schema.

References

1. Boguslavsky, I., Grigorieva, S., Grigoriev, N., Kreidlin, L., Frid, N.: Dependency treebank for Russian: Concept, tools, types of information. In: Proceedings of the 18th Conference on Computational Linguistics, vol. 2. Association for Computational Linguistics (2000)
2. Borschev, V., Partee, B., Paducheva, E., Testelets, Y., Yanovich, I.: Russian genitives, non-referentiality, and the property-type hypothesis (2007)
3. Bosco, C., Montemagni, S., Simi, M.: Converting Italian Treebanks: Towards an Italian Stanford Dependency Treebank. In: Proceedings of the 51st Annual Meeting of the Association for Computational Linguistics (2013)
4. Buchholz, S., Marsi, E.: CoNLL-X shared task on multilingual dependency parsing. In: Proceedings of the 10th Conference on Computational Natural Language Learning. Association for Computational Linguistics (2006)
5. Chang, P., Tseng, H., Jurafsky, D., Manning, C.: Discriminative reordering with Chinese grammatical relations features. In: Proceedings of the Third Workshop on Syntax and Structure in Statistical Translation (SSST-3) at NAACL HLT (2009)
6. Comrie, B., Corbett, G.: The Slavonic Languages. Taylor & Francis (2002)
7. de Marneffe, M., Manning, C.: Stanford typed dependencies manual (2008)
8. de Marneffe, M., Connor, M., Silveira, N., Borschev, V.: Genitives, relational nouns, and the argument-modifier distinction. In: ZAS Papers in Linguistics, vol. 17, pp. 177–201 (2000)
9. de Marneffe, M., Connor, M., Silveira, N., Bowman, S.R., Dozat, T., Manning, C.: More constructions, more genres: Extending Stanford Dependencies. In: Proceedings of the Second International Conference on Dependency Linguistics, DepLing 2013 (2013)
10. de Marneffe, M., Dozat, T., Silveira, N., Haverinen, K., Ginter, F., Nivre, J., Manning, C.: Universal Stanford Dependencies: A Cross-linguistic typology. In: Proceedings of LREC (2014)
11. Downing, A., Locke, P.: A university Course in English Grammar. Prentice Hall (1992)
12. Džeroski, S., Erjavec, T., Ledinek, N., Pajas, P., Žabokrtský, Z., Žele, A.: Towards a Slovene dependency treebank. In: Proceedings of the 5th International Conference on Language Resources and Evaluation (2006)

13. Hajič, J., Vidová-Hladká, B., Pajas, P.: The Prague dependency treebank: Annotation structure and support. In: Proceedings of the IRCS Workshop on Linguistic Databases (2001)
14. Hajič, J., Ciaramita, M., Johansson, R., Kawahara, D., Martí, M., Màrquez, L., Meyers, A., Nivre, J., Padó, S., Štepánek, J., Straňák, P., Surdeanu, M., Xue, N., Zhang, Y.: The CoNLL-2009 shared task: syntactic and semantic dependencies in multiple languages. In: Proceedings of the 13th Conference on Computational Natural Language Learning. Association for Computational Linguistics, Boulder, Colorado (2009)
15. Haverinen, K., Viljanen, T., Laippala, V., Kohonen, S., Ginter, F., Salakoski, T.: Treebanking Finnish. In: Proceedings of TLT9, pp. 79–90 (2010)
16. McDonald, R., Nivre, J., Quirmbach-Brundage, Y., Goldberg, Y., Das, D., Ganchev, K., Hall, K., Petrov, S., Zhang, H., Täckström, O., Bedini, C., Bertomeu, C.N., Lee, J.: Universal Dependency Annotation for Multilingual Parsing. In: Proceedings of the 51st Annual Meeting of the Association for Computational Linguistics (2013)
17. Nivre, J., Hall, J., Kübler, S., McDonald, R., Nilsson, J., Riedel, S., Yuret, D.: The CoNLL 2007 shared task on dependency parsing. In: Proceedings of EMNLPCoNLL (2007)
18. Partee, B., Borschev, V.: Genitives, relational nouns, and the argument-modifier distinction. In: ZAS Papers in Linguistics, vol. 17, pp. 177–201 (2000)
19. Paus, C.: Social and pragmatic conditioning in the demise of the Russian partitive case. Russian Linguistics 18(3), 249–266 (1994)
20. Petrov, S., Das, D., McDonald, R.: A universal part-of-speech tagset. In: Proceedings of LREC (2012)
21. Rappaport, G.: Numeral phrases in Russian: A minimalist approach. Journal of Slavic Linguistics 10(1-2), 327–340 (2002)
22. Seraji, M., Megyesi, B., Nivre, J.: Dependency Parsers for Persian. In: Proceedings of 10th Workshop on Asian Language Resources, COLING 2012, 24th International Conference on Computational Linguistics, Mumbai, India (2012)
23. Souček, M., Järvinen, T., LaMontagne, A.: Managing a Multilingual Treebank Project. In: Proceedings of the Second International Conference on Dependency Linguistics (2013)
24. Surdeanu, M., Johansson, R., Meyers, A., Màrquez, L., Nivre, J.: The CoNLL-2008 shared task on joint parsing of syntactic and semantic dependencies. In: Proceedings of CoNLL (2008)
25. Sussex, R., Cubberley, P.: The Slavic languages. Cambridge University Press, Cambridge (2006)
26. Tadić, M.: Building the Croatian Dependency Treebank: the initial stages. Suvremena Linguistica (2007)
27. Timberlake, A.: Reference conditions on Russian reflexivization. Language, 777–796 (1980)
28. Tsarfaty, R.: A Unified Morpho-Syntactic Scheme of Stanford Dependencies. In: Proceedings of ACL (2013)
29. Wardhaugh, R.: Understanding English Grammar: A Linguistic Approach. Blackwell (1995)
30. Wróblewska, A., Woliński, M.: Preliminary experiments in Polish dependency parsing. In: Bouvry, P., Kłopotek, M.A., Leprévost, F., Marciniak, M., Mykowiecka, A., Rybiński, H. (eds.) SIIS 2011. LNCS, vol. 7053, pp. 279–292. Springer, Heidelberg (2012)

Towards a Weighted Induction Method
of Dependency Annotation

Alina Wróblewska and Adam Przepiórkowski

Institute of Computer Science, Polish Academy of Sciences,
ul. Jana Kazimierza 5, 01-248 Warsaw, Poland
{alina,adamp}@ipipan.waw.pl
http://zil.ipipan.waw.pl/

Abstract. This paper presents a method of annotating sentences with
dependency trees which is set within the mainstream of the study on
dependency projection. The approach builds on the idea of weighted
projection. However, we involve a weighting factor not only in the process of projecting dependency relations (weighted projection) but also in
the process of acquiring dependency trees from projected sets of dependency relations (weighted induction). Using a parallel corpus, its source
side is automatically annotated with a syntactic parser and resulting dependencies are transferred to equivalent target sentences via an extended
set of word alignment links. Projected relations are initially weighted according to the certainty of word alignment links used in projection. Since
word alignments may be noisy and we should not entirely rely on them,
initial weights are thus recalculated using a version of the EM algorithm.
Then, maximum spanning trees fulfilling properties of well-formed dependency structures are selected from EM-scored directed graphs. An extrinsic evaluation shows that parsers trained on induced trees perform
comparably to parsers trained on a manually developed treebank.

Keywords: Dependency annotation, cross-lingual projection, weighted
induction.

1 Introduction

Supervised methods are very well-established in data-driven dependency parsing
and they give the best results so far. However, the manual annotation of training
data required by supervised frameworks is a very time-consuming and expensive
process. For this reason, intensive research has been conducted on unsupervised
grammar induction. However, performance of unsupervised dependency parsers
is still significantly below performance of supervised systems. Moreover, performance of unsupervised parsers is also substantially below performance of systems
based on cross-lingual projection methods [17].

The cross-lingual projection method has been successfully applied to various
levels of linguistic analysis and corresponding NLP tasks. An important area
of applying annotation projection is dependency tree projection and parser induction. Experiments with dependency projection were pioneered by [10], who

A. Przepiórkowski and M. Ogrodniczuk (Eds.): PolTAL 2014, LNAI 8686, pp. 164–176, 2014.

assume that dependencies in one language directly map to dependencies in another language. In order to acquire well-formed dependency trees Hwa and her colleagues apply additional smoothing techniques and aggressive filtering methods. Other research was conducted on projecting only reliable relations and training parsers on partial dependency structures [12,24]. There are also some constraint-driven learning approaches [8,22] which apply projected information to constrain estimation of dependency parsing models. Other related approaches consists in transferring delexicalised parsers between languages [31,17,23] or in multi-source cross-lingual transferring of parsers [17,23,18,26].

The cross-lingual dependency projection may be an alternative method of annotating sentences with dependency trees in less researched languages. The method builds on the assumption that a dependency tree encoding the predicate-argument structure of a sentence largely carries over to its translation since an integrated valency component determines both the number and the kind of complement slots of verbs, nouns, adjectives, etc., and these complement slots are relatively invariant across languages. Furthermore, dependency projection does not take into account the order of words, so it is thus perfectly suited for languages with different word orders.

The main idea behind dependency projection is to automatically parse source sentences and to project acquired dependency trees to equivalent target sentences. Since relations encoded in dependency trees connect tokens, projection of these relations may be sufficiently guided by word alignment which links corresponding tokens in parallel sentences. In the ideal case, projected dependencies constitute valid dependency structures of target sentences.

This paper describes a novel method of annotating Polish sentences with dependency trees which is set within the mainstream of the study on dependency projection. The approach builds on the idea of weighted projection [29]. However, we involve a weighting factor not only in the process of projecting dependency relations (*weighted projection*, Section 2.1) but also in the process of acquiring dependency structures from projected sets of dependency relations (*weighted induction*, Section 2.2). Using a parallel corpus, its English side is automatically annotated with a syntactic parser and resulting dependency relations are transferred to equivalent Polish sentences via an extended set of word alignment links. Projected relations are initially weighted according to the certainty of word alignment links used in projection. Since word alignments may be noisy and we should not entirely rely on them, initial weights are thus recalculated with the EM selection algorithm [7]. Then, maximum spanning trees fulfilling properties of well-formed dependency structures are selected from EM-scored directed graphs. An extrinsic evaluation shows that parsers trained on induced trees perform comparably to parsers trained on a manually developed treebank (Section 3). The novelty of the method proposed here consists in involving a weighting factor in the process of inducing dependency trees.

2 Weighted Induction Method

The weighted induction procedure consists of two successive processes – *projection* of dependency relations followed by *induction* of dependency trees from projected directed graphs (henceforth digraphs). Induced trees are treated as correctly built unlabelled dependency structures.

2.1 Weighted Projection

This section describes weighted projection which is the first step in the entire process of acquiring valid Polish dependency structures. According to the main idea behind weighted projection, arcs making up an English dependency tree are projected via an extended set of word alignment links between English and Polish tokens (*bipartite alignment graph*). Since we aim to project English relations which are restricted to sentence boundaries, only word alignment links within a pair of aligned parallel sentences are considered in projection. Projected arcs constitute initially weighted digraphs.

Bipartite Alignment Graph. Instead of projection only via automatic word alignment links, English dependencies are transferred via a set of links gathered from different automatic word alignments and extended with some additional links. This set of links constitutes a *complete bipartite alignment graph $BG :=$ $(V_{en} \cup V_{pl}, E)$*, for $E = V_{en} \times V_{pl}$. Vertices in BG are decomposed into two disjoint sets V_{en} and V_{pl} corresponding to English tokens with a ROOT node and Polish tokens with a ROOT node respectively. Every pair of vertices from V_{en} and V_{pl} is adjacent. Bipartite edges are weighted with the function $w :$ $E \to \{0, 1, 2, 3\}$. The weight w indicates the certainty of an edge and either corresponds to the number of occurrences of this edge in automatic alignment sets or is equal to 0 if it is not present in any alignment set. There are three word alignment sets: two unidirectional alignments and a set of bidirectional alignment links symmetrised with the *grow-diag-final-and* heuristic.[1] The edge between ROOT nodes, which is not present in any set of alignment links, is scored with 1.[2] Weighted bipartite alignment graphs built for each sentence pair are used to project English dependency arcs to Polish sentences.

Projection of Dependency Arcs. English dependency arcs are projected to corresponding Polish tokens via bipartite edges according the following procedure. The projection module takes as input a weighted bipartite alignment graph

---

[1] To improve the word alignment quality and overcome limitation of alignment scenarios, a method of unidirectional alignments symmetrisation was proposed by [19]. Next to union and intersection, there exist some more sophisticated symmetrisation concepts, e.g., the *grow-diag-final-and* symmetrisation method described in [14].

[2] Scoring the edge between a Polish ROOT and an English ROOT with 1 will guarantee the connectedness of projected digraph.

BG and an English dependency tree T_{en}. The English tree $T_{en} = (V_{en}, A_{en})$ consists of a set of vertices $V_{en} = \{u_0, u_1, ..., u_m\}$, where u_0 corresponds to the ROOT node, and a set of arcs $A_{en} \subseteq \{(u', u, gf)|u', u \in V_{en}, gf \in GF\}^3$ labelled with grammatical functions from GF.

Arcs of the English tree T_{en} are then iteratively projected to the Polish equivalent sentence. For each Polish lexical node v, its governor node v' is found in the following way. First, an English non-ROOT node u connected with v is looked for in the bipartite alignment graph BG. Then, the governor node u' of u is found in the English tree T_{en}. Finally, the Polish node v' which is connected with u' in BG is identified and recognised as the governor of v.

An arc (v', v, l) between Polish tokens v' and v, which is assigned the label l, can be added to the Polish digraph. The only restriction is that it is not possible to project arcs via bipartite edges which are both weighted with 0. The reason for this limitation is to avoid projection of arcs considered to be the most error prone. However, projection via two edges one of which is weighted with 0 is permitted in order to cover relations between English tokens one of which is not aligned with any Polish token in any of word alignment sets.

For each sentence pair the projection module outputs the set of Polish vertices V_{pl} and the set of arcs A_{pl} between these vertices. The set of vertices $V_{pl} = \{v_0, v_1, ..., v_n\}$ consists of an additional ROOT node v_0 and a set of lexical nodes $\{v_1, ..., v_n\}$, for each vertex v_i corresponding to the ith token of a Polish sentence $S = t_1, ..., t_n$. The vertices from V_{pl} are connected with arcs from the set $A_{pl} \subseteq \{(v_i, v_j, l)|v_i, v_j \in V_{pl}, l = (w_d, w_g, gf, f)\}$, for $w_d, w_g \in \{0, 1, 2, 3\}$, $gf \in GF, f \in \mathbb{N}_+$. These two sets constitute a Polish digraph in which each node directly or indirectly depends on the ROOT node and the ROOT node does not have any predecessor.

Any projected arc is assigned a label $l = (w_d, w_g, gf, f)$. The first element w_d refers to the weight of a bipartite edge connecting English and Polish tokens with the dependent status. The second element w_g refers to the weight of a bipartite edge connecting English and Polish tokens with the governor status. The third element gf indicates the label of the projected English dependency relation. The projection frequency f indicates the number of English relations labelled with the same grammatical function which are projected to the same two Polish tokens via equally weighted bipartite edges.

Intuitive Weighting of Projected Arcs. Intuitively, an arc between two tokens might be more important than arcs between other tokens if it is projected via bipartite edges with higher scores. Projected arcs are thus scored with initial weights that are estimated based on scores of bipartite edges (w_d and w_g) used in the projection of a particular arc and a projection frequency f. We define the following function $s(v_i, v_j, (w_d, w_g, gf, f)) = w_d + w_g + 2w_dw_gf$ scoring projected arcs. Initially weighted projected digraphs (or even multi-digraphs since English

---

[3] The arc (u', u, gf) indicates an edge directed from u' to u and labelled with the grammatical function gf.

arcs are projected via all possible pairs of bipartite edges) provide a starting point to induce final dependency trees.

2.2 Weighted Induction

This section presents weighted induction which is the second step in the process of acquiring Polish dependency structures. The main idea behind weighted induction is to identify the most likely arcs in initially scored projected digraphs and to assign them appropriate weights. Using methods of selecting maximum spanning trees from weighted directed graphs, final well-formed dependency structures, i.e., *maximum spanning dependency trees* (MSDTs), are inferred from weighted projected digraphs. A maximum spanning dependency tree $T = (V', A')$ extracted from a weighted projected digraph $G = (V, A)$, for $A' \subseteq A$ and $V' = V = \{v_0, v_1, ..., v_n\}$, where v_i corresponds to the ith token of a sentence $S = t_1, ..., t_n$ and v_0 is a ROOT node, corresponds to a valid dependency structure if v_0 is the root of T, i.e., $(v_i, v_0, l) \notin A'$, for $v_i \in V'$, $l \in L$, and v_0 has only one successor, i.e., if $(v_0, v_i, l) \in A'$, then $(v_0, v_j, l') \notin A'$, for $v_i \neq v_j$.

Arcs of projected digraphs are assigned initial weights calculated on the basis of weights of bipartite edges used in projection of these arcs. Weights of bipartite edges, in turn, result from automatic word alignment which is prone to errors. We therefore propose a heuristic of recalculating initial arc weights in projected digraphs. The recalculation applies the probability distribution over arcs in k-best MSDTs selected from initially weighted projected digraphs. The probability distribution over selected arcs identified by their feature representations is estimated using the EM-inspired selection algorithm. Projected digraphs with recalculated arc weights are used to induce final dependency structures. A schema of the weighted induction procedure is shown in Figure 1.

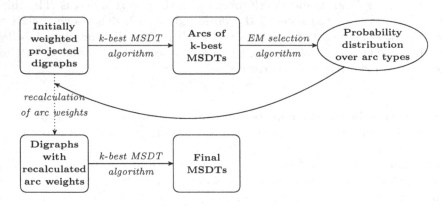

Fig. 1. Schema of the weighted induction procedure

K-best MSDTs. The induction procedure starts with the selection of k-best MSDTs from projected digraphs using a slightly modified version of the k-best MSTs selection algorithm by [4].[4] A list of k-best MSDTs of the digraph G is computed with the function *rank* of this algorithm. This function is slightly modified in relation to the original function *rank* which outputs a list of k-best MSTs (c.f., [4], p. 107). In our version, some additional conditions are imposed on candidate MSTs so that they meet properties of well-formed dependency trees. Since not all MSTs fulfil these properties, only valid MSDTs are taken into account in estimation of the probability distribution.

Feature Representations of Arcs. Arcs used in the EM training are represented with their features. Each of related nodes represents a token in a sentence and encodes information about the token's lemma, part of speech tag, and morphological features. Furthermore, any arc is assigned a label and an initial weight. The information available in the arc label and in the related nodes may be used in the feature representation j of this arc. The set of features identifying an arc is given with the function fr, i.e., $fr(v_h, v_i, (w_d, w_g, gf, f)) = j$.

Probability Distribution over Arc Types. The probability distribution over arc types is estimated with a version of the EM algorithm defined by [7]. This EM selection algorithm was originally designed to select the most probable valency frames from sets of valency frame candidates. Dębowski's algorithm is adapted for our purposes of identifying the most reliable arcs in sets of arcs in k-best MSDTs found in initially weighted projected digraphs.

Assume we have a training set $\mathcal{B} = \{B_1, ..., B_N\}$, where B_i is a set of arcs in k-best MSDTs coming into the ith vertex, for $i = 1, ..., N$ and N being a number of all nodes in k-best MSDTs. In this setting, the EM selection algorithm estimates model parameters $\theta_t = \left(p_j^{(t)}\right)_{j \in J}$, where t is the iteration number, for $t = 2, ..., T$, and j is a feature representation of an arc, for $j \in J$ and J being a set of all possible arc types in $\mathcal{B}$. The EM-inspired selection algorithm iterates over the formulae in (1) and (2) and defines a series of parameter values $\theta_2, ..., \theta_t$ until the last iteration. In the first step of each iteration, new parameter values $p_j^{(t)} = P(j|\theta_t)$ are estimated. The second step of each iteration is to estimate values $p_{ij}^{(t)} = P(j|B_i, \theta_t)$, for each $i = 1, ..., N$ and for each possible arc type $j \in J$. In the original version of the EM selection algorithm, the coefficient p_{ij} is a quotient of the probability value p_j of an arc with the type j and the sum of probability values $p_{j'}$ of all arcs in B_i. We modify the way of estimating the coefficient p_{ij} in order to take into account the initial weight $s(v_h, v_i, l)$ of the arc with the type j.

$$p_j^{(t+1)} = \frac{1}{N} \sum_{i=1}^{N} p_{ij}^{(t)} \qquad (1)$$

---

[4] The algorithm by [4] was used for other task related to some extent to our approach, e.g., for re-ranking of parses [9].

$$p_{ij}^{(t)} = \begin{cases} \dfrac{p_j^{(t)} \times s(v_h, v_i, l)}{\sum\limits_{j' \in B_i} p_{j'}^{(t)} \times s(v_{h'}, v_i, l')} & , \text{if } fr(v_h, v_i, l) = j \\ 0 & , \text{otherwise.} \end{cases} \tag{2}$$

The initial parameter values are set to 1, i.e., $p_j^{(1)} = 1$, as in the original approach by [7]. At each iteration, the new parameter values θ_t are calculated as a function of the previous parameter values θ_{t-1} and the training set $\mathcal{B}$. The EM-inspired selection algorithm iterates until the final iteration T is reached.

According to the original procedure by [7], the most likely arc would be selected from the set of possible arcs B_i. However, the most probable incoming arcs for each lexical node do not have to necessarily constitute a valid dependency tree (e.g., a resulting graph may contain a cycle). Therefore, our approach to recalculating weights does not build directly on the selected arcs but on the probability distribution $\left(p_j^{(T)}\right)$ over feature representations of arcs J estimated in the last iteration of the EM selection algorithm.

Recalculation of Arc Weights in Projected Digraphs. The new weight of an arc (v_h, v_i, l) with the feature representation j is calculated as the product of the square root[5] of the previous arc weight and the value p_j (see Equation 3).

$$s^* = \sqrt{s(v_h, v_i, l)} \times p_j, \qquad \text{for } fr(v_h, v_i, l) = j \tag{3}$$

If an arc is not present in any of k-best MSDTs, its probability value is equal to 0. Because there is a risk that some digraph arcs would be assigned 0 and they would have the same priority in the extraction of final MSDTs, their scores s^* are calculated as the product of the square root of the initial arc weight, the lowest value p_j in $\left(p_j^{(T)}\right)$ and an optimisation factor α which further decreases weights of unselected arcs (see Equation 4).

$$s^* = \sqrt{s(v_h, v_i, l)} \times \min_j p_j \times \alpha, \qquad \text{for some } 0 < \alpha < 1 \tag{4}$$

The main idea behind the recalculation is to reward arcs with the probability greater than zero by assigning them higher weights, and to penalise other arcs by assigning them lower weights. Arcs with higher weights are more likely to be selected as part of final dependency trees.

3   Experiments and Evaluation

To test the method outlined above, we conduct an experiment consisting in the projection of English dependency relations to Polish sentences and in the in-

---

[5] Arcs in projected digraphs are assigned initial weights from $\mathbb{N}_+$. In order to diminish the difference between initial weights and probability values, and therefore to raise the importance of relatively low probability values, initial weights are square rooted.

duction of Polish dependency trees. Since there is no Polish-English parallel corpus annotated with gold-standard dependency trees, we may evaluate neither the induction procedure itself nor the quality of induced trees. Instead, we perform an extrinsic evaluation to see to what extent induced trees affect performance of a parser trained on them.

3.1 Data and Preprocessing

The experiment is conducted on a large collection of Polish–English bitexts gathered from publicly available sources: *Europarl* [13], *DGT-Translation Memory* [25], *OPUS* [27] and *Pelcra Parallel Corpus* [21]. After tokenisation, sentence segmentation and sentence alignment, bitexts are used to produce automatic word alignment links using the statistical machine translation system MOSES [15]. Three sets of alignment links are generated: Polish-to-English, English-to-Polish and a set of links from both unidirectional alignments selected with the *grow-diag-final-and* method implemented as part of the MOSES system.

To parse the English side of the parallel corpus, we use the handcrafted wide-coverage English *Lexical Functional Grammar* [6,3], using the *Xerox Linguistic Environment* [5] as a processing platform.[6] The most probable LFG analyses are converted into dependency trees using a conversion procedure similar as in [20]. The conversion of permitted LFG analyses results in a collection of 4,946,809 English dependency trees, which constitute the subject matter of projection.

3.2 Automatic Induction of Polish Dependency Trees

Given three sets of word alignment links, English dependency trees, and Polish sentences enriched with morphosyntactic information using the *Pantera* tagger [1], the projection module (see Section 2.1) outputs 4,946,809 initially weighted digraphs. Then, the induction module (see Section 2.2) extracts 4,615,698 sets of k-best MSDTs (for $k = 10$) from the entire set of initially weighted digraphs, estimates the probability distribution over arc types in these k-best MSDTs within 10 iterations of the EM selection algorithm, recalculates initial arc weights in projected digraphs and acquires final MSDTs from these digraphs.

Since the final MSDTs are labelled with English grammatical functions, we treat them as unlabelled dependency structures at this point. Arcs in these

---

[6] Similarly as publicly available data-driven dependency parsers, the *XLE* parser for English may deal with some ungrammatical sentences. It applies the shallow parsing (or chunking) technique to identify well-formed chunks (constituents) in a problematic sentence and then composes them linearly into a FIRST-REST structure marked as FRAGMENTS. Hence the English *XLE* parser marks dubious sentences as FRAGMENTS in contrast to some other parsers which do not distinguish proper sentences from problematic strings of tokens. Since the aim of the current experiment is to build a bank of Polish dependency structures for strings of tokens considered as well-formed sentences or phrases, analyses marked as FRAGMENTS are not taken into account in projection.

unlabelled dependency trees are then assigned Polish dependency labels derived from projected English grammatical functions and morphosyntactic features of related Polish tokens using a set of predefined labelling rules. The entire induction procedure outputs a bank of 3,958,556 labelled dependency structures on which a Polish dependency parser may be trained.

3.3 Evaluation Experiment

We use the *Mate* system [2] in our evaluation experiment. The performance of the *Mate* parser trained on automatically induced trees is evaluated against a set of 822 dependency trees (*manual test*) taken from the Polish dependency treebank [28].[7] Furthermore, we provide a version of these test trees with automatically generated part of speech tags and morphological features (*automatic test*). In addition to these test sets, the parser is evaluated against a set of 100 relatively complex trees (*additional test*).[8]

Table 1 reports results of the *Mate* parser trained on induced dependency trees.[9] Parsing performance is measured with two evaluation metrics: *unlabelled attachment score* (UAS) and *labelled attachment score* (LAS) as defined by [16]. These results are compared with the performance of a *supervised* parser trained on a part of the Polish treebank.

A parser trained on automatically induced trees (*induced*) in one iteration[10] achieves 73.7% UAS if tested against the manual test set, 72.8% UAS if tested against the automatic test trees and 63.5% UAS if tested against the additional test trees. These results are significantly below the performance of a parser (*supervised*) trained on trees from the Polish dependency treebank.

---

[7] [28] provide a detailed description of the schema used to annotate Polish sentences with dependency tree representations.

[8] Additional test sentences were randomly selected from some Polish newspapers. The selected sentences are quite long and contain 15.3 tokens per sentence on average. They were first automatically tokenised, lemmatised and part of speech tagged, and then manually annotated with dependency trees by two experienced linguists. These linguists also corrected possible errors in lemmatisation and tagging, but not discrepancies in tokenisation.

[9] The reported experiment was preceded by a preliminary experiment. This experiment consisted in comparing performance of dependency parsers trained on two sets of MSDTs selected from a limited set of 1.1 million projected digraphs: (1) 1,000,797 MSDTs selected from projected digraphs with initially weighted arcs, and (2) 924,733 MSDTs selected from projected digraphs with EM-recalculated arc weights. According to the preliminary results the parser trained on the trees with EM-recalculated arc weights outperformed the baseline parser trained on the initially scored trees by 2.4 percentage points.

[10] Preliminary experiments show that the parsing performance decreases with the increasing number of iterations used to train the *Mate* parser. The decrease in parsing performance may be due to noise which is learnt in successive iterations. Therefore, we limit the number of *Mate* iterations to one.

Table 1. Performance of parsers trained with the *Mate* parsing system on the Polish dependency trees acquired with the weighted induction method (*induced*), induced and labelled (*labelled*), labelled and modified (*modified*), and labelled, modified and filtered (*filtered*). Settings of model training: one iteration, the heap size of 100 million features, the threshold of the non-projective approximation of 0.2. The *supervised* model is trained on 7405 trees from the Polish dependency treebank. Setting of supervised model training: 10 iterations, the heap size of 100M, the threshold of 0.2. Validation data sets: *Manual Test* – the set of 822 treebank trees; *Automatic Test* – the set of 822 treebank trees with automatic morphosyntactic annotations of tokens; *Additional Test* – the set of 100 sentences manually annotated with dependency trees.

Model	Data	Manual Test		Automatic Test		Additional Test	
		UAS	LAS	UAS	LAS	UAS	LAS
induced	3958556	73.7	–	72.8	–	63.5	–
labelled	3958556	74.6	69.4	74.0	68.1	63.7	58.3
modified	3958556	85.1	79.2	84.0	77.3	74.3	68.5
filtered	2352940	86.0	80.5	84.7	78.3	**76.1**	**70.3**
supervised	7405	**92.7**	**87.2**	**88.4**	**81.0**	76.0	69.5

Following Hwa's idea of improving automatically induced trees, we define 45 labelling rules and 31 correction rules.[11] Even if the induction process seems to be straightforward, there are still some Polish-specific morphosyntactic phenomena or linguistic structures diversely annotated in both languages the annotation of which may not result from the English dependency tree. The *Mate* parser trained on induced trees labelled with Polish dependency types (*labelled*) achieves 74.6% UAS and 69.4% LAS if tested against the manual test trees, 74% UAS and 68.1% LAS if tested against the automatic test trees and 63.7% UAS and 58.3% LAS if tested against the additional test trees. The *Mate* parser trained on induced dependency trees modified with predefined rules performs significantly better – 85.1% UAS and 79.2% LAS if tested against the manual test trees, 84% UAS and 77.3% LAS if test against the automatic test set and 74.3% UAS and 68.5% LAS if tested against additional test trees. These results are still below parsing performance of the supervised parser. Note, however, that in the third – more realistic – scenario on evaluating the parser on real data, the more useful measure LAS shows that the results of the semi-supervised procedure described here are directly comparably to the more costly supervised procedure.

Filtering is one of the most common optimisation techniques in projection-based approaches. Our results show that filtering of possibly incorrect trees does not contribute significantly to improving parsing performance since only two simple filtering criteria are used: percentage of non-projective arcs and percentage of arcs labelled with a default function *dep*. The best parsing results (*filtered*) are achieved if we reject trees with more than 30% of non-projective arcs and with more than 10% of *dep*-labelled arcs – 86% UAS and 80.5% LAS if tested against the manual test set, 84.7% UAS and 78.3% LAS if tested against

---

[11] Due to lack of space, a detailed presentation of all individual rules is not possible here. A general presentation of labelling and correction rules may be found in [30].

automatic trees and 76.1% UAS and 70.3% LAS if tested against the additional trees. The results of evaluation against the additional trees show that dependency parsers developed in an automatic way as described here may rival fully supervised – and, hence, more costly – parsers.

4 Conclusion

This paper presented a novel weighted induction method of obtaining Polish dependency structures. The weighted induction procedure consists of two main steps: projection of dependency relations and induction of well-formed dependency trees. The projection step resembles cross-lingual dependency projection pioneered by [10]. However, it is not required in our approach that projection results in dependency trees as in [10] or partial dependency structures as in [12] or [24]. Instead, all possible dependency arcs are projected and they constitute initially weighted digraphs. Previous approaches do not need any further steps after projection of dependency relations since projected trees (or tree fragments) are considered to be the final data for parser training. In our approach, projected digraphs may contain noisy arcs that should not be used in parser training. We thus proposed a method of recalculating initial arc weights and selecting the final MSDTs from the projected digraphs with recalculated arc weights. The weighted induction method allows to annotate most of sentences with proper dependency trees that could not necessarily be acquired in case of direct projection. Hence the aggressive filtering techniques are not applicable and weighted induction does not lead to a huge loss of data as in [11]. The well-formed induced MSDTs are thus presumably more appropriate than direct projections for parser training.

Results of an extrinsic evaluation consisting in training the *Mate* parser on induced trees are very encouraging. Even if they are mostly a little below the performance of the supervised parser, when tested on a homogenous set of rather short sentences from the treebank on which the parser was trained, a test against a small set of long and complex trees shows that a parser trained on so-induced trees may exceed the supervised upper bound. As this projection-based result was achieved with much less manual work than in the supervised scenario – construction of a few dozen labelling and correction rules as opposed to annotating thousands of sentences – we conclude that for the purpose of developing dependency parsers, the method described here rivals the supervised scenario.

While our experiment considered the Polish-English language pair, the weighted induction method may be applied to obtain dependency structures for other resource-poor languages which do not have any annotated data but have a reasonable number of sentences which are parallel with their translations in a resource-rich language. The weighted induction method was tested on the task of obtaining dependency structures, but it may also apply to other projection tasks, e.g., semantic role labelling or word sense disambiguation.

Acknowledgements. The presented research was supported by grant no POIG.01.01.02-14-013/09 from Innovative Economy Operational Programme co-financed by the European Union (European Regional Development Fund).

References

1. Acedański, S.: A morphosyntactic Brill tagger for inflectional languages. In: Loftsson, H., Rögnvaldsson, E., Helgadóttir, S. (eds.) IceTAL 2010. LNCS (LNAI), vol. 6233, pp. 3–14. Springer, Heidelberg (2010)
2. Bohnet, B.: Very High Accuracy and Fast Dependency Parsing is not a Contradiction. In: Proceedings of the 23rd International Conference on Computational Linguistics, pp. 89–97 (2010)
3. Bresnan, J.: Lexical-Functional Syntax. Blackwell, Oxford (2001)
4. Camerini, P.M., Fratta, L., Maffioli, F.: The K Best Spanning Arborescences of a Network. Networks 10, 91–110 (1980)
5. Crouch, D., Dalrymple, M., Kaplan, R., King, T., Maxwell, J., Newman, P.: XLE Documentation. Palo Alto Research Center (PARC), Palo Alto (2011)
6. Dalrymple, M.: Lexical-Functional Grammar. Syntax and Semantics, vol. 34. Academic Press (2001)
7. Dębowski, Ł.: Valence extraction using EM selection and co-occurrence matrices. Language Resources and Evaluation 43(4), 301–327 (2009)
8. Ganchev, K., Gillenwater, J., Taskar, B.: Dependency Grammar Induction via Bitext Projection Constraints. In: Proceedings of the 47th Annual Meeting of the ACL and the 4th International Joint Conference on Natural Language Processing of the AFNLP, vol. 1, pp. 369–377 (2009)
9. Hall, K.: k-best Spanning Tree Parsing. In: Proceedings of the 45th Annual Meeting of the Association of Computational Linguistics, pp. 392–399 (2007)
10. Hwa, R., Resnik, P., Weinberg, A., Cabezas, C., Kolak, O.: Bootstrapping Parsers via Syntactic Projection across Parallel Texts. Natural Language Engineering 11(3), 311–325 (2005)
11. Jiang, W., Liu, Q.: Automatic Adaptation of Annotation Standards for Dependency Parsing – Using Projected Treebank as Source Corpus. In: Proceedings of the 11th International Conference on Parsing Technologies, IWPT 2009, pp. 25–28 (2009)
12. Jiang, W., Liu, Q.: Dependency Parsing and Projection Based on Word-Pair Classification. In: Proceedings of the 48th Annual Meeting of the Association for Computational Linguistics, pp. 12–20 (2010)
13. Koehn, P.: Europarl: A Parallel Corpus for Statistical Machine Translation. In: Proceedings of the 10th Machine Translation Summit Conference, pp. 79–86 (2005)
14. Koehn, P.: Statistical Machine Translation. Cambridge University Press (2010)
15. Koehn, P., Hoang, H., Birch, A., Callison-Burch, C., Federico, M., Bertoldi, N., Cowan, B., Shen, W., Moran, C., Zens, R., Dyer, C., Bojar, O., Constantin, A., Herbst, E.: Moses: Open Source Toolkit for Statistical Machine Translation. In: Proceedings of the Annual Meeting of the Association for Computational Linguistics, pp. 177–180 (2007)
16. Kübler, S., McDonald, R.T., Nivre, J.: Dependency Parsing. Synthesis Lectures on Human Language Technologies, Morgan & Claypool Publishers (2009)
17. McDonald, R., Petrov, S., Hall, K.B.: Multi-Source Transfer of Delexicalized Dependency Parsers. In: Proceedings of the Conference on Empirical Methods in Natural Language Processing, pp. 63–72 (2011)

18. Naseem, T., Barzilay, R., Globerson, A.: Selective Sharing for Multilingual Dependency Parsing. In: Proceedings of the 50th Annual Meeting of the Association for Computational Linguistics: Long Papers, vol. 1, pp. 629–637 (2012)
19. Och, F.J., Ney, H.: A Systematic Comparison of Various Statistical Alignment Models. Computational Linguistics 29(1), 19–51 (2003)
20. Øvrelid, L., Kuhn, J., Spreyer, K.: Improving Data-Driven Dependency Parsing Using Large-Scale LFG Grammars. In: Proceedings of the 47th Annual Meeting of the ACL and the 4th International Joint Conference on Natural Language Processing of the AFNLP (Conference Short Papers), pp. 37–40 (2009)
21. Pęzik, P., Ogrodniczuk, M., Przepiórkowski, A.: Parallel and spoken corpora in an open repository of Polish language resources. In: Proceedings of the 5th Language & Technology Conference: Human Language Technologies as a Challenge for Computer Science and Linguistics, pp. 511–515 (2011)
22. Smith, D.A., Eisner, J.: Parser Adaptation and Projection with Quasi-Synchronous Grammar Features. In: Proceedings of the 2009 Conference on Empirical Methods in Natural Language Processing, pp. 822–831 (2009)
23. Søgaard, A.: Data point selection for cross-language adaptation of dependency parsers. In: Proceedings of the 49th Annual Meeting of the Association for Computational Linguistics: Human Language Technologies: Short Papers, vol. 2, pp. 682–686 (2011)
24. Spreyer, K.: Does It Have To Be Trees? Data-Driven Dependency parsing with Incomplete and Noisy Training Data. Ph.D. thesis, Universität Potsdam (2011)
25. Steinberger, R., Eisele, A., Klocek, S., Pilos, S., Schlüter, P.: DGT-TM: A freely Available Translation Memory in 22 Languages. In: Proceedings of the 8th International Conference on Language Resources and Evaluation, pp. 454–459 (2012)
26. Täckström, O., McDonald, R., Nivre, J.: Target Language Adaptation of Discriminative Transfer Parsers. In: Proceedings of the Conference of the North American Chapter of the Association for Computational Linguistics: Human Language Technologies, pp. 1061–1071 (2013)
27. Tiedemann, J.: Parallel Data, Tools and Interfaces in OPUS. In: Proceedings of the 8th International Conference on Language Resources and Evaluation, pp. 2214–2218 (2012)
28. Wróblewska, A.: Polish Dependency Bank. Linguistic Issues in Language Technology 7(1), 1–15 (2012)
29. Wróblewska, A., Przepiórkowski, A.: Induction of Dependency Structures Based on Weighted Projection. In: Nguyen, N.-T., Hoang, K., Jędrzejowicz, P. (eds.) ICCCI 2012, Part I. LNCS (LNAI), vol. 7653, pp. 364–374. Springer, Heidelberg (2012)
30. Wróblewska, A., Przepiórkowski, A.: Projection-based Annotation of a Polish Dependency Treebank. In: Calzolari, N., Choukri, K., Declerck, T., Loftsson, H., Maegaard, B., Mariani, J., Moreno, A., Odijk, J., Piperidis, S. (eds.) Proceedings of the Ninth International Conference on Language Resources and Evaluation, LREC 2014, pp. 2306–2312. ELRA, Reykjavík (2014)
31. Zeman, D., Resnik, P.: Cross-Language Parser Adaptation between Related Languages. In: Proceedings of the IJCNLP-08 Workshop on NLP for Less Privileged Languages, pp. 35–42 (2008)

Semantic and Syntactic Model of Natural Language Based on Non-negative Matrix and Tensor Factorization

Anatoly Anisimov, Oleksandr Marchenko,
Volodymyr Taranukha, and Taras Vozniuk

Faculty of Cybernetics, Taras Shevchenko National University of Kyiv, Ukraine
ava@unicyb.kiev.ua, rozenkrans@yandex.ua,
taranukha@ukr.net, taarraas@gmail.com

Abstract. A method for developing a structural model of natural language syntax and semantics is proposed. Factorization of lexical combinability arrays obtained from text corpora generates linguistic databases that are used for analysis of natural language semantics and syntax.

1 Introduction

Recently, the non-negative tensor factorization (NTF) method has become widely used in the natural language processing. From among numerous works in the area of particular interest are two works [1, 2]. They describe models for the tensor representation of the frequency for various types of syntactic word combinations in sentences. After non-negative factorization of tensors such a model allows for successful automatic extraction of specific linguistic structures from a corpus, such as selectional preferences [1] and Verb Sub-Categorization Frames [2], which combine data on syntactic and semantic properties of relations between verbs and their noun arguments in sentences.

The N-dimensional tensors contain estimates for frequency of word combinations sets in text corpora. The model takes into account syntactic positions of words. After large text corpora are processed and sufficient amounts of data are accumulated in the tensor, an N-way array is formed. It contains commutational properties of lexical items in the sentences of natural language. For the words presented in the tensor, the properties include: syntactic relations the word tends to be engaged into, other words in the tensor these relations point to, and frequencies of the corresponding relations. Moreover, these relations are multi-dimensional rather than binary, with N being the maximum number of possible dimensions. Then non-negative factorization for the obtained tensor is performed, which significantly transforms the presentation model. Originally, a multi-dimension tensor is sparse and extensive. Each of the N axes of the syntactic space contains tens of thousands or hundreds of thousands of points that represent words. After the tensor has been factorized, its data are represented as N matrices consisting of k columns (where k is much smaller than the number

A. Przepiórkowski and M. Ogrodniczuk (Eds.): PolTAL 2014, LNAI 8686, pp. 177–184, 2014.
© Springer International Publishing Switzerland 2014

of points in any of the tensor's N dimensions). Parameter k is a degree of factorization, the number of dimensions of the latent semantic space, and the number of attribute dimensions in it. In addition to a more compact data representation, the probability of every possible word combination can be estimated in different syntactic sentence structures. This can be done by calculating the sum of the products of the components for N k-dimensional vectors corresponding to the words chosen from the matrices corresponding, in turn, to their syntactic positions.

The number of dimensions in the tensor restricts the maximum length of sentences and phrases described by this model. Van de Cruys describes a three-dimensional tensor for modeling the syntactic combination: Subject – Verb – Object [1]. Van de Cruys and colleagues describe tensors of 9 and 12 dimensions to simulate up to twenty different types of syntactic relations [2]. The mere increase in the tensor dimension number, however, does not seem to be a good way of improving the model and handling more types of complex syntactic relations. It is quite reasonable, therefore, to look for other universal representation models for syntactical structures. The control spaces [3] have been chosen from among numerous time-tested classic formal models of language syntax representation owing to the fact that in this model an arbitrary complex structure is described using recursion through superposition of two basic syntactic relationships – binary syntagmatic and ternary predicative. The lexical and syntactic tensor model proposed here consists of a 3-dimensional tensor for ternary predicative relations (like Subject – Verb – Object) and a matrix for binary syntagmatic relations (like Noun – Adjective, Verb – Adverb, etc.). Sentences have two types of links: a ternary predicative relation and a closed cyclic dependency (binary syntagmatic). The use of control spaces appears to be an efficient means to reduce arbitrary n-ary syntactic relation to the superposition of binary and ternary relations.

Understanding natural language requires knowledge of language per se (vocabulary, morphology, syntax), and knowledge of the extralinguistic world. The tensor models include data on semantic and syntactic communicative properties only of the words from the texts already processed and only within the sentences and phrases in which these words are used. This paper proposes to use the hierarchical lexical database WordNet to generalize descriptions of communicative properties of words using implicit mechanisms of inheritance by taxonomy tree branches. Assuming a word A belongs to a synset S and has a certain property P, there is a high probability that the other words from S will also have the property P. Also, some words of the children synsets of S will almost certainly have P and words of the parent synsets of S are also likely to have P. These assumptions underpin the implementation of the generalization mechanism that describes communicative semantic and syntactic properties of words applying the principle of taxonomic inheritance.

The training set contains texts from The Wall Street Journal (WSJ) corpus, along with the English Wikipedia and the Simple English Wikipedia articles.

The latter two contain the definitions and basic information about concepts, which enhances semantics in the model.

2   Lexical-Syntactic Model of Natural Language

In order to construct a semantic-syntactic model of natural language, a method for automatic filling the three dimensional tensor F(for ternary predicative relations) and the matrix D (for cyclic binary dependencies) was designed. The method calls for the following steps:

- Sentences from a large corpus are taken and parsed by the Stanford Parser module, which generates the syntactic structures of sentences in the form of dependency trees and parse trees for phrase structure grammar [4, 5].
- The program examines the dependency tree and the CFG parse tree of the current sentence. It constructs the control space of the syntactic structure, analyzing relations between corresponding words to identify predicate combinations of length 3 (e.g., Subject – Verb – Object, etc.) and cyclic binary combinations of length 2 (Noun – Adjective, Verb – Adverb, etc.).
- In the control space of this sentence for every triad of points (i, j, k) connected with the ternary predicative sequence of links, in tensor F the cell $F[I, J, K]$ receives the value: $F[I, J, K] = F[I, J, K] + 1$. The coordinates I, J, K of the tensor cell correspond to pairs (w_i, A_i), (w_j, A_j) and (w_k, A_k), where w means words that are lexical values of the corresponding points (i, j, k), and A is a coded description of the characteristics of these words (part of speech, gender, number of lexical units, etc.).
- Similarly, in the control space of the syntactic structure of the current sentence for each pair of points (i, j) interconnected with the cyclic binary link, in matrix D the cell $D[I, J]$ is set to: $D[I, J] = D[I, J] + 1$.

The extremely large dimension and sparsity of matrix D and tensor F demand for non-negative matrix and tensor factorization in order to store the data in a more economical way. Matrix D is factored using Lee and Seung Non-negative Matrix Factorization algorithm [6] that decomposes matrix $D(N \times M)$ as a product of two matrices $W(N \times k) \times H(k \times M)$, where $k \ll N, M$. Tensor F is factored using the non-negative three-dimensional tensor factorization parallel algorithm PARAFAC [7]. The factorization yields corresponding matrices X, Y and Z.

After matrix D and tensor F factorization, the system forms a strong knowledge base which contains information about the syntactic framework of natural language sentences. Along with the description of general syntax that defines the structure of sentences in a general abstract form, the base also contains semantic restrictions that determine which words can form a syntactic connection of a certain type. To determine whether two words a and b form a cyclic binary relation, one has to take vector-row W_a from matrix W corresponding to the word a, and vector-column matrix H_b from matrix H which corresponds to the word b, and calculate the scalar product of vectors (W_a, H_b^T). If the product is greater than a certain threshold T, this relation is defined. In order to determine whether the three words a, b and c enter into predicative relation $(a \rightarrow b \rightarrow c)$, it is necessary to take vector X_a corresponding to the word a, vector Y_b corresponding to the word b, and vector Z_c corresponding to the word c and to calculate the value:

$$S_{abc} = \sum_{i=1}^{k} X_a[i] * Y_b[i] * Z_c[i]$$

If S_{abc} value is greater than a threshold, then this relation is defined. If not, it is considered undefined.

These matrices implicitly define a set of defined language clauses, which is specified with the input corpus. The vectors of words from the derived matrices implicitly describe their structural behavior. They define in which syntactic relation these words may join and which words they have joined. With the resulting matrix, one may parse sentences and generate the control space of their syntactic structures, using ascending algorithms such as Cocke – Younger – Kasami. The control space is built where possible.

3   Implementation

As the initial training text corpus, sets of articles from the English Wikipedia, the Simple English Wikipedia and the WSJ corpus are used. The texts are processed sequentially with the parser and with the program that constructs the control space of syntactic structures. First, the sentences are analyzed with the Stanford Parser yielding CFG parse trees (for phrase structure grammar) and dependency trees. Also, an algorithm has been developed to construct control spaces by converting the dependency tree and the CFG parse tree into the control space of a sentence. The algorithm is a recursive traversal from left to right of the sentence tree which creates points of the control space in each node of the CFG parse tree and performs conversion of corresponding relations of the dependency tree into connections of control space (either predicative or cyclic connections). Each point of the space is assigned a specific lexical value (a word or a phrase) and characteristics (part of speech, gender, number, etc.). At the outset every word is an isolated point in the control space. When points A and B are connected to form a new point S in the space, representing the relationship between A and B, this new point gains its own lexical value. This value can be inherited from the main element of the pair (A, B), e.g., the phrase *hot tea* consists of a pair *(hot, tea)* that has a Noun as the main word. Consequently, the new point will inherit value from *tea*. Also, the merger of two points may result in their lexical value forming a fixed collocation. For example, the combined value of point A *(Weierstrass)* and B *(theorem)* is the *Weierstrass theorem*, which is the lexical value of the new generated point C. Fixed collocations are obtained based on Wikipedia articles with corresponding titles.

The control space has been built, with matrix D and tensor F filled. 800,000 articles from the English Wikipedia and the Simple Wikipedia have been processed, along with the WSJ corpus. As the WSJ corpus is annotated manually and contains correct syntactic structures, a high number of quality syntactic structures control spaces are received. The processing yields a large matrix D for cyclic links (numbering approximately 2.3 million words × 2.3 million words, with up to 57 million non-zero elements) and the large three-dimensional tensor F for ternary predicative connections (consisting of approximately 2.3 million words × 52 thousand words × 2.3 million words, with up to 78 million non-zero elements). These arrays were factorized by the non-negative matrix factorization algorithm [6] and the non-negative tensor factorization parallel algorithm PARAFAC [7].

Factorized data sets allow for efficient computing of probability for cyclic binary relations between any two words using the scalar product of two corresponding vectors. To form ternary predicative relations among any three words the probability can be efficiently and easily calculated.

To investigate the applicability of this model for practical NLP tasks, a parser for the English language based on the obtained arrays of lexical-syntactic combinability has been implemented. This parser, based on the Cocke – Younger – Kasami algorithm, directly constructs the control space of a sentence.

The model describes only the relations among those words which actually occur in the corpora sentences and have been processed accordingly. When a pair of words A and B makes a cyclic binary link and has value in the array, the pair A_1 and B_1 (where A_1 is synonymous with A, and B_1 is synonymous with B) will not have the link if A_1 and B_1 are absent in the data. The same holds for ternary predicative relations. The matter can be easily dealt with by using synonym dictionaries. In the system we developed the WordNet is used to this end. We assume that if between A and B a relation exists, it also exists between an arbitrary pair of A_i and B_i, where A_i is any word from the synset that contains A, while B_i is any word from the synset that contains B. However, the question of homonymy arises when one word corresponds to several synsets in the WordNet. Every time a sentence is parsed, the point at issue is how to determine whether a pair or triplet of synsets is correct.

On the one hand, there are several standard approaches to solving this classic problem of ambiguous words (WSD). On the other hand, the two matrices W and H resulting from the non-negative matrix factorization of D can be considered powerful tools for determining the degree of semantic similarity between words according to the methods of latent semantic analysis.

So, to determine the presence of cyclic binary connections and to solve the problem of ambiguous words the following steps are carried out:

A: Take vector W_a corresponding to word a from term matrix W, vector column H_b which corresponds to word b from matrix H, and calculate the scalar product of the vectors (W_a, H_b^T). If the value $(W_a, H_b^T) > T$, then this link is **defined**. T is the threshold. The optimal value of T is found experimentally. If it fails:

B: Take synsets for words a and b from the WordNet. The set of synsets $\{A_i\}$ refers to word a, and the set of synsets $\{B_i\}$ refers to word b. Check the pairs of the words formed from the elements of $\{A_i\}$ and $\{B_i\}$. If there is a word a'_k from

$A_k \in \{A_i\}$ and a word b'_j from $B_j \in \{B_i\}$ such that scalar product of vectors $(W_{a'_k}, H^T_{b'_j}) > T$, then this link between a and b is **defined**. If not:

C: The set $\{A_i\}$ is expanded with synsets linked with nodes from $\{A_i\}$ with hyponym and hyperonym relations in the WordNet. The set $\{B_i\}$ is expanded in the same way. Check the pairs of words formed from elements of $\{A_i\}_{\exp}$ and $\{B_i\}_{\exp}$ (excluding the pairs already checked on step B). If there is a word a'_k from the synset $A_k \in \{A_i\}_{\exp}$ and a word b'_j from the synset $B_j \in \{B_i\}_{\exp}$ such that the scalar product of vectors $(W_{a'_k}, H^T_{b'_j}) > T$, then the link between a and b is **defined**. If it fails: expand $\{A_i\}_{\exp}$ and $\{B_i\}_{\exp}$ recursively 2 or 3 times and repeat step (C).

If it is always $(W_{a'_k}, H^T_{b'_j}) < T$, then the link **does not exist**.

During the expansion of $\{A_i\}$ and $\{B_i\}$ one should avoid adding synsets from the list of the concepts with the most general meanings from the top of the WordNet hierarchy. In expanding $\{A_i\}_{\exp}$ and $\{B_i\}_{\exp}$ with such concepts, the semantic similarity between a'_k and b'_j quickly deteriorates. Inheritance of properties through hyponymy/hypernymy is not correct for such synsets.

For the ternary predicative link, this algorithm works in the same way.

The taxonomic hierarchy of the WordNet lexical database together with the mechanism of inheritance allows us to generalize this representation model of syntactic and semantic relations of natural language. This turns the constructed system into a versatile tool for syntactic and semantic analysis of natural language texts.

4   Experiments

To form a robust syntactical and semantic relations base, it is crucial to have a huge corpus of correctly tagged texts. The WSJ corpus availability has a significant effect on assuring the quality of the resulting model. To construct tagged texts from the English Wikipedia and the Simple English Wikipedia, the Stanford Parser is used. It produces dependency and CFG parse trees. The accuracy of CFG parse trees is about 87%, while the accuracy of dependency trees is about 84%. As some of the trees are incorrect, it natural that they yield some inaccurate descriptions of the control spaces of the syntactic structures. The algorithm for converting CFG parse trees and dependency trees into control spaces of syntactic structures shows no errors on correct trees.

The development of the system for parsing and control spaces generation for natural language sentences based on created lexical and syntactic databases was followed by experiments. The accuracy was measured by computing control spaces of the syntactic structures. To generate test samples, 1,500 sentences were taken from the Simple English Wikipedia articles; 1,500 sentences – from the English Wikipedia articles (using the texts not included in the 800,000 items processed for constructing matrix D and tensor F).

The syntax trees of the sets of texts from the Wikipedia and the Simple Wikipedia that were processed with the Stanford Parser were automatically transformed into control spaces, applying conversion with the developed algorithm. The obtained control spaces were manually verified and corrected by experts. This annotated text corpus was formed for the purpose of checking the quality of parsing and generating syntactic structure control spaces for the Simple Wikipedia and the English Wikipedia texts.

The system for parsing and control spaces generation constructs control spaces of syntactic structures for sentences from the annotated corpus. Subsequently, the obtained control spaces were compared with the corresponding correct control spaces from the annotated test corpus.

Each cyclic binary link and each ternary predicative link that were found were automatically tested. The test was carried out with due regard for the algorithmic case in which a particular syntactic relation was found. Case **A** describes the identification of the direct link between words through the scalar product of their vectors; case **B** describes the usage of synonyms to compute the probability of the link. Case **C** describes the usage of the hyponym and hyperonym Word-Net connections for these words to find the probability of the link. The test was performed only for the sentences that had been successfully processed with the complete building of the syntactic structure control spaces (94.1% from 1500 sentences from the Simple English Wikipedia and 83.4% from 1500 sentences from English Wikipedia were successfully processed in the test set). Also, a test was performed on the WSJ corpus using cross-validation (when checking the quality of the system on 1 part of the corpus out of 10, the corresponding data obtained from the above mentioned part were temporarily excluded from the base of the model). The test on the WSJ corpus was performed automatically. 92.7% of sentences from the WSJ corpus obtained complete parse. The results are summarized in Table 1.

Table 1. Precision estimation of cyclic binary links and ternary predicative links on sentences from the Simple English Wikipedia, the English Wikipedia and the WSJ corpus

	Simple Wikipedia	Wikipedia	WSJ corpus
Cyclic binary links (case A)	95.17%	91.23%	93.71%
Cyclic binary links (case B)	91.29%	89.91%	91.05%
Cyclic binary links (case C)	89.17%	83.06%	85.07%
Ternary predicative links (case A)	96.17%	92.24%	94.37%
Ternary predicative links (case B)	93.21%	90.01%	91.33%
Ternary predicative links (case C)	91.03%	87.79%	89.79%

The precision estimates of the ternary predicative links are higher than the precision estimation of the cyclic binary links. It seems natural considering the positional stability for relations of type *Subject – Verb – Object* structure in the sentences. A certain small percentage of errors occurs even in case **A**. It indicates that errors must be present in the training set of control spaces of sentences that served as the base for constructing the cyclic links matrix D and the three-dimensional predicative relations tensor F. The model can be improved by checking and correcting the training set. The best estimates correspond to sentences from the Simple Wikipedia, which is quite understandable due to the simple and clear syntactic structure of its sentences. The English Wikipedia sentences are much more complicated, leaving more room for different interpretations of grammatical structures. Hence, the precision of processing for the WSJ corpus sentences is higher than that for the English Wikipedia sentences. It indicates that the high quality training data from the WSJ corpus allows for improving the model to a great extent.

5   Conclusions

The recursiveness of syntactic structures control spaces allows us to describe sentence structures of arbitrary complexity, length and depth. This enables the development of a semantic-syntactic model based on the single three-dimensional tensor and the single matrix instead of increasing the number of dimensions of connectivity arrays for lexical items. To investigate the applicability of this model for practical NLP tasks, a system for analysis and constructing syntactic structure control spaces has been developed on the basis of factorized arrays. It shows high quality and accuracy, thus proving the correctness and efficiency of the constructed model. The model is of high relevance both for theoretical and practical applications for computational linguistic systems.

References

1. Van de Cruys, T.: A Non-negative Tensor Factorization Model for Selectional Preference Induction. Journal of Natural Language Engineering 16(4), 417–437 (2010)
2. Van de Cruys, T., Rimell, L., Poibeau, T., Korhonen, A.: Multi-way Tensor Factorization for Unsupervised Lexical Acquisition. In: Proceedings of COLING 2012, pp. 2703–2720 (2012)
3. Anisimov, A.V.: Control Space of Syntactic Structures of Natural language. Cybernetics and System Analysis 3, 11–17 (1990)
4. Klein, D., Manning, C.D.: Accurate Unlexicalized Parsing. In: Proceedings of ACL 2003, pp. 423–430 (2003)
5. de Marneffe, M.-C., MacCartney, B., Manning, C.D.: Generating Typed Dependency Parses from Phrase Structure Parses. In: Proceedings of LREC (2006), http://nlp.stanford.edu/pubs/LREC06_dependencies.pdf
6. Lee, D.D., Seung, H.S.: Algorithms for Non-Negative Matrix Factorization. In: NIPS (2000), http://hebb.mit.edu/people/seung/papers/nmfconverge.pdf
7. Cichocki, A., Zdunek, R., Phan, A.-H., Amari, S.-I.: Nonnegative Matrix and Tensor Factorizations: Applications to Exploratory Multi-way Data Analysis and Blind Source Separation. J. Wiley & Sons, Chichester (2009)

Experiments on the Identification
of Predicate-Argument Structure in Polish

Konrad Gołuchowski

Institute of Computer Science, Polish Academy of Science, Warsaw, Poland

Abstract. This paper focuses on automatic methods of extracting a
predicate-argument structure in Polish. Two approaches to extract selected aspects of the predicate-argument structure are evaluated. In the
first experiment the multi-output version of the Random Forest classifier
is used to extract a valency frame for each predicate in a sentence. In
the second experiment the Conditional Random Fields classifier is used
to find syntactic heads of all arguments realised in a sentence. What is
more, the importance of various sources of features is presented, including shallow syntactic parsing, dependency parsing and a verb valency
information. Due to the lack of the high-quality syntactic parser, the
presented approach does not rely on the deep syntactic information.

Keywords: Argument identification, verb valency, predicate-argument
structure.

1 Introduction

Identifying a predicate-argument structure in a sentence is the initial step of
many natural language processing tasks, such as Semantic Role Labelling (SRL),
Information Extraction and verb clustering. Usual approaches to obtaining such
a structure heavily depend on deep parsing. [1,10] show that such information
is vital for the identification of argument boundaries. Although multiple efforts
have been put into building a syntactic parser for Polish [15,7], none of existing
tools gives satisfactory results yet. [2] shows that identification of exact argument
boundaries without syntactic parsing for Polish is a difficult task. However, in
many NLP tasks finding only selected aspects of a predicate-argument structure
(e.g. argument heads) instead of a full structure is useful.

For reasons given above, this paper presents two experiments that try to
identify two selected aspects of the predicate-argument structure without the
deep parsing information: heads of arguments and for each predicate, types of
all its arguments. The dependency parsing is used as an additional source of
features for argument types classification.

2 Predicate-Argument Structure

Predicate dependents are usually divided into two groups: arguments – that are
predicate-specific, and adjuncts – that can co-occur with almost any predicate.

A. Przepiórkowski and M. Ogrodniczuk (Eds.): PolTAL 2014, LNAI 8686, pp. 185–192, 2014.

This paper focuses only on the first group. Each argument may be labelled with a syntactic type, i.e. a syntactic information on how the argument is realised in a sentence. Argument types used in this paper were taken directly from the Składnica treebank [16,14] (see Sect. 4.1).

The sentence below presents a fairly simple example of the predicate-argument structure. Only one predicate *zjadł* (Eng. *ate*) is present. Its two core arguments, the subject (*subj*) and the nominal phrase in the accusative case (*np(bier)*), as well as an adjunct, have been annotated.

[SUBJMarek] **zjadł** [NP(BIER)kanapkę] [ADJUNCTwczoraj].
[SUBJMark] **ate** [NP(ACC)a sandwich] [ADJUNCTyesterday].

3   Related Work

In the semantic role labelling task, most algorithms for the argument identification start with syntactic parsing and then do a binary classification of each node in a parse tree whether it is an argument or not (see [1]). There are only a few examples that use the shallow syntactic information for the argument identification [2,3,13]. [1,10] both show the necessity of syntactic parsing to obtain argument boundaries in the semantic role labelling task. [6] uses the Markov logic to perform various tasks related to the identification of the predicate-argument structure. The presented Markov-logic-based classifier identifies predicates, arguments and senses at the same time. This approach does not rely on parse trees but uses features based on the dependency parsing. [5] treats assigning valency frame to a predicate as a verb sense disambiguation task. The set of possible verb senses is taken from VALLEX, a Czech lexicon of valency frames. A similar approach is presented in [12], which evaluates an attempt to assign a correct valency frame.

4   Resources and Tools

In subsequent subsections, the most important resources and tools are described.

4.1   Składnica – The Polish Treebank

Składnica is a Polish treebank [16,14]. It consists of 19998 sentences of which 8227 have manually corrected parse trees. This corpus is used both for training classifiers and for their evaluation. To assure that final solutions do not rely on any deep syntactic features, apart from the argument structure, only morphosyntactic features (such as a part-of-speech) were extracted from this corpus.

There are several argument types available in Składnica: 1) subjects, 2) nominal arguments with the case information, 3) prepositional phrases, 4) adjectival arguments, 5) information about reflexive pronoun *się*, 6) various dependent clauses, 7) infinitive arguments with the aspect information and 8) adverbial arguments. Whenever an adverbial argument was realised by a prepositional phrase, it was considered as a prepositional argument.

4.2   Walenty – The Valency Lexicon

Walenty [9] is the Polish Valency Lexicon that describes possible valency frames for almost 7000 verbs. Valency frame defines for a predicate what arguments in a sentence are expected and how these arguments are realised syntactically.

Some arguments in Walenty frames require certain words to occur inside them (e.g. in case of idiomatic expressions). Presence of such arguments is correlated with predicates, so for the purpose of experiments all such arguments were replaced with appropriate unlexicalized versions.

4.3   IOBBER – The Syntactic Shallow Parser

As mentioned before, reliable deep syntactic parsers are not yet available for Polish. However, tools performing shallow syntactic parsing were developed. One of such tools is IOBBER [11], which annotates a text with three types of syntactic groups: nominal and prepositional groups, verbal groups and adjectival groups. Moreover, IOBBER can find a syntactic head of each group.

4.4   The Polish Dependency Parser

The Polish dependency parser [18] was used to obtain a set of dependency relations between words. Relations between predicates and its dependents are the most interesting as they may indicate both presence of arguments and their type. A detailed description of dependency relations obtained from the Polish dependency parser can be found in [17].

5   Extracting Predicate-Argument Structure

In this paper, two aspects of extracting the predicate-argument structure are considered and presented in subsequent sections.

5.1   Experiment I: Extracting Arguments for Each Predicate

The goal of the first experiment was to obtain types of realised arguments for each predicate in a sentence. To achieve this goal, for each predicate in each sentence a set of features and a set of types of its arguments were extracted. All features were binary and stated that a word in nearest surrounding of the argument had some property (see Table 1). The multi-output version of Random Forest classifier from Scikit toolkit [8] was used because this classifier was able predict multiple arguments for a single input.

One approach to obtaining features for each predicate was to consider a window of neighbouring *chunks* surrounding the predicate. A *chunk* meant either a nominal or an adjectival group returned by the IOBBER or a single word if it was not in any group. This approach makes it possible to find most arguments that are in the nearest neighbourhood. Another approach to obtaining features for each predicate was to take advantage of predicate dependents given by the Polish dependency parser and extract the features for each dependent. Both these approaches, as well as the combined approach, were evaluated.

Table 1. Features used in Experiment I

Type	Feature
Morphosyntactic	presence of nouns, pronouns, adjectives with their case presence of conjunctions, complementizer and adverbs with their base forms, presence of noun and verb negation, presence of past participle, presence of reflexive pronoun *się*, presence of questions words (e.g. *what, which*, etc.), presence of words describing time (e.g. *yesterday, today*), presence of adverbs in base form (e.g. presence of the word *quickly*),
Walenty	all frame elements that can be realised with predicates from the sentence
Dependency parser-based	relation labels of selected dependents
Predicate-based	predicate part-of-speech

5.2 Experiment II: Extracting Argument Heads

The goal of the second experiment was to find which words represent syntactic argument heads in a sentence. Also, types of argument heads were determined. To achieve this goal, each word had the label assigned, either an argument type for an argument head, or not-argument-head label for all other words. In case of arguments that are coordination of phrases, the conjunction is considered as the argument head. Figure 1 presents a sample sentence with argument types labels. The not-argument-head label was denoted as _, the subject as *subj*, the nominal argument in the accusative case as *np(bier)* and the prepositional argument in the genitive case with the preposition *do* as *prepnp(do,dop)*.

Oni	często	chodzili	do		pubu	,	pili	piwo		, rozmawiali.
They	often	were going	to		a pub	,	drinking	a beer		, talking.
subj	_	_	*prepnp(do:gen)*	_		_ _		*np(bier)*	_ _	

Fig. 1. A sample sentence labelled with the argument heads

For each word in a sentence the following set of properties was extracted: 1) part-of-speech, 2) case (whenever applicable), 3) lemma of selected words (prepositions, complementizers, question words, *nie* and *się*), 4) information whether this word is the head of an IOBBER nominal or an adjectival group, 5) a matching Walenty argument if any, 6) a dependency relation if any. Then all properties in the window starting with two preceding words and one following word were considered as features for labelling a single word. Furthermore, a variant with the larger context (3 preceding and 3 following words) was evaluated.

To find the best possible labelling of words in a sentence the CRF++, the linear-chain Conditional Random Fields classifier, was used [4].

6   Evaluation

All evaluations were performed using 10-fold cross validation on the corpus extracted from the Składnica treebank. Standard measures of recall, precision and F-measure were used. In subsequent sections recall is understood as a correctly predicted fraction of arguments in the gold standard, precision – as a correctly predicted fraction of all predicted arguments and F-measure – as the harmonic mean of precision and recall. Correctness of the argument prediction is understood differently in each experiment. In the first experiment, the argument was predicted correctly if it was present in the valency frame of the considered predicate. In the second experiment, the predicted argument was considered correct when the argument head was recognised correctly.

6.1   Baseline Algorithms

In the first experiment, the baseline algorithm chooses the most frequent set of arguments for each predicate. For the second experiment – finding heads of arguments – the baseline algorithm chooses for each word the most common label for a concatenation of three features: its part-of-speech, its case (if applicable) and its base form.

6.2   Results of Experiment I: Predicate-Level Evaluation

The first experiment was performed in a few phases. Initially, only morphosyntactic features were extracted for each word in the sentence. In the later phases, features were extracted only for heads of syntactic groups. Next, the features extracted from the Walenty lexicon, as well as the features based on the Polish dependency parser, were added. In the best setting, the Random Forest classifier achieved the F-measure of 85.56% with the recall of 77.18% and the precision of 95.97%. Table 2 shows results achieved by the classifier in various setups. In this experiment the use of the IOBBER groups improves the recall but does not change the precision at all. The largest boost in the precision is obtained when the features based on the dependency parser relations are used.

This task is similar to assigning to a predicate a matching valency frame from the valency lexicon presented in [12]. However, [12] assigns not only arguments realised in a sentence as in experiment presented in Sect. 5.1 but also the full valency frame from VALLEX lexicon, To be able to approximately compare results from these two experiments, the accuracy of finding all arguments for each predicate is reported. The approach presented in this section achieved the accuracy on the level of 65.73% which is a slightly worse result than the accuracy presented in [12] (accuracy achieved by their best setup was 79.86%).

6.3   Results of Experiment II: Evaluation of Argument Head Extraction

In the task of finding argument heads using CRF-based classifier, most words are tagged with no-argument-head label. Including these tags in evaluation scores

Table 2. Results of finding arguments for each predicate in the sentence * – accuracy of finding the full frame for the predicate

Features	Precision (%)	Recall (%)	F (%)	Accuracy* (%)
Baseline	53.40	60.27	56.62	30.65
Window: 5 chunks				
Base features	88.87	74.03	80.77	59.45
+ Only IOBBER group heads	88.83	76.82	82.39	62.59
+ Walenty	89.74	76.57	82.63	63.32
+ Dependency-based relations	92.83	77.35	84.38	64.91
+ Dependency-based chunks	95.97	77.18	85.56	65.86
Window: 7 chunks				
Base features	91.17	73.71	81.51	60.30
+ Only IOBBER group heads	91.06	77.68	84.06	64.85
+ Walenty	92.28	77.13	84.02	65.06
+ Dependency-based labels	95.07	77.03	85.03	65.81
+ Dependency-based chunks	96.03	77.03	85.47	65.73
Dependency-based chunks only				
+ All features	94.00	74.65	83.21	62.22

Table 3. Results of finding heads of arguments * – accuracy of finding all types of arguments in the sentence

Features	Precision (%)	Recall (%)	F (%)	Accuracy* (%)
Baseline	71.31	70.21	70.75	37.68
Base features	85.21	80.90	83.00	57.54
+ IOBBER group head	85.99	84.11	85.05	61.69
+ Walenty	86.70	84.60	85.64	63.47
+ Dependency parser	91.17	88.91	90.02	73.06
+ Larger context	91.25	90.01	91.25	72.93

would lead to over-optimistic results and, in fact, would not reflect the real efficiency of finding argument heads. Therefore, only tags that reflect the actual argument types were user for scores calculations. Table 3 presents the results of this experiment.

Features based on the dependency relations gave the most noticeable improvement. Also, using information about syntactic heads improved the recall considerably. Detailed error analysis of the results showed that dependency relation-based features help to decrease three main sources of errors: misclassification of adverbial arguments, prepositional arguments and nominal arguments in the genitive case.

7   Conclusions and Future Work

This paper tackles the problem of finding two aspects of predicate-argument structure without the use of deep syntactic parsing. Two experiments were presented, as well as the impact of the various features types. All experiments gave reasonable results. Especially, in the experiment of finding argument heads, presented method achieved satisfying results, recognising correctly all arguments in 73% of sentences. Although the dependency parser outputs the predicate dependents, the first experiment showed that using a window around the predicate increases the recall of detected arguments.

In future work, it will be vital to merge both experiments into a single one, in order to obtain the predicate-argument structure, i.e. both argument heads and their governing predicates. Additional work may be necessary to improve the recall of finding valency frame for the predicate. Moreover, introducing semantic-based features (e.g. WordNet synset) should improve the distinction between adjuncts and arguments.

Acknowledgments. The paper is co-founded by the European Union from financial resources of the European Social Fund. Project PO KL "Information technologies: Research and their interdisciplinary applications".

References

1. Gildea, D., Palmer, M.: The necessity of parsing for predicate argument recognition. In: Proceedings of the 40th Annual Meeting on Association for Computational Linguistics, pp. 239–246. Association for Computational Linguistics (2002)
2. Gołuchowski, K., Przepiórkowski, A.: Semantic role labelling without deep syntactic parsing. In: Isahara, H., Kanzaki, K. (eds.) JapTAL 2012. LNCS (LNAI), vol. 7614, pp. 192–197. Springer, Heidelberg (2012)
3. Hacioglu, K., Pradhan, S., Ward, W., Martin, J.H., Jurafsky, D.: Semantic Role Labeling by Tagging Syntactic Chunks. In: Proceedings of CoNLL 2004, pp. 110–113 (2004)
4. Kudo, T.: CRF++: Yet another CRF toolkit (2005), Software available at http://crfpp.sourceforge.net
5. Lopatková, M., Bojar, O., Semecký, J., Benešová, V., Žabokrtský, Z.: Valency lexicon of czech verbs vallex: Recent experiments with frame disambiguation. In: Matoušek, V., Mautner, P., Pavelka, T. (eds.) TSD 2005. LNCS (LNAI), vol. 3658, pp. 99–106. Springer, Heidelberg (2005)
6. Meza-Ruiz, I., Riedel, S.: Jointly identifying predicates, arguments and senses using markov logic. In: Proceedings of Human Language Technologies: The 2009 Annual Conference of the North American Chapter of the Association for Computational Linguistics, pp. 155–163. Association for Computational Linguistics (2009)
7. Patejuk, A., Przepiórkowski, A.: Towards an LFG parser for Polish: An exercise in parasitic grammar development. In: Proceedings of the Eighth International Conference on Language Resources and Evaluation, LREC 2012, pp. 3849–3852. ELRA, Istanbul (2012)
8. Pedregosa, F., Varoquaux, G., Gramfort, A., Michel, V., Thirion, B., Grisel, O., Blondel, M., Prettenhofer, P., Weiss, R., Dubourg, V., Vanderplas, J., Passos, A., Cournapeau, D., Brucher, M., Perrot, M., Duchesnay, E.: Scikit-learn: Machine learning in Python. Journal of Machine Learning Research 12, 2825–2830 (2011)

9. Przepiórkowski, A., Hajnicz, E., Patejuk, A., Woliński, M., Skwarski, F., Świdziński, M.: Walenty: Towards a comprehensive valence dictionary of Polish. In: Calzolari, N., Choukri, K., Declerck, T., Loftsson, H., Maegaard, B., Mariani, J., Moreno, A., Odijk, J., Piperidis, S. (eds.) Proceedings of the Ninth International Conference on Language Resources and Evaluation, LREC 2014, pp. 2785–2792. ELRA, Reykjavík (2014), http://www.lrec-conf.org/proceedings/lrec2014/index.html

10. Punyakanok, V., Roth, D., Yih, W.T.: The importance of syntactic parsing and inference in semantic role labeling. Computational Linguistics 34(2), 257–287 (2008)

11. Radziszewski, A., Pawlaczek, A.: Large-scale experiments with NP chunking of Polish. In: Sojka, P., Horák, A., Kopeček, I., Pala, K. (eds.) TSD 2012. LNCS, vol. 7499, pp. 143–149. Springer, Heidelberg (2012)

12. Semecký, J.: Verb valency frames disambiguation. The Prague Bulletin of Mathematical Linguistics 88, 31–52 (2007)

13. Sun, W., Sui, Z., Wang, M., Wang, X.: Chinese semantic role labeling with shallow parsing. In: Proceedings of the 2009 Conference on Empirical Methods in Natural Language Processing, EMNLP 2009, vol. 3, pp. 1475–1483. Association for Computational Linguistics, Stroudsburg (2009)

14. Świdziński, M., Woliński, M.: Towards a bank of constituent parse trees for Polish. In: Sojka, P., Horák, A., Kopeček, I., Pala, K. (eds.) TSD 2010. LNCS (LNAI), vol. 6231, pp. 197–204. Springer, Heidelberg (2010)

15. Woliński, M.: An efficient implementation of a large grammar of polish. Archives of Control Sciences 15, 481–488 (2005)

16. Woliński, M.: Dendrarium – an open source tool for treebank building. In: Kłopotek, M.A., Marciniak, M., Mykowiecka, A., Penczek, W., Wierzchoń, S.T. (eds.) Proceedings of IIS 2010, pp. 193–204. Wydawnictwo Akademii Podlaskiej (2010)

17. Wróblewska, A.: Polish dependency bank. Linguistic Issues in Language Technology 7(1) (2012), http://elanguage.net/journals/index.php/lilt/article/view/2684

18. Wróblewska, A., Woliński, M.: Preliminary experiments in Polish dependency parsing. In: Bouvry, P., Kłopotek, M.A., Leprévost, F., Marciniak, M., Mykowiecka, A., Rybiński, H. (eds.) SIIS 2011. LNCS, vol. 7053, pp. 279–292. Springer, Heidelberg (2012)

Syntactic Approximation of Semantic Roles

Wojciech Jaworski[1,2] and Adam Przepiórkowski[1]

[1] Institute of Computer Science, Polish Academy of Sciences,
ul. Jana Kazimierza 5, 01-248 Warsaw, Poland
`adamp@ipipan.waw.pl`
[2] Institute of Informatics, University of Warsaw,
ul. Banacha 2, 02-097 Warsaw, Poland
`wjaworski@mimuw.edu.pl`

Abstract. The aim of this paper is to propose a method of simulating – in a syntactico-semantic parser – the behaviour of semantic roles in case of a language that has no resources such as VerbNet of FrameNet, but has relatively rich morphosyntax (here: Polish). We argue that using an approximation of semantic roles derived from syntactic (grammatical functions) and morphosyntactic (grammatical cases) features of arguments may be beneficial for applications such as text entailment.

Keywords: Thematic roles, parser, morphosyntax, LFG.

1 Introduction

There is a strong tradition in Slavic linguistics of relating morphosyntax to semantics, especially, of claiming that morphological cases have unified meanings. One of the most prominent proponents of this approach was Roman Jakobson (see, e.g., Jakobson 1971a,b), and it has been further developed by Anna Wierzbicka (e.g., Wierzbicka 1980, 1981, 1983, 1986), who claims that "cases have meanings and that this meaning can be stated in a precise and illuminating way" (Wierzbicka, 1986, p. 386).

While we do not fully subscribe to this tradition, we show that it turns out to be a useful approach in Natural Language Processing (NLP). In particular, we discuss the role of semantic roles in grammar engineering and argue that – in case of languages with rich morphosyntax but no manually created semantic role resources such as VerbNet or FrameNet – a relatively simple way of inferring an approximation of semantic roles from syntax and morphosyntax may be sufficient for some applications. In fact, it seems that even when a resource like VerbNet is available, this simpler approach to semantic-like roles may be beneficial.

The broad aim of the work partially reported here is to add a semantic component to the manually created LFG (Bresnan, 2001; Dalrymple, 2001) grammar of Polish (Patejuk and Przepiórkowski, 2012), implemented using the XLE platform (Crouch et al., 2011). Regardless of this particular context, we believe that the approach proposed in Sect. 3 has a wider applicability.

A. Przepiórkowski and M. Ogrodniczuk (Eds.): PolTAL 2014, LNAI 8686, pp. 193–201, 2014.

2   Semantic Roles in Grammar Engineering

The modern notion of semantic roles stems from the work of Gruber 1965 and Fillmore 1968, and it was brought to the foreground of linguistic research by Jackendoff 1972. Relatively small sets of semantic roles are commonly assumed in theoretical linguistics. For example, Fillmore 1968 distinguishes between Agentive, Dative, Instrumental, Factive, Locative, Objective, as well as Benefactive, Time and Comitative, and even fewer roles are assumed in LFG (Bresnan and Kanerva, 1989; Dalrymple, 2001). In more applicational or corpus-based work, much larger repertoires are adopted, e.g., 18 roles in the system of Sowa 2000 or 30 roles in VerbNet (Kipper *et al.* 2000; http://verbs.colorado.edu/~mpalmer/projects/verbnet.html).

Semantic roles are useful in those NLP tasks which use or produce semantic representations for the purpose of automatic reasoning, e.g., in text entailment or question answering. For example, instead of representing the sentence *Carrie ate pizza at Langley* naïvely as $\exists p \, pizza(p) \wedge eat(C, p, L)$, it may be a little less naïvely (but still ignoring tense, etc.) represented using semantic roles and the neo-Davidsonian approach (Parsons, 1990) as $\exists p, e \, pizza(p) \wedge eat(e) \wedge agent(e, C) \wedge patient(e, p) \wedge location(e, L)$. This latter representation makes the inference from *Saul ate pizza at Langley* to *Saul ate pizza*, represented as $\exists p, e \, pizza(p) \wedge eat(e) \wedge agent(e, C) \wedge patient(e, p)$, immediate – it's a matter of dropping the conjunct $location(e, L)$ in the semantic representation. On the other hand, on the more traditional approach, many meaning postulates would have to be formulated, including one relating the 3-argument *eat* predicate (as in $eat(C, p, L)$) to the corresponding 2-argument predicate (as in $eat(C, p)$).

Given the multiplicity of proposed systems of semantic roles, the question arises which one to use in a grammar engineering task. In Jaworski and Przepiórkowski 2014 we report the results of usability studies of the two systems mentioned above: Sowa's and VerbNet. The results are discouraging: the inter-annotator agreement is much too low to guarantee a reasonable quality of semantic role assignment – and, hence, the quality of any tools trained on corpora annotated with such semantic roles – and the investigation of disagreements reveals some internal inconsistencies in these systems.

On the basis of these experiments, as well as various remarks in the literature, we conclude that semantic role systems such as VerbNet or Sowa's are not really well-suited for the grammar engineering task and that other approaches must be explored. The one that we advocate here is to define 'semantic roles' on the basis of morphosyntactic information, including morphological cases, following the linguistic tradition referred to at the beginning of this section. This tradition is continued by Slavic linguists working within the Cognitive Linguistics paradigm, including Ewa Dąbrowska, whose view of the Polish dative reads like a definition of a semantic role: "the dative noun refers to an individual affected by a process or state which obtains in some part of his personal sphere, be it the sphere of potency, the sphere of empathy, the sphere of awareness, or the private sphere" (Dąbrowska 1997, p. 68; see also Dąbrowska 1994).

3 Syntactic Approximation of Semantic Roles

There are two general approaches to obtaining semantic representations in LFG-based parsing systems: co-description (CD) and description-by-analysis (DBA). The former, CD, is straightforward: lexical entries contain lexical semantic information, grammar rules or principles specify how meanings are composed, and semantic composition proceeds in parallel to syntactic parsing. This is the standard procedure in various formalisms and parsing platforms, including the HPSG (Pollard and Sag, 1994) English Resource Grammar (Copestake and Flickinger 2000; http://www.delph-in.net/erg/).

However, in LFG grammar engineering, the second approach, DBA, is common: semantic representation is obtained by analysing f-structures, i.e., non-tree-configurational syntactic representations (as opposed to more surfacy tree-configurational c-structures) containing information about predicates, grammatical functions and morphosyntactic features; this approach has been adopted for German (Frank and Erk, 2004; Frank and Semecký, 2004; Frank, 2004), English (Crouch and King, 2006) and Japanese (Umemoto, 2006).

In order to obtain representations employing semantic roles, resources external to the respective LFG grammars must be used in the process. Thus, in case of German, rules of transforming f-structures to semantic structures containing semantic role information were automatically acquired (Frank and Semecký, 2004) on the basis of a German treebank (Brants et al., 2002) annotated with FrameNet-like information, and subsequently generalised (Frank, 2004) to cover more unseen cases. For English, semantic roles were more directly transferred from VerbNet to the lexicon (Crouch and King, 2005) used in the system rewriting f-structures to semantic representations (Crouch and King, 2006). On the other hand, apparently no such external resources were used in case of Japanase (Umemoto, 2006), so the resulting representations use the names of grammatical functions such as subject and object, instead of true semantic roles.

Frank and Erk 2004 point out the benefits of adopting the DBA approach, especially at an early stage of developing a semantic module of an LFG parser, and we follow this advice here. However, there are currently no external resources for Polish that could supply information about semantic roles of particular predicates. But instead of falling back all the way to grammatical functions, as in case of the Japanese parser mentioned above, we capitalise on the fact that Polish has a relatively rich morphosyntactic system, with 7 morphological cases,[1] and a large number of preposition / morphological case combinations, many of which are highly correlated with specific semantic roles. In the remainder of this section we describe the procedure of assigning 'semantic roles' on the basis of morphosyntactic information; to constantly remind ourselves that they are just approximations of true semantic roles, they will be called R0, R1, etc., instead of Agent, Patient, etc., and the term 'semantic role' will be written in scare quotes.

How many roles do we need? We have seen above that too many roles cause classification problems, so we want as few different roles as possible. On the

---

[1] But only 6 of them are governable; the vocative is never governed.

other hand, there should be enough of them – and they should be sufficiently well differentiated – to minimise the probability of two arguments of the same predicate bearing the same role.[2] For the time being we settle on 11 core roles listed in Table 1 together with the meanings they are supposed to approximate (and, in some cases, with the usual names of such roles).

Table 1. Assumed 'semantic roles' and their approximate meanings

Role	Approximate description
R0	Actor of an action (Agent, Effector)
R1	Undergoer of an action (Patient, Theme, Product)
R2	Dative argument (Beneficiary, Recipient)
R3	Instrumental argument (Instrument)
R4	Adlative argument in both physical and abstract (functional, purposive) meaning (Destination, Recipient, Theme)
R5	Ablative argument in both physical and abstract (causal) meaning (Source)
R6	Locative argument in both physical and abstract meaning
R7	Perlative argument
R8	Topic of communication
R9	Temporal argument (point in time)
R10	Manner argument

The algorithm for assigning 'semantic roles' to arguments is rather simple. With one exception, the 'semantic role' is assigned on the basis of the grammatical function of the argument (as well as the voice of the verb; see below). The exception is the OBL(ique) argument – in the LFG grammar of Polish this is prototypically the grammatical function of various prepositional arguments. In this case, also the form of the preposition and the case of its object is taken into account. Tables 2 and 3 present the mapping from grammatical functions of arguments of an active form of the verb, and – in case of OBL – from particular preposition / case combinations, to 'semantic roles'.[3]

In case of passive forms, the deep object becomes the surface subject, so SUBJ maps to R1, and – conversely – the deep subject may be realised as a *by*-phrase (PRZEZ[acc]) bearing the OBL-AG grammatical function, so this function is mapped to R0.

---

[2] Note that it would be unrealistic to expect such situation never to happen; even VerbNet with its 25–30 roles needs roles such as Co-Agent, Co-Patient and Co-Theme. Moreover, in the experiments described in Jaworski and Przepiórkowski 2014, about 2.5–4.4% of verb occurrences had their arguments marked with duplicated roles (more precisely, 2.47% in case of VerbNet roles, 4.36% in case of Sowa's roles).

[3] The mapping given for OBL may also be used to assign 'semantic roles' to prepositionial adjuncts. Roles R6, R9 and R10 may also be used to indicate relations between a verb and its adverbial adjuncts. But if adjuncts are included in the 'semantic role' assignment, the problem of duplication of roles mentioned below becomes more serious and should be dealt with, e.g., by assigning adjuncts separate roles A2–A10, conventionally corresponding to R2–R10.

Table 2. Mapping of grammatical functions (with active verbs) to 'semantic roles'

Argument	Role
SUBJ	R0
OBJ	R1
OBJ-TH	R2
OBL-INST	R3
OBL-GEN	R1
OBL-STR	R1
OBL	see Table 3
XCOMP	R8
COMP	R8
XCOMP-PRED	R8

Table 3. 'Semantic roles' for OBL arguments

Preposition / morphological case	Role
DLA[gen], PRZECIW[dat], WOBEC[gen]	R2
DO[gen], KU[dat], MIĘDZY[acc], NA[acc], NAD[acc], PO[acc], POD[acc], POMIĘDZY[acc], PONAD[acc], POZA[acc], PRZED[acc], W[acc], ZA[acc]	R4
DZIĘKI[dat], OD[gen], SPOD[gen], SPOŚRÓD[gen], WSKUTEK[gen], Z[gen], ZZA[gen]	R5
KOŁO[gen], MIĘDZY[inst], NA[loc], NAD[inst], PO[loc], POD[inst], POMIĘDZY[inst], PONAD[inst], PONIŻEJ[gen], POZA[loc], PRZED[inst], PRZY[loc], U[gen], W[loc], WOKÓŁ[gen], WŚRÓD[gen], ZA[inst]	R6
BEZ[gen], POPRZEZ[acc], PRZEZ[acc], Z[inst]	R7
JAKO[nom], O[acc], O[loc]	R8
PODCZAS[gen]	R9
WEDŁUG[gen]	R10

This way of assigning 'semantic roles' conflates different grammatical functions while preserving the near-uniqueness of 'semantic roles'. First, normally only one of the grammatical functions OBJ (any passivisable argument, usually in the accusative), OBL-GEN (non-passivisable genitive argument) and OBL-STR (structurally cased, i.e., a usually accusative argument, which does not passivise) may appear in the f-structure of a given verb, so these – as well as the SUBJ of a passive verb – are uniformly mapped to R1. Second, in the valence dictionary for Polish mentioned below, there is only one rather special verb that has a valence schema with different arguments mapping to the grammatical functions of COMP (sentential complement) and XCOMP (infinitival complement), so it makes sense to translate both grammatical functions to R8, which approximates the Topic role. Less obviously, also XCOMP-PRED, which corresponds to the predicative argument of copula verbs, especially, BYĆ 'be' and ZOSTAĆ 'become', is translated to R8. It might at first seem that a sentence meaning *Brody is innocent* should be represented as, say, *innocent(b)*, but then there is no event that different tenses or modalities could modify. Without going into details of the envisaged semantic representation, let us assume that a proposition like *innocent(b)* – expressed by the XCOMP-PRED argument and its covert subject overtly realised as the subject of the copula – is the sole topical (R8) argument of the copula verb.

Lumping various grammatical functions into single 'semantic roles' should be contrasted with splitting OBL into different 'semantic roles', on the basis of the preposition and the case it governs (see Table 3). For example, R2 – the approximation of Beneficiary and Recipient – is assigned not only to dative arguments, but also to arguments headed by DLA[gen] 'for', etc. Similarly, DO[gen] 'to', KU[dat] 'towards', NA[acc] 'on(to)', etc., are reasonable indicators of the Adlative role, approximated here by R4.

This algorithm ensures high uniqueness of 'semantic role' assignment. Out of the total number of 24 170 morphosyntactic schemata in the September 2013 version of Walenty, a valence dictionary for Polish (http://zil.ipipan.waw.pl/Walenty; Przepiórkowski *et al.* 2014), only 343 (or 1.42%) contained two or more arguments which would be mapped to the same 'semantic role'.[4] In almost half of them, namely 162, R4 would be duplicated; this is because the relevant schemata contain a number of prepositional arguments of the same type. This is also a problem for the underlying LFG grammar, as all such arguments need to be mapped to essentially the same grammatical function, OBL. As this would violate LFG's coherence condition, the grammar introduces also OBL2.[5] Exactly the same problem occurs with R6, which is duplicated 69 times. In case of the 48 duplicates of R8, valence schemata contain a broadly verbal argument (COMP or XCOMP) and one of the prepositional arguments listed in the R8 row of Table 3. Moreover, OBJ co-occurs with OBL-GEN or OBL-STR 35 times, resulting in the duplication of R1; the other 29 duplication cases are less systematic. For some of these 343 cases duplication cannot easily be avoided, but for others a more sophisticated 'semantic role' assignment procedure can be devised; e.g., when OBL-GEN occurs next to OBJ, it should probably be mapped to the broadly ablative R5 rather than the thematic R1.

Obviously, the procedure just described is an engineering heuristic, and instances of 'wrong' decisions may be found. For example, OBL arguments of type z[inst] 'with' have at least two meanings, apart from the perlative (R7): thematic (R1) and co-agentive (R0); in fact, the sentence *Zrób z nim porządek*, lit. 'do with him order', is ambiguous between the two and may mean either 'Deal with him' (R1) or 'Clean up with him' (R0).

On the positive side, while we do not have any quantitative data on the effects of this approach to 'semantic roles' on tasks such as textual entailment or

---

[4] This should be contrasted with 2.47–4.36% of verb occurrences annotated with valence frames containing duplicated semantic roles in the experiments reported in Jaworski and Przepiórkowski 2014. As reported in that paper, on the same data the approach proposed here resulted in 1.73% of verb occurrences with valence frames containing duplicates.

[5] In fact, also OBL3 and OBL4. In the LFG valence dictionary, which was converted from the March 2014 version of Walenty, there are 19 787 schemata with OBL, 1843 (almost 10%) of them also mention OBL2, 45 of these 1843 include OBL3, and 2 of these 45 – also OBL4 (Agnieszka Patejuk, p.c.).

question answering,[6] we note that it makes various inferences immediate which would not be straightforward if arguments were marked only with grammatical functions, e.g., inferences of the b. sentences from the corresponding a. sentences below:

(1) a. *Janek pobił Tomka.* 'Janek beat Tomek up.'
    b. *Tomek został pobity.* 'Tomek was beaten up.'
(2) a. *Janek przesłał do Tomka książkę.* 'Janek sent a book to Tomek.'
        (lit. 'Janek sent to Tomek (a/the) book.ACC.')
    b. *Janek przekazał Tomkowi książkę.* 'Janek transferred a book to Tomek.'
        (lit. 'Janek transferred Tomek.DAT (a/the) book.ACC.')
(3) a. *Janek powiedział, że Tomek wygrał.* 'Janek said that Tomek had won.'
    b. *Janek mówił o Tomku.* 'Janek was talking about Tomek.'

4   Conclusions

Given various problems with the practical applicability of standard repertoires of semantic roles reported in this paper, and the fact that creating resources such as VerbNet or FrameNet takes a lot of time, money and expertise, we proposed an ersatz solution consisting in assigning approximations of semantic roles calculated on the basis of syntactic (grammatical functions) and morphosyntactic (case, preposition form) features of arguments. The algorithm presented above makes it possible to assign such 'semantic roles' to arguments almost uniquely, and the resulting neo-Davidsonian representations facilitate textual entailments well beyond what would be possible if arguments were marked with grammatical functions only.

References

Brants, S., Dipper, S., Hansen, S., Lezius, W., Smith, G.: The TIGER treebank. In: Hinrichs, E., Simov, K. (eds.) Proceedings of the First Workshop on Treebanks and Linguistic Theories (TLT 2002), Sozopol (2002)

Bresnan, J.: Lexical-Functional Syntax. Blackwell Textbooks in Linguistics. Blackwell, Malden (2001)

Bresnan, J., Kanerva, J.M.: Locative inversion in Chicheŵa: A case study of factorization in grammar. Linguistic Inquiry 20(1), 1–50 (1989)

Copestake, A., Flickinger, D.: An open-source grammar development environment and broad-coverage English grammar using HPSG. In: Proceedings of the Second International Conference on Language Resources and Evaluation, LREC 2000, Athens. ELRA(2000)

Crouch, D., King, T.H.: Unifying lexical resources. In: Proceedings of the Interdisciplinary Workshop on the Identification and Representation of Verb Features and Verb Classes (2005)

---

[6] To the best of our knowledge, no testing data for such tasks are available for Polish and Polish has never been included in evaluation initiatives of this kind.

Crouch, D., King, T.H.: Semantics via f-structure rewriting. In: Butt, M., King, T.H. (eds.) The Proceedings of the LFG 2006 Conference, Universität Konstanz, Germany. CSLI Publications (2006)

Crouch, D., Dalrymple, M., Kaplan, R., King, T., Maxwell, J., Newman, P.: XLE documentation (2011), http://www2.parc.com/isl/groups/nltt/xle/doc/xle_toc.html

Dalrymple, M.: Lexical Functional Grammar. Academic Press, San Diego (2001)

Dąbrowska, E.: Dative and nominative experiencers: two folk theories of the mind. Linguistics 32, 1029–1054 (1994)

Dąbrowska, E.: Cognitive Semantics and the Polish Dative. Cognitive Linguistics Research, vol. 9. Mouton de Gruyter, Berlin (1997)

Fillmore, C.J.: The case for case. In: Bach, E., Harms, R.T. (eds.) Universals in Linguistic Theory, pp. 1–88. Holt, Rinehart and Winston, New York (1968)

Frank, A.: Generalisations over corpus-induced frame assignment rules. In: Fillmore, C., Pinkal, M., Baker, C., Erk, K. (eds.) Proceedings of the LREC 2004 Workshop on Building Lexical Resources from Semantically Annotated Corpora, pp. 31–38. ELRA, Lisbon (2004)

Frank, A., Erk, K.: Towards an LFG syntax-semantics interface for frame semantics annotation. In: Gelbukh, A. (ed.) CICLing 2004. LNCS, vol. 2945, pp. 1–13. Springer, Heidelberg (2004)

Frank, A., Semecký, J.: Corpus-based induction of an LFG syntax-semantics interface for frame semantic processing. In: Hansen-Schirra, S., Oepen, S., Uszkoreit, H. (eds.) Proceedings of the 5th International Workshop on Linguistically Interpreted Corpora at COLING 2004, Geneva (2004)

Gruber, J.: Studies in Lexical Relations. Ph.D. thesis, Massachusetts Institute of Technology (1965)

Jackendoff, R.: Semantic Interpretation in Generative Grammar. The MIT Press, Cambridge (1972)

Jakobson, R.O.: Beitrag zur allgemeinen Kasuslehre. Gesamtbedeutungen der russischen Kasus. In: Selected Writings II, Mouton, The Hague, pp. 23–71 (1971a)

Jakobson, R.O.: Morfologičskie nabljudenija nad slavjanskim skloneniem. In: Selected Writings II, Mouton, The Hague, pp. 154–183 (1971b)

Jaworski, W., Przepiórkowski, A.: Semantic roles in grammar engineering. In: Proceedings of the 3rd Join Conference on Lexical and Computational Semantics (*SEM 2014), Dublin, Ireland (2014)

Kipper, K., Dang, H.T., Schuler, W., Palmer, M.: Building a class-based verb lexicon using TAGs. In: Proceedings of TAG+5 Fifth International Workshop on Tree Adjoining Grammars and Related Formalisms (2000)

Parsons, T.: Events in the Semantics of English: A Study in Subatomic Semantics. The MIT Press, Cambridge (1990)

Patejuk, A., Przepiórkowski, A.: Towards an LFG parser for Polish: An exercise in parasitic grammar development. In: Proceedings of the Eighth International Conference on Language Resources and Evaluation, LREC 2012, pp. 3849–3852. ELRA, Istanbul (2012)

Pollard, C., Sag, I.A.: Head-driven Phrase Structure Grammar. Chicago University Press / CSLI Publications, Chicago (1994)

Przepiórkowski, A., Hajnicz, E., Patejuk, A., Woliński, M., Skwarski, F., Świdziński, M.: Walenty: Towards a comprehensive valence dictionary of Polish. In: Calzolari, N., Choukri, K., Declerck, T., Loftsson, H., Maegaard, B., Mariani, J., Moreno, A., Odijk, J., Piperidis, S. (eds.) Proceedings of the Ninth International Conference on Language Resources and Evaluation, LREC 2014, pp. 2785–2792. ELRA, Reykjavík (2014)

Sowa, J.F.: Knowledge Representation: Logical, Philosophical, and Computational Foundations. Brooks Cole Publishing Co., Pacific Grove (2000)

Umemoto, H.: Implementing a Japanese semantic parser based on glue approach. In: Proceedings of the 20th Pacific Asia Conference on Language, Information and Computation, pp. 418–425. Tsinghua University Press, Huazhong Normal University (2006)

Wierzbicka, A.: The Case for Surface Case. Karoma, Ann Arbor (1980)

Wierzbicka, A.: Case marking and human nature. Australian Journal of Linguistics 1, 43–80 (1981)

Wierzbicka, A.: The semantics of case marking. Studies in Language 7, 247–275 (1983)

Wierzbicka, A.: The meaning of a case: A study of the Polish dative. In: Brecht, R.D., Levine, J.S. (eds.) Case in Slavic, pp. 386–426. Slavica Publishers, Columbus (1986)

Using Polish Wordnet
for Predicting Semantic Roles
for the Valency Dictionary of Polish Verbs

Natalia Kotsyba

Institute of Computer Science, Polish Academy of Sciences

Abstract. The paper describes a preliminary proposal of a method for
creating the semantic layer in a valency dictionary of Polish by enriching
it with information about semantic roles (meaning-related participant
types of predicate's arguments that ensure the same semantic properties
across their different syntactic realizations) taken from a Polish wordnet.
The peculiarities of organizing senses of verbs (in the form of synsets)
that can be helpful for extracting information about semantic roles from
the Polish wordnet are described. A working role set that is intended to
satisfy predicates with both abstract and concrete meanings within the
same role structure is proposed. Protoframes for selected verb classes are
presented.

Keywords: Semantic roles, valency dictionary, wordnet, FrameNet,
Polish.

1 Introduction

Semantic roles, introduced in the 1960s by Charles Fillmore [3], are meaning related participant types of predicate's arguments that ensure the same semantic
properties across their different syntactic realizations. They facilitate text mining and information extraction. Our current task, connected with the project
[11], is to enrich a valency dictionary of Polish, Walenty, with information about
semantic roles and selectional preferences (typical semantic groups of the arguments) of the arguments. Walenty will be further used i.a. for the construction of
grammars of Polish. It will cover ca. 12 000 verbs and 3 000 of predicative nouns
and adjectives.

Various semantic role sets have been used in a number of projects with shorter
or longer lifetimes. Most of them are based on the original set of roles proposed by
Fillmore [3,4] (AGENT, PATIENT, THEME, EXPERIENCER, GOAL, BENEFICIARY,
SOURCE, INSTRUMENT, LOCATIVE), but practically all applications of the theory
end up adding more roles, as those nine roles are not sufficient for covering the
whole range of verbal schemes. The number of roles used ranges from five, as
proposed by Gillian Ramchand [12], to ca. 1 200 in FrameNet [13], with most of
the projects using about 20–25 of them.

A. Przepiórkowski and M. Ogrodniczuk (Eds.): PolTAL 2014, LNAI 8686, pp. 202–207, 2014.

Apart from the quantity and quality of roles, sets of roles differ by their organization: from loose lists in most of the approaches to grouping into prototypes by David Dowty [2] and a strictly organized hierarchy by John F. Sowa [14].

2   Classes of Verbs in plWordNet and Their Features

A Polish wordnet called Słowosieć [10] seems to present a great potential to our task (we will use plWordNet for short further in the text). Even though modelled on Princeton WordNet for English, it features a number of its own original solutions, not only tailored for Polish, a language typologically different from English, but also aimed at an effective organization of the material within the resource.

Verbs are organized in plWordNet as a hierarchy of more abstract concepts at the top of the tree and more concrete ones at its branches and leaves. Where no appropriate hypernym was found for a group of verbs with specific meaning characteristics, an artificial synset was introduced, named by a short description in Polish and marked explicitly as *synset sztuczny* 'artificial synset'. The topmost artificial synsets roughly correspond to the verbal categories from a classification of verbs by Roman Laskowski [7], a derivation of Zeno Vendler's [15] system of aspectual classes that considers the characteristic features of Polish aspectology [9]. The classification covers basic ontological classes of verbs characterized by different values of dynamicity, telicity, change of state or its absence, volitionality, punctuality, and some other features, amounting to 12 classes, cf. Table 1 and [9]. The differences among those classes are of a lexical-grammatical nature, we will use the term LA (lexical aspect) axis further on to refer to them. Another axis according to which the verbal part of plWordNet can be represented as a rather stable structure is its grouping according to lexical-semantic criteria, further on referred to as LS classes. It includes eleven very general classes that are dispersed among the whole verbal part of plWordNet, with most of such classes appearing both above and inside of every LA class, identified by quite loose natural language descriptions. Table 2 presents the distribution of these data after some normalization. The following abbreviations are used in Table 2: abstract (abstract relations), body (reaction of organism or physiological activity),

Table 1. Features of LA classes

LA	dynamic	change	telic	intentional	perfective
act	+	+	−	+	+
activity	+/−	−	−	+	−
action	+	+	+	+	+/−
process	+	+	+	−	+/−
stative	−	−	−	−	−
accident	+	+	−	−	+
event	+	−	−	−	−
non-punctuals	+	+	−	+/−	+

Table 2. Distribution of verbs in plWordNet among LA and LS classes

LA	abstract	body	contact	loc	logic	phys	pos	prod	psych	soc	temp
act	85	127	263	665	616	147	392	–	642	839	15
activity	105	159	273	734	667	3	423	–	884	991	15
action	253	142	1305	1552	1125	202	126	318	208	519	–
process	244	19	–	95	82	1075	6	110	178	146	94
stative	109	–	–	63	–	62	–	–	313	–	–
accident	156	32	118	238	371	218	32	–	218	21	213
event	134	53	81	243	304	295	27	–	64	52	89

contact (physical contact), loc (location or locative relations), logic (situations connected with cause-effect relations), phys (physical phenomena), pos (change of possession), prod (producing something), psych (emotional, mental, intellectual states), soc (situation connected with social interaction), temp (temporal sequence of events).

3　Correlating Features of the Verb Classes with Roles

Using the grouping of verb senses described above allows making certain assumptions about their nature (some examples follow below) and assigning detected features automatically to the representatives of these groups. Among the considered parameters of the role frames are: the number of roles, their nature, and the character of interaction among them. While LAs are partly responsible for the former (States will tend to have one participant, while Actions and Processes, being telic, will most often have at least two of them), LS-specific classes tend to reflect the rest: social verbs will have humans or human organizations as participants, abstract ones will have processes and concepts as participants; the body class will frequently use the part-whole alternation, etc.

Information about LA and LS classes can be treated as clues for very abstract frames from which all other verbs inherit common features. It can be compared to the top FrameNet [13] frames, such as Event, State, Intentionally_act, Transitive_action, Objective_influence, Cause_emotion, Cause_bodily_experience, Communication, etc., that often have no direct lexical representatives of their own but are inherited by numerous frames with more concrete meanings.

For example, the Cooking_creation frame in FrameNet inherits from the following more abstract frames (the arrows show further inheritance): Intentionally_create → Creating → Transitive_action → Objective_influence → Intentionally_act → Event.

The imperfective and perfective verbs *przyrządzać* and *przyrządzić* 'prepare (food)' that are the most general cooking terms in plWordNet are hyponyms of the artificial synsets *CZASOWNIK – DZIAŁANIE (N)DK oznaczający wytworzenie czegoś 1* ('VERB – ACTION (IM)PERF meaning production of sth'), placed in the 'action-prod' cell in Table 2, which means the presence of the following features: Dynamic (+), Change (+), Telic (+), Intentional (+), production, cf. Table 1. The feature of intentionality allows an automatic assignment of the AGENT role to the subjects of verbs. The 'prod' feature presupposes the presence of an inanimate object that will be assigned the OBJECT role.

Similar reasoning allows formulating the most abstract scenarios, or proto-frames, that can be assigned to the subsets of plWordNet delimited by the cells in Table 2. We present below example protoframes grouped around LS classes without differentiating between LA subclasses or more detailed LS subclasses.

- Physical phenomena: CAUSE affects UNDERGOER; UNDERGOER gets affected;
- Production: INITIATOR produces RESULT (with the help of INSTRUMENT from MATERIAL);
- Psychological: EXPERIENCER has some feeling/is in some state (due to CAUSE); INITIATOR arouses a feeling/state in EXPERIENCER; COGNIZER interacts with CONTENT; COMMUNICATOR adresses ADRESSEE with CONTENT;
- Spatial: AGENT/THEME moves from SOURCE to DESTINATION; THEME moves/is moved from SOURCE to DESTINATION; INITIATOR moves THEME (from SOURCE to DESTINATION using INSTRUMENT in some MANNER).

The core role set can be structured as shown in Table 3. The working set of roles has to satisfy the needs of both the protoframes and frames proper, and needs to be more abstract for the former set. That is why the proposed inventory uses several levels of asbtraction for the roles. The most general ones (INITIATOR, UNDERGOER, PATH, RESULT(EE)) were taken from Ramchand's [12] proposal. INITIATOR and UNDERGOER are further divided into their volitional/non-volitional (AGENT vs CAUSE) or animate/inanimate (PATIENT vs OBJECT) representatives.

ENTITY and EVENT are used mostly with unary, existential predicates. Either can be futher divided into BACKGROUND and FOCUS that interact with abstract, non-symmentrical binary predicates, e.g. those of comparison.

SOURCE and GOAL are set apart as they are often optional arguments in more complex predicates. The left and rightmost roles inside, INITIATOR and RESULT, are those that do not undergo any changes. PATH refers to means by which INITIATOR operates to achieve the desired RESULT focusing on UNDERGOER, the affected group. RESULT appears if there is any change being a consequence of the process, and defines the essence of this change.

Roles with asterisks are LS specific, as can be seen by the possibility of group-ing them under more abstract concepts. They appear in the proposed role struc-ture model due to their presence in many role annotation schemes and popularity among specialists, for reference and/or, probably, further use. Their presence and degree of role granularity in general should be justified by the language material.

To conclude, on the one hand, such preliminary schemes and roles can be treated as working hypotheses to be manually checked by linguists; on the other hand, they can be useful for further consistency checks, and as the basic structure for the future framenet. For some of linguistic and/or technical tasks this level of the generalization of roles and frames may be useful in its present state as well.

Table 3. Inventory of semantic roles

Top-level categories (spanning the table):
COLLECTION / SCENARIO / TYPE — ENTITY — EVENT (PROCESS, STATE-OF-AFFAIRS) — ATTRIBUTE (LOCATION, TIME) — BACKGROUND, FOCUS — PART / PHASE / INSTANCE

| INITIATOR | | PATH | | UNDERGOER | | RESULT | | GOAL |
AGENT	CAUSE	MEDIATOR	INSTRUMENT	PATIENT	OBJECT	RESULTEE	PRODUCT	
COUNTERAGENT	*STIMULUS	BENEFACTOR	MANNER	*EXPERIENCER (psych)	*THEME	*RECIPIENT (pos)	*CONTENT (soc, psych)	*PURPOSE (logic)
*PROTAGONIST (soc)	EFFECTOR		*DURATION (temp)	*ADRESSEE (soc, psych)	*MATERIAL	*BENEFICIARY (loc)	*DESTINATION (loc)	
*COMMUNICATOR (soc, psych)			*TRACE (loc)	*PERCEIVER (psych)				
Cognizer (psych)			*ASSET (pos)					
SOURCE								
*REASON (logic)								
*CONDITION (logic)								

References

1. Apresjan, J.: Semantyka leksykalna. Synonimiczne środki języka (Lexical Semantics. Synonymic means of the language) (transl. Zofia Kozłowska i Andrzej Markowski), Zakład Narodowy im. Ossolińskich, Wrocław, Warszawa, Kraków (2000)
2. Dowty, D.: Thematic proto-roles and argument selection. Language 67(3), 547–619 (1991)
3. Fillmore, C.J.: The case for the case. In: Bach, E., Harms, R.T. (eds.) Universals in Linguistic Theory, pp. 1–88. Holt, Rinehard and Winston, New York (1968)
4. Fillmore, C.J.: The case for case reopened. Syntax and Semantics 8 (1977)
5. Hajnicz, E.: Automatyczne tworzenie semantycznych słowników walencyjnych (Automatic creation of semantic valency dictionaries), Akademicka Oficyna Wydawnicza EXIT, Warszawa (2011)
6. Kotsyba, N.: How light are aspectual meanings?: A study of the relation between light verbs and lexical aspects in Ukrainian. In: Robering, K. (ed.) Events, Arguments, and Aspects. Topics in the Semantics of Verbs. Studies in Language Companion Series, vol. 152, pp. 261–299 (2014)
7. Laskowski, R.: Kategorie morfologiczne języka polskiego — charakterystyka funkcjonalna (Morphological Categories of the Polish Language — a functional characteristics). In: Grzegorczykowa, R., Laskowski, R., Wróbel, H. (eds.) Gramatyka wspóczesnego języka polskiego. Morfologia, 2nd edn., vol. I. PWN, Warszawa (1998)
8. Levin, B.: English Verb Classes and Alternations: A Preliminary Investigation. The University of Chicago Press (1993)
9. Maziarz, M., Piasecki, M., Szpakowicz, S., Rabiega-Wiśniewska, J., Hojka, B.: Semantic relations between verbs in Polish Wordnet 2.0. In: Cognitive Studies (Etudes Cognitives), vol. 11, pp. 183–200. SOW Publishing House, Warsaw (2011)
10. Piasecki, M., Szpakowicz, S., Broda, B.: A Wordnet from the Ground Up, Oficyna Wydawnicza Politechniki Wrocławskiej, Wrocław (2009)
11. Przepiórkowski, A., Hajnicz, E., Patejuk, A., Skwarski, F., Woliński, M., Świdziński, M.: Walenty: Towards a comprehensive valence dictionary of Polish. In: Proceedings of the Ninth International Conference on Language Resources and Evaluation, LREC 2014, Reykjavík, Iceland, pp. 2785–2792 (2014)
12. Ramchand, G.: Verb Meaning and the Lexicon: A First Phase Syntax. Cambridge University Press (2008)
13. Ruppenhofer, J., Ellsworth, M., Petruck, M.R.L., Johnson, C.R.: FrameNet II: Extended Theory and Practice (2010),
    https://framenet.icsi.berkeley.edu/fndrupal/documentation
14. Sowa, J.F.: Knowledge Representation: Logical, Philosophical, and Computational Foundations. Brooks Cole Publishing Co., Pacific Grove (2000)
15. Vendler, Z.: Verbs and Times. In: Linguistics in Philosophy, ch. 4, pp. 97–121. Cornell University Press, Ithaca (1967)

Semantic Extraction with Use of Frames

Jakub Dutkiewicz, Maciej Falkowski, Maciej Nowak, and Czesław Jędrzejek

Institute of Control and Information Engineering, Poznań University of Technology,
Poznań, Poland

Abstract. This work describes an information extraction methodology
which uses shallow parsing. We present detailed information on the extraction process, data structures used within that process as well as the
evaluation of the described method. The extraction is fully automatic.
Instead of machine learning it uses predefined frame templates and vocabulary stored within a domain ontology with elements related to frame
templates. The architecture of the information extractor is modular and
the main extraction module is capable of processing various languages
when lexicalization for these languages is provided.

Keywords: Information extraction, event extraction, semantics, frames.

1 Introduction

Most of the data stored in the Internet today is unstructured and contaminated.
Methods of information extraction (IE) are often designed for a pure language.
We present a methodology for the extraction from contaminated data. There are
basically two methods of information extraction. The first method, open extraction systems based on statistical classifiers and machine learning are scalable,
but not very accurate. Rule based, domain specific systems that use linguistic
patterns are more accurate. The most important element of extraction is an
event, represented by a set of relations. Such an occurrent plays the central role
in describing a situation. An extraction of an event encompasses three layers:
syntactic, semantic and pragmatic. In our previous paper we have shown the
basic idea of how to use the shallow parsing method to extract events from natural language resources [1]. The first two layers of processing are based on the
process of recognizing the meanings of the extracted phrases within the sentence.
This paper extends our work with presentation of the detailed architecture and
methodology used in the extraction tool, including handling language ambiguity.
The method is novel for the Polish language. Comparison of our method with
other approaches for the English language will also be shown.

2 System Architecture

CAT IE Extractor is a system which has been developed to perform information extraction from natural language texts taken from the Internet. This type

A. Przepiórkowski and M. Ogrodniczuk (Eds.): PolTAL 2014, LNAI 8686, pp. 208–215, 2014.

of texts often lacks proper punctuation and may be grammatically and ortho-graphically incorrect. Also, the language used within these texts often contains many abbreviations and phrases (absent in official and literary language), which hinders their analysis. Methodology described in this section takes such realistic language properties into account.

The Extractor processes data in several steps. The extraction itself is the final step and is performed on the structures representing a parsed sentence. Figure 1 describes consecutive stages of the analysis (inner boxes) and the names of the tools that perform those stages for the Polish language (outer boxes). The IE system itself is capable of extracting information from texts of any language, providing set of preliminary processing tools is available. For the inital stages of processing the English language we use the Sundance [6] system. (The result for the English language will be published elsewhere.)

The architecture described here is quite typical for systems of this kind. For the Polish language we use the following tools:

- TaKIPI [2] – morphosyntactic tagger,
- Liner2 [3] – named entity recognizer,
- Spejd [4] – shallow parser.

Because errors are common in analyzed texts, the first step is to normalize input, mainly its punctuation. We use our own tool for this purpose (Text Segmenter), which attempts to fix proper markings at the end of sentences. It also adds the commonly skipped diacritic signs, replaces the phrases with common ortographi-cal errors and replaces commonly used abbreviations with their non-abbreviated equivalents. Before we proceed to the details of the IE system, it is crucial to introduce the key concepts and structures we use in the information extraction process, namely frames, frame instances and extraction patterns.

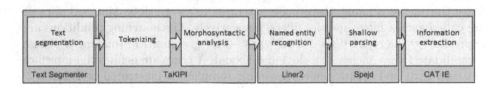

Fig. 1. Pipeline of architecture components

2.1 Frames

Data extracted as events from texts are expressed with artificial intelligence structures called frames. Frames can be considered as a domain data model. A frame is defined as a pair (R, S), where R is the semantic type of frame, and S is the set of its slots. Every slot in S is defined as a pair (R, T), where R is the relation, and T is a list of allowed semantic types of its object. An exam-ple of a frame of a type #Purchase, which describes the event of purchasing is presented in Table 1. The semantic types of frames and slots are expressed as ontology classes. This ensures richer semantics in the extracted data and also allows binding the vocabulary used to express slots with the ontology. This construction makes our system flexible and it is a step towards connecting existing lexical resources with our ontology (e.g. using a system similar to Lemon [5]).

Table 1. Data structures for the extraction of the *Purchase* event

Frame		Template		Instance	
Type: Purchase		Anchor: Active verb		bought	
Slots:		Slots:		Slots:	
Relation (attribute)	Allowed type	Case	prepositions	Value	Type
Buyer	#Person, #Organization	nominative	–	Jan Kowalski	#Man
Seller	#Person, #Organization	dative	–	–	–
Bought object	#Thing	accusative	–	new Fiat	#Car
Price	#Money	accusative	"za" ("for")	for a song	#Money
Place	#Place	locative	"w" ("in")	–	–
Date	#Date	locative	"w" ("in")	last year	#Time

2.2 Frame Extraction Template

A frame extraction template specifies a set of rules, because of which frames can be extracted from the text. A template consists of an anchor and a set of slots (attributes of an event). The extraction process starts with finding an anchor. An anchor is mostly a verb or a verb phrase, when it appears in a derivation (i.e. a syntactically parsed sentence), it triggers the process of recognition of an event. Once the anchor is found, the process attempts to assign slots. Matched phrases are called anchor fillers and slot fillers, respectively. The matching conditions are expressed by the restrictions of two types – first grammatical (making a phrase a candidate for an anchor or a slot) and lexical. A template defines the interface between texts and a data model. A sample template is illustrated in Table 1.

Grammatical conditions consist of the allowed grammatical cases of the phrase and the allowed prepositions at the beginning of the phrase. Note, that in Polish, each case of a noun has its own morphological form. If there are no prepositions defined in the template, prepositional phrases will not be taken into consideration. If there is at least one preposition in the template, phrases with no preposition will not be taken into consideration.

Once the grammatical conditions are fulfilled, the system attempts to find a semantic type of template element which corresponds to the textual value of the head of the phrase. The system uses dictionaries stored within the ontology to perform this task. The matching process has two stages. First, the system verifies the matching of at least one of the allowed semantic types of an anchor.

If the anchor matching is successful, the system next attempts to match slots. There are three possible outcomes of this process:

- Slot matched – The textual value of a phrase matches any allowed semantic type defined in the ontology.
- Slot not matched – The textual value of a phrase matches no allowed semantic type defined in the ontology.
- No match – The textual value of does not appear in the ontology; the phrase is either marked as a slot value candidate (aggressive mode), or it is rejected (conservative mode).

In this way determined template elements become an input in the evaluation process. Usually the conservative mode generates the correct extraction, but at the same time it rejects many correct slot matches, thereby making the extraction error. This is caused mainly by the quality of vocabulary stored in the ontology. On the other hand, the aggressive mode generates many more extractions than the conservative one, though, at the expense of their quality; in this mode incorrectly filled slots and incorrect frames are more common.

2.3 Frame Instance

A unit of information that is created as an effect of the extraction process is a frame instance (FI). A frame instance is created by filling slots of an appropriate frame with textual values including the semantic type of the slot. This semantic type can be obtained either from the Named Entity Recognition (NER) subsystem (if the phrase was recognized as a named entity) or from the ontology subsystem, which maps the phrases to ontological semantic types with its vocabulary. As an illustration in Table 1 of the process we used the following sentence: "Apparently Jan Kowalski bought a new Fiat for a song last year" *"Podobno Jan Kowalski kupił w zeszłym roku nowego Fiata za bezcen"*. The type of the extracted frame is #Purchase – it contains information describing the event of purchasing. Only a part of the possible slots are filled with values, since there were no remarks in the text about a place of purchase nor information about a seller.

2.4 Sentence Disambiguation

Let us define the sentence as a set of phrases.

$$\{p_i : p_i \in S\} \tag{1}$$

According to this definition, sentence contains a number of phrases. Due to the ambiguity of natural language, the result of shallow parsing does not give exactly one correct interpretation. Instead, it returns all available interpretations of phrases in the sentence as it is defined in (2). We assume that each phrase has the denoted, most probable phrase interpretation, as it is described in (3).

$$\{p_{i,j} : p_{i,j} \in p_i\} \tag{2}$$

$$\forall p_i \in S \; \exists p_{i,j} \in p_i, \; p'_i = p_{i,j} \tag{3}$$

The entire pool $I(S)$ of available sentence S interpretations could be interpreted as following:

$$I(S) = \{s_i : \forall p_j \in S \; \exists p_{j,k} \in p_j, \; p_{j,k} \in s_i\} \tag{4}$$

Unless the size of a set $I(S)$ exceeds a large number, the extraction process is performed on every single sentence interpretation. If that number exceeds predefined large number, extraction is performed on a sentence that consists solely of the most probable phrases. Out of all the extraction results one is chosen using the disambiguation process. The disambiguation process uses two arguments. The first one is the normalized number of changes between the most propable phrase interpretations and actual phrase interpretations in the sentence. That normalized number of changes $c(s_i, S)$ with the interpretation s_i of sentence S is defined in (5)

$$c(s_i, S) = \frac{|p'_i : p'_i \in s_i|}{|p_i : p_i \in S|} \tag{5}$$

The second argument of the disambiguation process is the actual interpretation of results. Currently we are using the aggressive approach – the more slots filled, the better. The $c(s_i, S)$ argument in the disambiguation process is secondary to the amount of extraction. If we use the quality function to reduce the number of interpretations to one, we remove the ambiguity of the sentences at this point. To denote that function let us define $|E(s_i)|$ as the a number of extracted slots from the interpretation s_i. The disambiguation process is described in (6) and (7).

$$s_i \in I(S') \iff \forall(s_j \in I(S), s_j \neq s_i) \; E(s_i) \geq E(s_j) \tag{6}$$

$$s_{\text{disamb}} = s_i \iff s_i \in I(S'), \forall(s_j \in I(S'), s_i \neq s_j) \; c(s_i, S') \geq c(s_j, S') \tag{7}$$

If the disambiguation process returns more than one sentence, the disambiguated sentence is chosen randomly out of all returned interpretations.

3   Evaluation

The evaluation process uses frame instances (not individual elements) with filled candidate anchors and slots. To be evaluated positively a template has to have the correctly assigned Agent and Patient (if it appears) thematic roles. Let us give examples for the #Killing event. A result depends on the ontology or named entity recognition along with annotation rules. Annotator decisions (as determined by "gold standard") are positive or negative. The system outcome can also be positive or negative. Examples are presented in Table 2. Case 1 – Alice was recognized by NER, *książka* is in the ontology. Case 2 – *Alice* was not recognized by NER, *książka* is not in the ontology. To evaluate the accuracy of described method we have downloaded data from the National Corpus of Polish (NKJP) [7]. We have randomly chosen 1000 sentences with the word "kill" and 1000 sentences with the word "purchase". We have manually annotated each of the 2000 sentences. If the event was present in the sentence, the annotator was

Table 2. Example evaluation results

Case	Polish clause	English clause	Assignment (aggressive mode)
Case 1/2	*Jan zabił Alę*	John killed Alice	True positive
Case 1	*Jan zabił ksiażkę*	John killed a book	True negative
Case 2	*Jan zabił ksiażkę*	John killed a book	False positive

Table 3. Precision and recall measures for the extraction process

	Recognizing events		Recognizing slots	
Event type	Precision	Recall	Precision	Recall
#Kill	33%	99%	48%	46%
#Purchase	52%	98%	48%	46%

meant to define all apparent slots for the event. Another restriction given to annotators was that, they annotate only #Kill and #Purchase events. Within the 1000 sentences with "kill" based words, annotators chose 268 with the event and have marked 424 slots. For the sentences with "purchase" based words, annotators chose 319 sentences and marked 599 slots within those sentences.

Exactly the same set of sentences was used to perform the extraction with the IE system. To succesfully run the experiment in the conservative mode, we would need to implement a large resource vocabulary to our ontology. The vocabulary we have implemented thus far lets us run the process in the aggressive mode. The system chose 603 sentences with the #Purchase event and 788 sentences with #Kill event. The set of 603 sentences contained all annotated sentences but 3. The set of 788 sentences contained all annotated sentences but 6. The extractor chose 277 slots correctly and 292 slots incorrectly for the #Purchase event: it chose 262 slots correctly and 184 slots incorrectly for #Kill event. The measures evaluated are presented in Table 3.

Table 4. Sample extraction

Sentence:	At the online auction, Heinz bought a cube for a pile of money.
Sentence (literally):	*Za grube pieniądze Heinz kupił kostkę na aukcji internetowej.*
Bought object:	a cube
Price:	for a pile of money
Place:	at the online auction

An example of the extraction is presented in Table 4. An error in the presented extraction highlights one of the major problems with the methodology. There is a high possibility, that a word of foreign origin will be marked with incorrect grammatical tags. This, low level error propagates trough the entire process. Propagation of errors is an issue, as a mistaken slot filler makes the extraction partially incorrect. Evaluation of the partially incorrect frame instances is summarised in Table 5. The *Precision*$_{er}$ measure is the precision of event recognition,

Table 5. Distribution of results regarding the number of errors within frames

Number of errors	Event type	Precision	Recall	$\frac{Precision}{Precision_{er}}$
0	#Kill	17%	50%	51%
<2	#Kill	28%	84%	85%
<3	#Kill	33%	91%	100%
0	#Purchase	17%	31%	33%
<2	#Purchase	36%	69%	69%
<3	#Purchase	47%	90%	90%

as it is presented in Table 3. The $\frac{Precision}{Precision_{er}}$ is the precision of extraction if we take only the frames which were correctly recognized as events.

4 Conclusions

In this work we improved our previous methodology [4] to enhance domain extraction. We created a mechanism that uses constraints described in the structualized data to perform extractions. We used the pattern based approach (with aggressive mode of matching). The next step would be to create a probabilistic model to find missing thematic roles outside of sentences with anchoring keywords. The methods described in this paper could be enhanced in several ways. The first, similar to the methods used in Boxer, would be based on the construction of a large ontology for most of verbs used in Polish. Such an ontology would look very similar to Verbnet. Valence dictionaries [4] would be helpful in this task. With a large enough ontology we could be able to extract and represent entire messages in a discourse representation compatible with ontological structures. This approach is scalable to a point: not only ontologies have to be enhanced but annotated corpora would likewise require enhancement. This way of improving our methodology would also require methods of representing the relations between events. On a deeper level the improvement of evaluation measures would also require enhancmenet of the shallow grammar parser used and neccesitate solving pronoun and coreference problems. Our method is similar to the Boxer [9] semantic parser, applied mostly to the English language. For selected events, used in this paper our thematic roles are more specific then for Verbnet derived eqiuvalent roles. For example, we use Perpetrator instead of Agent and Victim instead of Patient. There is a difference in the process of template matching for the Polish and English languages. Creating rule based matching templates for Polish is easier than for English, because of the greater expressive power of Polish with regards to its morphology. We plan to demonstrate this in our future work . That is why the recent work for template matching for English currently uses statistical methods [8].

Acknowledgements. This work was supported by the Polish National Centre for Research and Development (NCBR) No O ROB 0025 01 and 04/45/DSPB/0105 grants.

References

1. Dutkiewicz, J., Jędrzejek, C.: Ontology-based event extraction for the Polish language. In: LTC 2013 (2013)
2. Piasecki, M.: Polish Tagger TaKIPI: Rule Based Construction and Optimisation (2007)
3. Marcińczuk, M., Kocoń, J., Janicki, M.: Liner2 – A Customizable Framework for Proper Names Recognition for Polish. In: Bembenik, R., Skonieczny, Ł., Rybiński, H., Kryszkiewicz, M., Niezgódka, M. (eds.) Intell. Tools for Building a Scientific Information. SCI, vol. 467, pp. 231–254. Springer, Heidelberg (2013)
4. Przepiórkowski, A.: Powierzchniowe przetwarzanie języka polskiego, Akademicka Oficyna Wydawnicza EXIT, Warszawa (2008)
5. McCrae, J., Spohr, D., Cimiano, P.: Linking lexical resources and ontologies on the semantic web with lemon. In: Antoniou, G., Grobelnik, M., Simperl, E., Parsia, B., Plexousakis, D., De Leenheer, P., Pan, J. (eds.) ESWC 2011, Part I. LNCS, vol. 6643, pp. 245–259. Springer, Heidelberg (2011)
6. Riloff, E., Phillips, E.: An introduction to the sundance and autoslog systems (2004)
7. Lewandowska-Tomaszczyk, B., Bańko, M., Górski, R.L., Łazinski, M., Pęzik, P., Przepiórkowski, A.: Narodowy Korpus Języka Polskiego: geneza i dzień dzisiejszy (2010)
8. Huang, R., Riloff, E.: Modeling Textual Cohesion for Event Extraction. In: Proceedings of the 26th Conference on Artificial Intelligence, AAAI 2012 (2012)
9. Curran, J.R., Clark, S., Bos, J.: Linguistically Motivated Large-Scale NLP with C&C and Boxer. In: Proceedings of the ACL 2007 Demonstrations Session (ACL-07 demo), pp. 33–36 (2007)

Constraint Grammar-Based
Swedish-Danish Machine Translation

Eckhard Bick

Institute of Language and Communication, University of Southern Denmark,
Campusvej 55, DK-5230 Odense M, Denmark
eckhard.bick@mail.dk

Abstract. This paper describes and evaluates a grammar-based machine translation system for the Swedish-Danish language pair. Source-language structural analysis, polysemy resolution, syntactic movement rules and target-language agreement are based on Constraint Grammar morphosyntactic tags and dependency trees. Lexical transfer rules exploit dependency links to access contextual information, such as syntactic argument function, semantic type and quantifiers, or to integrate verbal features, e.g. diathesis and auxiliaries. Out-of-vocabulary words are handled by derivational and compound analysis with a combined coverage of 99.3%, as well as systematic morpho-phonemic transliterations for the remaining cases. The system achieved BLEU scores of 0.65-0.8 depending on references and outperformed both STMT and RBMT competitors by a large margin.

Keywords: Machine Translation, RBMT, Constraint Grammar.

1 Introduction

Over the last decade, riding on an exponential growth curve of computer processing power and corpus size, Statistical Machine Translation (STMT) has outpaced research into Rule-Based Machine Translation (RBMT), albeit there is a certain interest in hybrid systems, not least for languages with a rich morphology and a need for syntactic reordering (e.g. Hindi, Ahsan et al. 2010). Since STMT is a machine learning technique that depends on the availability of (bilingual or comparable) training data, it has enormously profited from big data techniques in general, but while regular parallel corpora like Europarl (Koehn 2005) have helped to develop the necessary methods, the largest public success has been achieved by the lords of big data, Internet giants Google and Bing, which can now be seen as bench marks for less widely known, and more specialized systems. Given the necessary data trove, STMT is a very cost-efficient method to produce machine translation for many language pairs and to harvest fluency from target language examples. However, STMT still suffers from certain more or less inherent problems:

1. For less-resourced languages, STMT may lack sufficient training data. This is particularly true if both languages in a translation pair are small. Even with English-bilingual data at hand, there will be quality loss in an English-mediated

transfer, because English as an interlingua[1] may disguise or create ambiguities relevant to the languages in question.

2. Without access to systematic linguistic analysis and generation, STMT on its own has difficulties in handling morphologically rich languages (Ahsan et al. 2010), because the individual inflexions, derivations and compounds are too rare. In addition, without syntactic-structural analysis long-distance features such as agreement between clause parts, pronominal anaphora and reflexivity are out of reach for n-gram based machine learning.

3. Because it is not possible to interfere directly with a phrase-based STMT core module, it is difficult to fix individual or systematic errors, even when identified, or to pass lexical knowledge such as word sense disambiguation on to an STMT system (Carpuat & Wu 2005), and it is not possible to systematically adapt the system to a domain different from the ones training data is available for.

4. Without dictionaries in a more traditional sense, STMT systems run the risk of semantic confusion of words that share the same context, such as currency words and antonyms.

Rule-based systems, on the other hand, can implement symbolic language models for small languages even in the absence of large corpus data (1), both for analysis and transfer, as argued for instance by Seiss & Nordlinger (2001) who use morphological finite-state transducers and manual transfer rules for Murrinh-Patha. Rule-based MT systems (as well as certain hybrid combinations) support deep module integration and have system-wide access to a full linguistic analysis, allowing them to take into account both analytical morphology and long-distance relations (2). Needless to say, rules can be changed or amended "locally" in order to handle errors, sense distinctions or domain migration (3). And finally, in a manual translation lexicon, currency words and antonyms will be listed individually without the risk of contextual confusion (4).

We therefore believe that rule-based MT systems should be given a second chance, not least for less-resourced languages, and though hybrid systems may be the ultimate solution, relatively "pure" systems can demonstrate strengths, weaknesses and evaluative insights that may guide later hybridization efforts. We are aware that phrase-based STMT can be improved by adding higher-order linguistic relations, for instance manual reordering rules using dependency relations for language pairs with SVO/SOV word order differences (Peng Xu et al. 2009), but it would be an added advantage if such pre- and postprocessing modules could be handled in one formalism, with a shared category set, shared tags and shared lexical information, with the possibility of easy cross-reference between modules, and an integrated system architecture. In this paper we present such a system for two small languages (Swedish to Danish), implementing a rule-based approach relying on high-quality

---

[1] English is more distant from both Swedish and Danish than the two languages are from each other. Rather, if the training data problem can be overcome through RBMT, it would make more sense to use one Scandinavian language as interlingua for another, as proposed by Bick & Nygaard (2007) for Norwegian and Danish.

source language dependency analysis (Constraint Grammar, Bick 2005) as a matrix for anchoring context-driven transfer rules, disambiguation, agreement and compounding. The general architecture of this approach is in principle language-independent, and inspired by similar work for the Danish-English machine translation (Bick 2007).

2   System Architecture

On the source language side, the core of the system is SweGram,[2] a Constraint Grammar (CG) parser using the CG3[3] formalism and rule compiler. SweGram was designed with robust corpus annotation[4] in mind, and provides the following information in a tag-based fashion, based on a 70,000-lemma lexicon and 8,500 tagging and disambiguation rules:

1. tokenization, including abbreviations, numerical and scientific expressions, complex function words and named-entity recognition (NER)
2. morphological analysis, including compound recognition, derivation and endings-based out-of-vocabulary heuristics
3. syntactic function tags (subject, object, predicative etc.)
4. dependency trees
5. semantic classification of common and proper nouns and valency classification of verbs

For instance, in the example analysis below, word #13, "snösmältningsmaskin" (snow-melting machine) is recognized as an allowed compound (with a fuge-s), and tagged as a noun (N) singular (S) in the common gender (UTR), nominative (NOM) and indefinite (IDF), functioning as the head of a direct/accusative object (@ACC) of word #10, the verb "inviga" (introduce, take into use).

De	[den] <*> ART nG P DEF @>N #1->3
senaste	[sen] <jtemp> ADJ SUP nG nN DEF NOM @>N #2->3
dagarnas	[dag] <dur> <temp> N UTR P DEF GEN @>N #3->4
snöstorm	[snöstorm] <event> <wea> <F:snö+storm> N UTR S IDF NOM @SUBJ> #4 >7
i	[i] <np-close> PRP @N< #5->4
New=York	[New=York] <civ> <*> PROP NEU S NOM @P< #6->5
fick	[få] <vt+INF> <mv> V IMPF AKT @FS-STA #7->0

[2] A demo version of the parser, as well as an overview of category and tag definitions, can be accessed at http://beta.visl.sdu.dk/visl/sv/parsing/automatic/

[3] CG3 has been developed as an open-source tool by the Danish language technology company GrammarSoft ApS in cooperation with the University of Southern Denmark, who maintain a documentation and download site at http://beta.visl.sdu.dk/constraint_grammar.html.

[4] SweGram annotated corpora are accessible at http://corp.hum.sdu.dk (CorpusEye)

myndigheterna [myndighet] <HH> <aci-subj> N UTR P DEF NOM @<ACC #8->10

att   [att] INFM @INFM #9->10

inviga   [inviga] <mv> V INF AKT @ICL-<OA #10->7

en   [en] ART UTR S IDF @>N #11->13

splitterny   [splitterny] <heur> <F:splitter+ny> ADJ UTR S IDF NOM @>N
                #12->13

snösmältningsmaskin [snösmältningsmaskin] <good-compound <N:snösmältning~s+maskin>
                <mach> N UTR S IDF NOM @<ACC #13->10

$.   [.] PU @PU #14->0

*[1-The 2-last 3-days' 4-snow storm 5-in 6-New York 7-got 8-the authorities 9-to 10-take into
use 11-a 12-brand new 13-snow melter]*

Note that this 13-word sentence contains 3 compounds, which is quite normal for
Swedish. Though only one ("snöstorm") was listed in the parser's lexicon, it assigned
correct lexical types (<wea> <event> weather event, and <mach> machine) to the
nouns. This will allow the translation system to construct plausible translations for the
compounds, and provide useful semantic context to other words in the sentence. The
parser also identified the special construction "få till att" (get sb to do sth), marking
"myndigheterna" (authorities) as both object of "få" and subject (<aci-subj>) of
"inviga", and allowing the translator to pick a reasonable translation for "få" (get)
which in Swedish is just as phrasally ambiguous as in English.

3   Lexical Transfer

For lexical transfer, Swe2dan exploits CG tags in two ways, as either local or contextual
distinctors, which are used to formulate transfer rules that help the system decide which
translation to pick, and can be used both for polysemy resolution and usage differences
("synonym picking"), with no statistical element needed. The idea is not to define, but
to distinguish meanings. While local distinctors refer to tags on the token itself (e.g. part
of speech, number, syntactic function, domain), contextual distinctors refer to features
of arguments, attributes, heads etc. of the word in question, tracing dependency links to
the second degree, or simply using relative positions left and right. The function word
"när" *(near, when)*, for instance, has 5–6 different translations, not counting fixed
expressions, which go in a separate lexicon. However, the different translations can be
reliably distinguished by word class (adverb ADV vs. conjunction KS), clause type
(interrogative vs. relative), syntactic function (pre-adject @>A), head verb tense (IMPF)
or immediate left context (P-1)[5].

- när_ADV :nær *[near]*; S=(<interr>) :hvornår *[when?]*; S=(<rel>) :når *[(at
  the time) when]*; S=(@>A) :næsten *[almost]*; P-1=('hart') :næsten *[almost]*
- när_KS :når *[when(ever)]*; H=(IMPF) :da *[when ..ed]*

---

[5] A reference to the head verb is clearly a long-distance distinctor, but even local distinctors
may depend on larger contexts – thus, syntactic function and clause type are local only in the
sense that the CG engine has created local tags for them.

In spite of the relatedness of Swedish and Danish, a one-on-one translation is possible in less than 50% of all tokens. Thus, lexicon entries with transfer rules account for only 4% of the ca. 100,000 lexemes, but for 53% in frequency terms. In other words, frequent lexemes are much more ambiguous, and more prone to usage variation than rarer ones. The structurally most important word class, verbs, stands for 40% of all contextual transfer rules. In the example, 5 translations for the verb "fräsa" are distinguished by specifying daughter-dependents (D) or dependents of dependents (granddaughters, GD) as subjects (@SUBJ) or objects (@ACC) with certain semantic features, such as human <H>, vehicles <V> or <food>. For closed-class items such as prepositions or adverbs (here: "åt", "iväg", "förbi"), it often makes sense to refer directly to word forms. Negative conditions are marked with a '!'-sign, optional conditions with a '?'.[6]

fräsa_V :hvæse *(to hiss like a cat)*;
   D1=("åt") GD1=(<H>) D2=(<H> @SUBJ) :vrisse *(to snap at sb)*;
   D=(<[HV].*> @SUBJ) D=("(iväg|förbi)") D!=(@ACC) :rase *(tear/speed along)*;
   D=(<food.*> @ACC) :stege, :brune, :brase, :lynstege *(to fry)*;
   D=(@ACC) D=(<H> @SUBJ) :fræse *(to mill, to cut a material or tool)*;

Sometimes a distinction depends on clues that are present in the overall context, but have not been explicitly tagged by the SweGram parser. We therefore introduced a separate CG, run after the parser and before translation, that adds the desired tags from context. Examples are reflexivity, article insertion or the propagation of number, definiteness and the +human feature to under-specified heads or dependents, or from anaphoric referents to pronouns. A second round of feature propagation is done in the translation program itself, after translations have been chosen. For instance, if a translated Danish noun differs from the Swedish original in gender (or sometimes, number), all other members of the noun phrase need to have changed their gender, too.

Finally, compounding and affixation can be used to assign different translations depending on whether a lexical item is used as first, last (second) or middle part, if necessary in combination with further conditions:

lock_N (25) :lok, :hårlok *[curl]*; S=(<second>) :låg *[cover]*; S=(NEU) :låg; S=(< first>) :lokke *[luring]*

4   Out-of-Vocabulary Words

Out-of-vocabulary words can occur both at parser level and translation level. Though we are primarily concerned with the latter here, it is relevant that the SweGram parser normalizes simple misspellings and provides a compound analysis for both known

---

[6] The formalism also allows reference to heads (H), heads of heads (grandmothers, GM), to numbered relative positions and their dependents, dependency direction, dependent n-grams and others.

and unknown compounds. As a first fallback for out-of-vocabulary words, the translator performs part-for-part translations following the parser's compound breakdown, using the above-mentioned rules for first and second parts. The second fallback is *transformation* rather than translation, i.e. heuristic translation based on partial translation (of word endings and affixes) and systematic letter replacement. For instance the definite plural noun ending '-orna' will be changed into Danish '-erne', the affixes '-ism' and '-skap' become '-isme' and '-skab', and Swedish 'ö'/'ä' will become Danish 'ø'/'æ'. Similar changes apply for participle and verb endings, as well as consonant gemination. In a way, the rationale behind these changes is treating Swedish as a kind of misspelled Danish, exploiting that a large portion of words has a common etymology.

In a newspaper test corpus with 144,456 non-punctuation tokens the parser classified 7,120 unknown non-name words as "good compounds" and 1,245 as outright heuristic analyses. Swe2dan came up with non-heuristic translations for 99.1% of the compounds and had ordinary lexicon entries for 62.1% of the heuristicals, leaving the rest to the transformation module. For ordinary, parser-sanctioned words the translation lexicon had a coverage of 99.71%, missing out on only 368 words, and bringing total coverage up to an impressive 99.33%. A breakdown of the 368 words that were known to the parser, but not to the translator, showed that roughly half (51.6%) of the Swedish words, many of them foreign, were left as-is and worked in Danish, too. 11.7 percent were transformed into the correct Danish word, and 36.7% produced wrong translations, either unchanged or with wrong or partial transformations. Very few good-compounds needed transformational translation, and in only 2 cases the transformation was wrong. 69% of the out-of-vocabulary words deemed "heuristic" by the parser were misspellings, non-letter characters and word fusions due to missing spaces. 11% had transformations (10% wrong, 1% correct), and 10% were correctly left as is. Including correct transformations and as-is translations, overall translation coverage was 99.62%.

5   Target Language Generation

Swedish and Danish both have 2 genders, but depending on the translation, noun genders may not match, not even for cognates. While finding the correct inflection for the noun itself is a simple lookup procedure, it is more difficult to propagate gender and number changes to the whole np, predicatives and complements which may be located far away in the sentence. Therefore, even the generator profits from the deep CG parse, propagating inflection changes along dependency links.

Evaluated on a 10,000 sentence newspaper text chunk, the coverage of the Danish generation lexicon was satisfactory. 94.49% of lookups were successful as complete lexemes, for 3.03% compound analyses provided by the parser were used for lookup of the inflectionally relevant last part of the compound, and for 0.78% the generator itself was able to create a (heuristic) compound analysis. As a last resort, unknown words were inflected following the most common Danish paradigm for the word class in question. This was necessary in 1.84% of cases (including 0.14% related to compound analysis), but caused virtually no inflection errors in the examined sample, probably because irregular forms tend to be frequent (and therefore lexicon-covered), while it is the regular paradigms that are productive and cover most of the Zipf curve tail. With a combined coverage of almost 100%, errors from the TL generation module are unlikely to be lexically caused, leaving only errors caused by wrongly assigned inflection tags, which are difficult to isolate from SL analysis and transfer.

6   Structural Transfer

Swedish and Danish are both East-Scandinavian languages and share basic sentence structure. However, there are certain important differences the treatment of which asks for a scope beyond n-gram matches:

1. Due to certain differences in the expression of definiteness, an np's left (article or pronoun) and right (noun inflection) edges have to be handled interdependently.
2. Adverb position differs on several accounts: Adverb particles of transitive phrasal verbs are placed after the object in Danish, but before it in Swedish.
3. Swedish has special supine verb forms, many of which are morphologically indistinguishable from active voice participles, which have to be translated with either past tense verbs or participle constructions in Danish.

At the level of individual words, our MT system can handle these cases in the lexical transfer rules themselves. For instance, the translation of the definite article (1) may be set to "nil" in a context of H=(N DEF), i.e. a definite head noun (Swedish double definiteness). Where more elaborate, or more global, conditions are needed, special CG rules are used. Like the underlying parser, these CG rules have access to virtually all tags and relations, and can change definiteness inflection, or insert articles, in preparation of the Danish translation. (2) can in principle also be handled by transfer rules, setting the translation of a phrasal adverb as "nil" while at the same time adding it to the object translation (a kind of indirect movement instruction):

packa_V D=("ut")_nil D=(@ACC)_[+ud] :pakke *[unpack – "pack out"]*

However, the lexical load is much bigger in this case. We therefore (also) use dependency-based movement rules like the following:

w(@MV<|@OA),g(<right> @ACC) -> 2,1

This rule changes the order of a word constituent (w) that is a verb particle (@MV<) or adverbial object complement (@OA) with a group constituent (g) that is a direct object (@ACC) to the right of its head verb. The rule will work independently of the size of the object np, because all dependents are automatically included in the movement. Similar adverb movement rules are also necessary for inverted (VS) clause order, where adverbs have to be moved out of the VS bracket (V A S → V S A), or for infinitive markers (insertion or adverb movement). All in all, the movement grammar contains 61 rules.

Supine forms, finally, (3) need structural information twice – first, the parser needs global context to disambiguate the form itself (it is ambiguous with ordinary, np-internal participles for all regular verbs), second, translation tense has to be chosen, with the possible insertion of an auxiliary and corresponding adverb or subject movements.

7   Evaluation

We evaluated the system on 100 random new sentences (ca. 1,500 words), taken from the Leipzig Wortschatz corpus collection,[7] comparing GramTrans' Sve2dan translations to those of three other systems, Google Translate,[8] Bing Translator[9] and Apertium,[10] all of which maintain open-access user interfaces. While Google Translate and Bing Translator rely on STMT, Apertium (Tyers et al. 2010) is an open-source RBMT system like GramTrans itself. However, where Apertium uses corpus-trained HMM taggers, GramTrans is rule-based also in the SL analysis modules.

First, we measured all systems against both an independent manual translation and best-case edited system translations, using the BLEU (Papineni et al. 2002) and NIST metrics.

Table 1. BLEU/NIST scores

	Manual reference (1)	Edited system reference	Multi-reference (all minus self)
GramTrans	**0.645 / 8.515**	**0.838 / 9.817**	**0.757 / 10.050**
Google	0.387 / 6.300	0.645 / 8.361	0.539 / 8.150
Apertium	0.390 / 6.391	0.516 / 7.361	0.468 / 7.418
Bing	0.342 / 6.006	0.600 / 8.064	0.492 / 7.793

In this comparison, GramTrans clearly outperformed all other systems. Apertium performed slightly better than the statistical systems, when measured against one manual translation, but came out last when measured against "self-edit" or "all others".[11] The statistical systems profited relatively more from the inclusion of self-edits, and in relative

---

[7] http://corpora.informatik.uni-leipzig.de/download.html

[8] Translations were taken from http://translate.google.com/ (15 March 2014)

[9] Accessed at http://www.bing.com/translator (15 March 2014)

[10] We used the demo at: http://www.apertium.org/?lang=n&lang=en (15 March 2014)

[11] In their own evaluation of Swedish-Danish Apertium and an early unpublished version of GramTrans, Tyers & Norfalk (2009), using word edit rates (WER) and editing distance, ranked their system (WER 30) below GramTrans (WER 26) and Google (WER 35), even when introducing an anti-GramTrans bias by only measuring GramTrans against post-edited Apertium-translations.

terms the difference between GramTrans and Google was bigger for BLEU than for NIST in all runs. Since NIST downplays the importance of short/common words and of small length difference, this finding may result from a particular strength of rule-base systems – function words, definiteness and inflexion/agreement, all of which result in the kind of "short" differences under-weighted by NIST.

In order to determine system similarity, we also measured systems against each other, using one system's edited translation as a BLEU-reference for another system:

Table 2. Cross-system BLEU scores

Reference: Test:	GramTrans edited	Google edited	Apertium edited	Bing edited
GramTrans	(0.838)	0.497	**0.666**	0.501
Google	0.387	(0.645)	0.384	**0.478**
Apertium	**0.426**	0.330	(0.516)	0.325
Bing	0.358	**0.446**	0.353	(0.600)

Here, GramTrans and Apertium score slightly better against each other's edited versions than against the manual standard, while dropping against Google and Bing edits, attesting to similar (rule-based) translation styles. Likewise, the statistical systems Google and Bing perform better against each other than against either the manual translation or edits of the rule-based systems.

Expecting even clearer correlations between results and RBMT/SMT system styles, we also performed an edit distance evaluation, using the TER metric (*Translation Error Rate,* Snover 2006) and again comparing system translations with both manual and edited translations. In TER, error rates can be interpreted as editing distances, covering insertion, deletion or substitution of a word, or the movement (shift) of a word or word chain. Fewer edits mean a lower TER and a better performance.

Table 3. TER distances

Reference: Test:	Manual	GramTrans edited	Google edited	Apertium edited	Bing edited
GramTrans	**20.84**	(8.57)	32.12	*19.77*	30.98
Google	45.05	44.40	(23.60)	45.20	*37.56*
Apertium	34.54	*31.13*	41.96	(24.51)	41.75
Bing	48.62	46.98	*40.70*	48.03	(28.05)

Again, GramTrans outperformed its competitors. Its relative advantage compared with the statistical systems was even bigger with TER than with BLEU, and it had the lowest editing cost (scores in parentheses), i.e. the GramTrans-translation needed fewest changes to become lexico-grammatically acceptable. In addition, GramTrans was also, for all combinations, the cheapest system to turn into the edited version of another system. Against the manual translation, GramTrans' relative TER advantage is shared by the other RBMT system, Apertium, which in the TER evaluation ranked second against this reference, while performing similar to Google with BLEU. Cross-system editing distances indicate that the statistical pair on the one hand (Google and Bing), and the rule-based pair on the other (GramTrans and Apertium) share inherent features even in the edited versions (bold italics).

A break-down of edit types corroborates the difference between the two system types:

Table 4. TER evaluation, edit types

	Inser-tions	Dele-tions	Substi-tutions	Shifts (word shifts)	Ins&del / subs	TER
GramTrans	20	11	103	5 (6)	0.40	8.57
Google	71	51	263	11 (11)	0.46	23.60
Apertium	21	33	333	11 (13)	0.16	24.51
Bing	87	74	290	18 (19)	0.56	28.05

The STMT systems have a relatively high need for deletions and insertions, compared to substitutions (second-last column), a finding that might be linked to problems with small function words and articles, possibly compound splitting caused by using English translations of Swedish compounds as a fall-back route into Danish (which compounds the same way Swedish does). Apertium has the opposite problem, with a high proportion of substitutions (in part due to its low lexical coverage), but a good score for insertions/deletions. Bing sticks out with the highest need for movements, indicating a poor syntactic engine.

A qualitative inspection of the data showed that GramTrans (and Apertium) completely avoided a number of error types typical of STMT systems:

- the confusion of "ontological sister terms"
  - Bing: "dollar/kroner"
  - Bing: "Per Wesslén/Wade", "Hale/Halestone", "Svensson/Smith"
  - Google: "Solbergaskolen"/"Solbergaleden"
- the literal translation of names caused by case folding
  - Bing: "Huge chockstartade" → "Kæmpe [= big/huge] chockstartade"

- errors caused by using a big-data training language as an intermediate step between small languages with insufficient direct training data,
  - ○ Google: "styrelse" (Swedish) → "board" (English) → "bord" [table] instead of "bestyrelse" (Danish)
  - ○ compounding: Google: "säkerhetsexpert" (Swedish) → "security expert" (English) → "sikkerhed ekspert" instead of "sikkerhedsekspert" (Danish)
  - ○ Google: "Se upp för elgen" (Swedish) → "Watch out for elg" (Danish)

Another difference were long-distance syntactic relations, such as the choice between reflexive ("sin") and non-reflexive ("hans") possessive pronouns, which were handled well by the RBMT systems, but badly by the STMT systems on all occasions. In the example sentence the reflexive noun phrase is correctly translated by GramTrans, while Bing introduces errors in both reflexivity (refl), number and compounding[12]. Google gets the number feature right, but not the other two, while Apertium simply retains the Swedish expression, including the reflexive.

Table 5. Qualitative differences: Reflexives

		refl	number	lex	compound
Swedish original	Nicole talar ut om sina viktproblem *Nicole speaks out about her weight problems*				
GramTrans	Nicole taler ud om sine vægtproblemer	ok	ok	ok	ok
Google	Nicole taler ud om hendes vægt problemer	err	ok	ok	err
Apertium	Nicole taler ud om sine viktproblem	ok	(ok)	err	(ok)
Bing	Nicole taler ud om hans vægt problem	err	err	ok	err

8   Conclusion and Outlook

We have shown, for the Swedish → Danish language pair, that a modular CG-based translation system with manual transfer rules can outperform bench mark systems such as Google Translate and Bing Translator, and discussed the architecture and performance of the individual modules. The system is publicly available at http://gramtrans.com, and being used in an integrated Swedish-Danish version of Wikipedia (http://dan.wikitrans.net). Future research goals include improved heuristics for out-of-vocabulary words, domain flags and improvements to a fledgling Danish → Swedish sister system – a task that for RBMT is by no means trivial since neither SL analysis nor transfer and disambiguation rules can be reused.

---

[12] Using the ordinary possessive forces systems to decide on possessor gender, and unlike Google, Bing gets that wrong, too.

References

1. Ahsan, A., Kolachina, P., Kolachina, S., Sharma, D.M.: Coupling Statistical Machine Translation with Rule-based Transfer and Generation. In: Proceedings of the 9th Conference of the Association for Machine Translation in the Americas, Denver, Colorado (2010), http://amta2010.amtaweb.org/AMTA/papers/2-06-AhsanKolachinaEtal.pdf
2. Bick, E.: Dan2eng: Wide-Coverage Danish-English Machine Translation. In: Maegaard, B. (ed.) Proceedings of Machine Translation Summit XI, Copenhagen, Denmark, September 10-14, pp. 37–43 (2007)
3. Bick, E., Nygaard, L.: Using Danish as a CG Interlingua: A Wide-Coverage Norwegian-English Machine Translation System. In: Nivre, J., Kaalep, H.-J., Muischnek, K., Koit, M. (eds.) Proceedings of the 16th Nordic Conference of Computational Linguistics, May 24-26, pp. 21–28. University of Tartu (2007)
4. Bick, E.: Turning Constraint Grammar Data into Running Dependency Treebanks. In: Civit, M., Kübler, S., Martí, M.A. (eds.) Proceedings of TLT 2005, pp. 19–27. Universitat de Barcelona (2005)
5. Carpuat, M., Wu, D.: Word Sense Disambiguation vs. Statistical Machine Translation. In: Knight, K., Ng, H.T., Oflazer, K. (eds.) Proceedings of ACL 2005, pp. 387–394. Association for Computational Linguistics, Stroudsburg (2005)
6. Koehn, P.: Europarl: A Parallel Corpus for Machine Translation. Machine Translation Summit, pp. 79–86 (2005)
7. Papineni, K., et al.: BLEU: A Method for Automatic Evaluation of Machine Translation. In: Proceedings of the 40th ACL, Philadelphia, pp. 311–318 (July 2002)
8. Seiss, M., Nordlinger, R.: An Electronic Dictionary and Translations System for Murrinh-Patha. In: Gimeno, A. (ed.) Proceedings of EUROCALL 2011, vol. 20(1) (2012); The Eurocall Review
9. Snover, M., Dorr, B., Schwartz, R., Micciulla, L., Makhoul, J.: A study of translation edit rate with targeted human annotation. In: Proceedings of Association for Machine Translation in the Americas, pp. 223–231 (2006)
10. Tyers, F.M., Sánchez-Martínez, F., Ortiz-Rojas, S., Forcada, M.L.: Free/open-source resources in the Apertium platform for machine translation research and development. The Prague Bulletin of Mathematical Linguistics (93), 67–76 (2010)
11. Tyers, F.M., Nordfalk, J.: Shallow-transfer rule-based machine translation for Swedish to Danish. In: Pérez-Ortiz, J.A., Sánchez-Martínez, F., Tyers, F.M. (eds.) Proceedings of the First International Workshop on Free/Open-Source Rule-Based Machine Translation, pp. 27–33. Universidad de Alicante, Alicante (2009)
12. Xu, P., Kang, J., Ringgaard, M., Och, F.: Using a dependency parser to improve SMT for subject-object-verb languages. In: Proceedings of NAACL 2009, pp. 245–253. Association for Computational Linguistics, Morristown (2009)

A Hybrid Approach to the Development of Bidirectional English-Oromiffa Machine Translation

Jabesa Daba[1] and Yaregal Assabie[2]

[1] Department of Computer Science, Wollega University, Ethiopia
jappy.db@gmail.com
[2] Department of Computer Science, Addis Ababa University, Ethiopia
yaregal.assabie@aau.edu.et

Abstract. This paper presents the development of bidirectional English-Oromiffa machine translation system using a hybrid of rule-based and statistical approaches. Since English and Oromiffa have different sentence structures, we implement syntactic reordering with the purpose of making the structure of source sentences similar to the structure of target sentences. Accordingly, reordering rules are developed for simple, interrogative and complex English and Oromiffa sentences. Two groups of experiments are conducted by using purely statistical approach and hybrid approach. The Oromiffa-English SMT yields a BLEU score of 41.50% where as English-Oromiffa SMT has a BLEU score of 32.39%. After applying local reordering rules, the system is improved to provide a BLEU score of 52.02% and 37.41% for Oromiffa-English and English-Oromiffa translations, respectively.

Keywords: English-Oromiffa machine translation, hybrid machine translation, local reordering.

1 Introduction

Machine translation is aimed at enabling a computer to transfer natural language expressions from one natural language (source language) into another (target language) while preserving the meaning and interpretation. The translation can be unidirectional or bidirectional [3]. In case of unidirectional translation, the system translates from the source language into the target language only in one direction. Bidirectional systems work in both directions in a way that one language can act as source language or a target language.

Different methods are currently used to develop machine translation systems and the basic approaches are rule-based, example-based, statistical and hybrid machine translation [10]. Rule-based machine translation relies on manually crafted linguistic rules for translation whereas example-based machine translations translate by decomposing a sentence into fragments, translating each of the fragments using the principle of analogy and then composing them properly. Statistical machine translation uses human produced translations known

A. Przepiórkowski and M. Ogrodniczuk (Eds.): PolTAL 2014, LNAI 8686, pp. 228–235, 2014.
© Springer International Publishing Switzerland 2014

as parallel corpus to extract linguistic knowledge. It is based on the idea that there is a possibility of every target sentence to be a translation of the source sentence. Hybrid approach uses the synergy effect of the strengths of the three previously mentioned approaches. More commonly, hybrid approaches combine statistical and rule-based translations using two methodologies [3]: *rules post-processed by statistics* and *statistics guided by rules*. Since word alignment is the basic and critical process in the statistical machine translation, efforts are directed aligning words in sentences of syntactically distant languages [4]. Due to the dominance of English as an international language, most of the studies on machine translation have focused on language pairs of English and other languages such as French [8], Spanish [7], Chinese [11], etc. However, there has been only a little published works on machine translation involving the major languages of Ethiopia such as Amharic, Oromiffa, Tigrigna, etc. [1], [2], [9]. This work is aimed at developing bidirectional English-Oromiffa machine translation system using hybrid approach.

The remaining part of the paper is organized as follows. Section 2 presents an overview of Oromiffa language along with its grammatical rules. The proposed hybrid machine translation system is presented in Sect. 3. Experimental results are reported in Sect. 4. Section 5 discusses conclusions and future works. We provide a list of references at the end.

2 Oromiffa Language

Oromiffa is one of the major languages spoken in Ethiopia and the total population speaking the language is currently estimated at 26 million [5]. It uses Latin-based alphabet known as *Qubee* [6] which has become the official script for writing since 1991. Qubee consists of twenty-eight basic letters, out of which five of them are vowels, the other five letters are double consonants (*Qubee dachaa*) and the rest are consonants. Double consonant letters are derived from a combination of two consonant letters.

Oromiffa and English have differences in their syntactic structure. In Oromiffa, the sentence structure is subject-object-verb (SOV) as opposed to the English (SVO). For example, in Oromiffa sentence *"Dagaagaan nyaata nyaate"*, *Dagaagaan* is subject, *nyaata* is object and *nyaate* is the verb. The sentence is translated into English as "Degaga ate food" where "Degaga" is subject, "ate" is verb and "food" is object. As compared to English, Oromiffa is morphologically complex language. In most cases, the derivation and inflection of words is accomplished through affixation. Oromiffa uses the same set of punctuation marks as English except for apostrophe (') which is used to represent a glitch sound known as *hudhaa*. It plays an important role in Oromiffa reading and writing system. For example, it is used to write a word in which most of the time two vowels appear together like *kaa'uu*.

3   The Proposed Machine Translation System

3.1   Reordering of Linguistic Structures

We introduced reordering of linguistic structures of the source text to reduce the complexities of the statistical translation process that arises as a result of differences in syntax of the two languages. The purpose of reordering is to make the local structure of the source text have the same structure as the target text. Accordingly, we apply reordering rules on English text for English-Oromiffa translation and Oromiffa text for Oromiffa-English translation. We classify reordering rules into four main categories: *reordering rules for local structures, simple sentences, interrogative sentences* and *complex sentences.*

Reordering Rules for Local Structures

The local structures that induce different order of words with respect to English and Oromiffa are *possessive pronouns, prepositional phrases, cardinal numbers, prepositions with cardinal numbers, verbs, verbs with auxiliary verbs,* and *present participle verbs with prepositions.*

Possessive Pronouns: In English-Oromiffa translation, we change the order of English possessive pronouns with their next noun phrases as shown in the example below. On the other hand, in Oromiffa-English translation, we apply local reordering on possessive pronouns in Oromiffa text.

Prepositional Phrases: For English-Oromiffa translation, we apply local reordering of prepositions on English text as shown in the following example. For Oromiffa-English translation, we change order of the prepositions in the Oromiffa text.

Cardinal Numbers: Cardinal numbers appear before noun phrases in English whereas they appears after noun phrases in Oromiffa. For English-Oromiffa translation, we change the order in which the cardinal number appears with noun phrases in English text as shown in the following example. For Oromiffa-English translation, we apply such reordering rules on the Oromiffa text.

Prepositions with Cardinal Numbers: In English, prepositions are placed before cardinal numbers whereas they are placed after the cardinals in Oromiffa. Thus, for English-Oromiffa translation, local reordering of prepositions with cardinal numbers is applied on English text as shown below. In Oromiffa-English translation, we perform local reordering on the Oromiffa text.

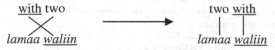

Present Participle Verbs with Prepositions: In English, phrases having present participle verbs and prepositions, the words are arranged in such a manner that the present participle verbs come before the preposition. However, in Oromiffa phrases, present participle verbs come after prepositions. In such cases, for English-Oromiffa translation, we apply local reordering on English text as shown in the following example. On the other hand, we apply local reordering on Oromiffa text in the case of Oromiffa-English translation.

Verbs with Auxiliary Verbs and Objects: In English, auxiliary verbs come before main verbs whereas auxiliary verbs come after main verbs in Oromiffa. Thus, in English-Oromiffa translation, the order of main verbs and auxiliary verbs of the English text is changed whereas the reverse process is applied on Oromiffa text in Oromiffa-English translation. Likewise, verbs and objects in the English text are reordered in English-Oromiffa translation whereas, in Oromiffa-English translation, we reorder objects and verbs in the Oromiffa text.

Reordering Rules for Simple Sentences

The reordering of simple sentences is achieved by recursively applying local reordering rules discussed above. An example of reordering procedures of a simple English sentence in English-Oromiffa translation is shown in Fig. 1. In the same way, we recursively apply local reordering rules on Oromiffa text in Oromiffa-English translation.

Reordering Rules for Interrogative Sentences

Interrogative sentences in Oromiffa and English do not have the same structures. English interrogative sentences constructed with interrogative verbs like "verb-to-be", "verb-to-do", "verb-to-have", and "auxiliary verbs" begin with the verbs themselves. In Oromiffa, the corresponding interrogative sentences take the form of simple sentences and followed by a question mark. Thus, in English-Oromiffa translation, interrogative verbs in English text are reordered so as to make the structure similar to that of Oromiffa simple sentences. Afterwards, reordering rules of local structures will follow to complete the reordering of English interrogative sentences. Likewise, the reverse process on Oromiffa interrogative sentence is performed for Oromiffa-English translation.

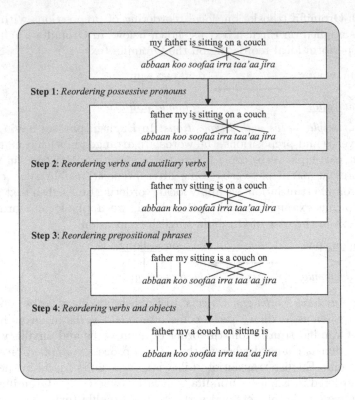

Fig. 1. An example of reordering English sentence in English-Oromiffa translation

Reordering Rules for Complex Sentences

Complex sentences are constructed from two or more simple sentences. We use divide-and-conquer strategy to reorder complex sentences where a complex sentence is divided into two or more simple sentences and those simple sentences are reordered independently. Finally, the reordered simple sentences are merged together to form a reordered complex sentence.

3.2   Language Model

Oromiffa language model is required for English-Oromiffa translation whereas English language model is required for Oromiffa-English translation. The language models are built on the original corpus before reordering rules are applied on the respective texts. Let us assume that e is an English text and o is its equivalent Oromiffa text. Thus, we compute $p(e)$ and $p(o)$ as language models for Oromiffa-English and English-Oromiffa translations, respectively. We used n-grams to compute language models with the help of IRSTLM tool. Smoothing was also employed to avoid zero probability for unseen n-grams.

3.3   Translation Model

The job of the translation model is to assign a probability that a given source language sentence generates target language sentence. Taking the previous notations that e is an English text and o is its equivalent Oromiffa text, let us also assume that $\acute{e}$ is the reordered form of e (whose order of words takes the form of o) and $\acute{o}$ is the reordered form of o (whose order of words takes the form of e). In our case, the source language would be the reordered English ($\acute{e}$) in English-Oromiffa translation and the reordered Oromiffa ($\acute{o}$) in Oromiffa-English translation. Thus, for English-Oromiffa translation, the best Oromiffa translation sentence $\hat{o}$ for a given source English text $\acute{e}$ is the one whose probability $p(o|\acute{e})$ is the highest and it is formulated as follows.

$$\hat{o} = \arg\max_{o} p(o|\acute{e}) \tag{1}$$

Bayes rule states that:

$$p(o|\acute{e}) = p(\acute{e}|o) * p(o)/p(\acute{e}) \tag{2}$$

where this is equivalent to the following.

$$\arg\max_{o} p(o|\acute{e}) = \arg\max_{o} p(\acute{e}|o) * p(o)/p(\acute{e}) \tag{3}$$

Since $p(\acute{e})$ is a given constant value, it could be ignored as it plays no role in finding the highest value in the right hand side of the equation. Then, substitution of (1) in (3) gives us the following *Noisy Channel Equation* for English-Oromiffa translation:

$$\hat{o} = \arg\max_{o} p(\acute{e}|o) * p(o) \tag{4}$$

where $p(\acute{e}|o)$ is the reordered English-Oromiffa translation model and $p(o)$ is the Oromiffa language model. By the same procedure, the *Noisy Channel Equation* for Oromiffa-English translation is computed as:

$$\hat{e} = \arg\max_{e} p(\acute{o}|e) * p(e) \tag{5}$$

where $p(\acute{o}|e)$ is the reordered Oromiffa-English translation model and $p(e)$ is the English language model.

4   Experiment

4.1   The Corpus

In this work, publicly available parallel corpus of English and Oromiffa languages is used. This parallel corpus includes some chapters of the Holy Bible, the Constitution of Ethiopia, the criminal code of Ethiopia, international conventions, Megeleta Oromia and a bulletin from Oromia Health Bureau. A total of 3000 parallel sentences of English and Oromiffa were used for experiment. Out of these parallel sentences, the training set includes 2700 parallel sentences whereas the rest were used for testing the system.

4.2   Test Results

The performance of the system was tested in terms of translation accuracy and the time it takes to translate a single English sentence to Oromiffa sentence and vice versa. BLEU score methodology was used to evaluate the performance of the translation and test results are summarized in Table 1. Although the time it takes to translate sentences significantly varies according to the variation in the complexity of sentences, the average time taken by English-Oromiffa translation was found to be greater than that of Oromiffa-English translation.

Table 1. Test results

Machine Translation	Approach	BLEU Score
English-Oromiffa	Statistical	32.39%
	Hybrid	41.50%
Oromiffa-English	Statistical	37.41%
	Hybrid	52.02%

4.3   Discussion

It can be seen from the test results that, by applying local reordering, the translation work becomes easier as if both languages were closely related in their syntax which ultimately eases the translation process. However, it should be noted that Oromiffa is morphologically more complex than English language. This morphological complexity would result in ambiguity when translating English sentences to Oromiffa. For example, there is a difference on marking gender in English and Oromiffa languages. As a result, "Kaku is running" was incorrectly translated by the system as "*Kakuun fiigaa dha*" rather than translating it correctly as "*Kakuun fiigaa jirti*". In this translation process "Kaku" was translated as "*Kakuun*" and "running" was translated as "*fiigaa*", which are both correct translations. But the system was unable to correctly translate the word "is" as "*jirti*", rather the system translated it as "*dha*" which was not a correct translation. The reason behind this is that the system was unable to identify to which gender "Kaku" belongs to. However, the Oromiffa sentence "*Kakuun fiigaa jirti*" was translated by the system as "Kaku is running" because "*jirti*" could be translated as without ambiguity as "is". Because of this reason the translation accuracy of Oromiffa-English translation was found to be better than English-Oromiffa translation. Moreover, Oromiffa-English translation on average took less computational time as compared to the reverse process. The same reason could justify the differences in computational time. In the case of English-Oromiffa translations, the system had to deal with a lot of translation ambiguities, an example of which is mentioned above. Dealing with such ambiguities demands as extra time as compared to Oromiffa-English translation.

5   Conclusion and Future Work

In this work, we proposed a bidirectional English-Oromiffa machine translation system. A hybrid of statistical and ruled based approaches was used to develop the machine translation system. English and Oromiffa have different morphological and syntactic structures which pose difficulty for statistical machine translation. A set of rules were identified and applied to reduce the differences in the syntactic structures of the two languages. Test results showed that reordering of English sentences improved the accuracy of English-Oromiffa translation whereas reordering of Oromiffa sentences improved the accuracy of Oromiffa-English translation. The effect of the morphological complexity of Oromiffa language was also observed in English-Oromiffa translations where verbs should agree with subject features (e.g. gender) in Oromiffa. In such cases, the translation process is subject to ambiguity affecting the accuracy and computational time of the translation. Thus, future work is directed at resolving such ambiguities by employing morphological analyzer for Oromiffa language.

References

1. Adugna, S.: English-Oromo Machine Translation: An Experiment Using a Statistical Approach. MSc Thesis, Addis Ababa, Addis Ababa University (2009)
2. Gasser, M.: Toward a Rule-Based System for English-Amharic Translation. In: The 4th Workshop on African Language Technology (2012)
3. Jurafsky, D., Martin, J.H.: Speech and Language Processing: An Introduction to Natural Language Processing, Computational Linguistics, and Speech Recognition. Prentice-Hall (2006)
4. Kumar, S., Byrne, W.: Local Phrase Reordering Models for Statistical Machine Translation. In: HLT-EMNLP, Vancouver, Canada (2005)
5. Lewis, P., Simons, F., Fennig, D.: Ethnologue: Languages of the World, 7th edn. SIL International, Dallas (2013)
6. Melbaa, G.: Oromia: An Introduction to the History of the Oromo People. Kirk House Publishers, USA (1999)
7. Nakov, P.: Improving English-Spanish Statistical Machine Translation: Experiments in Domain Adaptation, Sentence Paraphrasing, Tokenization, and Recasing. In: The Third Workshop on Statistical Machine Translation, pp. 147–150 (2008)
8. Schwenk, H., Fouet, J.-B., Senellart, J.: First Step Towards a General Purpose French/English Statistical Machine Translation System. In: The Third Workshop on Statistical Machine Translation, pp. 119–122 (2008)
9. Teshome, M.G., Besacier, L.: Preliminary Experiments on English-Amharic Statistical Machine Translation. In: The 3rd International Workshop on Spoken Languages Technologies for Under-resourced Languages (2012)
10. Tripathi, S., Sarkhel, J.K.: Approaches to Machine Translation. Annals of Library and Information Studies 57, 388–393 (2011)
11. Wang, C., Collins, M., Koehn, P.: Chinese Syntactic Reordering for Statistical Machine Translation. In: Joint Conference on EMNLP and CNLL, pp. 737–745 (2007)

Inflating a Training Corpus for SMT by Using Unrelated Unaligned Monolingual Data

Wei Yang and Yves Lepage

Graduate School of IPS, Waseda University,
2-7 Hibikino, Wakamatsu-ku, Kitakyushu, Fukuoka 808-0135, Japan
kevinyoogi@akane.waseda.ja, yves.lepage@waseda.ja

Abstract. To improve the translation quality of less resourced language pairs, the most natural answer is to build larger and larger aligned training data, that is to make those language pairs well resourced. But aligned data is not always easy to collect. In contrast, monolingual data are usually easier to access. In this paper we show how to leverage unrelated unaligned monolingual data to construct additional training data that varies only a little from the original training data. We measure the contribution of such additional data to translation quality. We report an experiment between Chinese and Japanese where we use 70,000 sentences of unrelated unaligned monolingual additional data in each language to construct new sentence pairs that are not perfectly aligned. We add these sentence pairs to a training corpus of 110,000 sentence pairs, and report an increase of 6 BLEU points.

Keywords: monolingual corpus, analogies, quasi-parallel corpus, machine translation.

1 Introduction

Sentence-level aligned parallel corpora are an extremely important resource as training data in statistical machine translation (SMT). The quantity of the parallel sentences is crucial, because the translation knowledge is acquired from these sentential parallel corpora [4]. The quality of the aligned parallel sentences is another important factor that impacts greatly the quality of the translation relations extracted between words or phrases between the source language and the target language. To summarize, translation quality depends on the quantity and the quality of the parallel corpus.

There exist numerous freely available bilingual or multilingual parallel corpora for language pairs that involve English, such as the Europarl parallel corpus [9]. The Europarl corpus was designed for research purposes in statistical machine translation and it has been used for multiple other research purposes, including for example word sense disambiguation. But the linguistic resources between languages like: Chinese, Japanese, Thai, Hindi or Bahasa Indonesian are relatively scarce. This does not mean that they are minority languages, as

A. Przepiórkowski and M. Ogrodniczuk (Eds.): PolTAL 2014, LNAI 8686, pp. 236–248, 2014.
© Springer International Publishing Switzerland 2014

all these languages have multimillion-strong speaker and writer bases and monolingual data are quite easy to collect. But bilingual sentence-aligned corpora in any of these language pairs are not so easily accessible. Manual construction of resources for such less-resourced language pairs is time consuming and costly. Researchers face many difficulties to extract parallel corpora from general texts [1] or from specialized texts like patent families [3]. For Chinese or Japanese, another important issue comes from copyright restrictions: most existing resources are not free due to copyright. For all the above reasons, we propose a novel method to combine small freely available aligned bi-corpora with approximately aligned sentence pairs generated from a reasonable size of monolingual unaligned data. We call such approximately aligned sentence pairs a quasi-parallel corpus, because it contains sentences that are translations to each other only to a certain extent. The degree of correspondence in translation is estimated by some similarity scores.

Our method to construct a quasi-parallel corpus is based on the notion of proportional analogy. This notion has already been applied to machine translation. For instance, [6] and [10] both address the problem of translating unknown words in a statistical machine translation framework and use analogy for that. Corpus-based analogical techniques have also been used to build a complete example-based machine translation system [13]. Analogy has also been used to cluster word pairs by semantic relations [2].

In this paper, we show how to cluster sentences to be used as rewriting models for new sentence generation and how to deduce translation relations between these newly generated sentences to construct a quasi-parallel corpus. We then report SMT experiments and compare a baseline system using a relatively small amount of parallel data and a system built on the baseline by adding the previous quasi-parallel corpus as additional training data. This new system performs significantly better. We also evaluate the quality of the quasi-parallel corpus in terms of language quality (grammaticality) and alignment quality (translation similarity).

2   Data Preparation for the Experiments

We perform our experiments on Chinese–Japanese which can be considered a less resourced language pair. We started by collecting parallel Chinese–Japanese data and monolingual data form the Web by using an in-house crawler.

2.1   Collecting Parallel Chinese–Japanese Data

We collected and aligned Chinese–Japanese sentences from the subtitles of movies and drama series, based on the time of the subtitles in two languages. Table 2.1 shows examples of Chinese and Japanese subtitles as translations to each other in a movie. We collected these data from the following websites: *Subscene.com* and *Opensubtitles.org*. Each text piece consists of one or two short sentences shown on the screen nearly every second in Chinese and Japanese. These sentences are short and simple for readers to easily understand them in limited time (one second). The subtitle corpus we used here comes from about 300 subtitle files. We obtained 106,310 pairs of aligned Chinese–Japanese sentence pairs after some cleaning.

Table 1. Examples for the Chinese–Japanese subtitle short sentence pairs

[Events] Format: Start, End, Name, Text
Dialogue: 0:02:11.99,0:02:14.66,cn.sub,不停船的话就击沉你们！快停！
Dialogue: 0:02:14.66,0:02:16.73,cn.sub,被抢劫啦是海盗哦
......
Dialogue: 0:02:11.99,0:02:14.66,jp.sub,止まらないと沈めるぞ！止まれー！
Dialogue: 0:02:14.66,0:02:16.73,jp.sub,さらわれるー 海賊だー
......

We also downloaded the JEC Basic Sentence Data by Kyoto U. and NICT with 5,304 Chinese–Japanese sentence pairs. We extracted 1,500 pairs of sentences for tuning and testing; the statistics about all this data will be given in Sect. 5.1. The rest of the data (3,804 pairs) will be combined with the subtitle corpus as the initial parallel corpus we used in this paper. Table 2.1 shows the statistics about the Chinese–Japanese initial parallel corpus we used in the experiments. Word segmentation has been done using the following toolkits: Mecab for Japanese and Urheen for Chinese.

Table 2. Statistics on the Chinese–Japanese subtitle data combined with a part of JEC sentences. This constitutes our initial parallel corpus (110,114 sentence pairs in total: 106,310 + 3,804).

	Language	# of different sentences (cleaned)	size of sentences in characters (mean ± std.dev.)	total characters	total words
Subtitle Corpus (106,310) + JEC (3,804)	Chinese	99,251	8.68 ± 3.59	861,723	589,757
	Japanese	90,406	11.99 ± 4.36	1,084,287	647,285

2.2 Collection of Monolingual Resources

To generate new quasi-parallel data, we use unrelated unaligned monolingual data. We collected monolingual Chinese and Japanese short sentences (less than 30 characters in size) mainly from the following websites: "Yahoo China", "Yahoo China News", "douban" for Chinese and "Yahoo! JAPAN", "Mainichi Japan" for Japanese. Table 3 gives the statistics of the cleaned 70,000 monolingual data we used in the experiments.

Table 3. Statistics on the cleaned Chinese and Japanese monolingual short sentences

	# of different sentences (cleaned)	size of sentences in characters (mean ± std.dev.)	total characters	total words
Chinese	70,000	10.29 ± 6.21	775,530	525,462
Japanese	70,000	15.06 ± 6.34	1,139,588	765,085

3 Construction of Analogical Clusters

3.1 Proportional Analogies

Proportional analogies establish a structural relationship between four objects, A, B, C and D: 'A is to B as C is to D'. An efficient algorithm for the resolution of analogical equations between strings of characters has been proposed in [11]. The algorithm relies on counting numbers of occurrences of characters and computing edit distances (with only insertion and deletion as edit operations) between strings of characters ($d(A, B) = d(C, D)$ and $d(A, C) = d(B, D)$).

Sentential Analogies. We gather pairs of sentences in Chinese and Japanese respectively, that constitute proportional analogies. For instance, the two following pairs of Japanese sentences:

紅茶が**飲み** たい。	あなたは紅茶が好 きですか。	::	ビールが**飲み**た い。	あなたはビールが 好きですか。
'*I'd like a* *black tea.*'	'*Do you like black* *tea?*'	::	'*I'd like a beer.*' :	'*Do you like beer?*'

are said to form an analogy, because the edit distance between the sentence pair on the left of '::' is the same as between the sentence pair on the right side: $d(A, B) = d(C, D) = 13$. The same must be true by exchanging the two sentences in the middle, B and C, so that $d(A, C) = d(B, D) = 5$. The relation on the number of occurrences of characters, which must be valid for each character, may be illustrated as follows for the character 飲: 1 (in A) − 0 (in B) = 1 (in C) − 0 (in D). We call any such two pairs of sentences a *sentential analogy*.

Analogical Cluster. When several sentential analogies involve the same pairs of sentences, they form a series of analogous sentences, and they can be written on a sequence of lines where each line contains one sentence pair and where any two pairs of sentences from the sequence of lines form a sentential analogy. We call such a sequence of lines an *analogical cluster*. The size of a cluster is the number of its sentential pairs. The following example in Japanese (English translation below) shows 6 possible sentential analogies. The size of this cluster is 4. In this analogical cluster, the box shows what remains the same on the left and on the right of the cluster, and the underline shows the changes on one sentence pair of the cluster. Analogical clusters like this can be considered as *rewriting models* to generate new sentences. The technique will be described in Sect. 4.

紅茶が 飲みたい。 ： あなたは 紅茶が 好きですか。
'I'd like a cup of black tea.' 'Do you like black tea?'

ビールが 飲みたい。 ：あなたは ビールが 好きですか。
'I'd like a beer.' 'Do you like beer?'

ジュースが 飲みたい。 ：あなたは ジュースが 好きですか。
'I'd like some juice.' 'Do you like juice?'

冷たいお水が 飲みたい。 ：あなたは 冷たいお水が 好きですか。
'I'd like some cold water.' 'Do you like cold water?'

3.2 Cluster Construction

In each language, independently, we construct analogical clusters from the unrelated monolingual data. The number of unique sentences used is 70,000 for both languages. Table 4 summarizes the statistics on the clusters constructed.

Table 4. Statistics on the Chinese and Japanese clusters constructed from our unrelated monolingual data independently in each language

	Chinese	Japanese
# of different sentences	70,000	70,000
# of clusters	23,182	21,975

3.3 Computing the Correspondence between Clusters

The core of our technique lies in the computation of the correspondence between clusters across languages. The goal of the technique described hereafter is to spot situations as the following one. Suppose that we have the two clusters in two different languages shown in Table 5. The lines in the clusters are not translations one of another. But the change between the left and the right columns are the same in both languages. In this case, we say that the two clusters correspond. The sequel of this section describes the needed computation to spot such correspondences.

Table 5. Two corresponding clusters. They do not have the same sizes and the sentences contained are not translations (i.e., here, the same). But the change between the left and the right columns ('I ... .' → 'Do you ... ?') is the same.

Language 1:

I like beer.:**Do you** like beer?

I like juice.:**Do you** like juice?

Language 2:

I study maths.:**Do you** study maths?
I watch movies.:**Do you** watch movies?
I read books.:**Do you** read books?

We extract corresponding clusters by computing the similarity between the changes in them using the following steps:

- First, for each sentence pair in a cluster, we extract the change between the left and the right sides by computing their longest common subsequence (LCS) [18].
- Then, we consider the changes between the left (S_{left}) and the right (S_{right}) sides in one cluster as two sets. We perform word segmentation on these changes so as to obtain minimal changes expressed in sets of words or characters.
- Finally, we compute the similarity between the left sets (S_{left}) and the right sets (S_{right}) of the Chinese and Japanese clusters. To this end, we make use of the EDR dictionary[1], a traditional-simplified Chinese variant table[2] and a Kanji–Hanzi Conversion Table[3] to translate all Japanese words into Chinese, or convert Japanese characters into simplified Chinese. We calculate the similarity between two Chinese and Japanese word sets according to a classical Dice formula:

$$Sim = \frac{2 \times |S_{\text{zh}} \cap S_{\text{ja}}|}{|S_{\text{zh}}| + |S_{\text{ja}}|} \tag{1}$$

S_{zh} and S_{ja} denote the minimal sets of changes across the clusters (both on the left or right) in both languages. The formula for computing the similarity between two Chinese and Japanese clusters is given in formula (2):

$$Sim_{C_{\text{zh}}-C_{\text{ja}}} = \frac{1}{2}(Sim_{\text{left}} + Sim_{\text{right}}) \tag{2}$$

The previous computation shows that the correspondence between two clusters does not rely on the translation correspondences between the entire sentences contained in the clusters, but only a small part of them. For this reason, correspondences between clusters have much higher chance to be found than translation between sentences. Figure 1 gives an example of the extracted corresponding clusters and the sets (S_{left} and S_{right}) of the changes (shown in underline) in the Chinese ($S_{\text{zh}_{\text{left}}} = \{$ 经典 $\}$ and $S_{\text{zh}_{\text{right}}} = \{$ 很, 不错 $\}$) and Japanese ($S_{\text{ja}_{\text{left}}} = \{$ クラシック $\}$ and $S_{\text{ja}_{\text{right}}} = \{$ この, は, とても, いい $\}$) clusters respectively, the similarity between the two clusters calculated according to formula (2) is 0.833. We set different threshold for $Sim_{C_{\text{zh}}-C_{\text{ja}}}$ and check the correspondence between these extracted clusters by sampling. While the $Sim_{C_{\text{zh}}-C_{\text{ja}}}$ threshold was set to 0.300, the acceptability of the correspondence between the extracted clusters were able to achieve to 80%. About 14,578 corresponding clusters were extracted ($Sim_{C_{\text{zh}}-C_{\text{ja}}} \geq 0.300$) by the above steps.

---

[1] http://www2.nict.go.jp/out-promotion/techtransfer/EDR
[2] http://www.unicode.org/Public/UNIDATA/
[3] http://www.kishugiken.co.jp/cn/code10d.html

Chinese cluster		Japanese cluster	
left part : right part		left part : right part	
经典游戏	：游戏很不错		
'classic game'	*'The game is very good.'*		
喜欢经典	：很不错喜欢	クラシック物語	：この物語はとてもいい
'I like classic.'	*'Very good, I like it.'*	*'classic narrative'*	*'The narrative is very good.'*
经典啊	：很不错啊	クラシック音楽	：この音楽はとてもいい
'Classic!'	*'Very good!'*	*'classic music'*	*'The music is very good.'*

Fig. 1. Two corresponding clusters constructed based on unrelated unaligned monolingual data. Chinese on the left of the figure, Japanese on the right of the figure. In both clusters, i.e., in both languages, the word meaning 'classic' on the left part of the cluster, is replaced on the right part of the cluster, by words meaning 'very good'.

4   Generation of New Sentences Using Analogical Associations

4.1   Generation of New Sentences

Analogy is not only a structural relationship. It is also a process [8] by which, given two related forms and only one form, the fourth missing form is built [5]. If the objects A, B, C are given, we may build an other object D according to the analogical equation $A : B :: C : D$. This principle can be illustrated with sentences:

紅茶が飲みたい。 ： あなたは紅茶が好きですか。 :: ビールが飲みたい。 ： x $\Rightarrow$ $x =$ あなたはビールが好きですか。

'I'd like a black tea.' ： *'Do you like black tea?'* :: *'I'd like a beer.'* ： x $\Rightarrow$ $x =$ *'Do you like beer?'*

In this example, the solution of the analogical equation is $D =$ "あなたはビールが好きですか。" (Do you like beer?). If we regard each sentence pair in a cluster as a pair $A : B$ (left to right or right to left), and any short sentence not belonging to the cluster as C (a *seed sentence*), the analogical equation $A : B :: C : D$ of unknown D can be forged. Such analogical equations allow us to produce new candidate sentences. Each sentence pair in a cluster is thus a potential rewriting template for the generation of new candidate sentences.

4.2   Experiments on New Sentence Generation and Filtering

We generate new sentences in Chinese and Japanese respectively and independently by using analogical associations based on the result of the clusters constructed and the initial parallel corpus (unique Chinese and Japanese sentences, described in Sect. 2.1) as the seed sentences. We generate new candidate sentences with each sentence pair in each cluster. Then we filter these candidate sentences by using the N-sequence method, a technique to assess the quality of outputs of NLP systems used in previous work [12]. The technique keeps the sentences where all N-sequences can be found in a reference corpus. Said the way round, we eliminate any sentence that contains an N-sequence not found in the reference corpus. We introduced begin/end markers to make sure that also the beginning and the end of a sentence are correct. We set N to 6 for Chinese and to 7 for Japanese. The size of the reference data we used are 1,059,985 for Chinese and 1,074,851 for Japanese. Table 6 shows the statistics on the new sentence generation.

Table 6. Statistics on new sentence generation in Chinese and Japanese. Q is the quality of the new candidate sentences or new valid sentences after filtering.

		Chinese		Japanese	
Initial data	# of seed sentences	99,251		90,406	
	# of clusters	23,182		21,975	
New sentence generation	# of candidate sentences	192,121,764		50,418,891	
		Q= 20%		Q= 50%	
Quality assessment (filtered)	# of new valid sentences	unique	seed–new–#	unique	seed–new–#
		34,230	105,537	142,820	191,409
		Q= 99%		Q= 99%	

Quality assessment was performed by sampling 1,000 sentences randomly and asking native speakers to check for grammaticality. The grammatical quality was at least 99%. This means that 99% of the Chinese and Japanese sentences may be considered as grammatically correct. For each valid sentence, we remember the corresponding seed sentence and the cluster identifier it was generated from. Table 6 shows the statistics of the filtering result. Table 7 shows the examples of new sentences generated from two given clusters (Fig. 1) and a pair of parallel seed sentences in Chinese and Japanese. In the Chinese example, we obtained two different valid sentences through the same cluster.

Table 7. The result of new sentence generation in Chinese and Japanese based on a pair of parallel seed sentences according to the clusters given in Fig. 1

A	:	B	::	C_{seed}	:	X_{new-zh}
经典游戏	:	游戏很不错				
喜欢经典	:	很不错喜欢	::	经典电影 'classic film'	⇒	电影很不错 'The film is very good.'
经典啊	:	很不错啊				很不错电影 'That's very good, the film.'

A	:	B	::	C_{seed}	:	X_{new-ja}
クラシック物語	:	この物語はとてもいい	::	クラシック映画 'classic film'	⇒	この映画はとてもいい 'The film is very good.'
クラシック音楽	:	この音楽はとてもいい				

4.3 Deduction of Translation Relations between New Generated Sentences

We deduce translation relations based on the initial parallel corpus and corresponding clusters between Chinese and Japanese. If the seeds of two new generated sentences in Chinese and Japanese are aligned in the initial parallel corpus, and if the clusters which they generated from are corresponding, we suppose that these two Chinese and Japanese newly generated sentences are translations of one another to a certain extent. In the example in Table 7, we obtain two pairs of quasi-parallel sentences: "电影很不错：この映画はとてもいい" and "很不错电影：この映画はとてもいい". Table 8 gives the statistics of the quasi-parallel corpus. Among the 76,151 unique Chinese–Japanese quasi-parallel sentences obtained, about 75% were found to be exact translations by manual check by sampling 1,000 pairs of sentences. This justifies our use of the term "quasi-parallel" for this kind of data.

Table 8. Statistics on the quasi-parallel corpus deducing

Chinese	Japanese	Chinese–Japanese		
seed–new–#	seed–new–#	Initial parallel corpus	Corresponding clusters	Quasi-parallel corpus
105,537	191,409	110,114	14,578	76,151

5 SMT Experiments

5.1 Experimental Protocol

To assess the contribution of the generated quasi-parallel corpus, we propose to compare two SMT systems. The first one is constructed using an initial parallel corpus. This is the baseline. The second one adds the additional quasi-parallel corpus obtained using analogical associations and analogical clusters.

Baseline. The statistics of the data used in the experiments are given in Table 9 (left). The training corpus consists of 110,114 sentences of initial Chinese–Japanese parallel corpus. The tuning set is 500 sentences from the JEC parallel corpus, and 1,000 sentences also from the JEC corpus were used for testing. We perform all experiments using the standard GIZA++/MOSES pipeline [15].

Adding Additional Quasi-parallel Corpus. The statistics of the data used in this second setting are given in Table 9 (right). The training corpus is made of 186,265 (110,114 + 76,151) sentences, i.e., the combination of the initial Chinese–Japanese parallel corpus used in the baseline and the quasi-parallel corpus.

Table 9. Statistics on the Chinese–Japanese corpus used for the training, tuning, and test sets in baseline (left) and baseline + quasi-parallel data (right). The tuning and testing sets are the same in both experiments.

Baseline	Chinese	Japanese	+ Quasi-parallel	Chinese	Japanese
sentences	110,114	110,114	sentences	**186,265**	**186,265**
words	637,036	721,850	words	1,147,098	1,318,747
mean ± std.dev.	5.94 ± 2.60	6.69 ± 2.94	mean ± std.dev.	6.06 ± 2.61	7.16 ± 3.08

(train)

Both experiments	Chinese	Japanese
sentences	500	500
words	3,582	5,042
mean ± std.dev.	7.15 ± 2.86	10.12 ± 3.39
sentences	1,000	1,000
words	7,285	10,126
mean ± std.dev.	7.28 ± 2.87	10.15 ± 3.30

(tune / test)

Experimental Results. Table 10 gives the evaluation results. We use the standard metrics BLEU [16], NIST [7], WER [14], and TER [17]. As Table 10 shows, significant improvement over the baseline is obtained by adding the quasi-parallel generated data.

Table 10. Evaluation results for Chinese–Japanese translation across two SMT systems (baseline and baseline + additional quasi-parallel data)

		BLEU	NIST	WER	TER
zh-ja	baseline	13.10	4.1732	0.7229	0.7344
	+ additional training data	**19.27**	**4.7013**	**0.6880**	**0.6933**
ja-zh	baseline	10.94	4.4028	0.7545	0.7621
	+ additional training data	**17.66**	**4.7989**	**0.7140**	**0.7214**

5.2   Analysis of the Results

We investigated the N (source length) × M (target length) distribution in phrase tables generated from the initial parallel corpus and the inflated training corpus by adding the quasi-parallel data. In Table 11 and 12, the statistics (zh→ja) show that the total number of phrase pairs generated by adding additional quasi-parallel corpus is larger than when using only the initial parallel corpus as training data. If we compare the number of entries, the number of phrase pairs (in Table 12) on the diagonal got a significant increase in the number of phrase pairs of similar length (except 1 × 1). Considering the correspondence between lengths in Chinese–Japanese translation, the increase in phrase pairs with different lengths (like 2 (zh) × 3 (ja) and 3 (zh) × 4 (ja)) is felicitous. This means that adding the additional quasi-parallel corpus for inflating the training corpus for SMT allowed us to produce much more numerous useful alignments. Table 13 gives samples of phrase pairs showing the same Chinese phrase aligned with different Japanese phrases. We obtained additional alignments compared to the use of the limited initial parallel corpus. The additional phrase pairs in this given samples are all correct by checking manually. By increasing the size of the training corpus by adding quasi-parallel data, we got more good alignment informations between Chinese and Japanese.

Table 11. Distribution of phrase pairs in the phrase translation table of GIZA++ (baseline zh→ja)

				Target language				
		1-grams	2-grams	3-grams	4-grams	5-grams	6-grams	7-grams	total
Source language	1-grams	23,789	35,494	25,069	13,670	6,568	2,982	1,257	108,829
	2-grams	34,865	52,596	38,612	22,429	12,300	6,413	3,113	170,328
	3-grams	18,904	33,116	39,633	29,465	19,262	11,881	6,617	158,878
	4-grams	8,097	15,948	24,779	28,160	23,629	17,628	11,495	129,736
	5-grams	3,235	7,020	12,656	18,532	21,166	19,277	15,072	96,958
	6-grams	1,195	2,860	6,027	10,405	14,537	16,470	15,245	66,739
	7-grams	466	1,223	2,615	5,196	8,395	11,003	12,239	41,137
	total	90,551	148,257	149,391	127,857	105,857	85,654	65,038	772,605

Table 12. Distribution of phrase pairs in the phrase translation table (baseline + additional training data: zh→ja). Compare with Table 13, increase in entry numbers in boldface.

				Target language				
		1-grams	2-grams	3-grams	4-grams	5-grams	6-grams	7-grams	total
Source language	1-grams	23,752	**38,758**	**29,920**	**18,025**	9,671	4,777	2,211	127,114
	2-grams	38,997	**56,814**	**44,647**	**28,149**	16,733	9,539	4,985	199,864
	3-grams	23,240	38,360	**45,596**	**35,724**	25,148	16,653	9,977	194,698
	4-grams	10,954	19,398	29,343	**33,124**	28,991	22,801	16,078	160,689
	5-grams	4,779	9,143	15,540	21,864	25,515	24,242	20,052	121,135
	6-grams	1,858	3,787	7,475	12,683	17,799	20,577	19,996	84,175
	7-grams	765	1,577	3,357	6,531	10,335	13,969	15,787	52,321
	total	104,345	167,837	175,878	156,100	134,192	112,558	89,086	**939,996**

6   Conclusion

Phrase-based statistical machine translation (PB-SMT) relies on the availability of parallel data to extract and align phrases and to estimate various features like translation probabilities. The quality of translation depends on the quantity of the available data and on its quality. The conventional approach for improving

Table 13. Samples of phrase alignments in zh→ja phrase table. The same Chinese phrase and corresponding Japanese phrases.

Baseline	只能这样了 これ で 行く しか ない
(Initial parallel corpus)	只能这样了 それ しか ない ん だ よ
	只能这样了 やる しか ない だろ
	只能这样了 これ で 行く しか なかっ た
Adding	只能这样了 これ しか ない
quasi-parallel data	只能这样了 それ しか ない ん だ
	只能这样了 やる しか ない
	只能这样了 やる しか なかっ た

performance of PB-SMT systems is to increase the size of the parallel corpus by adding new training data, usually extracted from comparable corpora.

We followed a slightly different path. Firstly, we chose to add more data that may be not so well aligned. We called such data quasi parallel data. Secondly, the quantity of data that we added was not so important as we expanded the training data by only two thirds (110,000 to 186,000). The main point in our view is the original method to generate the new quasi parallel data. These were obtained by structuring unaligned unrelated monolingual data according to analogical associations. These analogical associations are used as rewriting models to produce new sentences.

In experiments performed on Chinese–Japanese, by adding this kind of quasi-parallel data, no so large in quantity and not so good in quality (as translation is concerned), we were able to inflate the translation table in a rewarding way. On the same test set, the translation quality significantly increased over the baseline systems (more than 6 BLEU points).

The explanation why the method works may be as follows. The quasi-parallel data that we produced reflect changes, i.e., linguistic variations, that are attested in monolingual unrelated texts. This may greatly help to better cover the linguistic variations that may appear in a test set which, by definition, is unknown in advance. We claim that the analogical associations helped to capture and to reproduce in an efficient way these linguistic variations. We believe that, if we had used data from the same domain, our results may have been less convincing. Our method worked with a relatively small training corpus, and one may well think that, of course, its performance will decrease if really larger training data were used in the baseline.

An other interesting point in the method used is that the productivity of analogy allowed us to generate a reasonable quantity of new sentences. But it is also possible to use very simple techniques (attested n-gram method) to easily limit over-generation and to ensure the grammaticality of sentences. In our experiments, it was easy to reach grammaticality for 99% of the sentences generated. All this resulted in a significant increase in the number of entries in the phrase tables produced by the standard GIZA++/MOSES pipeline, especially for the lengths of phrases relevant for the languages at hand.

Acknowledgments. This work was supported in part by Foreign Joint Project funds from the Kitakyushu Foundation for the Advancement of Industry, Science and Technology (FAIS).

References

1. Abdul Rauf, S., Schwenk, H.: Parallel sentence generation from comparable corpora for improved SMT. Machine Translation 25(4), 341–375 (2011)
2. Biçici, E., Yuret, D.: Clustering word pairs to answer analogy questions. In: Proceedings of TAINN 2006, pp. 277–284 (2006)
3. Bin, L., Tao, J., Kapo, C., Benjamin, K.: T.: Building a large English-Chinese parallel corpus from comparable patents and its experimental application to SMT. In: Proceedings of LREC 2010, pp. 42–49 (2010)
4. Chu, C., Nakazawa, T., Kurohashi, S.: Chinese–Japanese parallel sentence extraction from quasi–comparable corpora. In: ACL 2013, pp. 34–42 (2013)
5. De Saussure, F.: Cours de linguistique générale. Payot, Paris (1916, 1995)
6. Denoual, E.: Analogical translation of unknown words in a statistical machine translation framework. In: MT Summit XI, Copenhagen pp. 10–14 (2007)
7. Doddington, G.R., Przybocki, M.A., Martin, A.F., Reynolds, D.A.: The NIST speaker recognition evaluation–overview, methodology, systems, results, perspective. Speech Communication 31(2), 225–254 (2000)
8. Itkonen, E.: Analogy as Structure and Process: Approaches in linguistics, cognitive psychology and philosophy of science 14 (2005)
9. Koehn, P.: Europarl: A parallel corpus for statistical machine translation. In: Proceedings of MT Summit, pp. 79–86 (2005)
10. Langlais, P., Patry, A.: Translating unknown words by analogical learning. In: EMNLP-CoNLL, pp. 877–886 (2007)
11. Lepage, Y.: Solving analogies on words: An algorithm. In: Proceedings of COLING-ACL 1998, Montréal, pp. 728–735 (August 1998)
12. Lepage, Y., Denoual, E.: Automatic generation of paraphrases to be used as translation references in objective evaluation measures of machine translation. In: IWP 2005, pp. 57–64 (2005)
13. Lepage, Y., Denoual, E.: Purest ever example-based machine translation: detailed presentation and assessment. Machine Translation 19, 251–282 (2005)
14. Nießen, S., Och, F.J., Leusch, G., Ney, H.: An evaluation tool for machine translation: Fast evaluation for machine translation research. In: Proceedings of LREC 2000, pp. 39–45 (2000)
15. Och, F.J., Ney, H.: A systematic comparison of various statistical alignment models. Computational Linguistics 29(1), 19–51 (2003)
16. Papineni, K., Roukos, S., Ward, T., Zhu, W.J.: BLEU: a method for automatic evaluation of machine translation. In: ACL 2002, pp. 311–318 (2002)
17. Snover, M., Dorr, B., Schwartz, R., Micciulla, L., Makhoul, J.: A study of translation edit rate with targeted human annotation. In: AMTA 2006, pp. 223–231 (2006)
18. Wagner, R.A., Fischer, M.J.: The string-to-string correction problem. Journal of the ACM 21, 168–173 (1974)

Uncovering Discourse Relations
to Insert Connectives between the Sentences
of an Automatic Summary

Sara Botelho Silveira and António Branco

University of Lisbon, Portugal
{sara.silveira,antonio.branco}@di.fc.ul.pt
http://nlx.di.fc.ul.pt/

Abstract. This paper presents a machine learning approach to find and classify discourse relations between two unseen sentences. It describes the process of training a classifier that aims to determine (i) if there is any discourse relation among two sentences, and, if a relation is found, (ii) which is that relation. The final goal of this task is to insert discourse connectives between sentences seeking to enhance text cohesion of a summary produced by an extractive summarization system for the Portuguese language.

Keywords: discourse relations, discourse connectives, summarization.

1   Motivation

An important research issue in which there remains much room for improvement in automatic text summarization is text cohesion. Text cohesion is very hard to ensure specially when creating summaries from multiple sources, as their content can be retrieved from many different documents, increasing the need for some organization procedure.

The approach presented in this paper aims to insert discourse connectives between sentences seeking to enhance the cohesion of a summary produced by an extractive summarization system for the Portuguese language [17]. Connectives are textual devices that ensure text cohesion, as they support the text sequence by signaling different types of connections or discourse relations among sentences. It is possible to understand a text that does not contain any connective, but the occurrence of such elements reduces the cost of processing the information for human readers, as they explicitly mark the discourse relation holding between the sentences, thus acting like guides in the interpretation of the text. The assumption in this work is that relating sentences that are retrieved from different source texts can produce a more interconnected text, and thus a more easier to read summary.

Marcu et al. (2002) noted that "discourse relation classifiers trained on examples that are automatically extracted from massive amounts of text can be used to distinguish between [discourse] relations with accuracies as high as 93%,

A. Przepiórkowski and M. Ogrodniczuk (Eds.): PolTAL 2014, LNAI 8686, pp. 249–261, 2014.
© Springer International Publishing Switzerland 2014

even when the relations are not explicitly marked by cue phrases" [9]. Following the same research line, this paper presents a machine learning approach relying on classifiers that predict the relation shared by two sentences. Considering two adjacent sentences, the final goal is to insert, between those sentences, a discourse connective that stands for the discourse relation found between them – including possibly the phonetically null one.

The procedure is composed by two phases. The first phase – *Null vs. Relations* – determines if two adjacent sentences share a discourse relation or not. If a relation has been found, the second phase – *Relations vs. Relations* – is applied, aiming to distinguish which is the discourse relation both sentences share. Based on this relation a discourse connective is retrieved from a previously built list to be inserted between those sentences. Consider for example the following sentences.

S_1 *O custo de vida no Funchal é superior ao de Lisboa.*
The cost of living in Funchal is higher than in Lisbon.
S_2 **No entanto**, *o Governo Regional nega essa conclusão.*
However, the Regional Government denies this conclusion.

These two sentences are related by the discourse connective *no entanto* ("however"), which expresses that the two sentences convey some adversative information. Hence, it is possible to say that these sentences entertain a relation of COMPARISON-CONTRAST-OPPOSITION, based on the discourse connective that relates them. The following example runs through the complete procedure.

1. Retrieve two adjacent sentences.
   O custo de vida no Funchal é superior ao de Lisboa
   The cost of living in Funchal is higher than in Lisbon.
   O Governo Regional nega essa conclusão.
   The Regional Government denies this conclusion.
2. Find the discourse relation.
   Apply model *Null vs. Relations* → "Yes" = both sentences share indeed a discourse relation.
   Apply model *Relations vs. Relations* → relation class = COMPARISON-CONTRAST-OPPOSITION
3. Look for the connective to insert.
   A random connective is obtained in the list for the class COMPARISON-CONTRAST-OPPOSITION → retrieved: *no entanto*.
4. Insert the discourse connective between the two sentences.
   O custo de vida no Funchal é superior ao de Lisboa
   The cost of living in Funchal is higher than in Lisbon.
   <u>*No entanto*</u>, *o Governo Regional nega essa conclusão.*
   However, the Regional Government denies this conclusion.

The remainder of this paper is structured as follows. Section 2 overviews previous works on finding discourse relations in text and details the approach pursued in this work; and Section 3 points out some future work directions, based on the conclusions drawn.

2 Uncovering Discourse Relations

The intent of the majority of the studies that address discourse relations is to recognize ([9], [5], [2], [7], [10], [19]) and classify discourse relations in unseen data ([20], [12], [11]).

Other works ([8], [1], [3]) approach this problem with different goals. Louis et al. aim to enhance content selection in single-document summarization [8]. Biran and Rambow are focused in detecting justifications of claims made in written dialog [1]. Feng et al. seek to improve the performance of a discourse parser [3].

Despite of their different goals, these studies follow a common approach to find and classify discourse relations in text, that is a machine learning techniques over annotated data. The task is to learn how and which discourse relations are explicitly – by means of cue phrases – or implicitly expressed on human annotated data. In the approach presented in this paper, this task is reverted. The classification of the discourse relation will be used to determine a discourse connective to be inserted between a given pair of adjacent sentences.

In order to build a classifier that decides which discourse relation is holding between two sentences, there are several decisions at stake: the initial corpus, the features to be used, the training and testing datasets, and the classification algorithm. The remainder of this section discusses these decisions.

2.1 Discourse Corpus

In order to feed the classifiers, a corpus that explicitly associates a discourse relation to a pair of sentences was created semi-automatically, relying on a corpus of raw texts and a list of discourse connectives.

The list of Portuguese discourse connectives was built by a human annotator who started by translating list provided by the English *Penn Discourse TreeBank (PDTB)* [14] [13]. After a first inspection to the raw corpus and taking into account the convenience of this task, some adjustments were made to this list, resulting in a final list that was used to create the discourse corpus. Table 1 shows an example of a connective for each class.

Prasad et al. (2008) state that discourse connectives have typically two arguments: ARG_1 and ARG_2. Also, they concluded that the typical structure in which the three elements are combined is ARG_1 <CONNECTIVE> ARG_2. The following example shows two sentences with this typical structure, where S_1 maps to ARG_1 and S_2 maps to ARG_2, with the connective "but" being included in ARG_2.

S_1 *Washington seguiu Saddam desde o início.*
  Washington followed Saddam from the beginning.
S_2 ***Mas** a certa altura as comunicações com Clinton falharam.*
  But at some point communications with Clinton failed.

Table 1. Examples of discourse connectives by class

Class	Connective	Translation
COMPARISON-CONTRAST-OPPOSITION	mas	but
COMPARISON-CONCESSION-EXPECTATION	apesar de	although
COMPARISON-CONCESSION-CONTRA-EXPECTATION	como	as
CONTINGENCY-CAUSE-REASON	pois	because
CONTINGENCY-CAUSE-RESULT	então	hence
CONTINGENCY-CONDITION-HYPOTHETICAL	a menos que	unless
CONTINGENCY-CONDITION-FACTUAL	se	if
CONTINGENCY-CONDITION-CONTRA-FACTUAL	caso	if
TEMPORAL-ASYNCHRONOUS-PRECEDENCE	antes de	before
TEMPORAL-ASYNCHRONOUS-SUCCESSION	depois de	after
TEMPORAL-SYNCHRONOUS	enquanto	until
EXPANSION-RESTATEMENT-SPECIFICATION	de facto	in fact
EXPANSION-RESTATEMENT-GENERALIZATION	em conclusão	in conclusion
EXPANSION-ADDITION	adicionalmente	additionally
EXPANSION-INSTANTIATION	por exemplo	for instance
EXPANSION-ALTERNATIVE-DISJUNCTIVE	ou	or
EXPANSION-ALTERNATIVE-CHOSEN ALTERNATIVE	em alternativa	instead
EXPANSION-EXCEPTION	caso contrário	otherwise

CETEMPúblico [16] is a corpus built from excerpts of news from *Público*, a Portuguese daily newspaper. This corpus was analyzed to find pairs of sentences complying with this structure. The composition of the discourse corpus is defined by triples as such (ARG$_1$, ARG$_2$, *DiscourseRelation*). So, after gathering the sentence pairs, a classification is required for the discourse relation holding between each pair of sentences. [13] argue that this typical structure is the minimal amount of information needed to interpret a discourse relation. Then, each pair was classified with the class of the discourse connective that links its sentences together. Also, the connective is removed from the sentence defined as ARG$_2$.

Finally, taking into account the goal of the task presented in this paper, when considering two adjacent sentences, those can share a discourse relation or not. Thus, pairs of adjacent sentences that do not have any discourse relation, that is that are not linked by any of the connectives considered, have also been retrieved. All the pairs that do not contain any connective linking them were classified with the NULL class, stating that there is no relation between the sentences.

This way a discourse annotated corpus has been built relating a pair of sentences and their respective discourse relation. This corpus was then used to create the datasets used to train and test the classifiers.

2.2 Experimental Settings

Experimental settings comprise the features, the datasets and the classification algorithms that were used to train the classifiers.

Features. Considering the task at hand, the features are expected to reflect the properties that could express the discourse relation holding between the two arguments in the relation (ARG_1 and ARG_2).

In order to find the best configuration, for the experiments, several features were tested. Considering the structure of the discourse corpus, the most straightforward approach would be to use both sentences (ARG_1 and ARG_2) to train the classifier.

Previous works ([9], [6], [20], [7], [8]) essayed different types of features to classify discourse relations, including contextual features, constituency features, dependency features, semantic features and lexical features. The presented approach is inspired in the one of Wellner et al. that reported high accuracy when using a combination of several lexical features [20].

In a sentence, the verb expresses the event so it can constitute a relevant information in helping to distinguish between different relations. Considering a specific relation, different pairs of sentences sharing that relation might have different verbs, although they could have the same discourse connective. This discourse connective typically requires the same verb inflections, not necessarily the same instance of the verb. Thus, instead of using the verb in each sentence, the verb inflections of each sentence were used.

Another feature is related to the context in which the discourse connective appears. Thus, a context window surrounding the occurrence of the discourse connective will be used. A six-word context window surrounding the location where the discourse connective occurs in the discourse relation is considered, where three words are the last three words of ARG_1 and the other three words are the first three words of ARG_2.

In addition, three more features were used to improve the identification of the tiny differences across discourse relations. These features include all the adverbs, conjunctions and prepositions found in each of the sentences. Conjunctions link words, phrases, and clauses together. Adverbs are modifiers of verbs, adjectives, other adverbs, phrases, or clauses. An adverb indicates manner, time, place, cause, or degree, so that it may help unveiling the grammatical relationships within a sentence or a clause. A non functional, semantically loaded preposition usually indicates the temporal, spatial or logical relationship of its object to the rest of the sentence. All these words can constitute clues to better identify the discourse relation between two unseen sentences, so they can help to enhance the accuracy of the classifier.

Datasets. The discourse corpus distribution indicates that it is highly uneven, containing some very big classes (e.g. NULL) and at the same time some very small ones (e.g. CONTINGENCY-CONDITION-FACTUAL). Taking this into account, all the experiments were based on even training datasets, that is the datasets always contain the same number of examples for each class.

Moreover, the training procedure was split in two phases. In the first training phase, the goal is to train a classifier that aims to identify whether the sentences share a discourse relation or not (named *Nulls vs. Relations*). Thus, the first dataset includes pairs from all the discourse classes – assigned as RELATION – and pairs assigned with the NULL class – assigned as NULL.

After uncovering that two sentences share indeed a discourse relation, the second training phase (named *Relations vs. Relations*) seeks to find which is that discourse relation. The second dataset will only include the pairs assigned with a specific discourse class (the NULL pairs are not included).

In what concerns the testing dataset, it will remain imbalanced as to reflect the normal distribution of discourse relations in a corpus.

Figure 1 illustrates the distribution of the testing dataset in the *Null vs. Relations* training phase, while Figure 2 shows the classes distribution in the *Relations vs. Relations* phase.

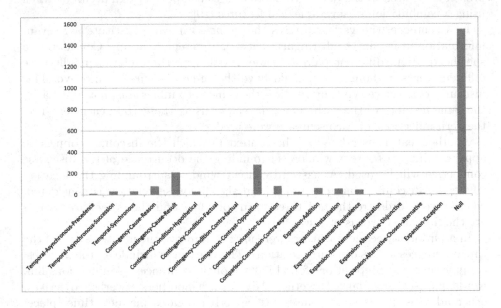

Fig. 1. Distribution of the classes in the testing dataset for *Null vs. Relations*

Classification algorithms. There are several algorithms that have been more frequently used in Natural Language Processing tasks.

Naïve Bayes [4] is a probabilistic classifier, which algorithm assumes independence of features as suggested by Bayes' theorem. Despite its simplicity, it achieves similar results obtained with much more complex algorithms.

C4.5 [15] is a decision tree algorithm. It splits the data into smaller subsets using the information gain in order to choose the attribute for splitting the data. In short, decision trees hierarchically decompose the data, based on the presence or absence of the features in the search space.

Finally, *Support Vector Machines (SVM)* [18] is an algorithm that analyzes data and recognizes patterns. The basic idea is to represent the examples as points in space, making sure that separate classes are clearly divided. *SVM* is a binary classifier, specially suitable for two-class in classification problems.

All these algorithms were used in the experiments reported in this paper.

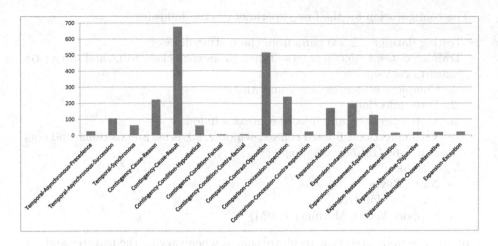

Fig. 2. Distribution of the classes in the testing dataset for *Relations vs. Relations*

2.3   Results

The classifiers trained aim to learn how to distinguish which is the discourse relation that two unseen sentences share, if any.

The first assumption of such a task would be to take into account all the classification classes at the same time. This means that, in this case, 19 classes would be taken into account, including all the 18 possible discourse classes plus the NULL class. However, as the corpus from which the datasets were built is highly imbalanced between all the 19 classes (as the distribution of the testing datasets suggest – cf. Figures 1 and 2), the classifier training procedure was divided in two phases, as already stated. *Null vs. Relations* determines if the sentences share a discourse relation or not. *Relations vs. Relations* discovers which is the discourse relation, already knowing that both sentences share one.

All the experiments reported were obtained using *Weka* workbench [21].

Null vs. Relations. The experiment procedure defines first the training and testing datasets used. The training dataset includes the same number of examples from all classes divided in a binary classification problem. The classifier will have then to decide if two sentences share a discourse relation ("yes") or if they do not ("no"). It contains 5,000 pairs evenly divided in both these classes. Yet, the testing dataset contains 2,500 pairs reflecting the normal distribution of the discourse relations in a corpus by remaining imbalanced (cf. Figure 1).

This first experiment aims mainly to identify which features improve the classifier accuracy, so that several combinations of features were used.

The results were then obtained using the most common classification algorithms (previously discussed).

The configuration for the first experiment is the following:

- Testing dataset – 2,500 pairs from the testing dataset
- Training dataset – 5,000 pairs split in half for each class, NULL and RELATION
- Features essayed:
  1. Complete sentences: ARG$_1$ and ARG$_2$
  2. Verb inflections
  3. Verb inflections and 6-word context window
  4. Verb inflections, 6-word context window, adverbs, prepositions, and conjunctions (*inf-6cw-adv-prep-cj*)
- Algorithms:
  - Naïve Bayes
  - C4.5 decision tree
  - Support Vector Machines (SVM)

In order to understand the results obtained when varying the features and the algorithms, a baseline for this task must be considered. The baseline assigns the most frequent class to all the instances. The most frequent class in the dataset is the NULL class. So, when assigning always the NULL class to the instances occurring in the test set, it is possible to achieve an accuracy of 61%, being this value the baseline to be overcome by a more sophisticated classifier.

The results for the first experiment are described in Table 2.

Table 2. Accuracy for each algorithm for the first experiment

Features	Naïve Bayes	C4.5	SVM
1 *sentences*	59.00 %	57.84 %	63.68 %
2 *verb inflections (inf)*	57.56 %	58.84 %	60.64 %
3 *+ 6-word context window (inf-6cw)*	59.12 %	60.16 %	61.32 %
4 *+ adverbs, prepositions and conjunctions (inf-6cw-adv-prep-cj)*	68.00 %	64.88 %	72.84 %

The results illustrated in the Table were obtained using the same training and testing datasets, so that several features and algorithms could be tested. Taking into account the goal of this task, the most straightforward approach would be to use the complete sentences – ARG$_1$ and ARG$_2$ – as features (feature#1). The first assumption when using this feature was that it might contain too much noise, as the complete sentences were being used. However, the sentences might also contain some singularities that help the classifier to achieve results close to the baseline. Feature#2 comprises the inflections of all the verbs in both sentences. By expressing an event, the verb can be a relevant source of information when regarding discourse relations and their specific inflections could help to identify the presence of a discourse relation or not. Despite empirically it could be a very relevant feature, by itself this feature achieves results below the baseline when considering all the algorithms tested.

As suggested by Wellner et al. (2006), we finally essayed to combine several features [20]. When combining the verb inflections with the 6-word context window – composed by three words in the end of ARG$_1$ and three words in the beginning of ARG$_2$ (feature#3) – we were able to improve the accuracy of all the three classifiers. With this configuration, it is even possible to overcome the baseline with *SVM*. Finally, feature#4 includes the combination of the previous features with all the adverbs, prepositions and conjunctions found in both arguments (results in the fourth line of the Table). Using this combination, we were able to significantly improve the results of the three classifiers, with all results scoring above the baseline.

Yet, when analyzing the behavior of each classifier, we can conclude that for all of them the combination of the features keeps enhancing their accuracy. Despite this, the results obtained using *SVM* are still the best ones, being more than 10 percentage points above the baseline.

After finding a combination of features that overcomes the baseline with all the algorithms, a new experiment was performed, varying only the size of the training dataset. Thus, the feature used was always the same (*inf-6cw-adv-prep-cj*), so as the algorithms and the testing dataset.

The second experiment aims then to verify if extending the training dataset would improve the accuracy of the classifiers. Table 3 reports the results for the training dataset extensions.

Table 3. Accuracy when extending the dataset for each algorithm

Number of pairs	Naïve Bayes	C4.5	SVM
5,000 pairs	68.00 %	64.88 %	72.84 %
10,000 pairs	67.80 %	67.88 %	75.20 %
20,000 pairs	67.68 %	69.76 %	76.72 %
40,000 pairs	67.08 %	70.56 %	76.80 %
80,000 pairs	67.52 %	72.96 %	78.68 %
160,000 pairs	66.96 %	70.20 %	78.92 %

The first line was obtained by training all the algorithms using a dataset containing 5,000 pairs. These are the final values reported in the previous experiment. The training dataset was then duplicated until the learning curve has reached a point where no relevant improvements were obtained. The learning curve is illustrated in Figure 3.

Note that all the training algorithms keep performing better when doubling the training dataset until the 80,000 pairs mark. At this point, there are slight improvements (case of *SVM*) or even worse performances (cases of *Naïve Bayes* and *C4.5*). In conclusion, the best performing algorithm – *SVM* – was used in a training dataset of 160,000 pairs to perform the first step of the connective insertion procedure: identify if two sentences enter a discourse relation or not.

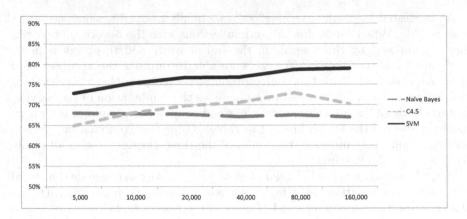

Fig. 3. Learning curve when extending the datasets

Relations vs. Relations. Once the previous classifier has determined that the sentences share a discourse relation, it is now time to identify which relation is that.

This is a multi-class classification problem, as we have 18 possible discourse relations to assign. Recall Figure 2 that shows the distribution of the classes in the testing dataset. Note that the most frequent discourse relation class is CONTIGENCY-CAUSE-RESULT. A baseline for this classification problem would assign the most frequent class to all the instances in the dataset, so that it would achieve an accuracy of 27%, being this value the lower boundary to be overcome by a more sophisticated classifier.

The first experiment takes together all the 18 classes in a all-vs-all approach. A training dataset containing 2,500 split unevenly through all the classes was used. In the same way, the testing dataset contains 2,500 pairs aiming to reflect the normal distribution of the discourse relations in a corpus. Results for this experiment are reported in Table 4.

Table 4. Accuracy for all the classes using a all-vs-all approach

Classes	*Naïve Bayes*	*C4.5*	*SVM*
ALL CLASSES	22.64 %	23.36 %	29.6 %

As these results point out, deciding between 18 different classes at the same time is a very hard task. Even though the best result (*SVM*) is slightly above the baseline, this is a very poor accuracy.

Hence, this problem was split into several problems, assuming a one-vs-all approach. Thus, for each class, we trained a classifier that aims to determine if a given pair of arguments share that relation or any of the other relation. This way, we turned a multi-class problem into a binary classification problem.

Taking into account that *SVM* had the best performance in the previous experiment, and that it is specially suitable for binary classification, this was the algorithm used in this experiment. Also the same combined features (*inf-6cw-adv-prep-cj*) – found in the first experiment – were used to train the classifiers. This experiment is based on training datasets containing 2,500 pairs divided in two: 1,250 from the specific relation and 1,250 from all the others. The goal was to build the training datasets with the same number of instances of the classes. However, we were unable to obtain in the corpus 1,250 for three classes (CONTINGENCY-CONDITION-FACTUAL, EXPANSION-ALTERNATIVE-DISJUNCTIVE and CONTINGENCY-CONDITION-CONTRA-FACTUAL). For each of these classes we built the training dataset by including the maximum number of instances for each class (66, 458 and 927, respectively) and the same number of instances of all the other classes. Thus, the training datasets for these three classes contained a total of 132, 916 and 1854, respectively.

Table 5 details the accuracy values obtained when training a single classifier for each class, using training datasets containing 2,500 pairs.

Table 5. Accuracy for each class using a one-vs-all approach

Classes	*SVM*
CONTINGENCY-CONDITION-FACTUAL	61.43 %
EXPANSION-ALTERNATIVE-DISJUNCTIVE	65.38 %
CONTINGENCY-CONDITION-CONTRA-FACTUAL	80.83 %
COMPARISON-CONCESSION-CONTRA-EXPECTATION	89.26 %
EXPANSION-EXCEPTION	74.13 %
EXPANSION-ALTERNATIVE-CHOSEN-ALTERNATIVE	64.87 %
EXPANSION-RESTATEMENT-GENERALIZATION	70.53 %
TEMPORAL-SYNCHRONOUS	62.23 %
TEMPORAL-ASYNCHRONOUS-PRECEDENCE	89.74 %
CONTINGENCY-CONDITION-HYPOTHETICAL	77.40 %
COMPARISON-CONCESSION-EXPECTATION	75.48 %
EXPANSION-RESTATEMENT-EQUIVALENCE	71.89 %
TEMPORAL-ASYNCHRONOUS-SUCCESSION	76.60 %
CONTINGENCY-CAUSE-REASON	60.51 %
EXPANSION-INSTANTIATION	75.80 %
EXPANSION-ADDITION	89.34 %
COMPARISON-CONTRAST-OPPOSITION	69.18 %
CONTINGENCY-CAUSE-RESULT	62.83 %

The results shown in the Table point out that all the classifiers performed significantly better when compared with the first experiment where the classes were considered altogether. Moreover, by using a one-vs-all approach we were able to create classifiers for each discourse class which are highly above the baseline and which are able to distinguish a specific class from all the other possible classes.

3 Final Remarks

This paper presents an approach to find discourse relations between sentences, in order to select a discourse connective to be inserted between those sentences. The procedure uses a sequence of classifiers, firstly to determine if there is any relation between the two sentences, and, afterwards, to distinguish which relation is that. By uncovering the discourse relation and selecting the corresponding connective, this work seeks to go a step forward in improving the quality of a text. The accuracy results of all the classifiers are very promising, suggesting that the probability of finding the correct connective is on average 72%.

The textual quality of a summary (e.g. fluency, readability, discourse coherence, etc.) has been repeatedly reported as the main flaw in current automatic summarization technology. Considering this, the ultimate goal of the procedure presented in this paper is to be included in a post-processing module of an automatic multi-document summarization system, that creates summaries using extraction methods. Post-processing is a module composed by three tasks executed in sequence – sentence reduction, paragraph creation and connective insertion. While sentence reduction aims to remove extraneous information from the summary, paragraph creation seeks to define topics of interest in the text. Yet, connective insertion is applied over the sentences in each paragraph by inserting between them the appropriate discourse connective (if any), creating interconnected text. Thus, the motivation behind this work is to seek for improvements in respect to the final quality of a summary built using extractive methods.

References

1. Biran, O., Rambow, O.: Identifying justifications in written dialogs by classifying text as argumentative. Int. J. Semantic Computing 5(4), 363–381 (2011)
2. Blair-Goldensohn, S., McKeown, K., Rambow, O.: Building and refining rhetorical-semantic relation models. In: Human Language Technologies 2007: The Conference of the North American Chapter of the Association for Computational Linguistics; Proceedings of the Main Conference, pp. 428–435. Association for Computational Linguistics, Rochester, Rochester (2007)
3. Feng, V.W., Hirst, G.: Text-level discourse parsing with rich linguistic features. In: Proceedings of the 50th Annual Meeting of the Association for Computational Linguistics: Long Papers, ACL 2012, vol. 1, pp. 60–68. Association for Computational Linguistics, Stroudsburg (2012)
4. John, G.H., Langley, P.: Estimating continuous distributions in Bayesian classifiers. In: Proceedings of the Eleventh Conference on Uncertainty in Artificial Intelligence, UAI 1995, pp. 338–345. Morgan Kaufmann Publishers Inc., San Francisco (1995)
5. Lapata, M., Lascarides, A.: Inferring sentence-internal temporal relations. In: HLT-NAACL, pp. 153–160 (2004)
6. Lee, A., Prasad, R., Joshi, A., Dinesh, N.: Complexity of dependencies in discourse: Are dependencies in discourse more complex than in syntax? In: Proceedings of the 5th International Workshop on Treebanks and Linguistic Theories, Prague, Czech Republic, p. 12 (December 2006)

7. Lin, Z., Kan, M.Y., Ng, H.T.: Recognizing implicit discourse relations in the Penn Discourse Treebank. In: Proceedings of the 2009 Conference on Empirical Methods in Natural Language Processing, EMNLP 2009: Empirical Methods in Natural Language Processing, vol. 1, pp. 343–351. Association for Computational Linguistics, Stroudsburg (2009)

8. Louis, A., Joshi, A., Nenkova, A.: Discourse indicators for content selection in summarization. In: Proceedings of the 11th Annual Meeting of the Special Interest Group on Discourse and Dialogue, pp. 147–156. Stroudsburg, PA, USA (2010)

9. Marcu, D., Echihabi, A.: An unsupervised approach to recognizing discourse relations. In: Proceedings of the 40th Annual Meeting on Association for Computational Linguistics, ACL 2002, pp. 368–375. Association for Computational Linguistics, Stroudsburg (2002)

10. Park, J., Cardie, C.: Improving implicit discourse relation recognition through feature set optimization. In: Proceedings of the 13th Annual Meeting of the Special Interest Group on Discourse and Dialogue, SIGDIAL 2012, pp. 108–112. Association for Computational Linguistics, Stroudsburg (2012)

11. Pitler, E., Nenkova, A.: Using syntax to disambiguate explicit discourse connectives in text. In: Proceedings of the ACL-IJCNLP 2009 Conference Short Papers, ACLShort 2009, pp. 13–16. Association for Computational Linguistics, Stroudsburg (2009)

12. Pitler, E., Raghupathy, M., Mehta, H., Nenkova, A., Lee, A., Joshi, A.: Easily identifiable discourse relations. In: Coling 2008: Companion Volume: Posters, pp. 87–90. Coling 2008 Organizing Committee, Manchester (2008)

13. Prasad, R., Dinesh, N., Lee, A., Miltsakaki, E., Robaldo, L., Joshi, A., Webber, B.: The Penn Discourse TreeBank 2.0. In: Proceedings of LREC (2008)

14. Prasad, R., Miltsakaki, E., Dinesh, N., Lee, A., Joshi, A., Robaldo, L., Webber, B.: The Penn Discourse Treebank 2.0 annotation manual. Tech. Rep. IRCS-08-01, Institute for Research in Cognitive Science, University of Pennsylvania (Dec 2007)

15. Quinlan, J.R.: Improved use of continuous attributes in C4.5. Journal of Artificial Intelligence Research 4(1), 77–90 (1996)

16. Rocha, P., Santos, D.: CETEMPúblico: Um corpus de grandes dimensões de linguagem jornalística portuguesa. In: 5th, pp. 131–140 (2000)

17. Silveira, S.B., Branco, A.: Combining a double clustering approach with sentence simplification to produce highly informative multi-document summaries. In: IRI 2012: 14th International Conference on Artificial Intelligence, Las Vegas, USA, pp. 482–489 (August 2012)

18. Vapnik, V.N.: The nature of statistical learning theory. Springer-Verlag New York, Inc., New York (1995)

19. Versley, Y.: Subgraph-based classification of explicit and implicit discourse relations. In: Proceedings of the 10th International Conference on Computational Semantics (IWCS 2013) – Long Papers, pp. 264–275. Association for Computational Linguistics, Potsdam (2013)

20. Wellner, B., Pustejovsky, J., Havasi, C., Rumshisky, A., Saurí, R.: Classification of discourse coherence relations: an exploratory study using multiple knowledge sources. In: Proceedings of the 7th SIGdial Workshop on Discourse and Dialogue, SigDIAL 2006, pp. 117–125. Association for Computational Linguistics, Stroudsburg (2006)

21. Witten, I.H., Frank, E.: Data Mining: Practical Machine Learning Tools and Techniques with Java Implementations. Morgan Kaufmann, San Francisco (2005)

Toward Automatic Classification
of Metadiscourse

Rui Correia[1,2], Nuno Mamede[1], Jorge Baptista[3], and Maxine Eskenazi[2]

[1] L²F – INESC-ID / Instituto Superior Técnico, Universidade de Lisboa, Portugal,
[2] Language Technologies Institute, Carnegie Mellon University, USA
[3] Universidade do Algarve, Portugal
{Rui.Correia,Nuno.Mamede}@inesc-id.pt, jbaptis@ualg.pt, max@cs.cmu.edu

Abstract. This paper describes the supervised classification of four metadiscursive functions in English. Training data is collected using crowdsourcing to label a corpus of TED talks transcripts with occurrences of *Introductions*, *Conclusions*, *Examples*, and *Emphasis*. Using decision trees and lexical features, we report classification accuracy.

1 Introduction

Commonly referred to as *discourse about discourse*, metadiscourse is composed of rhetorical patterns used to make the discourse structure explicit and guide the audience. Crismore et al. [4] define it as "linguistic material in texts, written or spoken, which does not add anything to the propositional content but that is intended to help the listener or reader organize, interpret and evaluate the information given". Some examples of metadiscursive acts include introductions ("I'm going to talk about..."; "In this paper we present..."), conclusions ("In sum,..."), or emphasis ("The take home message..."; "Note that...").

In this study we focus on metadiscourse as used in spoken communication. The main goal is to develop a unified platform of metadiscourse classification that differentiates strategies according to their function in discourse. The systematic analysis of metadiscourse in spoken language can be useful in areas such as discourse analysis, summarization, or teaching how to make effective presentations. This paper constitutes the first step toward that goal. Herein we discuss the selection of a taxonomy of metadiscourse, describe the task of building a corpus of the phenomenon using crowdsourcing to annotate texts extracted from presentations, and present the classification results for a limited set of categories in the taxonomy.

This paper starts with related work in Sect. 2. Section 3 explains the choice of a taxonomy of metadiscourse, describes the corpus of presentations, presents the labeling task, and comments on the quantity/quality of the collected data. Section 4 describes the experimental setup, which results are presented in Sect. 5. The discussion, conclusions and comments on future work are found in Sect. 6.

A. Przepiórkowski and M. Ogrodniczuk (Eds.): PolTAL 2014, LNAI 8686, pp. 262–269, 2014.

2   Related Work

The analysis of metadiscourse deals with interpretation of the explicit lexical clues that the speaker or writer intentionally uses to organize discourse. In the literature on discourse analysis we find studies that address function in discourse. The contributions of Miltsakaki et al. [16] to the Penn Discourse Treebank [14], reorganized discourse connectives according to their function, including categories such as giving examples (INSTANTIATION), making reformulations and clarifications (RESTATEMENT), or showing cause (REASON). Another example is the RST Discourse Treebank [13], a semantics-free theoretical framework of discourse relations based on Rhetorical Structure Theory [12], which also considers categories related to metadiscursive intentions, such as EXAMPLE or DEFINITION. Even though these projects explore function in discourse, they focus on written language and do not address the meta aspect of language.

Other studies focus specifically on metadiscourse-related topics. Wilson [22,23] focuses on the topic of metasemantics, analyzing the *use-mention paradigm* as defined by Lyons [9]. Wilson composes a list of 23 nouns and verbs that are "mention significant" (such as *meaning, name, say,* or *tell*), and uses it to retrieve a set of candidate sentences from Wikipedia articles. Experts then labeled each candidate sentence as containing metasemantic markers or not, and in the positive case, what their function is. Results show that annotators agree whether a sentence contains metasemantics or not, but that there is less of a consensus on how to functionally qualify them. Madnani et al. [11] explored the topic of shell language, used both to express claims and evidence ("The argument states..."), and to organize discourse ("In sum, the conclusion..."). The authors proved the complementary nature of rules and supervised learning to detect these elements in essays written by test-takers of a standardized test for graduate admissions. A final example is the work of Purver [17] and the follow-up by Rodríguez and Schlangen [18], where the focus goes to clarifications and repairs as used in dialog. The authors define a taxonomy of clarification according to the type of content being clarified (acoustic, lexical, etc.) and develop a system (CLARIE) capable of interpreting the most important types of clarifications in a dialog.

The absence of work targeted at metadiscourse in speech, which addressed the full spectrum of metadiscursive strategies, constituted the motivation to build a corpus of the phenomena and to develop automatic classification strategies.

3   Building a Corpus of Metadiscourse

The first step toward creating a corpus of metadiscourse in spoken language was to look into the theory and find a taxonomy that aligns with the above stated goal. The first approaches that address the spoken variety of metadiscourse [8,15,3] focused on the number of stakeholders involved, never discussing function. On the other hand, the work of Ädel [2], adopted in this paper, organizes metadiscourse functionally. Ädel's taxonomy is composed of 23 discourse functions and was built using MICUSP [19] – a corpus of academic papers – and MICASE [20] – comprised of university lectures.

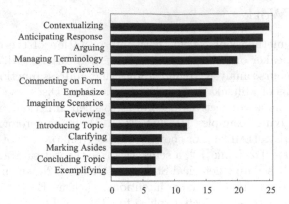

Fig. 1. Occurrences of the tags in ten TED talks

The next step was to select a source of data where such functions could be found. For this, the freely available corpus of TED talks[1] was chosen. The decision to use TED talks, instead of a corpus of university lectures (such as the MICASE), had to do with the crowdsourcing annotation strategy. Firstly, TED talks target a general audience, being absent of complex vocabulary that can hinder comprehension. Secondly, they are short, self-contained and typically do not need to be watched in a certain order to be understood. Finally, it is a multilingual resource, aligning with the goal of exploring other languages.

Having chosen a taxonomy and the TED talks, a small preliminary annotation task on ten randomly-chosen TED talks was carried out to test the suitability of this combination. Some of the 23 categories in Ädel's taxonomy were not found in the 10 talks (such as *repairs* or *clarifications*) due to the high degree of preparation of the talks. Figure 1 shows the distribution of the most frequent functions found in the ten talk sample. From these fourteen categories, a limited set of four was chosen to be explored in this first experiment, and used to annotate all 730 talks. Three criteria dictated the set of tags to use in this annotation task. We considered (a) the most frequent concepts in the literature on presentation skills, (b) the concepts that could be best explained to non-experts, and (c) the input from Carnegie Mellon's International Communications Center (entity that holds presentation skills workshops and is responsible for administering tests for non-native speakers applying for teaching assistant positions). The resulting set of categories are INTRODUCING TOPIC, CONCLUDING TOPIC, EXEMPLIFYING and EMPHASIZING. EXEMPLIFYING includes both EXEMPLIFYING and IMAGINING SCENARIOS, since they both consist of illustrating an idea.

The annotation was set up in the Amazon Mechanical Turk (AMT)[2] crowdsourcing platform. It has been shown that the quality of the results obtained using crowdsourcing, if the task is well-thought-out, can approach that of experts [5]. Four different tasks were uploaded, one per category, so as to lessen

---

[1] http://www.ted.com
[2] https://www.mturk.com/mturk/welcome

Table 1. Occurrences, self-reported confidence, and agreement for the categories IN-TRODUCING TOPIC, CONCLUDING TOPIC, EXEMPLIFYING and EMPHASIZING

Category	workers in agreement			conf	κ
	≥ 1	≥ 2	$= 3$		
INTRO	1,894	1,159	600	3.95	0.64
CONC	1,045	628	285	4.00	0.60
EXMPL	1,764	1,327	720	3.94	0.72
EMPH	3,450	2,580	750	3.99	0.58

the workers' cognitive load. The 730 TED talks were partitioned in segments of 300 words (rounded up to the nearest end of sentence). In each HIT (Human Intelligence Task – the smallest unit of work someone has to complete in order to receive payment), workers were presented with four random segments and asked to click on the words where the speaker used the metadiscursive act. Instructions included the motivation for this annotation, the definition of the concept at hand, and examples. For quality control purposes we (a) established training sessions with feedback, rejecting workers with unsuccessful training results, (b) compared answers to gold standards, (c) asked for a self-confidence report on a 5-point Likert scale for each segment, and (d) required 3 workers per segment.

Table 1 summarizes the annotation results. The column labeled ≥ 1 refers to the number of sentences that at least one worker marked as containing metadiscourse, and the column labeled $= 3$ indicates the number of sentences where all three workers agreed. Herein, annotators agree if the intersection of their selected words is not empty (for example, two workers agree when one selects "Today, I would like to say that" and the other misses some of the words, selecting only "I would like to say"). The last two columns refer to the self-reported confidence score on a 5-point Likert scale and inter-annotator agreement (κ).

Workers expressed roughly the same self-reported confidence for all categories. It is important to note that the number of INTRODUCING TOPIC and CONCLUDING TOPIC instances does not necessarily match the number of TED talks since speakers may introduce several topics throughout a single talk and do not always explicitly conclude them. The inter-annotator agreement is the highest for EXEMPLIFYING ($\kappa = 0.72$), and the lowest for EMPHASIZING ($\kappa = 0.58$). The latter result may be due to the fact that workers have different thresholds for considering that the speaker is emphasizing or not.

4 Experimental Setup

This experiment aims at finding how lexical information contributes to the detection of spoken metadiscourse. The crowd annotations are used as lexical training data to build decision trees for each category. The decision to train one classifier per category has to do with the fact that a single sentence may contain instances of several discourse functions. For instance, in "Let me finish with some examples" the speaker has both the intent of concluding and of exemplifying.

Table 2. Classifiers accuracy for the unigrams, bigrams and trigrams of POS, Lemmas and Words

Category	POS *n-grams*			Lemma *n-grams*			Word *n-grams*		
	1	2	3	1	2	3	1	2	3
INTRO	79.59	85.11	86.41	91.55	92.58	92.23	92.27	**92.71**	**92.71**
CONC	65.82	66.37	68.13	84.78	**86.93**	86.14	84.94	86.69	86.85
EXMPL	61.82	65.44	68.19	94.08	94.31	94.27	94.20	94.69	**94.92**
EMPH	67.31	68.15	68.86	77.87	79.58	79.75	79.42	79.52	**79.77**

This approach aligns with the aforementioned work by Madnani et al. [11] and Wilson [23] that showed the representative power of lexical information for metadiscourse-related phenomena. Similar lexical approaches are also found in areas such as sentiment analysis [1], and feedback localization [24].

This classification problem is formulated as: *For each sentence, decide if it contains occurrences of metadiscourse being used to introduce a topic, conclude a topic, exemplify or emphasize.* The features used in this experiment are *n*-grams of (a) parts-of-speech (POS) tags, extracted from the Stanford Parser [6], (b) word lemmas, and (c) the words themselves. It is important to point out that stop words were not discarded. This was the result of a small experiment where stop words proved to improve classification accuracy. Considering stop has also been successful in sentiment analysis and word sense disambiguation [7,21,10].

The amount of training data for each category corresponds to the numbers shown in Table 1 under column ≥ 2, which corresponds to the majority vote. Randomly chosen negative examples were added to the data, balancing the number of positive and negative instances. In this experiment, we used the C4.5 algorithm as implemented in WEKA.[3] Additionally, to address data sparsity issues, the feature reduction condition Information Gain > 0.0025 is applied.

5   Results

Table 2 reports the percentage of correctly classified items for each pair *feature-category* in a ten-fold cross-validation (bold-face representing the best settings).

For INTRODUCING TOPIC, the bigram and trigram word models achieved the best results (92.71%). These two setups were the only ones that significantly outperformed the lemma unigram model ($p < 0.05$). Regarding the false positive/negative rate (FP/FN), we registered a greater amount of FNs (8.8%) than FPs (5.8%). The features that are selected as more relevant include the word unigrams *I, to, about, talk,* and the bigrams *show you, tell you,* and *going to.*

The model that best predicted occurrences of CONCLUDING TOPIC was the lemma bigram model (86.9%). In this category, unigram models are consistently outperformed by higher-order models within the same feature type. FNs correspond to 19.8% of the negative examples, contrasting with 6.4% of FP occurrences.

---

[3] http://www.cs.waikato.ac.nz/ml/weka/

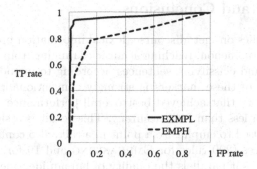

Fig. 2. ROC curves for two classifiers

The corresponding tree selects features such as the lemma unigrams *finally*, *end*, *last*, *conclude*, and the lemma bigrams *so in*, *so to*, and *my time*.

For the category EXEMPLIFYING the impact of the transition from POS to lexical features was the most accentuated, with an increase of 25% of overall performance. The best setting was the word trigram model (94.9%), the only that significantly surpassed the lemma unigram model ($p < 0.05$). Trigrams of words were sufficient to correctly classify 95% of the data, with a FP rate of 2.4% and a FN rate of 7.8%. The features that are selected for the tree include the word unigrams *example, imagine, instance, suppose*, the word bigrams *look at, give you, such as*, and the trigrams *if you were* and *think about the*.

Finally, for EMPHASIZING the word trigram model had the best performance (79.8%). Regarding the false positive/negative rates, again the classification errors are mostly due to FNs (27.1%) than FPs (12.4%). Unigrams such as *important, emphasize* and bigrams such as *to focus, point out, idea is* were selected.

Figure 2 shows the ROC curves for the best settings of the categories with highest and lowest accuracy. For EXEMPLIFYING the algorithm achieves a high TP rate maintaining a low FP rate, while for EMPHASIZING there is a compromise between these two measures.

Since the results from the previous experiments were inconclusive regarding the better adequacy of the different sets of features, we analyzed all the possible combinations of the previous settings (for example, POS and Lemmas; Lemmas and Words; or POS, Lemmas and Words together). The goal was to understand if the reason why using lemma and word features produced similar results was due to the fact that they both represent the same information, or if they are complementary. This only registered a statistically significant improvement for the category EMPHASIZING. Lemma and word trigrams together achieved an accuracy of 81.5%. Looking at the resulting tree, while the majority of rules are still based on word features, the algorithm generalized some cases using lemmas. These include the collapse of number and person variation in nouns and verbs (e.g. *that be really*, collapses the word trigrams *that is really* and *that are really*).

6   Discussion and Conclusions

As in previous studies on metadiscourse, lexical information proved to be representative of the phenomenon, reaching accuracies ranging from 80 to 95% on the task of detecting and classifying sentences according to metadiscourse. Results show that the use of these markers is strongly conventionalized. EXEMPLIFYING was the category that achieved best overall performance, contrasting with EMPHASIZING with less than 80% accuracy. This result is a sign of the greater lexical variability used to emphasize a point in a talk, also confirmed by the size of the resulting trees (200 rules for EMPHASIZING and 19 for EXEMPLIFYING). This is interesting as it parallels the quality of human identification observed.

As a first trial, this experiment addressed four of the 23 categories from the original taxonomy. Future work includes exploring additional functions, building a richer resource on the use of English spoken metadiscourse that we plan to release to the research community. In future experiments we plan to explore dependencies and dictionaries of discourse clues as features, analyzing their impact on the different metadiscursive markers. Another future work direction is to fine tune the classification to token-level, which implies reevaluating the crowd annotation with more informative metrics than majority vote. Finally, we aim at addressing the phenomenon in other languages, such as European Portuguese.

Acknowledgments. This work was supported by national funds through FCT – Fundação para a Ciência e a Tecnologia, under project PEst-OE/EEI/-LA0021/2013 and by FCT project CMU-PT/HuMach/0053/2008.

References

1. Abbasi, A., Chen, H., Salem, A.: Sentiment analysis in multiple languages: Feature selection for opinion classification in Web forums. ACM Transactions on Information Systems (TOIS) 26(3), 12 (2008)
2. Ädel, A.: Just to give you kind of a map of where we are going: A Taxonomy of Metadiscourse in Spoken and Written Academic English. Nordic Journal of English Studies 9(2), 69–97 (2010)
3. Auria, C.P.L.: Signaling speaker's intentions: towards a phraseology of textual metadiscourse in academic lecturing. English as a GloCalization Phenomenon. Observations from a Linguistic Microcosm 3, 59 (2006)
4. Crismore, A., Markkanen, R., Steffensen, M.S.: Metadiscourse in persuasive writing. Written Communication 10(1), 39 (1993)
5. Eskenazi, M., Levow, G.A., Meng, H., Parent, G.: Crowdsourcing for Speech Processing. John Wiley & Sons (2013)
6. Klein, D., Manning, C.D.: Accurate unlexicalized parsing. In: Proceedings of the 41st Annual Meeting on Association for Computational Linguistics, vol. 1, pp. 423–430. Association for Computational Linguistics (2003)
7. Lee, Y.K., Ng, H.T., Chia, T.K.: Supervised Word Sense Disambiguation with Support Vector Machines and Multiple Knowledge Sources. In: Senseval-3: Third International Workshop on the Evaluation of Systems for the Semantic Analysis of Text, pp. 137–140 (2004)

8. Luukka, M.R.: Metadiscourse in academic texts. In: Conference on Discourse and the Professions, Uppsala, Sweden, vol. 28, pp. 77–88 (1992)
9. Lyons, J.: Semantics, vol. 2. Cambridge University Press, Cambridge (1977)
10. Maas, A.L., Daly, R.E., Pham, P.T., Huang, D., Ng, A.Y., Potts, C.: Learning Word Vectors for Sentiment Analysis. In: Proceedings of the 49th Annual Meeting of the Association for Computational Linguistics: Human Language Technologies, vol. 1, pp. 142–150. Association for Computational Linguistics (2011)
11. Madnani, N., Heilman, M., Tetreault, J., Chodorow, M.: Identifying High-Level Organizational Elements in Argumentative Discourse. In: Proceedings of the 2012 Conference of the North American Chapter of the Association for Computational Linguistics: Human Language Technologies, pp. 20–28. Association for Computational Linguistics (2012)
12. Mann, W.C., Thompson, S.A.: Rhetorical Structure Theory: Toward a functional theory of text organization. Text 8(3), 243–281 (1988)
13. Marcu, D.: The Theory and Practice of Discourse Parsing and Summarization. The MIT Press (2000)
14. Marcus, M.P., Marcinkiewicz, M.A., Santorini, B.: Building a Large Annotated Corpus of English: The Penn Treebank. Computational Linguistics 19(2), 313–330 (1993)
15. Mauranen, A.: Reflexive academic talk: Observations from MICASE. In: Corpus linguistics in North America: Selections from the 1999 Symposium, pp. 165–178 (2001)
16. Miltsakaki, E., Robaldo, L., Lee, A., Joshi, A.: Sense Annotation in the Penn Discourse Treebank. In: Gelbukh, A. (ed.) CICLing 2008. LNCS, vol. 4919, pp. 275–286. Springer, Heidelberg (2008)
17. Purver, M.R.J.: The Theory and Use of Clarification Requests in Dialogue. Ph.D. thesis. Citeseer (2004)
18. Rodríguez, K.J., Schlangen, D.: Form, Intonation and Function of Clarification Requests in German task-oriented spoken dialogues. In: Proceedings of Catalog (the 8th Workshop on the Semantics and Pragmatics of Dialogue; SemDial 2004) (2004)
19. Römer, U., Swales, J.M.: The Michigan Corpus of Upper-level Student Papers (MICUSP). Journal of English for Academic Purposes (April 2009)
20. Simpson, R.C., Briggs, S.L., Ovens, J., Swales, J.M.: The Michigan Corpus of Academic Spoken English (2002)
21. Thelwall, M., Buckley, K., Paltoglou, G., Cai, D., Kappas, A.: Sentiment strength detection in short informal text. Journal of the American Society for Information Science and Technology 61(12), 2544–2558 (2010)
22. Wilson, S.: Distinguishing Use and Mention in Natural Language. In: Proceedings of the NAACL HLT 2010 Student Research Workshop, pp. 29–33. Association for Computational Linguistics (2010)
23. Wilson, S.: The Creation of a Corpus of English Metalanguage. In: Proceedings of the 50th Annual Meeting of the Association for Computational Linguistics: Long Papers, vol. 1, pp. 638–646. Association for Computational Linguistics (2012)
24. Xiong, W., Litman, D.: Identifying Problem Localization in Peer-Review Feedback. In: Aleven, V., Kay, J., Mostow, J. (eds.) ITS 2010, Part II. LNCS, vol. 6095, pp. 429–431. Springer, Heidelberg (2010)
25. Zaidan, O.F., Callison-Burch, C.: Crowdsourcing Translation: Professional Quality from Non-Professionals. In: Proceedings of the 49th Annual Meeting of the Association for Computational Linguistics: Human Language Technologies, vol. 1, pp. 1220–1229 (2011)

Detection of Nested Mentions
for Coreference Resolution in Polish

Maciej Ogrodniczuk[1], Alicja Wójcicka[1],
Katarzyna Głowińska[1,2], and Mateusz Kopeć[1]

[1] Institute of Computer Science, Polish Academy of Sciences
[2] Lingventa

Abstract. This paper describes the results of creating a shallow grammar of Polish capable of detecting multi-level nested nominal phrases, intended to be used as mentions in coreference resolution tasks. The work is based on existing grammar developed for the National Corpus of Polish and evaluated on manually annotated Polish Coreference Corpus.

1 Introduction

One of the numerous results of the National Corpus of Polish project[1] [1] was a formal shallow grammar of Polish, frequently referred to as *NKJP Grammar*, used by Spejd parser [2] to provide automated syntactic annotation [3] of the 1-billion-word corpus. The grammar was recently used by another project, CORE[2] for annotation of mentions — nominal groups referencing discourse-world objects in the Polish Coreference Corpus[3] [4], a 0.5-million-token manually annotated resource of general nominal coreference. Whereas in the former corpus the annotation of syntactic words and groups can be regarded as one of the target actions, in the latter one it is only the basis for subsequent identification of mentions (here: nominal constructs carrying reference to discourse-world objects). Therefore accuracy of this process and its compliance with mention representation (see Sect. 2) is crucial for the superior task of modelling coreference relations.

Nesting of nominal groups with disparate referents (see: *prezes firmy 'CEO of a company'*) has never been targeted by the NKJP grammar, therefore additional mechanisms have been implemented in the corpus to represent such inclusions (see Sect. 3). Section 4 reports on the process of the incorporation of the new rules into grammar while Sect. 5 evaluates the usefulness of the result to coreference resolution by contrasting mentions detected automatically with the new version of the grammar against manual annotation of mentions in the Polish Coreference Corpus.

---

[1] NKJP, Pol. Narodowy Korpus Języka Polskiego, see http://www.nkjp.pl
[2] *Computer-based methods for coreference resolution in Polish texts*, see http://zil.ipipan.waw.pl/CORE
[3] PCC, Pol. Polski Korpus Koreferencyjny, see http://zil.ipipan.waw.pl/PolishCoreferenceCorpus

A. Przepiórkowski and M. Ogrodniczuk (Eds.): PolTAL 2014, LNAI 8686, pp. 270–277, 2014.
© Springer International Publishing Switzerland 2014

2   PCC Mention Model vs. NKJP Grammar

Mentions in PCC are all nominal phrases (NGs) — syntactic groups[4] with nominal or pronominal heads (syntactic and/or semantic). In semantic annotation it is vital to preserve the deep structure of such phrases, e.g. to distinguish *a song* from *the song which was played when we first met* (in Polish even more evident due to absence of articles). A nested nominal phrase is marked as separate from the superior phrase when its syntactic/semantic head is other than the head of the superior phrase. Moreover, all potentially referential constructs are marked, because it is very difficult to define a clear-cut border between referentiality and non-referentiality, as in the following multi-word expression that usually is seen as non-referential:

Jedna jaskółka wiosny nie czyni. 'One <u>swallow</u> does not make a summer'.

Tą jaskółką było zniesienie cenzury. Ale to nie znaczy, że wprowadzono demokrację. 'A censorship abolishment was <u>this swallow</u>. But it does not mean that democracy was established.'

Since coreference resolution is a semantic task, the borderlines of nominal phrases are different from those in NKJP project, where, above all, syntactic criteria were taken into account. The PCC nominal phrase consists not only of adjectives, nouns, gerunds, conjunctions (coordinated groups) and subordinate numerals, but also of superordinate numerals (e.g., *trzy dziewczynki 'three girls'*), relative subordinate clauses (e.g., *kwiaty, które dostałam wczoraj 'the flowers, that I got yesterday'*), prepositional phrases, as well as adjectival participles. The complexity of the task is further increased by PP-attachment or by similar ambiguities involving potentially post-modifying adjectival participles.

The NKJP project was aiming for the creation of a 1-billion-word automatically annotated corpus of Polish, with a 1-million-word subcorpus annotated manually. Therefore, many decisions were influenced by the automatic annotation rules/process, and made in order to maintain a high level of consistency, whereas in the CORE project, the whole automatically pre-annotated corpus was verified and post-edited by the annotators. So some ambiguities could be solved by the linguists, e.g., PP-attachment ambiguities (*rozmowa o pogodzie 'conversation about the weather', rozmowa o piątej godzinie 'conversation at 5 o'clock'*), potentially post-modifying adjectival participles (*wierzba płacząca 'weeping willow', dziecko płaczące z wściekłości 'a child crying with rage'*).

Syntactic annotation in the National Corpus of Polish was limited to joining words together into constituents. Spejd grammar used in the PCC annotation was the modified version of the NKJP grammar, but due to the fact that NKJP nominal groups were different from the CORE nominal phrases, some modifications were made, e.g., the numeral groups were changed into nominal phrases.

The nominal groups in the NKJP project were extensive — they consisted of as many elements as possible, for e.g. in a phrase composed of consecutive

---

[4] A syntactic group is the longest possible sequence of syntactic words that satisfies certain conditions, i.e., match a Spejd rule or a description in the annotation guidelines.

nouns in the genitive case such as *propozycji wyznaczenia daty rozpoczęcia pro-cesu wprowadzania reformy ustroju*[5] *'proposal for setting the date of launching the process of introducing reform of the system'*, the whole phrase was the only de-tected nominal group despite the fact that seven other nested nominal phrases with distinct referents should have been detected.

3   Mention Detection Chain

MentionDetector (`http://zil.ipipan.waw.pl/MentionDetector`) is a tool that uses various information from several text processing applications to annotate Polish texts with mentions.

3.1   Preprocessing

The processing of a raw text begins with part-of-speech tagging with Pantera [6]. Then the text is shallow parsed with Spejd [2] and its morphological component Morfeusz SGJP [7]. The last step is to detect Named Entities, which is done by NER [8]. Information obtained from this step is then used to collect mention boundaries. Spejd has the biggest impact on mention detection, as it produces the largest number of noun groups and single-word nouns used as the mention candidates. With this respect, modifications of the Spejd grammar can bring the greatest benefit to the mention detection task.

3.2   Mention Detection Process

MentionDetector works in three steps:

1. It collects mention candidates from morphosyntactic, shallow parsing and/or named entity level (lack of any layer simply results in fewer mention candi-dates discovered) and also produces zero-anaphora candidates.
2. It removes redundant/unnecessary candidates.
3. It updates head information among mentions.

At the first stage of the process, mention candidates are extracted from the morphosyntactic level, taking all tokens with a noun (`subst|depr|ger`) or a per-sonal pronoun (`ppron3|ppron12`) tags assigned by the parser. From the shallow parsing level, all syntactic noun groups (with `NG.*` type) and syntactic words with noun or personal pronoun ctags (`Noun|Ppron.*`) are taken. Finally, from the named entity level, all named entities that contain at least one noun or pronoun token are also mention candidates. To enable zero subject processing, Mention-Detector marks each verb in sentences that do not contain any noun/pronoun token in the nominative case,[6] as a mention.

---

[5] Real NKJP example, see [5].
[6] Marking verbs instead of adding empty tokens representing zero subjects is just a technical measure implemented in PCC to maintain the original text unchanged.

At the second stage redundant mentions are detected by removing one of any two mentions having exactly the same boundaries, exactly the same heads, when one mention is the head of another mention or when two mentions intersect, but not in any way described as previous cases. For such pairs, a "less important mention" is selected for removal, which basically means removing the shorter mention or any mention in case of ties. For example in the following sentence:

Największa zagadka lotnictwa cywilnego musi zostać rozwiązana.
'The greatest mystery of civil aviation must be solved.',

preprocessing may produce the following mention candidates: (semantic heads of multi-word mentions are underlined)

- *lotnictwa 'aviation'* (based on a token tag or a syntactic word tag),
- *zagadka 'mystery'* (based on a token tag or a syntactic word tag),
- *lotnictwa cywilnego 'civil aviation'* (based on a syntactic noun group),
- *Największa zagadka lotnictwa cywilnego 'The greatest mystery of civil aviation'* (based on a syntactic noun group).

The task of the second stage is then first to remove all duplicates (e.g. *zagadka 'mystery'* could be found both as a token with a noun tag or a one-word noun group). Then finding mentions with the same heads will be followed by removing *lotnictwa 'aviation'*, as there is a broader mention of *lotnictwa cywilnego 'civil aviation'* with the same head. Similarly, *zagadka 'mystery'* will be removed for the analogous reason.

At the third stage of the process the first token is simply marked as the head of each mention, which does not have one detected automatically.

4   Towards the New Grammar

4.1   Change of Perspective

The original NKJP grammar detects nominal groups, but does not always reveal properly their internal structure. This is due to the order and structure of rules which are designed to detect the longest possible sequence irrespective of the fact whether the group is nested or not. For example the old version of the grammar detects the group: *bardzo małym druczkiem 'in very small print'*, consisting of two parts: adjectival group *bardzo małym 'very small'* and noun *druczkiem 'print'*; the structure of the group can be shown in this way: *bardzo małym druczkiem*. This division is not entirely correct, as the whole group should be interpreted as a group without children: *bardzo małym druczkiem*. The second interpretation, without nesting, is obtained by constructing a new version of grammar.

On the other hand, a nested group *usług firmy 'services of the company' (gen)* is interpreted as a group without children: *usług firmy* by the old version of grammar. The new version provides another interpretation; it detects the whole phrase 'usług firmy' and additionally preserves the information about the two smaller groups, which make up this group: *usług* (which is marked as syntactic and semantic head of the group) and *firmy*.

4.2 Rule Modification

In order to obtain such a result the structure of the section of rules detecting syntactical groups was modified.

First of all, rules for syntactic groups without nesting are in the new version of the grammar separated from rules for groups with nesting and are placed before them. The internal order of the first part of rules is based on two principles: the type of the group and length of the group. Generally speaking, more specialized rules (e.g. rule detecting addresses or dates) appear earlier in the grammar while the most frequent groups, nominal-adjective groups, are processed at the end. Within types, the rules are ordered from the broadest to the narrowest. The last group of rules corresponds to the creation of syntactic groups out of single nouns, adjectives and numerals.

Groups without nesting should contain only syntactic words (any syntactic group can be an element of such a group). In order to achieve such a result, rules describing groups without nesting are constructed in different ways from rules for groups with nesting. The main problem related to this part of grammar consists in the fact that even groups with complicated structure, containing e.g. adjectives and particles or numerals (as in a group: *kilka kolejnych filii szkolnych* 'a few other school branches') have to be built only from syntactic words. While designing rules, the recursiveness of adjective-nominal constructs has to be taken into consideration.

The most problematic group of rules in this part of the grammar is constituted by rules detecting nominal-nominal groups without nesting. Nominal-nominal groups in most cases are nested, but there are some exceptions, e.g. proper names of persons (*Jan Kowalski*) or appositions (*malarz pejzażysta* 'landscape painter'). The rules for these groups are quite restrictive in order to avoid for example a situation, where a nested group in the genitive is interpreted as an apposition in the genitive (in Polish the text *malarza pejzażysty* has two interpretations: 'a landscape painter (gen)' or 'a painter of a landscapist (gen)', the first is not nested, unlike the second). Our solution consists in making only nested groups from two subsequent nouns, if both are in the genitive and their orthographical forms begin with a small letter.

The second part of rules detecting syntactical groups — the part responsible for nested groups — is built in another manner. The only elements of these groups are other syntactical groups, nested or not nested. Recursiveness of such constructions cannot be achieved by a single rule with regular expressions; all parts of the grammar must be repeated. For example, if we have a group *przedłużenie terminu złożenia projektu budżetu* 'prolonging of the date of submitting the project of the budget', our aim is to detect the following structure: *przedłużenie terminu złożenia projektu budżetu*. In the first step the grammar detects a group *projektu budżetu*, in the second — *złożenia projektu budżetu*, in the third — *terminu złożenia projektu budżetu* and so on.

4.3 Nested Groups

There are four main types of nested groups: case-governed groups, prepositional groups, coordinated groups (conjunction governed groups) and relative clauses. Prepositional groups are excluded from this attempt since they are often very hard to distinguish — not only by parsers, but also by native speakers — between the two groups: the group with a preposition that is governed by a verb and a group governed by another nominal group. For example the text *Jaś obserwuje Marysię przy jedzeniu* can be interpreted as *'John is watching Mary while eating'* or *'John is watching how Mary eats'*.

Other types of groups are recognized by the new version of grammar. As mentioned above, in this part of the grammar, the proper order of repeated groups of rules is crucial. The problem arises that different types of groups with nesting can be embedded in all other types of groups (e.g., a coordinated group in a case-governed group and vice versa; a relative clause in a coordinated group and vice versa). Therefore the rules detecting various types of groups must be placed alternately. For example, the group *bandy partyzantów i terrorystów 'gangs of partisans and terrorists'* is made out of two smaller groups: the one-element group *bandy 'gangs'* and the coordinated group *partyzantów i terrorystów 'partisans and terrorists'*. If the rules detecting coordinated groups were placed first, the grammar would find the group *partyzantów i terrorystów* and in the second step the group *bandy partyzantów i terrorystów* would be created, which is the desirable result. However, there also exist groups such as: *naszego państwa oraz sposobu realizacji '(of) our state and way of realisation'*. The internal structure of the group is: *naszego państwa oraz sposobu realizacji*, so there is a group with nesting within the coordinated group. If the rules for coordinated groups where at the beginning of this part of the grammar, an incorrect group such as *państwa oraz sposobu 'our state and way'* would be created. Therefore the order of the rules is as follows:

1. the group of rules detecting case-governed groups, restricted only to the context without comma or conjuction on the right side of the given string (the group *bandy partyzantów* from *bandy partyzantów i terrorystów* is not found in the first step; on the other hand, the group *sposobu realizacji* being a part of *naszego państwa oraz sposobu realizacji* is detected)
2. the rules responsible for coordinated groups (the groups *partyzantów i terrorystów* and *naszego państwa oraz sposobu realizacji* are found)
3. the rules detecting case-governed groups, without the restriction mentioned above (the whole group *bandy partyzantów i terrorystów* is found)

The whole procedure is repeated by detecting longer groups and should be applied also to relative clauses (in the recent version of the grammar this method is used only by case-governed and coordinated groups).

5 Evaluation

Tables 1 and 2 present results of evaluation of the new grammar in two settings: setting 1 corresponds to real-life conditions, with best to-date mention detection, compensating potential grammar deficiencies with named entity recognition and zero-anaphora detection. Setting 2 intends to better illustrate gains resulting directly only from grammar improvements by including in the evaluation only groups detected by the grammar (without named entities etc.), i.e. NG, Noun and Ppron syntactic groups.

The evaluation has been carried out on a test set comprising of 530 texts (out of approx. 1,800) randomly selected from the Polish Coreference Corpus.

Table 1. Evaluation results, setting 1

		NKJP Grammar	New version
Mention statistics	Total gold mentions	53,407	53,407
	Total system mentions	51,217	51,750
	Total common mentions	33,839	34,176
Mention detection results	Precision	66.07%	66.04%
	Recall	63.36%	63.99%
	F1	64.69%	65.00%

Table 2. Evaluation results, setting 2

		NKJP Grammar	New version
Mention statistics	Total gold mentions	53,407	53,407
	Total system mentions	65,853	69,475
	Total common mentions	31,582	33,122
Mention detection results	Precision	47.96%	47.67%
	Recall	59.13%	62.02%
	F1	52.96%	53.91%

The difference in the number of system mentions between settings is a result of the second step of the mention detection algorithm, removing unnecessary mentions using simple heuristics.

Both settings show improvement of recall at the expense of precision (with F1 improved). Relatively low scores (in 50s–60s) results from the strict definition of mention match (exact boundaries) and the mention model itself, e.g. heavily dependent on relative clauses (difficult to access algorithmically).

6   Conclusions

The experiment showed slight improvement in absolute figures as far as mention detection is concerned, but should be regarded as the first step towards further reconstruction of NKJP grammar to enable nesting of different types of syntactic groups, not only the nominal ones. The feasibility of such a process has been confirmed.

In the mention detection chain some actions were taken in order to compensate grammar deficiencies. Now, with use of the new grammar, some of these deficiencies have been overcome.

Acknowledgments. The work reported here was carried out within the *Computer-based methods for coreference resolution in Polish texts (CORE)* project financed by the Polish National Science Centre (contract 6505/B/T02/2011/40). The work was also co-financed by the Polish Ministry of Science and Higher Education as a Investment in CLARIN-PL Research Infrastructure and by the European Union from resources of the European Social Fund, Project PO KL "Information technologies: Research and their interdisciplinary applications".

References

1. Przepiórkowski, A., Bańko, M., Górski, R.L., Lewandowska-Tomaszczyk, B. (eds.): Narodowy Korpus Języka Polskiego (Eng.: National Corpus of Polish). Wydawnictwo Naukowe PWN, Warsaw (2012) (in Polish)
2. Przepiórkowski, A., Buczyński, A.: Spejd: Shallow Parsing and Disambiguation Engine. In: Vetulani, Z. (ed.) Proceedings of the 3rd Language & Technology Conference, Poznań, Poland, pp. 340–344 (2007)
3. Głowińska, K.: Anotacja składniowa. [1], 107–127
4. Ogrodniczuk, M., Głowińska, K., Kopeć, M., Savary, A., Zawisławska, M.: Polish Coreference Corpus. In: Vetulani, Z., ed.: Proceedings of the 6th Language & Technology Conference: Human Language Technologies as a Challenge for Computer Science and Linguistics, Poznań, Poland, Wydawnictwo Poznańskie, Fundacja Uniwersytetu im. Adama Mickiewicza, 494–498 (2013)
5. Górski, R.L., Przepiórkowski, A., Łaziński, M., Lewandowska-Tomaszczyk, B.: Polski korpus. In: Academia. Polska Akademia Nauk, pp. 4–7
6. Acedański, S.: A Morphosyntactic Brill Tagger for Inflectional Languages. In: Loftsson, H., Rögnvaldsson, E., Helgadóttir, S. (eds.) IceTAL 2010. LNCS (LNAI), vol. 6233, pp. 3–14. Springer, Heidelberg (2010)
7. Woliński, M.: Morfeusz – a practical tool for the morphological analysis of Polish. In: Kłopotek, M.A., Wierzchoń, S.T., Trojanowski, K. (eds.) Proceedings of the International Intelligent Information Systems: Intelligent Information Processing and Web Mining 2006 Conference, Wisła, Poland, pp. 511–520 (June 2006)
8. Waszczuk, J., Głowińska, K., Savary, A., Przepiórkowski, A.: Tools and methodologies for annotating syntax and named entities in the National Corpus of Polish. In: Proceedings of the International Multiconference on Computer Science and Information Technology (IMCSIT 2010): Computational Linguistics – Applications (CLA 2010), Wisła, Poland, pp. 531–539. PTI (2010)

Amharic Anaphora Resolution
Using Knowledge-Poor Approach

Temesgen Dawit and Yaregal Assabie

Department of Computer Science, Addis Ababa University, Ethiopia
temesgen.dawit.ek@gmail.com, yaregal.assabie@aau.edu.et

Abstract. Building complete anaphora resolution systems that incorporate all linguistic information is difficult because of the complexities of languages. In the case of Amharic, it is even more difficult because of its complex morphology. In addition to independent anaphors, Amharic has anaphors embedded inside words (hidden anaphors). In this paper, we propose Amharic anaphora resolution system that also treats hidden anaphors, in addition to independent ones. Hidden anaphors are extracted using Amharic morphological analyzer. After anaphoric terms are identified, their relationships with antecedents are built by making use of the grammatical structure of the language along with constraint and preference rules. The system is developed based on knowledge-poor approach in the sense that we use low levels of linguistic knowledge like morphology avoiding the need of complex knowledge like semantics, world knowledge and others. The performance of the system is evaluated using 10-fold cross validation technique and experimental results are reported.

Keywords: Amharic anaphora resolution, knowledge-poor algorithm, hidden anaphors.

1   Introduction

Anaphora resolution is a technique or phenomenon of pointing back to an entity in a given text that has been introduced with more descriptive phrase in the text than the entity or expression which is referring back. The entity referred in the text can be anything like object, concept, individual, process, or any other thing [12]. The expression which is referring back is called *anaphor* whereas the previous expression being referred is called *antecedent* [2]. Anaphora can be categorized as pronominal anaphora, lexical noun phrase anaphora, noun anaphora, verb anaphora, adverb anaphora or zero anaphora based on the form of the anaphor or syntactic category of the anaphor. Based on the location of the antecedents in the text, anaphors can also be classified into *intrasentential* or *intersentential* [9], [16], [18]. Anaphora resolution is a complex task which needs various types of linguistic knowledge such as morphology, syntax, semantics, discourse and pragmatics. Furthermore, different disciplines such as logic, philosophy, psychology, communication theory and other related fields are also involved in the process of anaphora resolution [8]. Due to such complexities and

A. Przepiórkowski and M. Ogrodniczuk (Eds.): PolTAL 2014, LNAI 8686, pp. 278–289, 2014.

difficulties, most of the researches performed in the area of anaphora resolution were aimed at the resolution of pronouns or one type of anaphora. Though the resolution of anaphora is complex, it needs to be addressed strongly as it is used as an important component in many natural language processing (NLP) applications like information extraction, question answering, machine translation, text summarization, opinion mining, dialogue systems and many others. Depending on the inherent characteristics of languages, the process of anaphora resolution may pass through several steps. Among the most important steps are identification of anaphors, identifying the location of candidates and selection of an antecedent from the set of candidates [8]. Selection of an antecedent from the set of candidates is made by making use of constraint and preference rules [9], [14]. Constraint rules must be satisfied for any antecedent to be considered as candidate antecedent for the anaphora resolution. Gender, number and person agreement [6], [11], [18] and selectional restrictions [9], [18] are common constraint rules applied in most languages. Preference rules give more preference to antecedents when constraint rules are satisfied. They are applied on antecedents that passed constraint rules and when a certain anaphor has more than one antecedent as a choice [9], [14]. Some of the preference rules are definiteness [11], givenness [13], recency [5], frequency of mention [9], etc.

Based on the level of linguistic information required to resolve anaphora, two broad categories have emerged: *knowledge-rich* and *knowledge-poor* approaches. Knowledge-rich anaphora resolution approaches employ linguistic and domain knowledge in great detail starting from the low-level morphological knowledge to the very high level knowledge like world knowledge. As a result, morphological analyzers, syntactic analyzers, semantic analyzers, discourse processors and domain knowledge extractors are needed to deal with anaphora resolution using knowledge-rich approaches [15]. On the other hand, knowledge-poor approaches are formulated to avoid the complex linguistic knowledge used in anaphora resolution process. It is a result of looking for inexpensive solutions to satisfy the need of NLP systems in a practical way. The development of low-level NLP tools like part-of-speech (POS) tagger and morphological analyzer attracted the use of knowledge-poor approach. In general, knowledge-poor anaphora resolution approach depends on a set of constraint and preference rules [13], [14]. Depending on the characteristics of languages, several of the aforementioned rules, techniques and approaches have been used to develop anaphora resolution systems for various languages across the world such as English [11], Turkish [9], Norwegian [6], Estonian [15], Czech [17], etc. However, to our best knowledge, there was no any attempt to develop anaphora resolution system for Amharic. Thus, this work is aimed at developing anaphora resolution system for Amharic using knowledge-poor approach as the language is under-resourced with only limited number of NLP tools available.

The remaining part of this paper is organized as follows. Section 2 presents the characteristics of Amharic language with special emphasis on its morphology and personal pronouns. In Sect. 3, we present the proposed system for anaphora resolution. Experimental results are presented in Sect. 4, and conclusion and future works are highlighted in Sect. 5. References are provided at the end.

2 Linguistic Characteristics of Amharic

2.1 The Amharic Language

Amharic is the working language of Ethiopia having a population of over 90 million at present. Even though many languages are spoken in Ethiopia, Amharic is dominant and is spoken as a mother tongue by a large segment of the population and it is the most commonly learned second language throughout the country [10]. The language is categorized under the Semitic languages family and it is the second most spoken language among Semitic language families in the world, next to Arabic. Though the language is widely spoken or used as a working language in the country, it is still considered to be under-resourced since linguistic resources are not widely available. Amharic is written using its own script having 34 consonants (base characters), out of which six other characters representing combinations of vowels and consonants are derived for each character. This leads to have a total of 34x7 (=238) Amharic characters. In addition, there are about two scores of labialized characters used by the language for writing.

2.2 Amharic Morphology

Due to its Semitic characteristics, Amharic is one of the most morphologically complex languages. Amharic nouns and adjectives are marked for any combination of number, definiteness, gender and case. Moreover, they are affixed with prepositions [1], [19]. For example, from the noun *tämari* (student), the following words are generated through inflection and affixation: *tämariwoč* (students), *tämariw* (the student {masculine}/ his student), *tämariyä* (my student), *tämariyän* (my student {objective case}), *tämariš* (your {feminine} student), *lätämari* (for student), *kätämari* (from student), etc. Similarly, we can generate the following words from the adjective *fäṭan* (fast): *fäṭanu* (fast {definite, masculine, singular}), *fäṭanua* (fast {definite, feminine, singular}), *fäṭanoč* (fast {plural}), etc. Amharic verb inflections and derivations are even more complex than those of nouns and adjectives. Several verbs in surface forms are derived from a single verbal stem, and several stems in turn are derived from a single verbal root. For example, from the verbal root *sbr* (to break), we can derive verbal stems such as *säbr*, *säbär*, *sabr*, *säbäbr*, *täsäbabr*, etc. From each of these verbal stems, we can derive many verbs in their surface forms. If we take the stem *säbär*, the following verbs can be derived: *säbärä* (he broke), *säbäräč* (she broke), *säbärku* (I broke), *säbärkut* (I broke [it/him]), *alsäbärkum* (I didn't break), *alsäbäräčm* (she didn't break), *alsäbäräm* (he didn't break), *alsäbäräñm* (he didn't break me), *täsäbärä* ([it/he] was broken), *slätäsäbärä* (as [it/he] was broken), *kätäsäbärä* (if [it/he] is broken), *sisäbär* (when [it/he] is broken), etc. Amharic verbs are marked for any combination of person, gender, number, case, tense/aspect, and mood resulting in thousands of words from a single verbal root [1], [19]. As a result, a single word may represent a complete sentence constructed with subject, verb and object. For example, *yĭsäbräñal* ([he/it] will break me) is a sentence where the verbal stem *sbr* (will break) is marked for various grammatical functions as shown in Fig. 1.

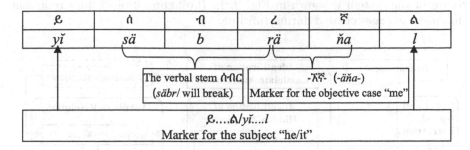

Fig. 1. Morphology of the word *yĭsäbräñal*

2.3 Amharic Personal Pronouns

Personal pronouns in Amharic language are *ĭne* (I), *ĭña* (we), *antä* (you {male, singular}), *anči* (you {female, singular}), *ĭnantä* (you {gender neuter, plural}), *ĭrswo* (you {gender neuter, polite}), *ĭrsu/ ĭsu* (he or it), *ĭrswa/ ĭswa* (she), *ĭrsačäw/ ĭsačäw* (he or she {polite}), *ĭnärsu/ ĭnäsu* (they). In addition, possessive and objective forms could be derived through prefixing and suffixing with *yä-* and *-n*, respectively. For example, *yäĭne* (my) and *ĭnen* (me) are derived from *ĭne* (I) through such affixation process. Amharic personal pronouns can exist in sentences in two forms: *independent* and *hidden*. Independent pronouns are those which appear independently in the text whereas hidden pronouns are embedded in other words. For example, in Fig. 1, the subjective pronoun *ĭsu* (he or it) and objective pronoun *ĭnen* (me) do not appear independently but are embedded in the word *yĭsäbräñal* ([he/it] will break me). Moreover, both independent and hidden personal pronouns are subject to complex affixation processes in Amharic.

3 The Proposed Amharic Anaphora Resolution System

3.1 System Architecture

A complete anaphora resolution system is a very complex process involving multidisciplinary concepts and requires different levels of linguistic analysis. Such requirements hinder the development of a complete anaphora resolution system for under-resourced languages like Amharic. Thus, the proposed anaphora resolution system focuses on resolution of pronominal anaphoric terms, which in many languages form the core of anaphora resolution systems. Accordingly, the proposed system has five major components: *low-level knowledge extraction, identification of independent anaphors, identification of hidden anaphors, identification of candidate antecedents* and *anaphora resolution.* In addition to the Amharic text to be processed, the system requires permanently stored list of *Amharic personal pronouns, constraint rules* and *preference rules* as inputs, and produces anaphors along with antecedent relations as an output. The architecture of the system is shown in Fig. 2. In the figure, dashed lines represent temporary storages created during runtime.

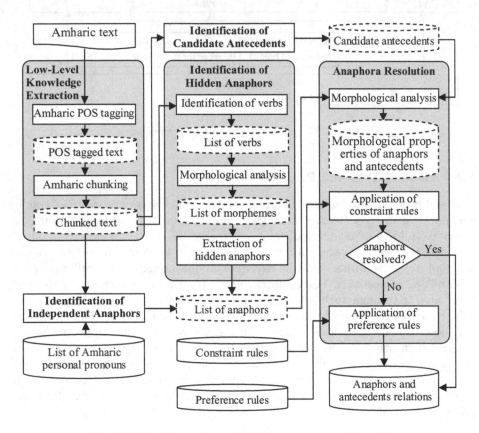

Fig. 2. Amharic anaphora resolution system architecture

3.2   Low-Level Knowledge Extraction

Independent anaphors can be easily identified by searching for personal pronouns in the text whereas hidden anaphors can be identified by analyzing the morphology of verbs. On the other hand, nouns and head nouns of noun phrases form antecedents. Thus, to identify independent anaphors, hidden anaphors and candidate antecedents, the Amharic text should be processed with the aim of labeling the text with POS tags and chunks which are considered to be low-level linguistic knowledge. POS tagging is a classification task with the goal of assigning word classes to the words in a text [3] and it helps identify antecedents and independent anaphors. Chunking is a task that consists of dividing a text into non-overlapping syntactically correlated words and it helps to get noun phrases from which candidate antecedents are identified. Furthermore, POS tagging and chunking are used to identify verbs from which we extract hidden personal pronouns. Because of the unavailability of a fully functional POS tagger for Amharic, we used a manually tagged Amharic text. POS tagged Amharic text is further processed to get chunks in the text. An Amharic chunker developed by Ibrahim and Assabie [7] was used to chunk POS tagged Amharic texts. The examples below show the results of low-level knowledge extraction process.

+ መስተዳድሩ ከፍተኛ ውጤት ላስመዘገቡ ተማሪዎች ሽልማት ሰጠ።
+ *mästädadĭru käfĭtäña wuṭet lasmäzägäbu tämariwoč šĭlĭmat säṭä፡፡*
+ The administration gave awards to students who scored high grade.
+ ('መስተዳድሩ' , 'N'), ('ከፍተኛ ADJ ውጤት N', 'NP'), ('ላስመዘገቡ', 'VPREP'), ('ተማሪዎች', 'N'), ("ሽልማት N ሰጠ V', 'VP')

+ ልጅቷ ትልቅ ወንድሟ እርሷ ከፍል ውስጥ እንዲያስጠናት እናቷን ጠየቀቻት፡፡
+ *lĭjĭtʷa tĭlĭq wendĭmʷa ĭrsʷa kĭfĭl wusṭ ĭndiyasṭänat ĭnatʷan ṭäyäqäčat፡፡*
+ The girl requested her mother so that her older brother would tutor her in her room.
+ ('ልጅቷ', 'N'), ('ትልቅ ADJ ወንድሟ N', 'NP'), ('እርሷ PERPRO ከፍል N ውስጥ PREP', 'PP'), ('እንዲያስጠናት' , 'V'), ('እናቷን NPREP ጠየቀቻት V' , 'VP')

3.3   Identification of Independent Anaphors

Personal pronouns in a given Amharic text are considered to be independent anaphors. Thus, to identify independent anaphors, we are searching for words tagged as personal pronouns in a POS tagged Amharic text. However, it is to be noted that Amharic personal pronouns are subject to various affixation processes. For example, many words can be derived through affixation processes from the personal pronoun *ĭrswa*, some of which are: *yäĭrswa, käĭrswa, bäĭrswa, ĭrswan, yäĭrswan, ĭrswanm, yäĭrswanm, ĭndäĭrswa,* etc. Thus, identification of independent anaphors is achieved by having a database of Amharic personal pronouns (in their root forms) and matching them against the sub-patterns of personal pronouns in the given POS tagged text. Identified independent anaphors are stored in a temporary storage which will be used later in the process of anaphora resolution.

3.4   Identification of Hidden Anaphors

The unique complexity of Amharic anaphora resolution lies in the identification of hidden anaphors. As discussed in Sect. 2.2, verbs are marked for combinations of several grammatical functions such as subject and object resulting in references to subjects and objects embedded in the verbs. Consequently, it is very common to construct Amharic sentences without the explicit use of independent anaphors. For instance, consider the following examples.

> ♦ ካሳ ጠንካራ ነው። መቶ ኪሎ ይሸከማል። የብረት ዱላም በእጁ ይሰብራል።
> ♦ *kasa ţänkara näw*። *mäto kilo yĭšäkämal*። *yäbĭrät dulam bäĭju yĭsäbral*።
> ♦ Kasa is strong. [He] can carry a hundred kilogram. [He] can also break an iron stick with his hand.

> ♦ ካሳ ጠንካራ ነው። እሱ መቶ ኪሎ ይሸከማል። እሱ የብረት ዱላም በእጁ ይሰብራል።
> ♦ *kasa ţänkara näw*። *ĭsu mäto kilo yĭšäkämal*። *ĭsu yäbĭrät dulam bäĭju yĭsäbral*።
> ♦ Kasa is strong. [He] He can carry a hundred kilograms. [He] He can also break an iron stick with his hand.

The verb *yĭsäbral* ([he] can break) is marked for the subject *ĭsu* (he) and implicitly shows the presence of the subject (hidden anaphor). The explicit use of *ĭsu* (he) as a subject results in redundancy of the subject as shown in the second example. Therefore, the use of Amharic independent anaphors is discouraged in such cases, and the first example is more common and natural than the second. These linguistic phenomena add complexity to the whole process of anaphora resolution for Amharic text. Thus, a successful identification of hidden anaphors is an important step in Amharic anaphora resolution. These hidden anaphors are identified from verbs in the chunked text by making use of morphological analyzer that indicates grammatical functions for which verbs are marked. Identification of hidden anaphors becomes more complex when verbs are marked for more number of grammatical functions. We used HornMorpho morphological analyzer developed by Gasser [4] to extract such information from verbs. For example, *yĭsäbral* ([he] can break) is analyzed as $yĭ + säbr + al$ where "$yĭ\ldots al$" would represent "he can", and *säbr* would be equivalent to "break". Accordingly, the hidden anaphor *ĭsu* (he) can be identified from the verb *yĭsäbral* ([he] can break) though morphological analysis. Identified hidden anaphors are stored in a temporary storage along with independent anaphors.

3.5   Identification of Candidate Antecedents

Identification of candidate antecedents is a process whereby a list of words is prepared from which the correct antecedents referred by anaphors are selected. All nouns and head nouns of noun phrases are used as possible candidate antecedents. We used chunked texts to identify nouns and head nouns in a given text. For instance, let us take the chunked sentence.

> ('መስተዳድሩ', 'N'), ('ክፍተኛ ADJ ውጤት N', 'NP'), ('ላሰመዘገቡ', 'VPREP'), ('ተማሪዎች', 'N'), ('ሽልማት N ሰጠ V', 'VP').

From the sentence, መስተዳድሩ/*mästädadru* (the administration), ውጤት/*wuṫet* (grade), ተማሪዎች/*tämariwoč̃* (students) and ሽልማት/*s̃ilĭmat* (award) are nouns or head nouns of noun phrases. As a result, they are candidate antecedents for the anaphor ("he" or "it") hidden inside the verb ሰጠ/*sätä* ([he/it] gave).

3.6 Anaphora Resolution

Now that anaphors and candidate antecedents are identified, the final step is to establish relevant relations between them. This relation is built by applying a stored knowledge of constraint and preference rules which are developed by taking the inherent characteristics of Amharic language into consideration. The constraint rules we applied in this work are *gender*, *number* and *person* agreements where we use morphological analyzer to extract the gender, number, and person properties of nouns and head nouns of noun phrases. Constraint rules are applied on both anaphors and candidate antecedents to select correctly related antecedents that agree with the anaphors. The antecedent-anaphor relationship is built and stored in a database if only one plausible antecedent is left after the constraint rules are applied. If two or more candidate antecedents compete for a single anaphora, then preference rules are applied. Preference rules give more preference to antecedents that satisfy the rule while give nothing or negative value to those that do not satisfy the rule.

The criteria to establish preference rules used in this work are *subject place*, *definiteness, recency, mention frequency*, and *boost pronoun*. Nouns at subject position of a sentence are more preferred than nouns not placed at subject positions. Definite nouns are more preferred as relevant antecedents than nouns which are not definite. On the other hand, nouns which are most recent to the anaphor are given more weight than others which are not recent. The antecedent mentioned repetitively is also given more priority. Boost pronouns are pronouns used as antecedents for anaphors and the pronouns themselves can be considered as possible candidates. The preference of one criterion over the other is determined through values calculated using the probability distribution they have in Amharic language. The probability distribution is calculated by statistically analyzing Amharic text corpus. Table 1 shows the probability distribution of preference rules (with percentage from 5).

Table 1. Values of preference rules for Amharic text

Preference criteria	Percent from 5
Subject place	+ 2.88
Definiteness	+ 0.94
Recency	+ 0.89
Mention frequency	+ 0.02
Boost pronoun	+ 0.12

Each candidate antecedent receives a value from each preference criterion. If a candidate antecedent fulfills a preference criterion, it would receive the corresponding value for that criterion. Otherwise, it would receive zero value. Finally, a candidate antecedent that has the highest aggregate score is selected as an antecedent for a given anaphor. For example, let us take the following sentence.

> ◆   መስተዳድሩ ከፍተኛ ውጤት ላስመዘገቡ ተማሪዎች ሽልማት ሰጠ፡፡
> ◆   *mästädadïru käfïtäña wuṭet lasmäzägäbu tämariwoč šïlïmat säṭä*፡፡
> ◆   The administration gave awards to students who scored high grade.

In the example, the verb ሰጠ/*säṭä* ([he/it] gave) has the hidden personal pronoun *ïsu* (he). መስተዳድሩ/*mästädadru* (the administration), ውጤት/*wuṭet* (grade), ተማሪዎች/*tämariwoč* (students) and ሽልማት/*šïlïmat* (award) are nouns or head nouns of noun phrases which can be candidate antecedents. Gender, number and person information of the verb ሰጠ/*säṭä* ([he/it] gave) is male, singular and third person, respectively. On the other hand, gender, number and person information of candidate antecedents is shown in Table 2.

Table 2. Gender, number and person information of candidate antecedents

Candidate antecedent	Gender	Number	Person
መስተዳድሩ (*mästädadïru*/the administration)	[male]	singular	third
ውጤት (*wuṭet*/ grade)	[male]	singular	third
ተማሪዎች (*tämariwoč*/ students)	neuter	plural	third
ሽልማት (*šïlïmat*/award)	[male]	singular	third

Among the candidates, ተማሪዎች/*tämariwoč* (students) will be discarded due to the constraint on its number which is plural as opposed to that of the hidden anaphor. The remaining three candidates pass to the next step where preference rules are applied. Table 3 shows the respective values (along with their aggregate scores) that the remaining candidate antecedents receive from each preference criterion. From the three candidate antecedents, the candidate having largest aggregate score is መስተዳድሩ/*mästädadru* (the administration). As a result, it is selected as the antecedent being referred by the personal pronoun *ïsu* (he) hidden inside verb ሰጠ/*säṭä* ([he/it] gave).

Table 3. Preference values given to candidate antecedents

Candidate antecedent	Subject place	Definiteness	Recency	Mention frequency	Boost pronoun	Aggregate score
መስተዳድሩ	+2.88	+0.94	0	0	0	+3.82
ውጤት	0	0	0	0	0	0
ሽልማት	0	0	+0.89	0	0	+0.89

4   Experiment

4.1   The Corpus

To evaluate the performance of the system, two sets of Amharic text corpora were collected from various sources. The first set of the corpus was used to establish preference rules as discussed in Sect. 3.6. A total of 495 sentences were statistically analyzed to compute the probability distribution of preference criteria. Another set of corpus was used to test the performance of the system. We used 311 sentences that were generated by the Amharic chunker. The text was collected from Walta Information Center and an Amharic grammar book [19]. From these sentences, we identified a total of 315 verbs which contain hidden personal pronouns. It was observed that the number of independent anaphors in the sentences was very minimal. Thus, we collected an additional text from the Amharic Bible where we selected a total of 163 sentences having 110 independent personal pronouns.

4.2   Test Results

We used 10-fold cross-validation technique to test the performance of the system. The original sample is randomly partitioned into 10 equal size subsamples and the cross-validation process is then repeated 10 times. Accordingly, we obtain 10 results from the folds which can be averaged to produce a single estimation of the overall performance of the system. The evaluation methods we used to test the performance of the system are *Success Rate* (*SR*) and *Critical Success Rate* (*CSR*) which are computed as follows.

$$SR = \frac{Number\ of\ correctly\ resolved\ anaphors}{Number\ of\ all\ anaphors} \tag{1}$$

$$CSR = \frac{Number\ of\ correctly\ resolved\ anaphors}{Number\ of\ anaphors\ with\ more\ than\ one\ antecedent} \tag{2}$$

Equation (2) is used when an anaphor has more than one antecedent after the application of gender, number and person constraint rules. The overall performance of the proposed system also depends on the performances of Amharic POS tagger, chunker, and morphological analyzer. The accuracy of the Amharic chunker is 93.75% [7] and the accuracy of the morphological analyzer, using our tests, is 92.7%. With these limitations, the overall performance of our proposed anaphora resolution system is shown in Table 4.

Table 4. Test result for Amharic anaphora resolution system

Type of anaphors	Success rate	Critical success rate
Hidden	81.79%	76.07%
Independent	70.91%	57.58%

5   Conclusion

Amharic is one of the most morphologically complex and less-resourced languages. Despite the efforts being undertaken to develop various Amharic NLP applications, only few usable tools are publicly available at present. Researchers and developers frequently mention that the morphological complexity of the language is a feature that hinders the development of natural language processing applications for the language. Amharic anaphora resolution also suffers from this problem since anaphoric terms are subject to affixation processes and are also embedded in other words. In this work, we tried to overcome this problem by POS tagging, chunking, and analyzing the morphology of words. Due to unavailability of high-level linguistic tools for the language, we follow knowledge poor anaphora resolution approach that uses the results of low-level linguistic analyses. With the limitations on Amharic chunker and morphological analyzer we used as components in the system, tests have shown that the proposed Amharic anaphora resolution system yields promising results. The performance of the system can be further enhanced by improving the effectiveness of the chunking and morphological analysis modules. Thus, future work is recommended to be directed at improving the chunking and morphological analysis components of the system. Moreover, the preference rules can be refined better by statistically analyzing a large amount of corpus.

References

1. Amare, G.: Zemenawi Ye'amarigna Sewasew Beqelal Aqerareb (Modern Amharic Grammar in a Simple Approach). Addis Ababa University, Addis Ababa (2010)
2. Deoskar, T.: Techniques for Anaphora Resolution: A Survey. Cornell University, USA (2004) (Unpublished)
3. Fissaha, S.: Part of Speech Tagging for Amharic using Conditional Random Fields. In: Proceedings of the ACL Workshop on Computational Approaches to Semitic Languages, Addis Ababa, Ethiopia, pp. 47–54 (2005)
4. Gasser, M.: HornMorpho: A system for morphological processing of Amharic, Oromo, and Tigrinya. In: Proceedings of Conference on Human Language Technology for Development, Alexandria, Egypt (2011)
5. Ho, H., Min, K., Yeap, W.-K.: Pronominal anaphora resolution using a shallow meaning representation of sentences. In: Zhang, C., Guesgen, W.H., Yeap, W.-K. (eds.) PRICAI 2004. LNCS (LNAI), vol. 3157, pp. 862–871. Springer, Heidelberg (2004)
6. Holen, G.: Automatic Anaphora Resolution for Norwegian (ARN). PhD thesis, University of Oslo, Oslo, Norway (2006)
7. Ibrahim, A., Assabie, Y.: Hierarchical Amharic Base Phrase Chunking Using HMM with Error Pruning. In: Proceedings of the 6th Language and Technology Conference, Poznan, Poland (2013)
8. Kabadjov, M.: A Comprehensive Evaluation of Anaphora Resolution and Discourse Classification. PhD thesis, University of Essex, UK (2007)
9. Kucuk, D.: A knowledge-poor pronoun resolution system for Turkish. PhD thesis, Middle East Technical University (2005)

10. Lewis, P., Simons, F., Fennig, D.: Languages of the World, 17th edn. SIL International, Dallas (2013)
11. Liang, T., Wu, D.: Automatic Pronominal Anaphora Resolution in English Texts. Computational Linguistics and Chinese Language Processing 9(1), 21–40 (2004)
12. Liddy, E.: Anaphora in Natural Language Processing and Information Retrieval. Information Processing and Management 26(1), 39–52 (1990)
13. Mitkov, R.: Robust Pronoun Resolution with limited knowledge. In: Proceedings of the 36th Annual Meeting of the Association for Computational Linguistics and 17th International Conference on Computational Linguistics, Montreal, Canada, pp. 869–875 (1998)
14. Mitkov, R.: Anaphora Resolution: The State of the Art. In: Working paper (Based on the COLING 1998/ACL 1998 Tutorial on Anaphora Resolution) (1999)
15. Mutso, P.: Knowledge-Poor Anaphora Resolution System for Estonian. Master's thesis, University of Tartu, Tartu, Estonia (2008)
16. Nemcik, V.: Anaphora Resolution. Master's thesis, Masaryk University, Brno, Czech (2006)
17. Nemcik, V.: Anaphora Resolution for Czech. PhD thesis, Masaryk University, Brno, Czech (2007)
18. Nobre, N.: Anaphora Resolution. Master's Thesis, Portugal, Universidade Tecnica de Lisboa (2011)
19. Yimam, B.: Ye'amarigna Sewasew (Amharic Grammar). Eleni Printing Press, Addis Ababa (2000)

The First Resource
for Bengali Question Answering Research

Somnath Banerjee, Pintu Lohar,
Sudip Kumar Naskar, and Sivaji Bandyopadhyay

Department of Computer Science and Engineering,
Jadavpur University, India
{s.banerjee1980,pintu.lohar}@gmail.com,
{sudip.naskar,sivaji_cse_ju}@yahoo.com

Abstract. This paper reports the development of the first tagged resource for question answering research for a less computerized Indian language, namely Bengali. We developed a tagging scheme for annotating the questions based on their types. Expected answer type and question topical target are also marked to facilitate the answer search. Due to scarcity of canonical documents in the web for Bengali, we could not take the advantage of web as the resource and the major portion of the resource data was collected from authentic books. Six highly qualified annotators were involved in this rigorous work. At present, the resource contains 47 documents from three domains, namely history, geography and agriculture. Question answering based annotation was performed to prepare more than 2250 question-answer pairs. The inter-annotator agreement scores measured in non-weighted kappa statistics is satisfactory.

Keywords: Bengali resource, corpus development, question answering corpus, question answering.

1 Introduction

In computational linguistics, a corpus is a large and structured set of texts that can be used to perform statistical analysis and hypothesis testing, checking occurrences or validating linguistic rules within a specific language territory. A significant landmark in corpus development was achieved with the publication of the Brown corpus in 1960. Following that many notable monolingual and multilingual corpora have been developed across many languages.

In recent years, research in question answering (QA) has grown rapidly with the explosion of information in the internet. The first standard corpus for QA research (in English) was made available with the organization of TREC (1992). Since then new QA resources are being released in TREC each year. Since 2003, QA@CLEF has been providing monolingual and multilingual QA corpora. QA4MRE@CLEF also provides documents and multiple choice questions from various domains. NTCIR has been providing similar QA resources in Japanese since 1999. Besides three aforesaid

A. Przepiórkowski and M. Ogrodniczuk (Eds.): PolTAL 2014, LNAI 8686, pp. 290–297, 2014.

major contributors, some notable corpora development works [1–4] on QA research can also be found in the literature.

In spite of the sound importance of QA in NLP, research in Bengali QA has achieved very less attention than their western and middle-eastern counterparts. Though some work [5–7] has been reported on QA, but no published work on corpus development in the area of QA is available yet.

2   Data Collection

2.1   Challenges

The practice of the web as a corpus for teaching and research has been proposed a number of times [8–11]. Particularly, the Wikipedia as web resource has become a promising corpus and a big frontier for researchers as it contains documents in various fields such as Arts, Geography, History, Science, Sports, Games, etc. However, unfortunately for Bengali the scenario is different. The content of the page is not present or the page contains only a few sentences for majority of documents particularly for history and agriculture fields. Moreover, being a less computerized language, the availability of authentic and well formatted documents is rare. Therefore, for history and agriculture documents we had no choice but to manually prepare the data from authentic government text books of pre-college standard.

2.2   Document Acquisition

Due to the difficulties described in the previous subsection, we were unable to prepare large collection of data. Data were collected for three domains, namely History, Agriculture and Geography. For history domain, 33 documents were selected from authentic text books. Table 2 details the statistics for acquired documents. For agriculture domain, out of 4 documents 2 documents were obtained from Wikipedia and 2 documents from text books. For geography domain, all the 10 documents were acquired from Wikipedia. Thus, a total of 47 documents were collected for corpus preparation. Around 81 hours of rigorous manual typing work were performed to prepare soft copies of 35 documents out of 47 documents.

2.3   Question Acquisition

Question acquisition is a much more challenging task than the acquisition of text data. The question-answer pairs must be bound to the associated documents. In this regard, the following issues were addressed.

- The answer to a question can be either directly extracted or inferred from the document.
- The question size; i.e., number of questions in corpus, has to be reasonable.
- The prepared questions in the corpus should be diverse and of broad coverage. The task should involve as much as question setters as possible to reduce biasness.

To address the second and forth issues, a cloud based service was built and re-quests were sent to students. After accepting the request, the authenticity credentials were sent to the interested students. 40 students agreed to provide us volunteer ser-vice. Two groups with 20 students were formed, namely Gr-A and Gr-B. Different documents were provided to each member of the two groups and they were asked to submit at least 10 questions for each document. The first and third issues were addressed by binding them in question submission constraints. While preparing ques-tions, the students had to follow the given constraints – the answer to the asked ques-tion must be present directly in the document or can be inferred from the source document, and use at least 5 different interrogatives of their own choice. The submit-ted questions were stored in our web server along with the document information. After receiving these questions we analyzed for their overlapping nature. Hence, we introduced a term: Question Overlapping Frequency (QOF) that represents the num-ber of members who formed exactly the same question text. For example, QOF=3 implies that 3 members have generated the same question. Subsequently, the students belonging to Gr-A were requested to answer the questions submitted by the students of Gr-B and vice versa. After collecting the question-answer pairs, we took only the questions having QOF value of 5; i.e., at least 5 students submitted the same question without any interaction among themselves. This also increased the validity of the questions. It was observed from the submitted question set that the questions having low frequencies suffered from typographic errors and sometimes failed to satisfy the question submission constraints. However, there was another possibility that the ques-tions with lower QOF value might not be the erroneous but rare and informative one. To deal with this situation, we sorted out the questions with QOF=1, i.e., only one of the students submitted the question and the validity of such questions was verified with human intervention. Through this survey, we found 12 such questions and only 2 of them finally resulted in valid questions with the help of human intervention. Thus, we collected 2033 valid question-answer pairs.

3   Proposed Template

For document management and storing, the XML technology was chosen because of its popularity and ease of understanding. The corpus documents were stored in XML format using the tagset defined in Table 1. The tags in the tagset were used for three

Table 1. Corpus tagset

Tag	Definition	Tag	Definition
Doc ID	Document ID number	Domain	Domain of the document
Lang	Language of the document	Title	Topic name
Text	Text part of the topic	Paragraph	Paragraph of text part
Questions	Questions answering part	Question	Single question information
Qid	Question ID number	Qtype	Question class
Qstr	Question	EAT	Expected answer type
QTT	Question topical target	Ans	Answer of the question

purposes – document information, text annotation and question answering annotation. The proposed template for each document is depicted in Fig. 1.

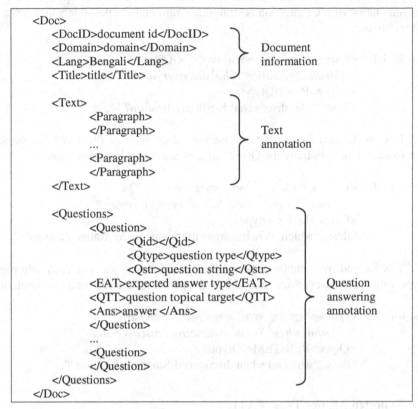

Fig. 1. Document template

4   Question Answering Annotation

Question answering based annotations have been described in this section, in particular the Question Type, the Expected Answer Type, and the Question Topical Target.

4.1   Question Type

The set of question categories (classes) is referred to as question type taxonomy or question ontology. Although different question taxonomies have been proposed for different languages, the taxonomy proposed by [5] is the only existing standard taxonomy for Bengali. Therefore, for Question Type (Qtype) annotation task we followed the taxonomy proposed by [5] which is a single layer taxonomy with nine coarse-grained classes: PER (person), ORG (organization), LOC (location), TEM (temporal), NUM (number), METH (method), REA (reason), DEF (definition) and

MISC (miscellaneous). However, the Qtype annotation is not straightforward for Bengali. The Bengali questions may take interrogatives of three categories - Unit Interrogative (UI), Dual Interrogative (DI), and Compound Interrogative (CI). The Qtype annotation of a UI question is straightforward and the possible value is the one of the 9 classes.

Qtype for UI: <Qstr>সিন্ধু¹ সভ্যতা² কে³ আবিষ্কার⁴ করেন⁵ ?⁶</Qstr>
          [*Sindh*¹ *civilization*² ***who***³ *discover*⁴ *did*⁵ *?*⁶]
          <Qtype>PER</Qtype>
          Gloss: "who discovered Sindh civilization?"

All DIs are formed by two UIs. However, since both of the UIs take the same Qtype value, so for simplicity the Qtype value is annotated by single value.

Qtype for DI: <Qstr>কোন¹ কোন² অঞ্চল³ বাবরের⁴ অধিকারে⁵ আসে⁶ ?⁷</Qstr>
          [***which***¹ ***which***² *area*³ *Babar*⁴ *control*⁵ *come*⁶ *?*⁷]
          <Qtype>LOC</Qtype>
          Gloss: "which were the areas that came under Babar's control?"

A CI is formed by combining two different interrogatives. For each interrogate, a Qtype value is required. So, a total of two Qtype values are required for annotation.

Qtype for CI: <Qstr>কে¹ কবে² সিন্ধু³ সভ্যতা⁴ আবিষ্কার⁵ করেন⁶ ?⁷</Qstr>
          [***who***¹ ***when***² *Sindh*³ *civilization*⁴ *discover*⁵ *did*⁶ *?*⁷]
          <Qtype>PER-TEM</Qtype>
          Gloss: "who and when discovered Sindh civilization?"

4.2   Expected Answer Type (EAT)

Prager [12] defined the Expected Answer Type (EAT) as the class of object (or rhetorical type of sentence) required by the question; in other words, it is the semantic category associated with the desired answer, chosen from a predefined set of labels. The named entity (NE) category of the answer to the question can be used as EAT. The tagset defined in IJCNLP-08 NERSSEAL shared task [14] was used as the NER tagset. However, the EAT is not applicable (NA) when the Qtype is of type METH, REA and DEF; the reason being the answer is not an NE in such cases. For other types of Qtype, i.e., PER, ORG, LOC, TEM, NUM, and MISC, the answer is single word and NE of the answer is the EAT.

EAT for LOC: <Qstr>কোন¹ কোন² অঞ্চল³ বাবরের⁴ অধিকারে⁵ আসে⁶ ?⁷</Qstr>
             [***which***¹ ***which***² *area*³ *Babar*⁴ *control*⁵ *come*⁶ *?*⁷]
             <Qtype>LOC</Qtype>
             <EAT>NEL</EAT>
             Gloss: "which were the areas came under Babar's control?"

EAT for Qtype DEF: <Qstr>বেদ^1 কি^{2} ?^{3}</Qstr>
            [*Beda^1 what^2 ?^{3}*]
            <Qtype>DEF</Qtype>
            <EAT>NA</EAT>
            Gloss: "what is Beda?"

4.3   Question Topical Target (QTT)

Question Topical Target (QTT) is a part of the question that describes the entity about which the information request has been made. The QTT is sometimes referred to as question focus [13], or question topic [12]. QTT corresponds to a noun or a noun phrase that is likely to be present in the answer. But unlike English questions, the interrogatives of a Bengali question can appear at the beginning, within or at the end of the question text [5]. Therefore, it is very difficult to separate the noun phrase when the interrogative appears within the question text. To follow the same strategy for all cases, the named entities in the question text are considered as QTT. So in the present work, QTT is a comma separated named entity list.

Example-8:   QTT with single NE
          <Qstr> সূর্যের^1 ভর^2 কত^3 ?^{4}</Qstr>
             [*Sun^1 mass^2 what^3 ?^{4}*]
          <QTT>সূর্য</QTT>
          Gloss: "what is the mass of the Sun?"

Example-9:   QTT with multiple NEs
          <Qstr> সূর্যকে^{1} প্রদক্ষিণ^2 করতে^{3} পৃথিবীর^4 কত^5 সময়^{6} লাগে^{7} ?^{8}</Qstr>
             [*Sun^1 orbit^2 do^3 Earth^4 how^5 time^6 require^7 ?^{4}*]
          <QTT>সূর্য,পৃথিবী</QTT>
          Gloss: "how long does it take for the Earth to orbit the Sun?"

5   Annotation Aggrement

Prior to the initiation of this study, six language specialists gathered to discuss the question type, EAT and QTT in details to minimize disagreement. They divided the whole annotation task in in three subparts, namely QType annotation, EAT annotation and QTT annotation. Each subpart was evaluated by two independent language specialists. Inter-annotator agreement was evaluated in non-weighted kappa coefficients. The kappa coefficient for Qtype, EAT and QTT annotation tasks are 0.95, 0.91 and 0.89 respectively.

6   Corpus Statistics

Annotation and distribution of questions in the corpus for three different domains, namely history, geography and agriculture, are given in Table 2. Also Table 2 details the document source and statistics for individual domains. The distribution of different question types for individual domains is shown in Table 3.

Table 2. Questions statistics

Domain	#Documents	Source	#Questions
History	33	Books	1922
Geography	10	Wikipedia	251
Agriculture	4	Wikipedia (2) + Books (2)	84

Table 3. Qtype statistics

Type	History	Geography	Agriculture	Overall
Person	777	22	0	799
Organization	54	4	6	64
Location	137	6	11	154
Temporal	171	47	13	231
Numerical	35	60	15	110
Methodical	139	3	10	152
Reason	128	17	8	153
Definition	28	17	7	52
Miscellaneous	430	65	9	504
Person-Temporal	8	3	0	11
Person-Location	6	0	0	6
Temporal-Location	4	4	3	11
Numerical-Miscellaneous	5	3	2	10

7   Conclusion and Future Work

The development of the first annotated dataset for question answering research has been presented in this paper. We were unable to take the full advantage of web as a resource and the major portion of the dataset was collected from authentic textbooks. In the corpus, 2257 questions were annotated according to three question answering based levels, namely Question Class, Expected Answer Type and Question Topical Target. The corpus statistics show that questions of type person (799 questions) appear more frequently than the questions of other types.

We are under no illusions that the corpus is fully comprehensive and without anomalies. Further revision of the annotation scheme and of the corpus is inevitable. However, we believe that this corpus forms a significant contribution to the study of Bengali question answering research, and is a significant step for corpus linguistics in the direction of the semantic study of languages.

In the future, more documents from other domains will be included to increase the volume of the corpus.

Acknowledgements. We acknowledge the support of the Department of Electronics and Information Technology (DeitY), Ministry of Communications and Information Technology (MCIT), Government of India funded project *"CLIA System Phase II"*.

References

1. Verberne, S., Boves, L., Oostdijk, N., Coppen, P.A.J.M.: Data for question answering: the case of why. In: Proceedings of the International Conference on Language Resources and Evaluation (LREC 2006), 5th edn., Genoa, Italy (2006)
2. Inoue, M., Akagi, T.: Collecting humorous expressions from a community-based question-answering-service corpus. In: Proceedings of LREC, pp. 1836–1839 (2012)
3. Cabrio, E., Coppola, B., Gretter, R., Kouylekov, M., Magnini, B., Negri, M.: Question answering based annotation for a corpus of spoken requests. In: Proceedings of the Workshop on the Semantic Representation of Spoken Language, Salamanca, Spain (2007)
4. Louis, A., Nenkova, A.: A corpus of general and specific sentences from news. In: Proceedings of LREC, pp. 1818–1821 (2012)
5. Banerjee, S., Bandyopadhyay, S.: Bengali Question Classification: Towards Developing QA System. In: Proceedings of the 3rd Workshop on South and Southeast Asian Natural Language Processing (SANLP), COLING, India, pp. 25–40 (2012)
6. Banerjee, S., Bandyopadhyay, S.: An Empirical Study of Combining Multiple Models in Bengali Question Classification. In: Proceedings of International Joint Conference on Natural Language Processing (IJCNLP), Japan, pp. 892–896 (2013)
7. Banerjee, S., Bandyopadhyay, S.: Ensemble Approach for Fine-Grained Question Classification in Bengali. In: Proceedings of 27th Pacific Asia Conference on Language, Information, and Computation (PACLIC), Taiwan, pp. 75–84 (2013)
8. Rundell, M.: The biggest corpus of all. Humanising Language Teaching 2(3) (2000)
9. Fletcher, W.H.: Concordancing the Web with KWiCFinder. In: Proceedings of the Third North American Symposium on Corpus Linguistics and Language Teaching, Boston, MA (2001)
10. Robb, T.: Google as a Corpus Tool? ETJ Journal 4(1) (2003)
11. Fletcher, W.H.: Making the Web more useful as source for linguists corpora. In: Conor, U., Upton, T.A. (eds.) Applied Corpus Linguists: A Multidimensional Perspective, pp. 191–205. Rodopi, Amsterdam (2004)
12. Prager, J.: Open-Domain Question-Answering. In: Foundations and Trends in Information Retrieval. Now Publishers (2007)
13. Monz, C.: From Document Retrieval to Question Answering. Ph.D. thesis, University of Amsterdam (2003)
14. Singh, A.K.: Named Entity Recognition for South and South East Asian Languages: Taking Stock. In: Proceedings of the IJNLP 2008 Workshop on NER for South and South East Asian Languages, Hyderabad, India, pp. 5–16 (2008)

Computer-Assisted Scoring of Short Responses: The Efficiency of a Clustering-Based Approach in a Real-Life Task

Magdalena Wolska[1], Andrea Horbach[2], and Alexis Palmer[2]

[1] LEAD Graduate School, Eberhard Karls Universität Tübingen, Germany
magdalena.wolska@uni-tuebingen.de
[2] Saarland University, Saarbrücken, Germany
{andrea,apalmer}@coli.uni-saarland.de

Abstract. We present an extrinsic evaluation of a clustering-based approach to computer-assisted scoring of short constructed response items, as encountered in educational assessment. Due to their open-ended nature, constructed response items need to be graded by human readers, which makes the overall testing process costly and time-consuming. In this paper we investigate the prospects for streamlining the grading task by grouping similar responses for scoring. The efficiency of scoring clustered responses is compared both with the traditional mode of grading individual test-takers' sheets and with by-item scoring of non-clustered responses. Evaluation of the three grading modes is carried out during real-life language proficiency tests of German as a Foreign Language. We show that a system based on basic clustering techniques and shallow features yields a promising trend of reducing grading time and performs as well as a system displaying test-taker sheets for scoring.

1 Introduction

A *constructed-response* item is a type of open-ended test question which elicits free production, rather than specifying a closed set of answers from which a test-taker selects (as in multiple-choice or true-false tests). *Short constructed responses* – so-called *short answers* – which we address here, may range from a single word or phrase through a few sentences. While selected-response tests can be easily scored automatically, constructed responses need to be scored by humans because of their open-ended nature. The scoring process is thus time-consuming, potentially prone to errors of inconsistency, and more expensive. Automated scoring, exploiting natural language processing (NLP) techniques, offers reduction of the time and cost of scoring constructed responses and has been successfully deployed in the essay grading scenario; see [15] for an overview.

Most current NLP research on short answers has focused on fully automated scoring. The techniques employed are based on rule-based matching of responses with concepts and model answers [2,9], probabilistic learning for concept matching [16], classification [13], or alignment of linguistic units, syntactic or semantic [12,11,5,7], to mention just a few. Existing short answer scoring systems have

A. Przepiórkowski and M. Ogrodniczuk (Eds.): PolTAL 2014, LNAI 8686, pp. 298–310, 2014.
© Springer International Publishing Switzerland 2014

been shown to agree highly with human-assigned scores, yet they never reach perfect accuracy. It is disputable whether, from the test-takers' point of view, any amount of grading error attributable to machine-scoring would be admissible even in low-stakes exams, let alone in high-stakes critical assessments.

In this work, we pursue a related, but different approach to short answers: rather than automatically assigning a grade, we exploit language processing techniques to streamline *manual* scoring. The "traditional" scenario in this case consists of teachers scoring test sheets on paper. This is done on a *by test taker*-basis – teachers score test sheets of individual test takers – and may result in scoring inconsistencies, that is, the same responses given by different persons may be scored differently, especially if the set of test sheets is large and the time for scoring limited. The least that computer-assisted scoring can offer with little effort is *by item*-scoring of preprocessed unique responses: rather than presenting test-takers' sheets, a system can present all responses to a particular item. The key question we investigate in this context is whether the order in which responses are presented to the raters can contribute to making scoring faster. Specifically, we explore the idea of aggregating similar responses into groups using clustering methods. The premise here is that similar responses are likely to be graded with the same scores, thus presenting them together has the potential of making scoring faster and more consistent.

The same idea has been developed in parallel to our work by Basu et al. [3]. "Powergrading" uses more sophisticated clustering techniques and has been shown to yield promising results, but has been so far evaluated only in a simulated setting. By contrast, our work has been set in a real-life scenario of foreign language testing for course placement. We use actual learner test data, and our evaluation is conducted as part of the placement process. In [8] we showed in a simulated setting that teacher's workload can be reduced to labeling only 40% of all responses to a question, while still maintaining a grading accuracy of more than 85%. In the present work, we compare time-efficiency of clustering-based grading in the same context to two baseline systems: one reproducing the traditional scoring mode, familiar to the teachers, that of grading individual test-takers' sheets, and the other displaying all responses to individual questions ordered by frequency. Our results show that, in a real-life task, clustering-based scoring leads to reductions in scoring time over frequency-based ordering and can be at least as efficient as the traditional, by test-taker scoring mode.

2   Data

All linguistic and scoring data used in this study were acquired during *actual* placement tests for German as a Foreign Language (GFL) semester and term-break courses taught at Saarland University. Test takers were students who enrolled for the language courses and needed to be placed at the appropriate level. Scoring was done by the GFL teachers responsible for assigning students to courses.

Short Answer Data. The placement tests were administered via a custom web-based platform we developed. The listening comprehension part of the tests consisted of three sections with pre-recorded audio prompts and up to 24 short answer comprehension questions in total. Most questions elicited a single piece of information (as in "Where does X come from?"), but some explicitly asked for enumeration of up to three pieces of information (e.g. "What information was provided about X? (name 3 things)"). Test-takers could read the questions before listening to the audio. Responses were type-written. There were no restrictions on the linguistic form nor on the time allowed for completing the listening task.

Scoring Data. The tests were scored via a web-based scoring interfaces which displayed sets of responses. We will refer to sets of responses displayed for scoring as *answer sheets* (or simply *sheets*), by analogy with manual scoring.

Most of the single-information items were scored on a [0,1] scale. The multi-part items were scored on a 2- or 2.5-point scale; partial credits were given at .5-score points. Scores were entered manually into text fields. Empty responses were automatically scored at 0 and excluded from manual scoring. Scores were assigned strictly based on content; grammar, spelling or other errors were not taken into account as long as a response could be semantically interpreted. Responses matching exactly one of the target answers provided by the teachers were scored automatically. Grading was performed by Saarland University GFL teachers. The majority of items were graded by two teachers. At this time we did not control for scoring consistency, however, in the future we are planning to test both within- and between-grader scoring reliability. Quantitative information about the data and details on the different scoring modes evaluated in this paper will be elaborated in Sect. 4 when we describe the evaluation setup.

3 Response Clustering

Test-takers' responses were preprocessed in a standard way. Basic preprocessing included: removing intra- and inter-sentential punctuation, lemmatization using TreeTagger [14], and lowercasing the response strings. Multiple instances of the resulting wording patterns are collapsed into one observation. Three types of features are used for clustering: n-gram, question material, and keyword-features.

N-grams. N-gram features include word n-grams and character n-grams. Word unigams, bigrams, trigrams and skip-bi- and tri-grams (pairs and triples of words with an arbitrary number of words between them) are extracted from lemmatized strings. Prior to lemmatization, character bi- to four-grams are extracted from lowercased strings in order to account for misspellings.

Keywords. Keywords represent the most relevant concepts (main points) in target answers and correspond roughly to the most compact correct answers per question. For the most part, in our case, these consist of the nouns present in the target answers and were created specifically for each question. Based on the defined keywords, we generate a binary feature (KW) specifying whether the keyword(s) is/are present within a learner's response.

Question Material. The question material feature (QM) models the fact that the part of a response which reiterates the question's topic (what is *given*) does not contain the answer. Excluding question material from the response has been shown to improve classification accuracy in automated scoring [11]. As Meurers et al. point out, the issue is not about the question material as such, but rather about distinguishing between the topic and focus part of the response [10]. Here, as in [11], we do not perform topic-focus analysis, but approximate it by removing from the response strings the lexical material which appears in the question, based on unigram matching. As a result, for instance, after lowercasing, both responses "they live in berlin" and "in berlin" to the question "Where do they live?" would be reduced to a single observation "in berlin".

Clustering Method And Parameters. For the present study, we use a variant of the simple but efficient single pass clustering method [6]: vectors are iterated over in a sequence. If a new vector is within a specified similarity threshold to a centroid of an existing cluster, it is added to the cluster; otherwise, a new cluster is formed. Vector cosine, a standard measure of vector similarity derived from the Euclidean dot product, is used as a similarity measure.

Four clustering models have been tested. The conditions differ in the combination of keyword and question material features: +KW denotes that the keywords feature has been included and +QM denotes that the question material has been excluded. This results in the following four feature combinations (models): +KW +QM, +KW −QM, −KW −QM, −KW +QM. All models use the same n-gram features. As a simple method of giving the keyword feature greater weight, it is repeated one hundred times in the feature vectors.

Baselines. In the main evaluation (Sect. 5), we compare clustering-based scoring to two baseline scoring modes. The first baseline interface replicates the traditional scoring routine of grading individual test-takers' sheets: the interface displayed all responses from one test participant in the order in which questions appeared in the test, that is, grading was on *per test-taker* basis. We will refer to this data set and system as **TT**. In this scoring condition, responses are not preprocessed, and multiple token-identical responses from different test-takers have to be scored multiple times, as in the pen-and-paper setting.

The second baseline interface displayed responses on *per question* basis, providing a basic workload reduction by collapsing multiple token-identical responses into one entry to be scored. Collapsing token-identical responses already results in, on average, around 30% reduction of the number of responses to be scored and consequently reduction in total scoring time. Within one set of responses for scoring, responses were sorted *by frequency* of occurrence in decreasing order. We will refer to this data set and system as **Q_F**.

Note that the first scoring mode is the one the teachers were familiar with and originally favoured. Thus, we expected that this scoring mode would prove efficient. The choice of frequency over random ordering for the second baseline system was motivated by its potential logistic benefit: with the most frequent responses being scored first, the set of completely scored test-taker sheets is likely to grow faster than in the latter case. Thus, assuming that among the frequent responses are many correct ones (advanced learners), in the placement testing scenario, advanced-level groups could be formed earlier, resulting in overall reduction of the time needed to form groups based on the placement test results.

4   Evaluation Setup

The evaluation consists of two parts: first, we select a model and parameter setting for the timing experiment. We do this by pre-testing the performance of the four models on a data set which has been scored using one of our baseline systems, in order to identify the best overall model-parameter combination for clustering (intrinsic evaluation of clustering-based scoring). Then, we evaluate the selected clustering-based system in a real-life setting on a time-on-task criterion (extrinsic evaluation). We will refer to the data and systems which involved *per question scoring of clustered responses* as **Q_C**. Before we describe the model and parameter selection method, we present the data sets used in the evaluation.

4.1   Evaluation Data

As mentioned in Sect. 2, all our data stems from actual tests and has been collected as part of placement testing for GFL courses. Data from five tests have been used for the analysis: two tests have been scored using baseline scoring systems (**TT** and **Q_F**) and three using clustering-based scoring (**Q_C 1–3**).

We did not influence the selection of the test materials used by the GFL centre nor did we restrict the number of participants. This has two consequences which affect the evaluation: first, the five sets of test-taker responses correspond to different listening comprehension items (different material used for the placement tests). This means that we cannot tune the performance for specific items. Second, the experimental design is unbalanced (different numbers of test participants – and thus also responses – for each test). Figure 1 shows the distribution of the number of responses per test for the five chronologically ordered test times, before and after preprocessing (see Sect. 3). The reason for the large differences in the total number of responses per condition is that the number of participants in the term-break courses was much smaller than in the semester courses.

4.2   Model and Parameter Selection

The particular model, as well as the similarity threshold, for the clustering-based systems to be timed on task were selected based on data scored using the **Q_F** system. Even though this data set is smaller than the **TT** data set, its size

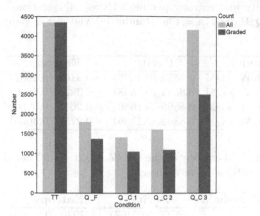

Fig. 1. Distribution of number of responses per scoring mode; All: total number of responses, Graded: number of scored responses (after preprocessing)

corresponds more closely to the size of the data set we expected to obtain for the first cluster-based scoring time (less chance of overfitting). After excluding answer sheets which were scored multiple times (see Sect. 4.5), 21 answer sheets scored using the **Q_F** system were used to select a model and a threshold setting.

Single pass clustering, as described in Sect. 3, was performed on the **Q_F** data with similarity thresholds in the range from 0.1 to 0.9 at 0.1 increments. (Threshold values of 0 and 1 yield one cluster and as many clusters as data points, respectively. Neither of these is a desired result, so we do not consider them here.) Since these data have already been scored, we use standard cluster evaluation measures of purity and entropy with respect to the response scores – treated here as cluster labels – to select the feature combination for the model to be used in the timing experiment. Purity and entropy both measure a clustering method's accuracy in identifying existing classes. For a set of labels $\{l_i\}$ and clusters $\{c_j\}$ they are defined as follows:

$$Purity = \frac{1}{n} \sum_j \max_i n_{ij} \qquad Entropy = \sum_j \frac{n_j}{n} \left(- \sum_i \frac{n_{ij}}{n_j} \log_2 \frac{n_{ij}}{n_j}\right)$$

where n is the number of elements in the distribution, n_{ij} is the number of elements labelled i in the cluster j, and n_j is the total number of elements in the cluster j.[1]

Ideally, we would like to obtain clusters of responses which should receive the same grades, therefore it is precision-oriented evaluation measures that are most relevant. Recall that the questions in the task-based evaluation are different from the ones used for parameter selection. Thus, we choose a feature combination which yields high mean purity/low mean entropy and low variance over all

---

[1] See, for instance, [1] for a comparison of various cluster evaluation measures.

Table 1. Mean purity and entropy per model over all questions (and all similarity thresholds) for the **Q_F** data ($n = 189$; standard deviations in parentheses)

Model	Purity	Entropy
+KW +QM	0.8553 (0.1696)	−0.3169 (0.2887)
+KW −QM	0.8634 (0.1658)	−0.3228 (0.2866)
−KW +QM	0.8896 (0.1649)	−0.2615 (0.3200)
−KW −QM	0.8862 (0.1586)	−0.2758 (0.3060)

Table 2. Mean purity and entropy of the +KW +QM model per similarity threshold ($n = 21$; standard deviations in parentheses)

Threshold	Purity	Entropy
0.1	0.6505 (0.2467)	−0.4940 (0.2243)
0.2	0.7929 (0.1621)	−0.4776 (0.2977)
0.3	0.8163 (0.1550)	−0.4517 (0.3313)
0.4	0.8423 (0.1383)	−0.3841 (0.2964)
0.5	0.8539 (0.1344)	−0.3769 (0.2887)
0.6	0.8900 (0.1281)	−0.2932 (0.2385)
0.7	0.9121 (0.1298)	−0.2147 (0.2268)
0.8	0.9575 (0.0696)	−0.1143 (0.1614)
0.9	0.9825 (0.0377)	−0.0459 (0.0984)

questions and similarity thresholds, in order not to bias the selection toward best performance on the specific responses in the Q_F data set.

Table 1 shows the means and standard deviations of cluster purity and entropy by feature combination; n denotes the number of observations per cell (here: 21 questions * 9 thresholds). Violation of normality assumption was found, however, because group variances were equal (as per Levene's test) and the groups are of moderate and equal sizes, ANOVA was deemed appropriate. Two one-way ANOVAs were used to test if purity and entropy were different for the four clustering models. Differences were not statistically significant at $p < 0.05$. The +KW +QM model has been selected for the clustering systems because this combination of features is most commonly used and performs well in existing automated scoring systems. However, in follow-up work, we will analyze the effect of the quality of the keywords and accuracy of question material extraction in more detail and perform analogous analysis on other data sets in order to test whether the +KW +QM model indeed performs best across test questions.

Table 2 shows the means and standard deviations of cluster purity and entropy for the +KW +QM model for each of the thresholds for the 21 questions (n). We select a mid-range similarity threshold of 0.4 in order to allow for larger differences in wording and spelling (n-gram features) within clusters. All the **Q_C** systems (**Q_C** 1–3) were based on the same model and similarity threshold.

4.3   Procedure

The in-use evaluation of the scoring interfaces was performed as part of course placement process. The system displayed sets of responses on web pages and recorded time elapsed from the point of opening the page to the point of clicking submit. In the case of the **TT** system, each sheet contained all responses by a particular test-taker. In the case of the **Q_F** and **Q_C 1–3** systems, each sheet contained all responses to a particular question. In the **Q_C** conditions responses were grouped according to clusters, in the order of decreasing cluster size. Groups were not made explicitly visually distinct and the teachers were not told that responses were being displayed by similarity. However, it was of course possible to notice the grouping (which the teachers did and reported the observation). The interface allowed the teachers to open a sheet multiple times and review or revise already given scores; this feature had been explicitly requested by the teachers who performed the grading. This feature also lead to certain response sheets being excluded from analysis (see Sect. 4.5).

4.4   Evaluation Measures

The main question we investigate in this work is whether clustering makes manual scoring of constructed responses more efficient in terms of reductions in time-on-task. Because the evaluation was conducted in a real-life setting rather than as a laboratory experiment, we did not have control over external factors which affect time measurements, such as third-party interference during scoring or graders' fatigue. Therefore, we do not use total scoring time of all sheets as an evaluation measure. However, we do use *per sheet scoring time*, calculated as the difference between the submit time and opening time in seconds, to remove outliers (see below). As the main evaluation measure we use *mean per response scoring time* (in seconds), calculated for each sheet as total scoring time per sheet divided by the number of responses on the given sheet. Moreover, we assess the relationship between the amount of content per answer sheet (expressed as number of tokens) and the time it takes to grade the content.

4.5   Anomalous Data and Outliers

Because our evaluation was conducted as a real-life task, the data contains anomalies which lead us to exclude data points as follows: since it is not clear how to treat sheets which have been scored multiple times (scores were revised or deleted, possibly by mistake, and retyped), in the final evaluation we include only two types of answer sheets: (i) those which have been opened (and scored) only once and (ii) of those opened multiple times, those which also were completely scored the first time. In the latter case, we include only the first scoring time and assume that subsequent opening served merely to review the scores.

As shown in Fig. 2 the distribution of the number of responses per sheet is heavily unbalanced across the scoring conditions.[2] In particular, all sheets in the

---

[2] All box plots throughout the paper are Tukey box plots. Means are marked within boxes and linked.

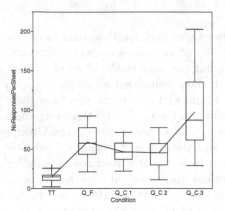

Fig. 2. Number of responses per sheet

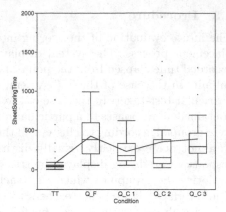

Fig. 3. Per-sheet scoring time

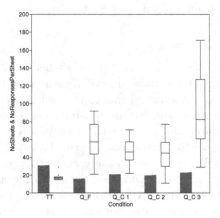

Fig. 4. Number of sheets (bars) and number of responses per sheet (box-plots) included in the analysis

TT condition (scored by test-taker) could contain at most 30 responses (total number of pieces of information elicited). Response sheets of low proficiency learners would contain only a fraction of that number (e.g. 14 sheets contained fewer than 10 responses). In order to obtain a closer match to the other conditions (cf. Fig. 2), we include only those sheets from the **TT** condition which contain at least half of the elicited responses.

Finally, we exclude from all conditions the answer sheets with anomalously long scoring times. Figure 3 shows the distribution of per sheet scoring times across conditions. The 11 marked outliers have been manually removed from the data set. Figure 4 shows the final number of sheets and the distribution of number of responses per sheet included in the analysis.

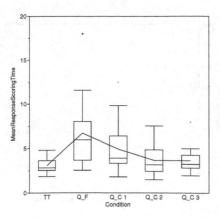

Table 3. Descriptive information on mean per response scoring times (grand means; standard deviations in parentheses)

Model	n	Mean	Median
TT	31	3.0709 (0.8795)	2.8125
Q_F	16	6.7192 (4.1427)	5.9565
Q_C 1	21	4.9128 (2.7282)	3.9048
Q_C 2	20	3.6454 (1.5932)	3.1645
Q_C 3	23	3.6434 (1.2667)	3.2414

Fig. 5. Mean per response scoring time (cell sizes are not equal)

5   Analysis and Results

In this study, we are mainly interested in two questions. First, is clustering-based scoring faster than scoring non-clustered responses (in our case, the **TT** and **Q_F** modes)? Second, what is the relationship between the *amount of material* presented for scoring and the time required for the task, and further, can the time needed to grade a given amount of material be reliably estimated? The first thing to note is that the properties of our timing data do not meet parametric assumptions. Because for some measurements the violation of assumptions is substantial, rather than presenting detailed statistical analysis, we only show the observed tendencies and leave statistical analysis for a controlled experiment.

We first compare the average time needed to grade a response in each of the scoring conditions. Table 3 and Fig. 5 show the grand means and the distribution of scoring time per condition (*n* denotes the number of sheets; Table 3 shows the grand means: means of per answer sheet means). There are several outliers, the scoring times are not normally distributed, and the variances are not equal (confirmed with Shapiro-Wilk and Levene's tests). We see that the shortest and the most stable scoring time per sheet is in the **TT** condition and the longest in the **Q_F** condition (twice as long as in **TT**). In the clustering conditions the mean scoring duration decreases with each iteration, and **Q_C 3** reaches values almost in the range of the **TT** condition. The fact that in the **TT** condition, on average, scoring proceeds the fastest can possibly be explained by the fact that the material is always presented to the raters in smaller chunks (answer sheets never contain more than 30 responses). Among the clustering conditions we see a trend of average scoring durations decreasing over time. This is a promising observation that possibly suggests a learning effect, that is, it may be that teachers needed to get used to the new scoring mode. The Kruskal-Wallis test conducted on the data showed statistical differences among the conditions,

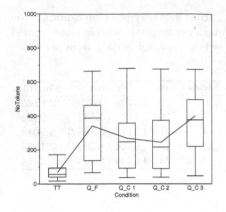

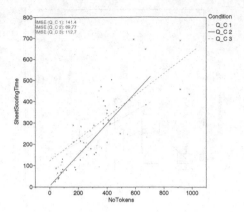

Fig. 6. Number of tokens per answer sheet (cells are not equal in sizes)

Fig. 7. Linear fit of sheet scoring time as a function of the number of tokens

however, pairwise comparisons yield different groups depending on the post hoc test used (Wilcoxon or Steel-Dwass). The statistical differences found by both post hoc tests (at $\alpha = 0.05$) are between the **TT** condition and the **Q_F** and **Q_C 1** conditions and between the **TT** condition and the **Q_C 2** condition; the difference between **TT** and **Q_C 3** is marginally statistical ($p = 0.0707$) according to the Steel-Dwass test. Again, these results need to be taken with caution due to the properties of the data. Overall the most interesting observation is the trend of the average time needed for scoring decreasing with consecutive uses of the clustering mode.

Now, while the comparison by answer sheet intuitively corresponds to the setup of the scoring task, it of course confounds the fact that scoring sheets contain different amounts of material and material of different natures. Figure 6 shows the distribution of the number of tokens per sheet by scoring condition. In the **TT** scoring mode, the raters have to remember the order of questions in the test in order to keep track of which response answers which question, but the average number of responses to score per sheet is smaller than in the other conditions. In the **Q** modes, depending on the question, the number of different responses may be small or quite large (cf. Fig. 4), and the length and linguistic complexity of the responses may differ. We hypothesized a linear relationship between the amount of material to be scored and the scoring time. To test the hypothesis, we applied linear fit to the time data as a function of the length of an answer sheet in tokens. Figure 7 shows a linear fit for scoring time for the clustering conditions. Linear fit was not statistically significant for the **TT** and **Q_F** conditions. The best fit is found for the **Q_C 2** condition. The fact that no fit is found for the **TT** and **Q_F** conditions and the large differences in RMSE show that the relationship is strongly dependent on the responses. In other words, individual questions may require much more or less scoring time than the average. Again, this might have to do with the complexity of the responses or the learner errors (in particular, spelling errors might contribute to longer scoring times). We plan to investigate this relationship further by annotating learner errors and correlating the error types with the present results.

6   Conclusions

We have presented results of a pilot experiment in which we investigate the effectiveness of grouping responses to constructed response items of a listening comprehension task. A simple clustering technique and shallow features show promising results: scoring clustered responses is shown to proceed faster than scoring responses to a specific question when they are not clustered. Moreover, scoring in the clustering mode is as efficient as scoring sheets of individual learners, the mode that is most familiar to the teachers in our scenario. It is worth noting that our results are comparable to the recently (March 2014) presented results obtained with the Powergrading system in terms of the general tendency of the clustering system being more efficient than a non-clustering system [4]. While our scoring times in the clustering mode are longer than those presented by Brooks at al., our experiment is conducted in the course of a real-life task without any control on our part over external factors which might affect scoring. Moreover, our interface itself is much less sophisticated than the interface used in Brooks at al. experiments. Indeed, it is most likely a combination of the presentation mode (clustered vs. non-clustered) and a convenient interface that would most contribute to faster scoring of constructed responses. We plan to investigate both of these aspects of streamlining the scoring process in our further work.

Acknowledgments. We would like to thank Dr. Kristin Stezano Cotelo and Anja Koch, M.A., from the Saarland University International Office, as well as the other International Office DaF teachers who participated in the placement testing, for their collaboration on test scoring using our platform. We also thank the three anonymous reviewers for their helpful comments.

This work was partially funded by the Cluster of Excellence "Multimodal Computing and Interaction" of the German Excellence Initiative. Magdalena Wolska's work is supported by the Institutional Strategy of the University of Tübingen (Deutsche Forschungsgemeinschaft, ZUK 63).

References

1. Amigó, E., Gonzalo, J., Artiles, J., Verdejo, F.: A comparison of extrinsic clustering evaluation metrics based on formal constraints. Information Retrieval 12(4), 461–486 (2009)
2. Bachman, L.F., Carr, N., Kamei, G., Kim, M., Pan, M.J., Salvador, C., Sawaki, Y.: A reliable approach to automatic assessment of short answer free responses. In: Proceedings of the 19th International Conference on Computational Linguistics, pp. 1–4 (2002)

3. Basu, S., Jacobs, C., Vanderwende, L.: Powergrading: a clustering approach to amplify human effort for short answer grading. Transactions of the Association of Computational Linguistics 1, 391–402 (2013)
4. Brooks, M., Basu, S., Jacobs, C., Vanderwende, L.: Divide and correct: Using clusters to grade short answers at scale. In: Proceedings of the 1st Conference on Learning at Scale, pp. 89–98 (2014)
5. Hahn, M., Meurers, D.: Evaluating the meaning of answers to reading comprehension questions: A semantics-based approach. In: Proceedings of the 7th Workshop on Innovative Use of NLP for Building Educational Applications, pp. 326–336 (2012)
6. Hill, D.: A vector clustering technique. In: Proceedings of the FID/IFIP Joint Conference: Mechanised Information Storage, Retrieval and Dissemination, pp. 225–234 (1968)
7. Horbach, A., Palmer, A., Pinkal, M.: Using the text to evaluate short answers for reading comprehension exercises. In: Proceedings of *SEM 2013: 2nd Joint Conference on Lexical and Computational Semantics, pp. 286–295 (2013)
8. Horbach, A., Palmer, A., Wolska, M.: Finding a tradeoff between accuracy and rater's workload in grading clustered short answers. In: Proceedings of the 9th International Conference on Language Resources and Evaluation, pp. 588–595 (2014)
9. Leacock, C., Chodorow, M.: C-rater: Automated Scoring of Short-Answer Questions. Computers and the Humanities 37(4), 389–405 (2003)
10. Meurers, D., Ziai, R., Ott, N., Kopp, J.: Evaluating Answers to Reading Comprehension Questions in Context: Results for German and the Role of Information Structure. In: Proceedings of the TextInfer Workshop on Textual Entailment, pp. 1–9 (2011)
11. Mohler, M., Bunescu, R.C., Mihalcea, R.: Learning to grade short answer questions using semantic similarity measures and dependency graph alignments. In: Proceedings of the 49th Annual Meeting of the Association for Computational Linguistics: Human Language Technologies, pp. 752–762 (2011)
12. Mohler, M., Mihalcea, R.: Text-to-text semantic similarity for automatic short answer grading. In: Proceedings of the 12th Conference of the European Chapter of the Association for Computational Linguistics, pp. 567–575 (2009)
13. Pulman, S.G., Sukkarieh, J.: Automatic short answer marking. In: Proceedings of the 2nd Workshop on Building Educational Applications Using NLP, pp. 9–16 (2005)
14. Schmid, H.: Probabilistic part-of-speech tagging using decision trees. In: Proceedings of the International Conference on New Methods in Language Processing, pp. 44–49 (1994)
15. Shermis, M.D., Hamner, B.: Contrasting state-of-the-art automated scoring of essays. In: Handbook of Automated Essay Evaluation: Current Applications and New Directions, pp. 313–346 (2013)
16. Sukkarieh, J.Z., Blackmore, J.: c-rater: Automatic content scoring for short constructed responses. In: Proceedings of the 22nd International Conference of the Florida Artificial Intelligence Research Society, pp. 290–295 (2009)

Pinpointing Sentence-Level Subjectivity through Balanced Subjective and Objective Features

Munhyong Kim and Hyopil Shin

Department of Linguistics, Seoul National University, Korea
{likerainsun,hpshin}@snu.ac.kr

Abstract. The sentence-level subjectivity classification is a challenging task. This paper pinpoints some of its unique characteristics. It argues that these characteristics should be considered when extracting subjective or objective features from sentences. Through various sentence-level subjectivity classification experiments with numerous feature combinations, we found that balanced features for both subjective and objective sentences help to achieve balanced precision and recall for sentence subjectivity classification.

Keywords: sentence-level subjectivity analysis, subjective and objective features, balanced features.

1   Introduction

Sentiment analysis is one of the most active research areas in the field of natural language processing due to the need to analyze opinions on products, political issues, and so on. Distinguishing between subjective and objective information, called subjectivity classification, is a challenging task, independent from polarity or stance classification of opinions. Subjectivity classification can be performed at the document level, sentence level, or phrasal level depending on the need. These three types of subjectivity classification are co-related; phrasal-level subjectivity analysis helps to infer sentence-level subjectivity, while sentence-level subjectivity analysis in turn helps to infer document-level subjectivity.

As the size of web text data is continually growing, extracting only factual or opinionated sentences is becoming more necessary than ever. Moreover, the ability to discriminate between subjective and objective sentences could be used to ensure an objective voice which is required in such sources as Wikipedia or newspaper articles.

The goal of this research is to examine the particular characteristics or aspects of sentence-level subjectivity classification task and to discover more balanced feature combinations for both subjective and objective sentence classification.

In this study, the problem of defining a subjective sentence is pointed out to clarify the sentence-level subjectivity classification task. Then, some characteristics of the task are introduced to help understand the design of the experiments. Though various experiments on sentence subjectivity classification, we argue that the use of balanced feature combinations benefits the classification of subjective and objective sentences

A. Przepiórkowski and M. Ogrodniczuk (Eds.): PolTAL 2014, LNAI 8686, pp. 311–323, 2014.
© Springer International Publishing Switzerland 2014

and shows that increasing feature space dimensions is not necessary. Our approach proves to be robust even for problematic data.

This paper is organized as follows. Section 2 briefly introduces the previous studies. Section 3 analyzes the characteristics of the sentence-level classification task. Section 4 describes the experiment design in detail. Section 5 presents and discusses the results of the experiments. Finally, Sect. 6 summarizes the study's findings and offers conclusions.

2   Previous Studies

[2] started to build a subjectivity corpus, containing 1,001 sentences annotated for their subjectivity. [11] performed sentence-level subjectivity classification using a Naïve Bayes Model, based on the presence or absence of certain syntactic classes (pronouns, adjective, cardinal numbers, modal verbs and adverbs), punctuation marks, and sentence position in a document. These features are subjective indicators for the sentence-level subjectivity task. According to [15], the features of [11] with Naïve Bayes classifier show a distinctively superior performance in finding subjective sentences than objective sentences (Subjectivity Prec.: 69%, Rec.: 97, Objectivity Prec.: 38%, Rec.: 3%). This result indicates that the features are only effective for finding subjective sentences. Although [15] used other features, such as counts of semantically oriented words [5], counts of the polarities of sequences of semantically oriented words and counts of parts-of-speech tags combined with polarity information, the performance of subjectivity and objectivity classification was still the asymmetrical.

[12] built various sentence subjectivity classifiers for unannotated text. First, a rule-based classifier that used subjective clues from a lexicon and automatically extracted subjective or objective patterns was developed for comparison with a statistical classifier. Second, a Naïve Bayes classifier was proposed that used subjective clues, extracted patterns, and word lists (the syntactic class expressions from earlier works) as features. This classifier showed 74.2% (subjective) and 72.7% (objective) F1 scores and a 73.8% accuracy. The second statistical classifier achieves relatively balanced recall and precision for both subjective and objective sentences, compared to those of the rule-based classifier. Even though sentence-level subjectivity classification experiments have been conducted by many researchers in various ways, some crucial characteristics of this task have not been specifically and fully addressed.

3   Sentence-Level Subjectivity Classification Task

This section clarifies the notion of a subjective sentence that varies depending on the researcher and introduces some of the unique characteristics of the subjectivity classification task.

3.1 Subjective Sentences

To differentiate between subjective and objective sentences, a definition of a subjective sentence is required. However, defining a subjective sentence is not simple because of the vagueness of sentence subjectivity. It is even hard for a human to tell whether a sentence is subjective or objective. Consider the following example:

(a) The conference opened with fierce demonstrations by the opponents of economic globalization, who maintained that what is being concocted by globalization is in fact poverty.

The sentence above is annotated as an objective speech event in Multi-Perspective Question Answering (MPQA) [13], but it can be also interpreted as a subjective sentence due to the impression produced by such words as *fierce*.

Thus, the definition of a subjective sentence can vary. [14] adopted the notion of a private state from [9] for referring to a mental and emotional state that is not open to objective observation or verification. This term can be referred to as subjectivity in linguistics [1]. In [14], a sentence is considered to be subjective "if it contains one or more private state expressions". This definition is not sufficient, because not all sentences containing private states are subjective and objective sentences can also include private state expressions.

[15] did not explicitly define a subjective or objective sentence, but collected gold standard data by asking human annotators to assign one of the following labels: "fact", some opinion type labels, and "uncertain". The "uncertain" label indicates that defining a subjective sentence is difficult even for a human.

A survey study on sentiment analysis [7] tried to formalize and clarify the concept of sentence subjectivity. [7] categorized opinions as explicit or implicit; explicit opinions are expressed in subjective sentences and implicit opinions are implied in objective sentences. Then, an opinionated sentence is defined to include both implicit and explicit opinions:

(b) The voice quality of this phone is amazing.   <explicit>
(c) The earphone broke in two days.   <implicit>

In accordance with [7]'s classification, the notion of a subjective sentence in this study matches with that of an opinionated sentence. Although they provide a definition for sentence subjectivity, this definition alone is not sufficient due to its vague nature.

Before defining a subjective sentence, two important properties for subjective sentence judgment should be pointed out. First, it is important to narrow down the source of subjectivity in a sentence. The sentence's subjectivity should not be judged in relation to any opinion holders other than its writer. The other opinion holders should only be considered during phrasal subjectivity classification.

Second, although a sentence is usually written with a main purpose – expressing one's personal feelings and beliefs or reporting an objective fact – these two could appear together in a single sentence. For instance, a journalist can report an event unconsciously or consciously choosing expressions reflecting his or her judgment. In this case, it is impossible to make a clear distinction between a subjective and objective sentence. It is reasonable to rely on a reader's intuition on sentence subjectivity. Therefore, we suggest the following definition of a subjective sentence based on human intuition.

A sentence is subjective, if the reader perceives the writer's subjectivity towards something from explicit words or implicit contextual information of any kind.

3.2   Characteristics of Sentence-Level Subjectivity Classification Task

Due to the nature of subjective and objective sentences, the sentence subjectivity classification task has unusual characteristics compared to other classification tasks: the asymmetry of features between the two classes and the fact that subjective words do not equally contribute to the sentence subjectivity.

Consider a two-class document classification task, for instance, a task in which documents are classified as related to either cars or fishing. If we use bag-of-words features for each class, there must be words that appear only in one class and their count in each class is likely to be roughly equal (provided the texts in both classes are of the same size).

Yet, this is not the case for sentence subjectivity classification. Objective words can appear in either a subjective or an objective sentence but subjective words are only likely to occur in subjective sentences.

Therefore, using a subjective expression lexicon to find subjective sentences might result in finding only subjectivity features and no objectivity features unless the absence of a specific subjective word is counted.

This asymmetry of subjective and objective features should be considered when extracting sentence subjectivity features. Therefore, this study is focused on finding balanced feature sets for both classes (balanced and imbalanced features will be explained with examples in Sec. 4.2).

Another aspect of the sentence subjectivity task is that the sentence subjectivity is not determined by all subjective words in the sentence. Rather, it depends on only a few words in the sentence; main predicates, adverbs of predicates, conjunctions and auxiliary verbs are often strong indicators for sentence subjectivity. See the below examples.[1]

(d) The Bonn Conference <u>however</u>, which dealt with the global environment and examined the condition of world climate, <u>could</u> be **appraised as an exception** and **a relative success**.

---

[1] For better understanding, we show English examples from Multi-Perspective Question Answering (MPQA) [13], but it is originally an observation found in KOSAC.

Though the words in bold in (d) are sentiment words, only the underlined words, *however* and *could*, make the sentence subjective. Therefore, considering this aspect of sentence subjectivity could help to extract better features from sentences.

4   Experiment

This section introduces the experiments, conducted to show how subjective sentences could be discriminated from the objective ones, together with the collected corpora and extracted features.

4.1   Data and Task

Three corpora were used for experiments. The Korean Sentiment Analysis Corpus (KOSAC)[2] [6] was used for the first experiment. KOSAC contains a total of 7,713 sentences (subjective: 2,658, objective: 5,055). The sentence subjectivity judgments in KOSAC were validated by cross-checking of three annotators.

KOSAC includes direct quotations as sentences, so their subjectivity was modified according to the opinion holder of the quotation, not the writer who quotes it, because these quotations were uniformly annotated as objective in KOSAC.

The sentence subjectivity classification task is performed with a linear SVM model [3,4] implemented in Scikit-learn [8] and a 10-fold cross-validation: 1/10 test and 9/10 training set.

To validate the results of the first experiment, another corpus was collected from the Sejong part-of-speech tagged corpus (SEJONG)[3]. A total of 900 sentences were collected and annotated (500 subjective and 400 objective). Sentences were collected from various sections of two newspapers, Chosun and Hankyoreh (culture, opinions, sports, editorials, columns, TV, women), six magazines, and five novels, ensuring that the types of corpus text are balanced.

Furthermore, two corpora that were composed only of objective sentences were collected to determine how well the selected feature set performed on a certain type of data. Specifically, the objective corpora included war and disease related sentences taken from Korean Wikipedia[4] (NEG-WIKI) and reporting sentences that were not included in KOSAC (REPORT-SEJONG): 308 sentences from Wikipedia and 50 reporting sentences from the Sejong Corpus.   NEG-WIKI was collected because war or disease-related Wikipedia articles contain negative expressions. Sentences with such negative expressions are likely to be misclassified as subjective. Also, REPORT-SEJONG sentences were collected since reporting objective sentences often include subjective expressions. These corpora can be used to verify the robustness of the features that this study suggests. The same SVM model experiments were executed for the SEJONG and objective corpora, but the 10-fold validation was omitted.

---

[2] http://word.snu.ac.kr/kosac
[3] http://www.sejong.org
[4] http://ko.wikipedia.org

4.2   Features

Features were automatically extracted from seed expressions (KOSAC-DICT) included in KOSAC.[5] Before introducing each feature set, the meaning of balanced features needs to be explained. As mentioned in 3.2, the sentence subjectivity classification task is asymmetrical; the presence of subjective expressions can be a strong indicator for subjective sentences, but the absence of them cannot be a feature for objective sentences. Therefore, balanced features are those that can help recognize subjective as well as objective sentences. The features consisted of sentiment expression frequency features, probability features, expression type features, semantic type features, and intensity features.

Expression Frequency Features (SF, N-SF, T-SF). The most naive assumption of sentence subjectivity is that the sentence subjectivity is determined by the number of subjective expressions in the sentence. For this group, frequency (**SF**), frequency normalized by the sentence length (**N-SF**) and three-valued (0, 1 or > 2) frequency (**T-SF**) were extracted. These are not balanced features since objective sentences can include many subjective expressions though subjective sentences are likely to include a larger number of subjective expressions.

Probability Features (PES, PPOSS, PSS, PSS-POS). This group included four types of feature sets and four features that were derived from these sets. The four feature sets were the probability of expression subjectivity (**PES**), the probability of part-of-speech sequence subjectivity (**PPOSS**), the probability of subjectivity of the sentence that the expression belongs to (**PSS**), and the probability of subjectivity of the sentence that the part-of-speech sequence belongs to (**PSS-POS**). (1) The PES feature indicates how likely is that the expression is subjective. It is the number of expression occurrences as subjective divided by the total number of occurrences. The PPOSS is the same measure as PES except that it is calculated with only the part-of-speech tag sequence, not the expression. (2) The PSS feature measures how likely is that the sentence the expression belongs to is subjective based on the training corpus. The meaning of PSS-POS is the same as PSS, but PSS-POS only considers the POS sequence, not the expression. These four measures are considered to be feature sets because each of them includes all entries from the KOSAC subjective expression dictionary (KOSAC-DICT). Two features, PES and PSS, are summarized as follows.

$$PES = \frac{S_{eo}}{N_{eo}} \times \frac{N_{eo}}{N_{eo}+K} \quad ^6 \tag{1}$$

$$PSS = \frac{S_{ss}}{N_s} \times \frac{Ns}{Ns+K} \quad ^7 \tag{2}$$

---

[5]   Seed expressions in KOSAC are subjectivity indicating expressions, annotated according to Korean Subjectivity Markup Language (KSML) annotation guideline [10].

[6] N_{eo}: the total number of the expression occurrences; S_{eo}: the number of the expression occurrences as subjective; $N_{eo}/(N_{eo}+ K)$: smoothing term for compensating the effect of low frequent words; K: a parameter that determines the degree of smoothing. K is set to 1.

[7]   Ns: number of sentences that the seed belongs to; Sss: number of subjective sentences that the subjective expression belongs to.

The remaining four probability features (AVG-PES, AVG-PPOSS, AVG-PSS, AVG-PSS-POS) are the averaged values of the feature sets explained above: PES, PPOSS, PSS, and PSS-POS feature values for all recognized potentially subjective words in a sentence are respectively summed up and averaged. It is important to note that the four averaged features can be calculated only when the sentence contains at least one subjective expression. If no such expression exists then the averaged features are set to 0, indicating that they can better reflect the objective aspect of sentences.

The PES, PPOSS, PSS and PSS-POS are imbalanced feature sets which are meaningful only for subjective sentences since the feature entries of these sets correspond to the subjective word entries in the dictionary. On the contrary, the four averaged features are balanced for subjective and objective sentences because their values respectively indicate how likely a sentence is to be subjective or objective. Subjective sentences are likely to include subjective words that make the entire sentence subjective, and objectives sentences are less likely to have them.

Intensity Features (CAT-INT, CONT-INT, AVG-CONT-INT). This feature group was composed of two feature sets and one averaged feature: categorical and continuous intensity feature sets (**CAT-INT, CONT-INT**) and one averaged continuous intensity value of all subjective expressions for a sentence (**AVG-CONT-INT**). The first two feature sets were dictionary-entry features, just like PES, PPOSS, PSS, and PSS-POS, thus they were imbalanced. On the other hand, the AVG-CONT-INT is a balanced feature.

The categorical intensity value for each expression comes from the most frequent intensity value of the expression in the KOSAC: high-3, medium-2, or low-1. The continuous intensity value was calculated based on the frequency of each categorical intensity value for each expression as below.

$$\text{CONT} - \text{INT} = \frac{((3 * \text{freq\_high}) + (2 * \text{freq\_medium}) + (1 * \text{freq\_low}))}{\text{total\_freq\_except\_none}} \quad (3)$$

Expressive Type Features (EXP-TYPE-FREQ, EXP-TYPE-NORM, EXP-TYPE-THREE). In KOSAC-DICT, the expressive types are direct-speech, direct-action, direct-explicit, indirect, and writing-device. The three direct types are subjective, indicating main predicates in the sentence. Indirect type expressions are usually nouns and adverbs, modifying adjectives. The writing-device type expressions are sentential adverbs, auxiliary verbs, modals, particles, and conjunctions that in most cases critically determine the sentence subjectivity. Three feature sets were available for this group: the pure frequency (**EXP-TYPE-FREQ**), the frequency normalized by the sentence length (**EXP-TYPE-NORM**), and the three-valued frequency (**EXP-TYPE-THREE**) of each expressive type. For example, the EXP-TYPE-FREQ features for a sentence could be "direct-action: 1, direct-speech: 0, direct-explicit: 0, indirect: 4, writing-device: 3". Thus, five expressive type features were extracted for each set. If there were no expressions of a given type, it was assigned a value of 0, which enabled the features to reflect sentence objectivity.

Semantic Type Features (SEM-TYPE-FREQ, SEM-TYPE-NORM, SEM-TYPE-THREE). Each expression in KOSAC-DICT is also assigned a semantic category: *agreement, argument, emotion, intention, judgment, speculation,* and *others.*

This feature group also has three feature sets, similar to expressive type features: **SEM-TYPE-FREQ, SEM-TYPE-NORM**, and **SEM-TYPE-THREE**. Six semantic type features except the *others* type belonged to each set.

Bag-of-Words Features (BOW). All sentences in both classes were tokenized into unigrams and counted. Thus, a sentence was represented as a set of unigrams and their frequency in the sentence. This feature is simple and naive but was used here because it might include some objectivity-indicating features that we were not aware of as well as subjective features.

5   Experiment Result

This section introduces and discusses the results of sentence subjectivity classification experiments with all collected corpora and findings in the balanced features.

5.1   Feature Testing Experiment in KOSAC

To see how effective each feature was, the features were tested separately using KOSAC data. Table 1 shows features with an accuracy above 60% except one-feature results, (1)–(3). The AVG-PSS and AVG-PSS-POS features show lower accuracy (55.88% and 55.79%) than (1)–(3), and thus are not included in Table 1. Precision, recall, and F1 are given as A / B, where A and B represent subjective and objective sentences correspondingly.

First, expression frequency features were tested alone. Of the three frequency features, N-SF was used as a baseline to which the other features were added. Although the accuracy of SF was higher than that of N-SF, N-SF was chosen because N-SF is more intuitively robust as a seed frequency measure.

When using only the N-SF feature set, the results for subjective and objective sentences seemed to be opposites of each other. This indicates that N-SF is not a balanced feature for recognizing objective sentences. Also, N-SF is not a suitable feature for subjective sentence classification because of the low recall.

It should be noted that the number of features used in PES, PSS, CAT-INT, or CONT-INT was much larger than the number used in AVG-PES, EXP-TYPE-FREQ, or SEM-TYPE-FREQ. Thus, the accuracy from these two groups cannot be compared directly.

Compared to the N-SF feature set, features from (4) to (10) allowed to find subjective and objective sentences in a more balanced way. Therefore, more numerous combinations were tested as a next step.

Table 2 summarizes the results of the feature combination tests. From (2) and (3), it can be said that AVG-PES and AVG-PPOSS helped to significantly increase the overall accuracy. Specifically, the recall dropped by 7% but the precision increased

Table 1. Separate Feature Examination within KOSAC Corpus

N.	Feature Combination	Precision	Recall	F_1	Acc.
(1)	SF	54.2 / 74.1	90.7 / 26.0	67.9 / 38.4	57.7
(2)	N-SF	53.4 / 84.8	96.6 / 18.7	68.8 / 30.6	57.0
(3)	T-SF	53.0 / 80.5	95.5 / 18.0	68.1 / 29.4	56.1
(4)	N-SF + PES	55.8 / 84.4	94.8 / 27.3	70.2 / 41.2	60.4
(5)	N-SF + PSS	56.3 / 81.9	93.1 / 30.2	70.1 / 44.0	61.0
(6)	N-SF + AVG-PES	57.8 / 78.3	89.4 / 37.0	70.2 / 50.2	62.8
(7)	N-SF + EXP-TYPE-FREQ	56.5 / 81.6	92.7 / 31.2	70.2 / 45.1	61.4
(8)	N-SF + SEM-TYPE-FREQ	56.6 / 81.1	92.4 / 31.5	70.1 / 45.3	61.4
(9)	N-SF + CAT-INT	61.3 / 71.9	78.9 / 51.8	68.9 / 60.1	**65.1**
(10)	N-SF + CONT-INT	61.0 / 72.4	80.1 / 50.7	69.2 / 59.5	**65.0**

Table 2. Feature Combination Examination within KOSAC Corpus[8]

N.	Feature Combination	Precision	Recall	F_1	Acc.
(1)	N-SF	53.44/84.87	96.57/18.71	68.77/30.61	56.94
(2)	N-SF + AVG-PES* + AVG-PPOSS*	57.90/78.26	89.33/37.24	70.23/50.42	62.83
(3)	N-SF + AVG-PSS* + AVG-PSS-POS*	52.72/83.40	96.63/16.28	68.18/27.18	55.73
(4)	(2) + CAT-INT	65.10/69.15	70.39/63.55	67.53/66.08	66.89
(5)	(4) + CONT-INT	64.23/68.34	69.99/62.22	66.85/64.96	66.00
(6)	(4) + PSS	64.13/70.00	73.10/60.44	68.23/64.75	66.63
(7)	(4) + EXP-TYPE-FREQ* + SEM-TYPE-FREQ*	61.10/76.55	84.69/47.92	70.93/58.84	65.96
(8)	(2) + EXP-TYPE-FREQ*	64.39/71.93	75.99/59.37	69.66/64.99	67.52
(9)	(2) + SEM-TYPE-FREQ*	66.36/70.13	71.38/64.96	68.71/67.37	68.11
(10)	ALL-FEATURES	57.33/69.03	80.01/69.03	66.67/52.20	60.85

---

[8] The asterisk (*) indicate balanced features that appeared to help sentence subjectivity classification in the experiment.

by 4% for subjective sentence classification. Moreover, for objective sentence classification, this combination almost doubled the F1 score. Then, in (4) and (5), CAT-INT and CONT-INT feature sets were added one by one. This resulted in a more balanced performance between precision and recall and an increased accuracy for finding subjective sentences. For objective sentence classification, the recall substantially increased. Though AVG-CONT-INT was expected to work as a balanced feature, it turned out this feature was not effective for differentiating subjective sentences from objective ones. It shows that the intensity of subjective expressions in a sentence may not be directly related to its subjectivity.

We can assume CONT-INT does not benefit much from adding CAT-INT since the features represented the same thing, differing only in their values. By replacing the categorical value with a continuous one, one would expect a better result but adding CONT-INT instead of CAT-INT did not show any significant difference (65.5%-CAT-INT, 65.9%-CONT-INT).

The effect of the PSS feature set on accuracy was checked in (6). However, it only increased the recall by 3%. In (7), EXP-TYPE-FREQ and SEM-TYPE-FREQ were added to feature combination (4). Though the accuracy did not noticeably change, the recall raised by about 14%. Also, the recall of objectivity classification dropped by about 16%. The effect of other features could have been diminished because of the relatively high dimension of CAT-INT features.

Hence, in (8) and (9), CAT-INT was removed. EXP-TYPE-FREQ and SEM-TYPE-FREQ were respectively added to feature combination (2). It seemed to be the combinations which showed the most well-balanced and highest accuracy for both subjectivity and objectivity classification. Also, this indicated that only four or five feature sets could be sufficient to classify the subjective and objective sentences without any dictionary-entry features.

The bag-of-words features could be more balanced for subjective and objective sentences as explained earlier. With only bag-of-words features, the result was the following.

Pr.: 72.39 / 72.27, Rec.: 72.25 / 73.04, F_1: 72.29 / 72.6, Acc.: 72.80 / 72.02

Surprisingly, the bag-of-words feature set was better than any of the other feature combinations in Table 2. For this reason, bag-of-words features (BOW) were included and tested with the feature combinations from Table 1. The best accuracy (73.57%) was obtained by the feature combination (9) of Table 1 + BOW. The major difference from the experiment without BOW is that the feature combination (2) in Table 1 with BOW (Acc.: 73.15%) shows as high accuracy as (9) of Table 1 + BOW (Acc.: 73.57%). This result indicates that N-SF, AVG-PES, and AVG-PPOSS alone could be enough to classify sentence subjectivity with the balanced bag-of-words features. It is important notice that AVG-PES and AVG-PPOSS are one of balanced features. Also, EXP-TYPE-FREQ and SEM-TYPE-FREQ are more balanced features than just frequency features, such as N-SF, since their types are related to sentence subjectivity and the zero counts of such types reflect the sentence objectivity.

5.2   Validation of Feature Combinations with Other Corpora

The result of the feature combination tests above could have been valid only for KOSAC, so it needed to be tested with another independent corpus, SEJONG.

Each separate feature set was tested in the same way as in Table 1. Classification with the only starting point feature (N-SF) showed a 59.67% accuracy. The separate features with accuracies above 62% differ somewhat within SEJONG from those shown in Table 1. Thus, the different combinations of these features were examined in Table 3.

Table 3. REPORT-SEJONG / NEG-WIKI Classification Task

N.	Feature Combination	Pr.	Rec.	F_1	Acc.
(1)	BOW + N-SF	68.10/76.49	88.22/48.12	76.87/59.08	70.44
(2)	(1) + AVG-PES*	72.57/68.00	76.05/63.91	74.27/65.89	70.67
(3)	(2) + EXP-TYPE-FREQ*	70.00/74.48	85.23/54.14	76.87/62.70	**71.44**
(4)	(2) + SEM-TYPE-THREE*	73.99/61.38	69.26/68.92	71.55/64.94	67.00
(5)	(2) + EXP-TYPE-FREQ* + SEM-TYPE-THREE*	66.62/69.84	88.82/64.41	76.13/67.01	**71.89**
(6)	(5) + CAT-INT	67.13/72.08	87.23/55.64	75.87/62.80	70.78
(7)	(5) + CONT-INT	66.67/72.97	88.22/54.14	75.95/62.16	70.78
(8)	(5) + CAT-INT + CONT-INT	67.07/73.17	87.82/52.63	76.06/61.22	70.44
(9)	(5) + PES	64.56/71.74	89.82/57.89	75.13/64.08	**71.22**
(10)	(5) + PSS	64.37/71.52	89.42/55.39	74.85/62.43	70.44
(11)	(5) + PSS-POS	64.88/73.40	91.82/57.39	76.03/64.42	**71.89**

PSS-POS is a new addition as effective feature (Pr.: 61.70 / 68.13, Rec.: 88.42 / 31.08, F1: 72.68 / 42.69, Acc.: 63.0). Moreover, SEM-TYPE-FREQ changed to SEM-TYPE-THREE (Pr.: 61.42 / 72.85, Rec.: 91.82 / 27.57, F1: 73.60 / 40.0, Acc.: 63.33).

Even though combinations (5) and (11) show the highest accuracies, the objectivity classification F1 of (5) was higher than (11). Also, (3) and (9) showed relative high accuracies compared to other feature combination results. This indicates that, without increasing feature dimensions by adding CAT-INT, CONT-INT, PES, PSS, or PSS-POS sets, subjective and objective sentences can be classified with only BOW, N-SF, AVG-PES, EXP-TYPE-FREQ, and possibly SEM-TYPE-THREE.

The best feature combination within KOSAC was BOW + N-SF + AVG-PES + AVG-PPOSS + SEM-TYPE-FREQ. Though the feature combination result is a bit different, it consistently shows that dictionary-entry features that lead to an enormous increase of feature space dimension are not necessary for a robust classifier.

However, this does not mean that we do not need a list of seed words since AVG-PES and EXP-TYPE-FREQ are calculated from them.

5.3 Objective Corpus Experiment

The two objective corpora, NEG-WIKI and REPORT-SEJONG, were tested in Table 4 with the feature combinations found in the previous experiments.

The REPORT-SEJONG sentences on average had 51.38 words with 18.9 subjective expressions. NEG-WIKI sentences included 7.17 subjective expressions out of 37.87 words per sentence on average.

Table 4. REPORT-SEJONG / NEG-WIKI Classification Task

N.	Feature Combination	Pr.	Rec.	F_1	Acc.
(1)	BOW* + N-SF + AVG-PES* + AVG-PPOSS* + SEM-TYPE-FREQ*	100%	78/64%	87/78%	78/64%
(2)	BOW* + N-SF + AVG-PES* + AVG-PPOSS* + CAT-INT	100%	74/76%	85/86%	74/76%

Despite the high proportion of subjective expressions in the sentence, the feature combinations in Table 4 showed a fairly high recall. The reason behind the 100% precision is that all test data is objective. These results suggest that such feature combinations allow problematic objective sentences to be correctly found with fairly high accuracy.

6   Conclusion

This research introduced interesting characteristics of the sentence-level subjectivity classification task and found that balanced features considering those characteristics can make the performance balanced for precision and recall for both subjective and objective sentences. The sentence-level subjectivity task has asymmetric nature in the number of features for each class. Also, it is important to notice that the sentence subjectivity is determined by not all, but a few subjective expressions.

Through experiments, we discovered that extracting more balanced features for both classes helped to increase the precision and recall of objectivity classification and improved the overall performance of subjectivity classification. Also, the feature combinations indicate that it is possible to ensure good classification performance even without the enormous increase of feature dimensions incurred by adding all dictionary entries.

Acknowledgments. This work was supported by the National Research Foundation of Korea grant funded by the Korean government (NRF-2011-327-A00322).

References

1. Banfield, A.: Unspeakable sentences: Narration and representation in the language of fiction, p. 256. Routledge & Kegan Paul, Boston (1982)
2. Bruce, R., Wiebe, J.: Recognizing subjectivity: A case study in manual tagging. Natural Language Engineering 5(2) (1999)
3. Cortes, C., Vapnik, V.: Support vector machine. Machine Learning 20(3), 273–297 (1995)
4. Fan, R.E., Chang, K.W., Hsieh, C.J., Wang, X.R., Lin, C.J.: LIBLINEAR: A library for large linear classification. Journal of Machine Learning Research 9, 1871–1874 (2008)
5. Hatzivassiloglou, V., Wiebe, J.: Effects of adjective orientation and gradability on sentence subjectivity. In: Proceedings of the 18th Conference on Computational Linguistics, pp. 299–305. Association for Computational Linguistics, Stroudsburg (2000)
6. Jang, H., Kim, M., Shin, H.: KOSAC: A Full-fledged Korean Sentiment Analysis Corpus. In: Proceedings of the 27th Pacific Asia Conference on Language, Information and Computation (2013)
7. Liu, B.: Sentiment analysis and subjectivity. In: Handbook of Natural Language Processing, vol. 2, pp. 627–666 (2010)
8. Pedregosa, F., Varoquaux, G., Gramfort, A., Michel, V., et al.: Scikit-learn: Machine learning in Python. The Journal of Machine Learning Research 12, 2825–2830 (2011)
9. Quirk, R., Greenbaum, S., Leech, G., Svartvik, J.: A Comprehensive Grammar of the English Language. Longman, New York (1985)
10. Shin, H., Kim, M., Jo, Y.M., Jang, H., Cattle, A.: Annotation Scheme for Constructing Sentiment Corpus in Korean. In: Proceedings of the 26th Pacific Asia Conference on Language, Information and Computation, pp. 181–190 (2012)
11. Wiebe, J.M., Bruce, R.F., O'Hara, T.P.: Development and use of a gold standard dataset for subjectivity classifications. In: Proceedings of the 37th Annual Meeting of the Association for Computational Linguistics, pp. 246–253 (1999)
12. Wiebe, J., Riloff, E.: Creating subjective and objective sentence classifiers from unannotated texts. In: Gelbukh, A. (ed.) CICLing 2005. LNCS, vol. 3406, pp. 486–497. Springer, Heidelberg (2005)
13. Wiebe, J., Wilson, T., Cardie, C.: Annotating expressions of opinions and emotions in language. Language Resources and Evaluation 39(2-3), 165–210 (2005)
14. Wilson, T.A.: Fine-grained Subjectivity and Sentiment Analysis: Recognizing the intensity, polarity and attitudes of private states. Doctoral Dissertation, Univ. of Pittsburgh (2008)
15. Yu, H., Hatzivassiloglou, V.: Towards answering opinion questions: separating facts from opinions and identifying the polarity of opinion sentences. In: Proceedings of the 2003 Conference on Empirical Methods in Natural Language Processing, pp. 129–136. Association for Computational Linguistics, Stroudsburg (2003)

Using Graphs and Semantic Information
to Improve Text Classifiers

Nibaran Das[1], Swarnendu Ghosh[1], Teresa Gonçalves[2], and Paulo Quaresma[2,3]

[1] Dept. of Computer Science and Engineering, Jadavpur University, Kolkata, India
[2] Dept. of Computer Science, School of S&T, University of Évora, Évora, Portugal
[3] L2F – Spoken Language Systems Laboratory, INESC-ID, Lisbon, Portugal
nibaran@ieee.org, swarbir@gmail.com, {tcg,pq}@uevora.pt

Abstract. Text classification using semantic information is the latest
trend of research due to its greater potential to accurately represent text
content compared with bag-of-words (BOW) approaches. On the other
hand, representation of semantics through graphs has several advantages
over the traditional representation of feature vector. Therefore, error
tolerant graph matching techniques can be used for text classification.
Nevertheless, very few methodologies exist in the literature which use
semantic representation through graphs. In the present work, a methodology has been proposed to represent semantic information from a summarized text into a graph. The discourse representation structure of a
text is utilized in order to represent its semantic content and, afterwards,
it is transformed into a graph. Five different graph matching techniques
based on Maximum Common Subgraphs (mcs) and Minimum Common
Supergraphs (MCS) are evaluated on 20 classes from the Reuters dataset
taking 10 docs of each class for both training and testing purposes using
the k-NN classifier. From the results it can be observed that the technique has potential to perform text classification as well as the traditional
BOW approaches. Moreover a majority voting based combination of the
semantic representation and a traditional BOW approach provided an
improved recognition accuracy on the same data set.

1 Introduction

The text classification task has been the focus of much research work in the last
years [3,11]. State of the art approaches typically represent documents as vectors
(bag-of-words) and use a machine learning algorithm, such as k-NN or SVM, to
create a model and classify new documents. However, and in spite of being able
to obtain good results, these approaches fail to represent the semantic content
of the documents, losing much information and limiting the tasks that can be
implemented over the document representation structure. Taking into account
these limitations some research has been done aiming to use and evaluate more
complex knowledge representation structures [10,1].

A. Przepiórkowski and M. Ogrodniczuk (Eds.): PolTAL 2014, LNAI 8686, pp. 324–336, 2014.

In this paper we propose a new approach which integrates a deep linguistic analysis of the documents with a graph-based representation for the purpose of classification. As a result of the linguistic analysis, the semantic content of the texts is represented by a discourse representation structure (DRS) [8], which is transformed by our system into a graph structure. Then, we proposed, applied, and evaluated several graph distance measures [9] and integrated them in a k-NN classifier. The proposed approach shown is able to obtain results at least as well as the ones obtained by traditional bag-of-words approaches.

This paper has the following structure. The next section briefly presents the theoretical background required by our approach (discourse representation theory, graph representation, k-NN classifiers, graph measures). Section 3 describes our system and its modules. Section 4 exposes the performed experiments and discusses the obtained results. In the final section conclusions and future work are presented.

2   Theory and Algorithms

2.1   Brief Description of DRS

In order to extract and represent the information conveyed by texts several approaches can be applied, varying from deep linguistic symbolic analysis to statistical based ones. In the context of this paper we will use a "typical" deep linguistic processing sequence: lexical, syntactic, and semantic analysis.

One of the most relevant work on semantic analysis is the Discourse Representation Theory (DRT) was proposed by Kamp and Reyle [8]. In DRT, the aim is to associate sentences with expressions in a logical language which represents their meaning. These expressions are composed of, a) a set of referents, which are the entities that have been introduced into the context, b) a set of conditions, which are the predicates that hold between the referents. In DRT, each sentence is viewed as an update of an existing context, having a new context as a result (an example is presented in Sect. 3).

DRT provides a very potent platform for the representation of the semantic structure of sentences including relations like implications, propositions and negations. It is also able to separately analyze almost every kind of events and find out their agent and patient.

The main component of DRT is the Discourse Representation Structure (DRS). A DRS consists of two parts: the set of discourse referents and the set of conditions. The discourse referents are variables representing all of the entities in the DRS. The conditions are the logical statements about these entities. Figure 1 shows an example of a DRS representation for the sentence "He throws the ball.".

2.2   Brief Description of GML

Graph Modeling Language (GML) [7] is a simple and efficient way to represent weighted directed graphs. A GML file is basically a 7-bit ASCII file and, as such,

```
[ x1, x2, x3:
    male(x1), ball(x2), throw(x3),
    event(x3), agent(x3, x1), patient(x3, x2)
]
```

Fig. 1. DRS representation for the sentence "He throws the ball."

can be easily read, parsed, and written. Several open source applications[1] are available which enable viewing and editing GML files.

Graphs are represented by the keys viz. node and edge. The basic structure is modeled with the node's *id* and the edge's *source* and *target* attributes. The *id* attribute assigns numbers to nodes, which are then referenced by *source* and *target*. Weights can be represented by the *label* attribute.

2.3   k-NN Classifier

The k-nearest-neighbour is a well known example of an example based classifier and is amongst the simplest machine learning algorithms. This kind of classifiers do not build an explicit class representation and depend only on the class labels of the training examples that are similar to the test example. These methods are also known as lazy learners since they delay the class generalization until the testing phase.

The similarity between examples is usually determined using a distance measure; the closer, the more similar. In the basic version of the k-NN algorithm, a new example is simply classified according to the majority vote of its neighbours (train examples that are most similar to it).

The construction of the classifier also involves the determination of the number of neigbours (value of k) that produce the best results. Usually, this number is determined experimentally.

2.4   Distance Measures for Graphs

It has already been mentioned that the objective of the present work is to develop a comparative assessment among different distance measures for text classification task. Here, five different distance measures taken from [9] are utilized for this purpose. They are popularly used in object recognition task, but for text categorization they were not yet utilized. For two graphs G_1 and G_2, if $d(G_1, G_2)$ is the dissimilarity/similarity measure, then this measure would be a distance if d has the following properties:

1. $d(G_1, G_2) = 0$, iff $G_1 = G_2$
2. $d(G_1, G_2) = g(G_2, G_1)$
3. $d(G_1, G_2) + d(G_2, G_3) \geq d(G_1, G_3)$

---

[1] http://en.wikipedia.org/wiki/Graph_Modelling_Language

The measures used in the present work follow the above rules and the corresponding equations are:

$$d_{\text{mcs}}(G_1, G_2) = 1 - \frac{|\text{mcs}(G_1, G_2)|}{\max(|G_1|, |G_2|)} \tag{1}$$

$$d_{\text{wgu}}(G_1, G_2) = 1 - \frac{|\text{mcs}(G_1, G_2)|}{|G_1| + |G_2| - |\text{mcs}(G_1, G_2)|} \tag{2}$$

$$d_{\text{ugu}}(G_1, G_2) = |G_1| + |G_2| - 2 * |\text{mcs}(G_1, G_2)| \tag{3}$$

$$d_{\text{MMCS}}(G_1, G_2) = |\text{MCS}(G_1, G_2)| - |\text{mcs}(G_1, G_2)| \tag{4}$$

$$d_{\text{MMCSN}}(G_1, G_2) = 1 - \frac{|\text{mcs}(G_1, G_2)|}{|\text{MCS}(G_1, G_2)|} \tag{5}$$

In the equations $\text{mcs}(G_1, G_2)$ and $\text{MCS}(G_1, G_2)$ denote maximum common subgraph and minimum common supergraph of the two graphs G_1 and G_2. Theoretically $\text{mcs}(G_1, G_2)$ is the largest graph in terms of edges that is isomorphic to a subgraph of G_1 and G_2. $\text{mcs}(G_1, G_2)$ has been formally defined in the work of Bunk et al. [5]. As stated earlier, it is a NP complete problem and actually, the method of finding the mcs is a brute force method which finds all the subgraphs of both the graphs and select the maximum graph which is common to both.

To make the program computationally faster, it was modified to an approximate version of $\text{mcs}(G_1, G_2)$ using the fact that the vertices which exhibit greater similarity in their local structures among two graphs have a greater probability of inclusion in the mcs graph. In this way it ensures that the approximation version exhibits most of the properties of a mcs, while keeping the complexity in a polynomial time. The two pass approach used in the present work is as follows:

1. All the vertex pairs (one from each graph) are ranked according to their similarity of local structures. In the present case, the number of matching self-loops is used for similarity measures.
2. The mcs is built by including each vertex pair (starting with the one with the highest similarity) and considering it a common vertex; then including the rest of the edges (non self-loop edges) which satisfy the existing common self loops in both the graphs.

The minimum common supergraph (MCS) [1] is formed using the union of the two graphs, i.e. $\text{MCS}(G_1, G_2) = G_1 \bigcup G_2$.

The distance measures of (1), (2) and (5) were used directly without any modifications; the ones of (3) and (4) were divided by $|G_1| + |G_2|$ and $|\text{MCS}(G_1, G_2) + \text{mcs}(G_1, G_2)|$ respectively to make them normalized, keeping the value within the range [0,1].

2.5   Tools

In order to process texts C&C/Boxer [6,4] a well-known open source tool available for download at http://svn.ask.it.usyd.edu.au/trac/candc was used. The tool consists of a combinatory categorial grammar (CCG) [6] parser and outputs the semantic representations using discourse representation structures (DRS) of Discourse Representation Theory (DRT) [8].

3 Methods

The method described in the present work, is mainly divided into three major components. The first is the creation of the DRS with the semantic interpretation of the summarized text; the second is the construction of graphs in GML from the obtained DRS using some predefined rules; the third one is the classification phase where the different graph distances are assessed using a k-NN classifier [11]. The entire diagram of the system is shown in Fig. 2. It is worthy to mention here that the computation of traditional mcs and MCS is NP complete. To minimize the complexity, summarized text is used for the present work instead of entire one.

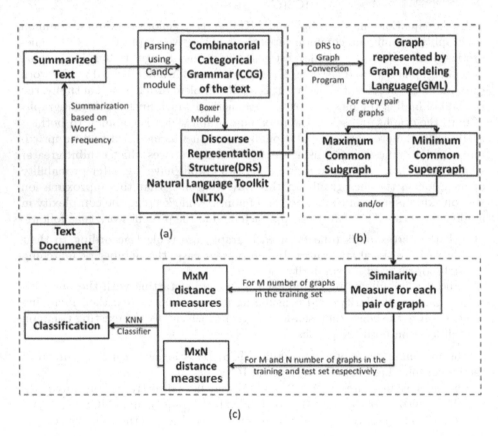

Fig. 2. Block diagram of the system: a) Semantic information extraction, b) Generation of graphs, c) Classification using k-NN

3.1 Semantic Information Extraction

Though the traditional bag-of-word approach provides good results for most of the text classification tasks, it suffers from the lack of ability to understand and differentiate the intrinsic relations between various subjects and predicates in a piece of text. Words may be the building block of language but they mean nothing in the absence of a firm semantic structure. Hence, it is essential to explore the semantic level analysis of the language and DRT in such a framework.

However, extensive use of DRSs demands a computationally efficient data structure for representation of the semantic relations. Graphs are such a structure and present a major advantage in case of similarity analysis. Various graph similarity measures can be developed using the maximum common subgraph and the minimum common supergraph for a pair of graphs. Hence, a robust system may be built which possess the ability to minutely observe and analyze complex semantics of natural language and efficiently categorize them.

As mentioned earlier, summarized text is used for the present work instead of entire one to compute mcs and MCS within polynomial time bound. Summarization is done on the basis of frequency of words; the sentences are chosen whose words occur with greatest frequency over a particular class. The sentences are ranked in order to easily choose the best ones for summarization. The summarization is done using the tool described in the url[2] below. The summarized text is then sent to the C&C parser [6] to identify the CCG derivation, POS tags, lemmas and named entity tags which are then used by Boxer [4] to produce the DRSs based on the inherent semantic interpretation of the sentence.

3.2 Formation of Graphs

To represent a graph in GML from DRSs, several rules are formed in the present work. As described earlier, Boxer is able to produce different kind of complex semantic relationships such as implications, negations and propositions of a sentence. In general Boxer represents a sentence through some discourse referents and conditions based on the semantic interpretation of the sentence.

In the graph, the referent is represented by vertex after resolving the equity among different referents of the DRS; and a condition is represented by an edge value between two referents. The condition of a single referent is represented as a self-loop of the referent (source and destination referents are same.) Special relationships such as proposition, implication etc. are treated as edge values between two referents; Agent and patient are also treated as conditions of discourse, hence represented by the edge values of two referents. An example of a sentence and its transformations (syntactic, semantic and graph representation) is shown in the Fig. 3.

To measure the distance between two graphs, the approximate $mcs(G_1, G_2)$ is constructed based on the steps described in Sect. 2.4. It is then for the creation of $MCS(G_1, G_2) = G_1 + G_2 - mcs(G_1, G_2)$ to make it computationally faster. Figure 4 shows the mcs and MCS of two graph sentences.

---

[2] `git://github.com/amsqr/NaiveSumm.gitmaster`

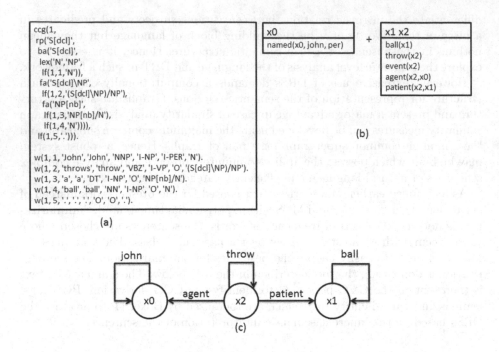

Fig. 3. The transformations for the sentence "`John throws a ball.`": a) C&C output, b) Boxer output and c) the corresponding graph

3.3 Evolution of Different Distance Measures Using k-NN Classifier

It has already been mentioned that the different distance measures (see (1)–(5)) were calculated based on mcs and MCS. The values of mcs and MCS can be represented by the number of similar vertices or the number of similar edges. Thus, ten different distances were calculated based on (1)–(5).

For classification purposes and for each measure two matrices were calculated (one for edges, other for vertex similarity) keeping the distance between each test example with each of the training examples. Thus, for M training and N testing examples, a matrix of $N \times M$ is formed. The results were used to evaluate the performance of each distance on the dataset.

4   Experimental Results and Discussion

In order to evaluate the method, a subset of Reuters-21578, the most well-known corpus for text classification has been used. For this purpose, 20 documents (10 for training, 10 for testing) of each of 20 different classes were selected. The 20 selected classes were: `cpi`, `acq`, `carcass`, `bop`, `alum`, `gas`, `ipi`, `copper`, `cocoa`, `earn`, `gold`, `dlr`, `coffee`, `barley`, `grain`, `cotton`, `interest`, `corn`, `fuel` and `crude`.

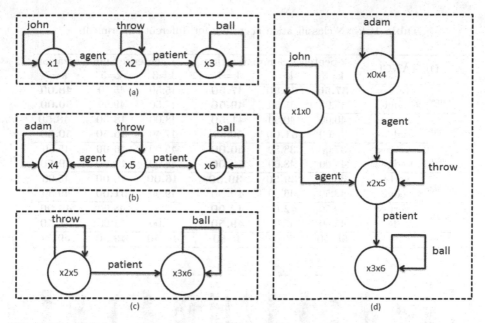

Fig. 4. Graphical overview of mcs and MCS: a), b) graph representation of sentences meaning "John throws a ball." and "Adam throws a ball." ; c) maximum common subgraph; d) minimum common supergraph

As stated, a summarization technique based on word frequencies has been applied and this work reports results for two and three sentences summarization. The summarized texts are then passed through NLTK toolkit [2] where C&C and Boxer are used to generate the DRSs. The generated DRSs are further used to generate the graphs. Then, the ten distance measures (see (1)–(5)) are calculated based on the generated graphs and afterwards, used for classification purpose using k-NN classifiers. The accuracies observed for the test dataset for 3, 5 and 7 nearest neigbours (k value) are shown in Table 2, along with a result for the traditional BOW approach. It is worthy to mention here that we have discarded all the stop words during creation of the bag of word vectors.The highest recognition accuracies among 3, 5, 7 nearest neighbours are highlited in the table by bold face.

From Table 1 and Fig. 5a it could be observed that all the edge based distances perform better than their vertex counterpart. So, the DRS conditions, which are represented by edge values, play an important role in the classification job. From Table 1 it can also be observed that average recognition accuracies of two sentence are lower than that of the three sentence summarization techniques. This can be easily visualized in Fig. 5b.

The maximum accuracy observed for the present work is 51.50% for edge based experiment for (5) with 3 sentence summarization for k=7; the minimum accuracy observed is 35.50% for vertex based experiment for (4) with 2 sentence summarization for k=3.

Table 1. k-NN classification accuracy for different experiments

DISTANCE		2 Sentence Summarization			3 Sentence Summarization		
		k=3	k=5	k=7	k=3	k=5	k=7
d_{mcs}	vertex	**37.50**	**37.50**	**37.50**	45.50	45.50	**48.00**
	edge	45.00	45.50	**49.50**	47.50	49.50	**50.00**
d_{wgu}	vertex	40.00	40.50	**41.50**	48.00	**48.50**	48.50
	edge	45.00	**51.00**	50.50	47.50	**50.50**	50.50
d_{ugu}	vertex	37.00	38.50	**40.00**	45.00	**46.00**	45.50
	edge	44.00	48.00	**50.00**	48.00	**49.50**	49.50
g_{MMCS}	vertex	35.50	**39.50**	**39.50**	**46.00**	**46.00**	45.00
	edge	42.50	46.50	**49.50**	49.00	**51.00**	49.50
d_{MMCSN}	vertex	39.50	**42.00**	**42.00**	47.50	48.00	**49.00**
	edge	45.00	49.00	**49.50**	47.00	51.00	**51.50**
bow		**50.50**	50.00	48.50	**49.50**	**49.50**	49.00

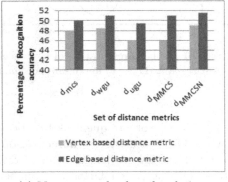

(a) Vertex vs. edge based technique

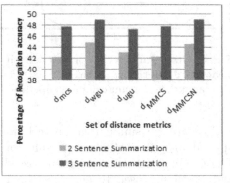

(b) 2 vs. 3 sentence summarization

Fig. 5. Accuracy comparison for vertex vs. edge based measures and for two vs. three sentence summarization

The average of 3 sentence summarization accuracy over k, observed for the five different distances with edge based calculations are $49.00\pm1.00\%$, $49.50\pm1.41\%$, $49.00\pm0.87\%$, $49.83\pm0.85\%$ and $49.83\pm2.01\%$. From the result it is observed that both distance measures based on (4) and (5) provide the same average accuracy on the test dataset. The overall accuracy on 3 sentence summarized texts and calculated using edge based formulae averaged over k is $49.43\pm0.38\%$, which denotes that the five distances are more or less comparable based on the observed recognition accuracies.

To analyze the result further, precision, recall and F_1 measures were calculated for the bag-of-words and the best graph distance (d_{MMCSN}). Precision is the ratio of the number of documents that belong to the class to the total number of documents classified as belonging to the class ($prec = TP/(TP + FP)$); recall is the ratio of the number of documents classified as belonging to the class to the total number of documents that belong to the class ($rec = TP/(TP + FN)$); F_1 is the harmonic mean of precision and recall ($F_1 = 2 * rec * prec/(rec + prec)$). The comparative assessment of the two approaches is shown in Table 2.

Table 2. Recall, precision and F_1 for the bag-of-words and semantic approaches

CLASS	RECALL		PRECISION		F_1	
	BOW	GRAPH	BOW	GRAPH	BOW	GRAPH
cpi	0.80	0.60(−0.2)	0.38	0.60(+0.22)	0.52	0.60(+0.08)
acq	0.10	0.60(+0.5)	0.20	0.60(+0.40)	0.13	0.60(+0.47)
carcass	0.90	0.30(−0.6)	1.00	0.33(−0.67)	0.95	0.32(−0.63)
bop	0.50	0.80(+0.3)	0.46	0.67(+0.21)	0.48	0.73(+0.25)
alum	0.50	0.30(−0.2)	1.00	0.38(−0.62)	0.67	0.33(−0.34)
gas	0.30	0.30(+0.0)	1.00	0.33(−0.67)	0.46	0.32(−0.14)
ipi	0.90	0.80(−0.1)	0.75	0.80(+0.05)	0.82	0.80(−0.02)
copper	0.30	0.50(+0.2)	0.60	0.31(−0.29)	0.40	0.39(−0.01)
cocoa	0.60	0.60(+0.0)	0.86	0.86(+0.00)	0.71	0.71(+0.00)
earn	0.40	0.80(+0.4)	1.00	0.89(−0.11)	0.57	0.84(+0.27)
gold	0.20	0.60(+0.4)	0.33	0.80(+0.47)	0.25	0.71(+0.46)
dlr	0.30	0.80(+0.5)	0.75	0.62(−0.13)	0.43	0.70(+0.27)
coffee	0.50	0.50(+0.0)	0.46	0.46(+0.00)	0.48	0.48(+0.00)
barley	0.60	0.50(−0.1)	0.46	0.50(+0.04)	0.52	0.50(−0.02)
grain	0.60	0.30(−0.3)	0.40	0.43(+0.03)	0.48	0.35(−0.13)
cotton	0.80	0.40(−0.4)	0.47	0.27(−0.20)	0.59	0.32(−0.27)
interest	0.50	0.30(−0.2)	0.28	0.60(+0.32)	0.36	0.40(+0.04)
corn	0.40	0.20(−0.2)	0.80	0.40(−0.40)	0.53	0.27(−0.36)
fuel	0.30	0.50(+0.2)	0.19	0.39(+0.20)	0.23	0.44(+0.21)
crude	0.70	0.60(−0.1)	0.54	0.43(−0.11)	0.61	0.50(−0.11)
average	0.51	0.52	0.60	0.54	0.51	0.51

There, it can be observed that sometimes the graph distance provides significantly better results than bag-of-words approach. In the case of the `carcass` class the BOW approach provides very satisfactory result due to having simple words like "beef" or "pork" which are enough to uniquely identify the category. On the other hand, the `gold` class shares some common words with other classes; the word "gold" itself can be found in `copper` and `alum` classes; other common words like "reserves" occur in many other classes. Moreover, there are words like "ounces" or "carat" which are overlooked in the BOW approach due to their comparatively low no. of occurrences. The use of the semantic approach enables the binding of words like "gold", "reserves", "carat" and "ounce" in such a way that they are highly unique for the `gold` class, giving better results.

Keeping the above factors in consideration a majority voting based combination methodology was implemented using the graph based and bag-of-words representations. During the process the confidence value (probability of a class) for each class is taken into consideration.The confidence values are obtained using k-NN classifier of the Weka machine learning toolkit[3]. Generally the class with the highest confidence value is selected as the class of the corresponding sample by the k-NN classifier. If B_j^i is the confidence value for i-th class of j-th sample for the BOW approach and G_j^i the confidence value for i-th class of j-th sample for the graph based approach then the new combined confidence value $C_j^i = (B_j^i + G_j^i)/2$. Now, the sample is assigned to the class having maximum C_j^i value.

Table 3. Accuracy for the combination of graph and BOW representations

distance		2 sent summarization			3 sent summarization		
		graph	graph + BOW	inc	graph	graph + BOW	inc
d_{mcs}	vertex	37.50	48.00	10.50	48.00	**54.00**	6.00
	edge	49.50	52.50	3.00	50.00	**55.00**	5.00
d_{wgu}	vertex	41.50	51.00	9.50	48.50	**55.00**	6.50
	edge	51.00	52.50	1.50	50.50	**55.50**	5.00
d_{ugu}	vertex	40.00	49.00	9.00	46.00	**54.50**	8.50
	edge	50.00	52.50	2.50	49.50	**56.00**	6.50
g_{MMCS}	vertex	39.50	49.50	10.00	46.00	**54.50**	8.50
	edge	49.50	51.50	52.00	51.00	**56.00**	5.00
d_{MMCSN}	vertex	42.00	51.00	9.00	49.00	**54.50**	5.50
	edge	49.50	52.00	2.50	51.50	**55.00**	3.50
bow		50.50			49.50		

As it can be seen in Table 3, classification using three sentence summarization obtains the best results and the combination of the BOW and the graph representations is able to significantly increase the individual results. With three sentence summarization we have observed average recognition accuracies of 54.5% and 55.5% after combining different distance measures with BOW for vertex and edge respectively; the minimum accuracy observed after increment is 54% which is also higher than any individual approach. And the maximum recognition accuracy observed is 56% which is much higher than the result of BOW approach.Therefore we can infer that the present approach has some mutually exclusive information over BOW representation. Hence, it can be strongly established that the graph distance based approach provides significant information and can highly contribute to a better recognition rate for textual data with semantically coherent information.

5   Conclusions and Future Work

In this paper a new approach has been proposed to the text classification task which integrates a deep linguistic analysis of the documents with a graph-based

---

[3] http://www.cs.waikato.ac.nz/ml/weka/

representation. In the linguistic analysis, discourse representation structures (DRS) are used to represent text semantic content and, afterwards, these structures are transformed into graphs. We have evaluated existing graph distance measures and proposed some modifications which are more adequate to calculate graph distances between graph-drs structures. Finally, we integrated graph-drs structures and the proposed graph distance measures into a k-NN classifier. The obtained results allow us to state that the proposed approach is able to obtain at least as good results as the standard BOW approach (in some cases it was able to outperform the standard approach). Moreover, the combination with BOW representation provides improved recognition accuracy which supersedes the accuracies achieved by the individual approaches (graph and BOW). This result is a good indicator of the adequacy of using semantic information to represent texts and text content.

As future work we intend to explore further the impact of the summarization module in our system and evaluate the use of more sentences or other summarization algorithms to represent each text. We also intend to evaluate our proposal with other machine learning algorithms, such as support vector machines using a graph kernel.

References

1. Angelova, R., Weikum, G.: Graph-based text classification: Learn from your neighbors. In: Proceedings of the 29th Annual International ACM SIGIR Conference on Research and Development in Information Retrieval, SIGIR 2006, pp. 485–492. ACM, New York (2006)
2. Bird, S., Klein, E., Loper, E.: Natural Language Processing with Python: Analyzing Text with the Natural Language Toolkit. O'Reilly, Beijing (2009), http://www.nltk.org/book
3. Bleik, S., Mishra, M., Huan, J., Song, M.: Text categorization of biomedical data sets using graph kernels and a controlled vocabulary. IEEE/ACM Trans. Comput. Biology Bioinform. 10(5), 1211–1217 (2013)
4. Bos, J.: Wide-coverage semantic analysis with Boxer. In: Proceedings of the 2008 Conference on Semantics in Text Processing, STEP 2008, pp. 277–286. Association for Computational Linguistics, Stroudsburg (2008)
5. Bunke, H., Foggia, P., Guidobaldi, C., Sansone, C., Vento, M.: A comparison of algorithms for maximum common subgraph on randomly connected graphs. In: Caelli, T.M., Amin, A., Duin, R.P.W., Kamel, M.S., de Ridder, D. (eds.) SSPR&SPR 2002. LNCS, vol. 2396, pp. 123–132. Springer, Heidelberg (2002)
6. Curran, J.R., Clark, S., Bos, J.: Linguistically Motivated Large-Scale NLP with C&C and Boxer. In: Carroll, J.A., van den Bosch, A., Zaenen, A. (eds.) ACL. The Association for Computational Linguistics (2007)
7. Himsolt, M.: GML: A portable Graph File Format. Tech. rep. Universität Passau, 94030 Passau, Germany (1999), http://www.infosun.fim.uni-passau.de/Graphlet/GML/gml-tr.html

8. Kamp, H., Reyle, U.: From Discourse to Logic: Introduction to Model-theoretic Semantics of Natural Language, Formal Logic and Discourse Representation Theory. Studies in Linguistics and Philosophy, vol. 42. Kluwer, Dordrecht (1993)
9. Riesen, K., Bunke, H.: Graph Classification and Clustering Based on Vector Space Embedding. World Scientific Publishing Co., Inc., River Edge (2010)
10. Wang, Z., Liu, Z.: Graph-based knn text classification. In: 2010 Seventh International Conference on Fuzzy Systems and Knowledge Discovery (FSKD), vol. 5, pp. 2363–2366 (2010)
11. Zhang, L., Li, Y., Sun, C., Nadee, W.: Rough set based approach to text classification. In: IEEE/WIC/ACM International Conference on Web Intelligence and Intelligent Agent Technology, vol. 3, pp. 245–252 (2013)

Exploring the Traits of Manual E-Mail Categorization Text Patterns

Eriks Sneiders, Gunnar Eriksson, and Alyaa Alfalahi

Stockholm University, Department of Computer and Systems Sciences,
Forum 100, SE-164 40, Kista, Sweden
{eriks,gerik,alyalfa}@dsv.su.se

Abstract. Automated e-mail answering with a standard answer is a text categorization task. Text categorization by matching manual text patterns to messages yields good performance if the text categories are specific. Given that manual text patterns embody informal human perception of important wording in a written inquiry, it is interesting to investigate more formal traits of this important wording, such as the amount of matching text, distance between matching words, n-grams, part-of-speech patterns, and vocabulary in the matching words. Understanding these features may help us better design text-pattern extraction algorithms.

Keywords: Text patterns, part-of-speech patterns, POS patterns.

1 Introduction

Text pattern matching has been used in Information Extraction [1], automated question [2] and e-mail [3] answering, and e-mail filtering [4]. Any text recognition task that involves regular expressions means matching surface text patterns. To the best of our knowledge, however, no one has studied what actually happens when text patterns do match pieces of text being recognized.

In this paper, we explore matching of manual text patterns to e-mail messages in order to assign standard answers, i.e., in order to put a message into a specific text category. The syntax of the text patterns resembles regular expressions. Content-wise the text patterns embody informal human perception of important wording for particular e-mail inquiries. In a text pattern, we can define a number of synonyms that designate a concept, we can define the order of the words and the distance between the words. Words are represented by word stems that match different inflections of a word. The text patterns are convenient for matching compound words, which are common in Germanic languages. A complete description of the text patterns is available in [3].

When a text pattern matches a piece of query text, it leaves a footprint. The difference between a text pattern and its footprint is the same as the difference between a regular expression and the bits of text it matches. We restrict a footprint to one sentence, which makes it easier to establish a representative number of words and

A. Przepiórkowski and M. Ogrodniczuk (Eds.): PolTAL 2014, LNAI 8686, pp. 337–344, 2014.

the distance between the words in the footprints, as well as establish part-of-speech patterns in the footprints. In the following example, footprints are the underlined words: "I applied for housing allowance on 13 January 2009. When will my money come?" The two footprints are "applied housing allowance" and "when money come".

We explore the footprints made by manual text patterns in e-mail messages during the process of automated e-mail answering. We explore by asking:

- How many words in query text do need to match a text pattern for successful matching outcome, i.e. correct message categorization?
- Are these words organized in n-grams or spread all around the sentence?
- How domain specific are the matching words?
- Do footprints form any part-of-speech (POS) patterns?

The language of the text being explored is Swedish, but we believe that similar conclusions would apply also to text in English.

The structure of this paper is straightforward. Section 2 describes the data that has created the footprints. Section 3 answers the above questions, and Sect. 4 summarizes the conclusions.

2   Experiment Data

The original collection was 9663 e-mail messages sent by citizens to the Swedish Social Security Agency. For our experiment, we selected 1909 messages, written in Swedish, that were correctly answered by an automated e-mail answering system, i.e. they were correctly assigned a standard answer, i.e., they were placed in a correct text category. 1882 messages belong to one text category while 27 messages belong to two categories (i.e. they have two standard answers), which makes 1936 message-category pairs. The categories and the number of selected messages in them are Cat1 (330), Cat2 (269), Cat3 (174), Cat4 (103), and Cat5 (1060). We ignored incorrect message categorization instances.

The size of the messages varied. The minimum, maximum, average, and median number of words per message were 4, 321, 45.5, and 35. The minimum, maximum, average, and median number of sentences were 1, 45, 5.2, and 4.

154 text patterns were used in order to categorize the messages. Because the text patterns were created manually, they are subjective, and so are their footprints. In order to give a somewhat objective picture of the footprints we describe them by the correctness of e-mail message categorization. The 154 text patterns categorized e-mail messages with precision about 90% and recall about 60% in the joint set of Cat1–Cat5. (The description of the measurements lies outside the scope of this paper.) We assume, without any proof, that footprints left by a different set of text patterns that achieve the same level of correctness of e-mail categorization, these different footprints would be similar.

Although we have 1936 message-category pairs, the system reproduced only 1918 of them because the system placed 18 messages, which belonged to two categories, into only one category.

3   Experiment Results

3.1   Amount of Matching Text

In total, 2273 sentences in the 1918 categorization instances matched one of the 154 text patterns, i.e. we have 2273 footprints. In 1577 categorization instances the message has one footprint (only one sentence in the message matched a text pattern), in 327 categorization instances there are 2 footprints, and in 14 categorization instances 3 footprints. The total number of words in the 2273 footprints is 12 108.

The overwhelming majority – 82% – of the messages in our experiment have only one matching sentence. We should bear in mind, however, that our messages lie within five specific text categories and have explicit information needs. Most information needs that were captured by the system happened to be stated in one sentence.

Table 1 shows the number of matching words per sentence (i.e., the size of a footprint) and per message, and the number of corresponding sentences and messages. The number of sentences in the table is independent from the number of messages. The majority of the messages – 1338 or 70% of the total – have 5 to 7 matching words. It is 5 to 7 times less than the median number of words per message, which is 35. For 19 messages, a correct answer was assigned on the grounds of only 3 matching words; these were very specific 3 words. In half of the messages, the footprints occupied less than 18% of the text.

Table 1. Number of matching words per sentence and per message

Matching words	Num. sentences	Num. messages	Matching words	Num. sentences	Num. messages
1	43	0	8	141	207
2	30	0	9	66	123
3	212	19	10	17	49
4	420	162	11	6	12
5	565	473	12	0	7
6	505	501	15	0	1
7	268	364			

3.2   Gaps between Matching Words

Knowing the number of matching words is not enough for having a complete idea of what the footprints look like. The distribution of the words across the sentence is another characteristic.

A gap between two matching words is a number of non-matching words between them. A gap of size 0 means that two matching words follow each other; a gap of size 1 means that there is a random word in between. Table 2 shows the gap sizes between matching words across all the footprints and the frequency of these gap sizes. In about 84% of the cases the gap is no larger than 1, which means that the matching words prefer staying in a cluster instead of evenly spreading across the sentence.

Table 2. Gap sizes across all the footprints

Gap size	Num.	%	Gap Size	Num.	Gap Size	Num.
0	6092	61.9	4	218	8	31
1	2155	21.9	5	106	9	17
2	728	7.4	6	76	More	56
3	309	3.1	7	47	Total	9835

An n-gram is a sequence of *n* words that follow each other; there is no gap in between. Table 3 shows the number of n-grams in all the footprints, as well as the number of distinct n-grams extracted from the footprints. The distinct n-grams are made of word lemmas and do not contain duplicates. The fourth column shows how many distinct n-grams could be decomposed into extracted distinct bi- and trigrams, while the fifth column permits also larger n-grams in the decomposition. Please observe that decomposition is not overlap. Our 9-grams could not be decomposed, but they certainly overlap with other n-grams.

Table 3. N-grams in footprints and extracted from footprints

Size	Num. in footprints	Num. distinct	Split into 2-3-grams	Split into 2-to-7-grams	
				Num.	%
1-gram	2981	428	n/a	n/a	n/a
2-gram	1374	544	n/a	n/a	n/a
3-gram	823	451	n/a	n/a	n/a
4-gram	468	282	62	62	22
5-gram	232	157	61	61	38.9
6-gram	102	74	7	15	20.3
7-gram	25	23	7	8	34.8
8-gram	8	8	1	5	62.5
9-gram	3	3	0	0	0
Total	**6016**	**1970**	**138**	**151**	

The number of distinct 2-to-9-grams is half the total number of 2-to-9-grams, which means the frequency of individual n-grams in the footprints is not high and they are not statistically representative. Because the text patterns, which made these n-grams, operate a rich vocabulary of synonyms, we have a good reason to believe that the n-grams would become more statistically representative if they were made of concepts, which cover different synonyms, rather than word lemmas. This means that text pattern extraction from e-mail messages is likely to be more successful if these patterns are made of concepts, not individual words.

Our initial assumption was that larger n-grams would be easily decomposable into smaller n-grams, and we could cover the footprints with uni-, bi-, and trigrams. Our n-grams made of word lemmas show results different from what we hoped for.

3.3   Part-of-Speech Patterns

The POS pattern of a footprint is a sequence of POS attributes of the words in the footprint disregarding the gaps between the words. Two footprints may share the same POS pattern while containing different words. In order to discover the POS patterns, we did automatic POS-tagging of the entire email text.

Table 4. Most frequent POS patterns

No.	Swedish POS patterns	N	Examples translated to English	N
1	vb nn nn	51	sent application housing-allowance	17
2	pn ab vb nn	43	I/we not got [payment]	22
3	pn vb nn	41	I applied housing-allowance	19
4	nn	37	[domain specific concepts]	37
5	ha vb ps nn	35	when comes my [payment]	33
6	pn vb nn pp nn	27	I need form about parental-allowance	3
7	ha vb nn	26	when comes [payment]	10
8	vb ha pn vb nn	24	wonder when I get [payment]	18
9	vb nn ab vb nn	23	applied housing-allowance not got [reply]	22
10	pn ab vb nn pp nn	23	I not got [payment] in/for [period]	22
11	vb ha jj nn vb ab pp	22	wonder how many [days] left over for	22
12	ha vb pn ps nn	20	when get I my [payment]	18
13	vb pn vb nn	20	can/would you send form	14
14	pn vb vb nn	19	I want order form/brochure	14
15	vb ha ps nn vb	17	wonder when my [payment] comes	12
16	vb ab vb nn	17	have not got [domain-dependent-noun]	17
17	vb nn pp nn	16	sent papers about [allowance]	8
18	pn vb pn vb nn	15	I want you send form/brochure	13
19	vb ha nn vb	15	wonder when [payment] comes	12
20	vb ha jj nn vb vb pl pp	14	wonder how many [days] have taken out for	12

Our 2273 footprints host 942 POS patterns. Table 4 shows 20 most popular ones. The parts-of-speech in the table are verbs (vb), nouns (nn), pronouns (pn), question adverbs (ha), adverbs (ab), prepositions (pp), adjectives (jj), possessives (ps), and particles (pl).

The frequency of individual POS patterns drops quickly. 600 of the 942 POS patterns (63.7%) occur only once. Table 4 shows also the most representative example of each POS pattern. The examples are word lemmas translated to English. A lemma in square brackets stands for a number of close synonyms that represent the concept. For example, [days] means different wordings for child-care days paid by the state.

The POS patterns are dominated by individual expressions. For example, the pattern no. 11 occurs 22 times, and always with the same expression. The domination of individual expressions weakens the role of the POS patterns, detached from the text patterns, as representative features of the text categories.

Table 5. Most frequent lemmas

POS	Lemma	English	N	POS	Lemma	English	N
pn	jag	I	841	pp	för	for	198
vb	få	get	724	nn	*ersättning*	*compensation*	159
ha	när	when	467	vb	kunna	can	157
ab	inte	not	448	Jj	många	many	149
vb	undra	wonder	386	vb	vilja	want	147
vb	ha	have	371	nn	*dag*	*day*	142
ha	hur	how	334	vb	vara	be	137
vb	*skicka*	*send*	294	pp	på	on	126
nn	*blankett*	*form*	293	nn	*ansökan*	*application*	122
nn	*peng*	*money*	274	pn	det	this/that	111
ps	min	my	258	vb	ta	take	111
nn	*bostads-bidrag*	*housing allowance*	246	nn	*föräldrapenning*	*parental allowance*	157
vb	komma	come	244	pl	ut	out	94
nn	*utbetalning*	*payment*	226	nn	*pension*	*pension*	88
pn	ni	you	223	vb	*betala*	*pay*	82

Table 5 shows the most frequent words (their lemmas in Swedish and translation into English) in the footprints. Domain-specific words, in italic, designate what the e-mail inquiry was about. For example, "day" is essential for the concept of child-care days; "pay" is essential in expressions about payments.

Somewhat surprisingly, seven most frequent are common language words, not domain words. The footprints mirror manual text patterns crafted to embody the essence of an inquiry in an e-mail message. People do not communicate through sets of keywords; the words that help formulate intelligible sentences are as important as domain-specific keywords. Khosravi and Wilks [5] have observed that the share of nouns, verbs, adjectives, and adverbs in e-mail messages is about 40%, close to that in spoken language. More than half are words that support the communication.

3.4   POS Patterns across Text Categories

The distribution of the POS patterns in the footprints across Cat1–Cat5 is uneven (see Table 6). We are curious how representative the POS patterns are in the entire e-mail text across Cat1–Cat5. If certain POS patterns do stick out, we could use them in text categorization outside the scope of our text patterns matching. In order to judge representativeness of the POS patterns, we have to normalize their number per category with respect to the size of the category. Normalized POS pattern frequencies (not included here because of space limitations) show that most POS patterns are somewhat evenly distributed across Cat1–Cat5. Some irregularities are marked in Table 6: weak overrepresented in bold, strong overrepresented in bold underlined, underrepresented in italic underlined. POS pattern no. 7 covers a variety of expressions (see Table 4), and it is underrepresented in Cat1. The other six POS patterns with irregularities cover mostly one expression, their content lacks diversity. Interesting is Cat3: POS patterns no. 5, 12, 16, and 19 are not represented in the footprints but are overrepresented in the rest of the text.

Table 6. Number of most frequent POS patterns (referenced by their sequence number in Table 4) across text categories

No.	In the footprints, Cat1–Cat5					In the entire text, Cat1–Cat5				
	1	2	3	4	5	1	2	3	4	5
1	3	37	0	1	10	1643	1238	699	607	4585
2	12	15	0	0	16	406	396	166	127	1412
3	9	21	0	7	4	1197	982	529	381	3522
4	5	15	7	1	9	3270	2437	1420	1198	9377
5	0	0	0	0	35	_25_	65	**97**	22	326
6	15	5	0	1	6	603	494	294	205	1652
7	1	7	0	7	11	_97_	189	171	88	774
8	0	4	0	0	20	187	234	216	108	785
9	0	22	0	0	1	326	365	151	134	1230
10	0	0	0	0	23	165	145	84	57	553
11	0	0	22	0	0	20	_4_	**_143_**	_2_	45
12	0	2	0	0	18	13	25	**29**	11	127
13	15	4	0	0	1	1056	812	477	382	2944
14	17	0	0	1	1	857	672	357	260	2438
15	0	7	0	0	10	601	614	251	223	2220
16	0	0	0	0	17	47	39	**55**	**34**	237
17	2	1	0	0	13	1286	939	545	468	3349
18	15	0	0	0	0	460	391	240	158	1431
19	0	0	0	0	15	_109_	144	**288**	77	718
20	0	0	14	0	0	16	_3_	**73**	2	15

4   Conclusions

We have researched text pattern matching in order to categorize e-mail messages into specific text categories, where all messages in one category share the same standard answer. We have explored the traces of manual text patterns in the text of almost two thousand e-mail messages.

In 70% of all categorization instances, only 5–7 words in a message were used in order to decide the right text category for the message. Half of the messages were categorized using less than 18% of their text.

A gap between two words in an e-mail message that match a manual text pattern is the number of non-matching words between the matching words. About 84% of such gaps are no larger than 1; representative words tend to stick together.

About 75% of all matching words lie in n-grams of size 2 to 9. The number of distinct 2-to-9-grams, where duplicates are removed, is half of the total number of the 2-to-9-grams found in the message text, which means that individual n-grams are not statistically representative. We believe that n-grams made of concepts, which enclose a number of synonyms, rather than word lemmas would be more representative.

A POS pattern covers the words in one sentence that have correctly matched a manual text pattern. Such POS patterns are not representative features of the text categories. We do believe, however, that mixing text patterns and POS patterns in text pattern matching could increase the recall of categorization.

Unlike one would expect, most matching words are common language words. It is the combination of common words and a few domain keywords that is representative for an inquiry, not the single words themselves. If we were to extract inquiry-specific text patterns automatically, focusing on domain keywords or inverted document frequency would not help, except maybe for extracting seed terms the way Downey et al. [1] did.

References

1. Downey, D., Etzioni, O., Soderland, S., Weld, D.S.: Learning text patterns for web information extraction and assessment. In: Proc. AAAI 2004 Workshop on Adaptive Text Extraction and Mining, pp. 50–55 (2004)
2. Sneiders, E.: Automated FAQ Answering with Question-Specific Knowledge Representation for Web Self-Service. In: Proc. 2nd International Conference on Human System Interaction (HSI 2009), Catania, Italy, May 21-23, pp. 298–305. IEEE (2009)
3. Sneiders, E.: Automated Email Answering by Text Pattern Matching. In: Loftsson, H., Rögnvaldsson, E., Helgadóttir, S. (eds.) IceTAL 2010. LNCS (LNAI), vol. 6233, pp. 381–392. Springer, Heidelberg (2010)
4. Wang, J.H., Chien, L.F.: Toward Automated E-mail Filtering – An Investigation of Commercial and Academic Approaches. In: TANET 2003 Conference, pp. 687–692 (2003)
5. Khosravi, H., Wilks, Y.: Routing email automatically by purpose not topic. Natural Language Engineering 5, 237–250 (1999)

Relation Extraction for the Food Domain without Labeled Training Data – Is Distant Supervision the Best Solution?

Melanie Reiplinger, Michael Wiegand, and Dietrich Klakow

Spoken Language Systems, Saarland University, D-66123 Germany
{melanie.reiplinger,michael.wiegand,dietrich.klakow}@lsv.uni-saarland.de

Abstract. We examine the task of relation extraction in the food domain by employing distant supervision. We focus on the extraction of two relations that are not only relevant to product recommendation in the food domain, but that also have significance in other domains, such as the fashion or electronics domain. In order to select suitable training data, we investigate various degrees of freedom. We consider three processing levels being argument level, sentence level and feature level. As external resources, we employ manually created surface patterns and semantic types on all these levels. We also explore in how far rule-based methods employing the same information are competitive.

1 Introduction

In view of the large interest in food in many parts of the population and the ever increasing amount of new dishes, there is a need of automatic knowledge acquisition, especially relation extraction. Some relations, such as food items that can be served together or relations that express that two food items can be substituted by each other, are highly relevant to product recommendation systems or virtual customer advice. Unfortunately, such knowledge does not exist in conventional (structured) knowledge bases, yet it has been found that domain-specific corpora, i.e. unstructured natural language texts, contain an abundance of such relations [1].

In this paper, we focus on distant supervision [2] for relation extraction, where training data (i.e. sentences with mentions of particular relation instances) are automatically generated with the help of a relation database. Such a database contains argument pairs representing instances for relations that one wants to extract (e.g. *<rice pudding, fruit salad>* is an entity pair that expresses the relation that the two food items can be served together). Distant supervision rests on the assumption that sentences with mentions of those entity pairs are genuine instances of the relation (Sentence (1)). This way of producing labeled training data (given a relation database) is considerably faster than manually labeling sentences from a corpus in which food items occur. Of course, this approximation is not guaranteed to produce correct mentions of relation instances as exemplified in Sentence (2), so one also needs to devise methods to remove spurious relation mentions.

A. Przepiórkowski and M. Ogrodniczuk (Eds.): PolTAL 2014, LNAI 8686, pp. 345–357, 2014.

(1) For tonight, I planned to have rice pudding with fruit salad.
(2) Other types of food I like are pizza, falafel, rice pudding and fruit salad.

The aim of this investigation is to examine the different degrees of freedom in the design of a relation extraction classifier for the food domain based on distant supervision. Among the different aspects of such a classifier, we consider different kinds of knowledge (semantic types and surface patterns) and apply them on different processing levels. We also want to show that one has to take into consideration special properties of the relations to be extracted. For different relations, different kinds of classifier configurations may be suitable. We also want to critically assess whether predictive types of knowledge sources actually require a distantly supervised classifier, or whether a simple incorporation of such knowledge into rule-based classification already produces comparable results.

Our experiments are carried out on German data, but our findings should carry over to other languages since the issues we address are language universal. For general accessibility, all examples are given as English translations.

2   Data and Annotation

The relations that are to be extracted are *SubstitutedBy* and *SuitsTo* as illustrated in Table 1. We focus on these two relations because we consider them not only relevant to customer advice/product recommendation in the food domain, but also to similar applications in other types of domains. Customers want to know which items suit together, be it two food items that can be used as a meal, two fashion items that can be worn together, or two electronic devices that can be combined/connected in some way. Substitutes are relevant in all situations in which item A is out of stock but item B can be offered as an alternative.

Table 1. The different relations and their distribution in our gold standard

Relation	Description	Example	Perc.
SuitsTo	food items that are typically consumed together	My kids love the simple combination of fish fingers with mashed potatoes.	60
SubstitutedBy	similar food items commonly consumed in the same situations	We usually buy margarine instead of butter.	9
Other	other relation *or* co-occurrence of food items is co-incidental	On my shopping list, I've got bread, cauliflower, ...	31

As a gold standard, we randomly extracted from a domain-specific corpus about 2200 sentences in which at least two food items co-occur and manually labeled them with their pertaining relation. As a corpus, we chose a crawl of *chefkoch.de* [3], the largest German web platform dealing with food-related issues. This corpus also serves as an unlabeled dataset from which training data for distant supervision are to be extracted. In addition to the two relations from

Table 2. Coverage of the relation database on the unlabeled food corpus

Relation	Argument Pairs	Matched Sentences
SuitsTo	1,374	44,692
SubstitutedBy	781	34,771
None	62,191	1,187,101

above, we introduce the label *Other* for cases in which either another relation between the target food items is expressed, or where their co-occurrence is co-incidental. The class distribution in our gold standard is also shown in Table 1 (last column). On a subset of 400 sentences, an interannotation agreement of $\kappa = 0.78$ was measured which can be considered *substantial* [4].

For distant supervision, a relation database of entity tuples representing different relation instances is required. We make use of the resource introduced in [5] that has been specifically designed for distant supervision experiments. Table 2 lists the size of this database[1] together with the corresponding amount of sentences in our corpus where these food pairs match. By *None*, we understand all pairs that are neither contained in *SuitsTo* nor in *SubstitutedBy*.

The resource was created by giving two annotators partially instantiated relations, such as *SuitsTo(broccoli, ?)* or *SubstitutedBy(beef roulade, ?)*. The annotators then produced lists of food items that fit those relations, e.g. {*potatoes, fillet of pork, mushrooms, ...*} (food items that suit to *broccoli*) and {*goulash, braised meat, rolled pork, ...*} (food items that can be substituted by *beef roulade*). The annotators were allowed to consult various information sources for research, such as the internet. However, in order to obtain unbiased results, they were specifically asked not to focus on a particular source, e.g. a particular website. Note that the resource also contains other relations than the ones considered in this paper (for instance, relations that describe for what event a particular food item is suited, or which food items are recommended/not recommended for people with a particular health condition). It would be beyond the scope of this paper to examine all those different relations for distant supervision. We focus on the two relations due to their relevance to other domains (see discussion above).

Table 2 shows that there is a huge number of potentially negative data. However, one should keep in mind that such pairs may also contain positive pairs, since our relation database (i.e. the pairs representing instances of *SuitsTo* and *SubstitutedBy*) is not exhaustive.

3   Method

Figure 1 presents the most important aspects of the relation extraction system examined in this paper. The figure can be read in the following way. There are two main *knowledge* sources that are examined in this paper being *patterns* and

---

[1] We excluded any food pairs from our manually labeled test set (i.e. gold standard).

types. The sources can be harnessed by either of the two classifiers, *distant supervision* and *rule-based classification*. With regard to distant supervision, there are also three different processing levels to be considered. The final classifiers are to extract either the relation *SuitsTo* or *SubstitutedBy*. Due to the different nature of the two different relations, the different knowledge sources, classification methods and processing levels may have a different impact on extraction performance. In the following, we discuss these issues in more detail.

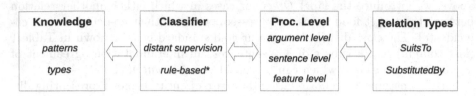

Fig. 1. The most important dependencies examined in this task (*: this classifier cannot be paired with the different *processing levels*)

3.1   Knowledge

Surface Patterns. The most straightforward approach to relation extraction is a classification based on surface patterns. We exclusively rely on the set of manually-designed surface patterns introduced in [3] as illustrated in Table 3. A pattern is a lexical sequence comprising the words between two food items.

Table 3. Illustration of the manually designed surface patterns

Relation	#Patterns	Examples
SuitsTo	8	FOOD and FOOD; FOOD as well as (a) FOOD; FOOD go with FOOD; FOOD with FOOD; FOOD fit to FOOD
SubstitutedBy	8	FOOD or (even) FOOD; FOOD (FOOD; FOOD instead of FOOD; FOOD in place of FOOD; FOOD , resp. FOOD

Type Constraints. Another important source of information for relation extraction is intrinsic relation argument information. In the context of the food domain, this corresponds to food type information (since all arguments of the relations we consider are food items). Following [6], we consider the food categorization according to the Food Guide Pyramid [7] (Table 4). In all our experiments, we make use of the best food type model from [6] that had been induced in a semi-supervised manner.

Certain type combinations may be indicative of certain relations. This information could be particularly exploited in situations in which contextual information is inconclusive. For example, Sentences (3) and (4) both contain food items that match the same surface pattern, i.e. *FOOD and FOOD*. Yet the two

Table 4. The different food types

Category	Description	Size	Perc.
MEAT	meat and fish (products)	394	20.87
BEVER	beverages (incl. alcoholic drinks)	298	15.78
VEGE	vegetables (incl. salads)	231	12.24
SWEET	sweets, pastries and snack mixes	228	12.08
SPICE	spices and sauces	216	11.44
STARCH	starch-based side dishes	185	9.80
MILK	milk products	104	5.51
FRUIT	fruits	94	4.98
GRAIN	grains, nuts and seeds	77	4.08
FAT	fat	41	2.18
EGG	eggs	20	1.06

Table 5. Type assumption for the two target relations

Relation	Rule
SuitsTo	Is the type pair of the form <x,y>, e.g. <MEAT,STARCH> for <*fish, chips*>?
SubstitutedBy	Is the type pair of the form <x,x>, e.g. <FAT,FAT> for <*butter, margarine*>?

sentences convey different relations. A simple type-based rule may help to disambiguate this context (Table 5): If the food items possess different types, then they are more likely to represent instances of *SuitsTo*. However, if the food items are of the same food type, then they are likely to be instances of *SubstitutedBy*.

(3) I very often eat fish$_{MEAT}$ and chips$_{STARCH}$. *(Relation: SuitsTo)*
(4) For these types of dishes you can offer both Burgundy wine$_{BEVER}$ and Champagne$_{BEVER}$. *(Relation: SubstitutedBy)*

3.2 Classifiers

Apart from the distantly supervised classifier, we also examine rule-based classification. We designed two different classifiers that exclusively make usage of the two knowledge resources that can also be utilized outside the context of distant supervision, i.e. the manually designed surface patterns and the food type information. The resulting classification algorithms are straightforward. For the **pattern-based classifier ($RB_{pattern}$)**, we assign to a sentence the label *SuitsTo* or *SubstitutedBy*, in case one of their respective patterns fires. If no pattern fires, then neither of these two relations holds. For the **type-based classifier (RB_{type})**, we use the type rules from Table 5 to predict either of our two target relations. Finally, we also include a **combined classifier (RB_{comb})** which only predicts a target relation if both the type assumption for the respective relation and one of its patterns fires.

3.3 Processing Levels

In the following, we discuss the three different processing levels we consider for distant supervision. The two knowledge sources, surface patterns and food types, can be applied on all of these levels. We also want to investigate whether the same knowledge resource can have a different impact depending on which level it is applied. Table 6 lists the different methods for the different processing levels.

Table 6. Methods on different processing levels for distant supervision

Level	Method	Description
argument	random	select argument pairs from relation database at random
	frequency	sort argument pairs according to frequency (of co-occurring in the text corpus)
	pmi	sort argument pairs according to *pointwise mutual information (pmi)*
	patt$^+$	sort argument pairs according to pmi of food items and the surface patterns pertaining to the target relation in *descending* order
	patt$^-$(a)	sort argument pairs according to pmi of food items and the surface patterns pertaining to the target relation in *ascending* order
	patt$^-$(b)	sort argument pairs according to pmi of food items and the surface patterns pertaining to the contrast relation in *descending* order
	type	sort type pairs (e.g. <MEAT,STARCH>) according to pmi and consider their actual food instantiations as arguments
	wup	sort argument pairs according to Wu-&-Palmer [8] similarity in GermaNet [9]
sentence	pattern	only include sentences in which target food items co-occur with surface pattern from target relation
	type	only include sentences in which type rule for the pertaining relation (Table 5) is fulfilled
feature	pattern	include all surface patterns as additional features
	type	include features indicating the types of the target food items, e.g. <MEAT,STARCH> for <*fish,chips*>
	standard	standard features directly extracted from training data without external knowledge resources (see Table 7)

Argument-Level Filtering. By argument-level filtering, we understand methods by which we **select the arguments** (i.e. entity pairs) from the relation database to represent typical positive (and negative) relation instances. All these methods have in common that they produce a ranking of argument pairs where the higher ranked pairs are considered more suitable than the lower ranks. For instance, one can sort the argument pairs by their frequency of co-occurrence in our unlabeled text corpus.

We also employ the argument-level filtering for the selection of negative training data, i.e. training instances considered not to convey the target relation. As *negative* argument pairs, we consider the union of food pairs of *None* (Table 2) and the pairs of the contrast relation (i.e. *SubstitutedBy* for *SuitsTo* and vice versa[2]). For generating a ranking of argument pairs for negative training data with the help of the surface patterns, we explore two different methods: $patt^-$ *(a)* considers the ranking produced by the patterns of the target relation,[3] however, the pairs are sorted according to pointwise mutual information in *ascending order*. By that, we mostly aim for food pairs not strongly correlating with the target relation. $patt^-$ *(b)*, on the other hand, considers the ranking produced by the patterns of the contrast relation (in *descending order*), so, in this case, we aim for food pairs correlating with the contrast relation.

Another measure that is less self-explanatory is the Wu-&-Palmer [8] similarity computed from GermaNet [9], the German version of WordNet [10] (i.e. *wup*). In principle, this measure indicates the semantic distance of two food items (more precisely, their synsets) in the GermaNet hypernymy graph. It is considered a good measure for detecting (near-)synonyms. We employ it since typical entity pairs for *SubstitutedBy* are similar to (near-)synonyms.[4]

Sentence-Level Filtering. By sentence-level filtering, we understand methods by which we **filter the sentences** that match entity pairs from the relation database to represent training data. For example, if we consider a pattern-based sentence filter (i.e. *pattern*), we only include those sentences in which the arguments are connected via a surface pattern of the pertaining target relation (Table 3). For negative training data, we simply *exclude* those sentences that match a particular condition. We apply these filtering methods on the best respective argument-level selection method for the respective relation.

Feature-Level Processing. By feature-level processing, we understand the traditional form of feature engineering for supervised learning. Apart from features that make use of either surface patterns or food types, we also include a large set of standard features that are commonly applied to relation extraction tasks. All these features have in common that they are directly extracted from the sentences which are to be classified. They do not depend on other knowledge resources. The individual standard features are listed in Table 7.

---

[2] Note that we have two different sets of negative training data depending on what target relation we consider as positive class, i.e. *SuitsTo* or *SubstitutedBy*. We will train two binary classifiers, one for each target relation (versus the remaining instances).

[3] Even though the food pairs do not include instances of the target relation (acc. to the relation database), some of those pairs may still match the patterns of that relation.

[4] Wu-&-Palmer similarity is only used at the argument level. This is due to the fact that the methods at the other processing levels are binary functions rather than continuous functions like Wu-&-Palmer similarity. We did not find an intuitive binary function based on that similarity.

Table 7. Standard features, i.e. features not employing external knowledge

Feature	Description
word-left-window	a window of 2 words to the left of *arg1*
word-right-window	a window of 2 words to the right of *arg2*
word-window	the word sequence between *arg1* and *arg2*
left-lemma-window	a window of 2 words to the left of *arg1* as lemmas
right-lemma-window	a window of 2 words to the right of *arg2* as lemmas
lemma-window	the word sequence between *arg1* and *arg2* as lemmas
bow	all words in the sentence
lemma-bow	lemmas of w_{i-1}, w_{i+1} where $w_i \in \{arg1, arg2\}$
lemma-bigrams	all bigrams $< w_i, w_j >$ between *arg1* and *arg2* as lemmas
pos-left-window	part-of-speech tags of words in a window of 2 words to the left of *arg1*
pos-right-window	part-of-speech tags of words in a window of 2 words to the right of *arg2*
pos-window	part-of-speech sequence between *arg1* and *arg2*
pos-unigrams	part-of-speech tags of w_{i-1}, w_{i+1} where $w_i \in \{arg1, arg2\}$
pos-bigrams	all part-of-speech bigrams $< t_i, t_j >$ between *arg1* and *arg2* using lemmas

3.4   The Two Target Relations

We want to point out that due to the different properties of our two target relations, the demands for a good classifier may vary. By different properties, we specifically mean the notable difference in occurring in our corpus (Table 1). *SubstitutedBy* is a typical *minority* class, while *SuitsTo* is the *majority* class. We assume that for *SuitsTo*, an appropriate classifier needs to focus on precision. With a proportion of 60%, even poor classifiers equivalent to guessing are likely to extract a reasonable amount of true positives, yet precision will be low. With a proportion of only 9%, classifiers for *SubstitutedBy* equivalent to guessing may produce not a single true positive. So, recall also seems important to this relation.

4   Experiments

Even though we explore different methods of producing labeled training data via distant supervision, the training sets always have the same size (10,000 labeled instances). As a supervised classifier, we use Support Vector Machines. As an implementation, we chose SVM^{light}.[5]

For each relation we want to detect (i.e. *SuitsTo* and *SubstitutedBy*), we will build a separate binary classifier, where the positive class is the target relation to be detected, and the negative class are all remaining instances (including instances of the respective contrast relation). We always enforce the class distribution from our gold standard (Table 1). Given a particular ranking of argument

---

[5] http://svmlight.joachims.org

pairs for the positive and the negative class, respectively, we randomly sample up to 100 sentences from each pair (where those two food items co-occur),[6] starting from the top of the rankings until the 10,000 instances have been obtained.

4.1   Performance of Argument-Level Filtering

We first examine the different argument-level filtering methods. For these experiments, no sentence-level filtering is employed. Moreover, we only train classifiers using only the standard features (Table 7). Table 8 presents the results. All filtering methods are separately applied for the creation of positive training data (all those training data that represent instances of the respective target relation, i.e. *SuitsTo* or *SubstitutedBy*) and negative training data (all those training data that comprise instances not representing instances of the respective target relation). The reason for allowing different filtering methods for those two types of training data is that we must not assume that one single filtering method is optimal for the creation of both positive and negative training data.

Table 8. F-scores (macro-average) of different argument-level filtering methods (*positive*: filtering methods are applied for the creation of positive training data; *negative*: filtering methods are applied for the creation of negative training data)

| | SuitsTo | | | | | | SubstitutedBy | | | | | |
| | *positive* | | | | | | *positive* | | | | | |
negative	rand.	freq	pmi	wup	patt$^+$	type	rand.	freq	pmi	wup	patt$^+$	type
random	41.8	45.0	49.2	46.0	45.3	42.1	61.8	59.0	63.0	59.8	65.1	60.0
freq	40.1	44.1	50.1	41.4	44.8	40.4	61.0	58.1	61.7	59.0	64.0	59.0
pmi	42.3	45.1	50.7	43.5	47.8	43.0	62.8	58.9	64.2	61.0	64.8	59.3
wup	42.4	45.8	50.3	44.4	45.8	42.0	59.3	57.4	60.8	57.6	64.7	55.3
patt$^-$(a)	43.7	47.2	52.2	45.4	47.7	42.0	64.8	62.6	**67.0**	62.9	66.8	63.5
patt$^-$(b)	55.0	56.6	**60.4**	56.7	54.9	54.5	52.5	53.1	57.8	54.7	62.2	50.3
type	42.4	43.8	49.2	44.3	45.9	41.0	61.5	59.3	63.3	59.0	64.6	61.1

The table shows that indeed there is a difference in effectiveness of the methods between the different relations. Moreover, there is also a difference in effectiveness of the methods depending on whether they are applied to rank positive or negative training data. For ranking positive data for *SuitsTo*, *pmi* is the most effective method. For ranking positive data for *SubstitutedBy*, patterns are slightly more effective than *pmi*. For ranking negative data, for both relations, the best ranking is produced by applying patterns, however, the form of pattern-based ranking is different. That is, for *SubstitutedBy* the best negative ranking comprises arguments that do not correlate with the target relation (*patt$^-$(a)*),

---

[6] In fact, we construct 3 random samples per configuration and report the averaged results. Having 3 individual results per configuration also allows us to apply a paired t-test for statistical significance testing.

while for *SuitsTo* it is those arguments that strongly correlate with the contrast relation ($patt^- (b)$). We assume that the reason for this lies in the intrinsic predictiveness of the surface patterns for the different relations (we particularly mean *precision* here, as it is the most relevant evaluation measure to rankings). For *SubstitutedBy*, patterns are more effective than for *SuitsTo* (its precision is 77.72 compared to 55.40 for *SuitsTo* (Table 11:$RB_{pattern}$)). This is why, patterns from *SubstitutedBy* may also serve as negative cues for *SuitsTo*.

In summary, argument-level filtering is very relevant to both relations; for *SuitsTo*, we achieve an increase in F-score by 18.2% points and for *SubstitutedBy* by 5.2% points over the standard random argument selection.

4.2 Performance of Sentence-Level Filtering

Table 9 displays the performance of the different sentence-level filtering methods. We use for each relation the best respective result from argument-level filtering (Table 8). We still use only the standard feature set and apply the filtering methods for the creation of positive and negative training data separately.

Table 9. F-scores (macro-average) of different sentence-level filters (*positive*: filtering methods are applied for the creation of positive training data; *negative*: filtering methods are applied for the creation of negative training data)

	SuitsTo				SubstitutedBy		
	positive				*positive*		
negative	no filter	pattern	type	*negative*	no filter	pattern	type
no filter	**60.40**	43.22	60.66	no filter	**67.00**	64.52	66.05
pattern	47.44	49.19	47.51	pattern	28.34	63.66	27.05
type	60.70	43.15	60.90	type	67.53	64.50	66.15

The table shows that sentence-level filtering only degrades performance, no matter which combination of methods is chosen. We assume that filtering sentences too drastically reduces the instance diversity for a particular relation. For example, the surface patterns may capture some instances of the pertaining relation, however, if only sentences matching those patterns are included as training data, then many other (representative) instances of that class are excluded. Moreover, the supervised classifier may finally end up learning only the knowledge that is encoded in the surface patterns and nothing beyond it.

4.3 Performance of Feature-Level Processing

Table 10 shows the impact of adding pattern and type information as features. It also displays the impact of argument-level and sentence-level filtering and contrasts it with the performance of different feature sets. For all classifiers under *feature level*, we use the best respective configuration from argument-level and sentence-level filtering.

Table 10. F-scores (macro-average) of different processing levels including feature level; *: significantly better than *sentence-level best* at $p < 0.05$ (paired t-test)

	argument level		sentence level	feature level		
	random	best	best	+pattern	+type	all
SuitsTo	41.78	60.40	60.90	60.58	61.81*	**61.89***
SubstitutedBy	61.75	67.00	67.53	67.78	70.37*	**70.50***

The table shows that on feature level, type information is beneficial while patterns are not. This is completely opposite to argument-level filtering. This finding suggests that different types of knowledge sources have varying impact depending on the processing level on which they are applied.

4.4   Comparing Distant Supervision with Other Classifiers

Table 11 compares distant supervision with a majority classifier (always predicting the majority class), rule-based classification and, as an upper bound, a supervised classifier applying 10-fold cross-validation on our gold standard dataset. Like DS_{best}, the supervised classifier is trained on the best feature set (*all* from Table 10). Even though for distant supervision we increased extraction performance notably by employing argument filters and appropriate feature engineering, for *SuitsTo*, the rule-based classifier exploiting type information produces virtually the same performance as the best distantly supervised classifier.

Table 11. Comparison of majority classifier, distant supervision, rule-based classifier and supervised classifier (upper bound). We report macro-average precision, recall and F-score.

		Major	DS_{random}	DS_{best}	$RB_{pattern}$	RB_{type}	RB_{comb}	super
SuitsTo	Acc	60.00	60.84	**68.05**	45.49	68.03	44.73	77.58
	Prec	30.00	53.74	**67.45**	55.40	66.91	59.03	81.32
	Rec	50.00	50.59	62.18	53.13	**62.69**	53.66	72.49
	F	37.50	41.78	61.89	42.60	**62.65**	40.00	73.48
SubstitutedBy	Acc	91.00	86.39	86.86	91.47	79.28	**91.56**	92.54
	Prec	45.50	61.28	67.29	77.72	60.55	**81.72**	78.87
	Rec	50.00	62.33	**77.59**	60.28	72.13	57.99	77.31
	F	47.64	61.75	**70.51**	63.97	62.02	61.26	77.77

As already indicated in Sect. 3.4, for the relation *SuitsTo* a precision-oriented classifier is most important. Indeed, the best performing classifiers (i.e. DS_{best} and RB_{type}) have a higher precision than the remaining classifiers.

For *SubstitutedBy*, the situation is different. Here, the distantly supervised classifier outperforms all rule-based classifiers. Obviously, for this relation, which represents a minority class, various kinds of information are necessary in order to produce a good classifier. The two basic rule-based classifiers, i.e. $RB_{pattern}$

and RB_{type}, both produce similar F-scores (even though based on different recall/precision levels). From that, we conclude that contextual information *and* types are equally informative. We pointed out in Sect. 3.4 that we assume that for *SubstitutedBy*, recall also plays a significant role. Indeed, the classifier with the highest performance, i.e. DS_{best}, exceeds all other classifiers in terms of recall.

In summary, a distantly supervised classifier can produce reasonable performance, however, in some cases, much simpler classifiers, such as RB_{type} for *SuitsTo*, may produce competitive results. Moreover, the best distantly supervised classifier still performs notably worse than a fully supervised upper bound. This suggests that there is still room for improving distantly supervised classification.

5 Related Work

The food domain has recently received some attention in the NLP community. Different types of classification have been explored including ontology mapping [11], part-whole relations [12], recipe attributes [13], dish detection and the categorization of food types according to the Food Guide Pyramid [6]. Relation extraction tasks have also been examined. While a strong focus is on food-health relations [14,15], relations relevant to customer advice have also been addressed [1,3,6]. Beyond that, sentiment information was related to food prices with the help of a large corpus consisting of restaurant menus and reviews [16]. Moreover, actionable recipe refinements have been extracted [17]. To the best of our knowledge, we present the first work to investigate the usefulness of distant supervision for relation extraction in the food domain.

6 Conclusion

We examined relation extraction in the food domain by employing distant supervision. We focused on the extraction of two relations that are not only relevant to product recommendation in the food domain, but that also have significance in other domains, such as the fashion or electronics domains. We examined various degrees of freedom in order to select suitable training data. We considered three processing levels being argument level, sentence level and feature level. While argument-level filtering and feature-level processing help to increase classification performance, sentence-level filtering turned out to be not effective. As external resources, we examined manually created surface patterns and semantic types on all these levels. Their effectiveness varies depending on the processing level onto which they are applied. Patterns are effective for argument-level filtering while types can be harnessed on the feature level. We showed that a careful selection of training data and appropriate feature design substantially improves classification performance of distantly supervised classifiers, however, in some cases, similar performance can also be achieved by much simpler rule-based methods.

Acknowledgements. Michael Wiegand was funded by the German Federal Ministry of Education and Research (BMBF) under grant no. 01IC12SO1X. The authors would like to thank Stephanie Köser for annotating the newly created dataset presented in this paper.

References

1. Wiegand, M., Roth, B., Klakow, D.: Data-driven Knowledge Extraction for the Food Domain. In: Proc. of KONVENS (2012)
2. Mintz, M., Bills, S., Snow, R., Jurafsky, D.: Distant Supervision for Relation Extraction without Labeled Data. In: Proc. of ACL/IJCNLP (2009)
3. Wiegand, M., Roth, B., Klakow, D.: Web-based Relation Extraction for the Food Domain. In: Bouma, G., Ittoo, A., Métais, E., Wortmann, H. (eds.) NLDB 2012. LNCS, vol. 7337, pp. 222–227. Springer, Heidelberg (2012)
4. Landis, J.R., Koch, G.G.: The Measurement of Observer Agreement for Categorical Data. Biometrics 33(1), 159–174 (1977)
5. Wiegand, M., Roth, B., Lasarcyk, E., Köser, S., Klakow, D.: A Gold Standard for Relation Extraction in the Food Domain. In: Proc. of LREC (2012)
6. Wiegand, M., Roth, B., Klakow, D.: Automatic Food Categorization from Large Unlabeled Corpora and Its Impact on Relation Extraction. In: Proc. of EACL (2014)
7. U.S. Department of Agriculture, H.N.I.S.: The Food Guide Pyramid. Home and Garden Bulletin 252, Washington, D.C., USA (1992)
8. Wu, Z., Palmer, M.: Verbs semantics and lexcial selection. In: Proc. of ACL (1994)
9. Hamp, B., Feldweg, H.: GermaNet - a Lexical-Semantic Net for German. In: Proc. of ACL Workshop Automatic Information Extraction and Building of Lexical Semantic Resources for NLP Applications (1997)
10. Miller, G., Beckwith, R., Fellbaum, C., Gross, D., Miller, K.: Introduction to WordNet: An On-line Lexical Database. International Journal of Lexicography 3, 235–244 (1990)
11. van Hage, W.R., Katrenko, S., Schreiber, G.: A Method to Combine Linguistic Ontology-Mapping Techniques. In: Gil, Y., Motta, E., Benjamins, V.R., Musen, M.A. (eds.) ISWC 2005. LNCS, vol. 3729, pp. 732–744. Springer, Heidelberg (2005)
12. van Hage, W.R., Kolb, H., Schreiber, G.: A Method for Learning Part-Whole Relations. In: Cruz, I., Decker, S., Allemang, D., Preist, C., Schwabe, D., Mika, P., Uschold, M., Aroyo, L.M. (eds.) ISWC 2006. LNCS, vol. 4273, pp. 723–735. Springer, Heidelberg (2006)
13. Druck, G.: Recipe Attribute Detection Using Review Text as Supervision. In: Proc. of the IJCAI-Workshop on Cooking with Computers (2013)
14. Miao, Q., Zhang, S., Zhang, B., Meng, Y., Yu, H.: Extracting and Visualizing Semantic Relationships from Chinese Biomedical Text. In: Proc. of PACLIC (2012)
15. Kang, J.S., Kuznetsova, P., Luca, M., Choi, Y.: Where Not to Eat? Improving Public Policy by Predicting Hygiene Inspections Using Online Reviews. In: Proc. of EMNLP (2013)
16. Chahuneau, V., Gimpel, K., Routledge, B.R., Scherlis, L., Smith, N.A.: Word Salad: Relating Food Prices and Descriptions. In: Proc. of EMNLP/CoNLL (2012)
17. Druck, G., Pang, B.: Spice it up? Mining Refinements to Online Instructions from User Generated Content. In: Proc. of ACL (2012)

Semantic Clustering of Relations between Named Entities

Wei Wang[1], Romaric Besançon[1], Olivier Ferret[1], and Brigitte Grau[2]

[1] CEA, LIST, Vision and Content Engineering Laboratory
91191 Gif-sur-Yvette Cedex, France
`wei.wang@lip6.fr`, {`romaric.besancon,olivier.ferret`}`@cea.fr`
[2] LIMSI, UPR-3251 CNRS-DR4, Bat. 508,
BP 133, 91403 Orsay Cedex, France,
`brigitte.grau@limsi.fr`

Abstract. Most research in Information Extraction concentrates on the extraction of relations from texts but less work has been done about their organization after their extraction. We present in this article a multi-level clustering method to group semantically equivalent relations: a first step groups relation instances with similar expressions to form clusters with high precision; a second step groups these initial clusters into larger semantic clusters using more complex semantic similarities. Experiments demonstrate that our multi-level clustering not only improves the scalability of the method but also improves clustering results by exploiting redundancy in each initial cluster.

Keywords: unsupervised information extraction, relation extraction, clustering.

1 Introduction

Unsupervised Information Extraction (UIE) differs from standard Information Extraction (IE) approaches by opening IE systems to unknown information structures. Such approaches allow to discover non-predefined relations between entities [12], which helps handling the heterogeneous relation types found in open-domain [2,9] and proves useful in application contexts such as strategic or competitive intelligence. A very light form of supervision can also be taken into account in such approaches by enabling users to delimit a topical context, such as in the *On-Demand Information Extraction* paradigm [23].

Most of the work in UIE is dedicated to the extraction of relations and less to their organization. Based on a statistical classifier (e.g., TEXTRUNNER [2]), on bootstrapping (e.g., OLLIE [16]) or on patterns and rules (e.g., REVERB [9] or [1,11]), such systems concentrate on guaranteeing the validity of each extraction rather than on the organization of relations. However, structuring the extracted relations is important both to characterize the type of relations and facilitate the access to information for the end-users.

In [14], a clustering of relations is performed but misses a more semantic dimension in relation similarity and fails to group synonyms or paraphrases. A

A. Przepiórkowski and M. Ogrodniczuk (Eds.): PolTAL 2014, LNAI 8686, pp. 358–370, 2014.

more semantic similarity is proposed by [7] but is evaluated only on a small corpus. Large-scale semantic clustering of extracted relations is still a challenge, even if some approaches have been proposed such as [18] or [19]. However, [18] applies semantic criteria mainly for finding equivalent entities, whereas we focus on similar relation expressions, and [19] exploits the specific context of Wikipedia.

We present in this article an efficient and effective multi-level clustering procedure that succeeds in grouping relations that are expressed either by similar expressions or synonymous phrases and can be applied at a large scale. We focus only on relations between named entities because, in an applicative context of strategic or competitive intelligence, relations of interest are mostly oriented by named entities. Experiments demonstrate that our multi-level clustering not only improves the scalability of the method, but also improves the clustering results by exploiting redundancy in each initial cluster.

2   Related Work

In UIE approaches based on clustering [12,22,23], clustering methods play a dual role since they provide a cluster structure to relation instances at the same time these instances are extracted. In [12], each cluster is likely to contain semantic variations of the same relation, including synonyms, since it results from the merging of sets of relation instances based on the co-occurrence of the same pair of named entities. [23] creates an off-line base of paraphrases, relying on shared named entities to align sentences from multiple newspapers reporting the same event: relation patterns linking the same pair of entities are placed in the same pattern set. In general, these clustering methods are designed to detect reliable relation patterns while we are interested in a method that focuses specifically on finding synonymous patterns.

[14] proposes a method for extracting high-level relations and concepts from the relations of TEXTRUNNER through a co-clustering method based on Markov Logic that simultaneously generates classes of arguments and classes of relations. However, as it doesn't exploit any lexical semantic resources, it generally fails to group synonyms or paraphrases. In the same way, [4] performs co-clustering but applies it on a dual representation of relations, either as entity pairs (extension) or as lexical patterns (intension). This co-clustering relies on a matrix of co-occurrence between entity pairs and lexical patterns and requires, to be effective, a good connectivity in the entity relation graph. Similarly, the effectiveness of the inference procedure of generative models in UIE [21,27] relies on the connectivity of the whole entity relation graph, which is not very scalable. Our multi-level clustering approach addresses this issue by limiting the use of semantic similarity measures to small sets of relation instances. [18] also clusters both the relations and their arguments but adopts a less integrated approach and relies on lexical semantic resources built from corpora. Its clustering method globally takes advantage of redundant information from a first clustering so that more equivalent entities can be detected. However, our objective is more focused on the detection of equivalent expressions of relations.

Concerning semantic similarities, [7] exploits lexical information from Word-Net but its evaluation is done on a small corpus whereas we target large-scale approaches. Moreover, it only exploits verbs, whereas we want to include nouns as well, and we rely on an initial clustering step that provides a more robust base to our semantic clustering.

3   A Multi-level Clustering

We propose a multi-level clustering procedure for relation organization that groups the instances of relations extracted from a large corpus into clusters by relying on their semantic similarity.

Each relation instance is extracted from the co-occurrence of two named entities in a sentence and then filtered according to the two-step method defined in [24]: a filtering based on simple heuristics (such as a threshold on the number of words between the entities) is first applied to throw away a large number of incorrect relations with a good precision, followed by a second, more fine-grained, filtering based on a statistical classifier. Relation instances are then characterized by a pair of named entities and a linguistic form, composed of the normalized part of the sentence between these entities, called *Cmid*.

In unsupervised IE tasks, the number of extracted relation instances can be very large, which makes the direct search for semantic similarities among these instances too costly. At the same time, we need to deal with the diversity of these instances both in terms of types and forms of expression. We also observed that this diversity can be decomposed into several levels and more particularly, that a part of the extracted relation instances can be described by the same keyword, with slight variations, as illustrated in Table 1. In this table, each line refers to the same relation, expressed with different linguistic forms.

Table 1. Examples of variations of the linguistic form of relations

Category	Relations, grouped by form, with normalized words	
ORG − ORG	create the, who create, ...	establish the, who establish the, ...
ORG − LOC	base in, a company base in, ...	locate in, which be locate in, ...
ORG − PER	a group found by, which be found by, ...	
PER − ORG	who be the head of, become head of, ...	

The large-scale constraint, the variability of the expressions of the relations and this observation motivate our multi-level approach: by first clustering relations with very similar linguistic expressions (such as *create the* and *who create*), we can efficiently group the syntactic paraphrases of a relation into small but precise initial clusters. Then, a second level of clustering takes into account more complex semantic similarities to further group the initial clusters into larger semantic clusters. The cost of using semantic criteria is limited by the fact that the initial clusters are far less numerous than the extracted relations. Both initial clustering and semantic clustering are applied within each relation category, characterized by the type of the named entities linked by the relation.

3.1   Initial Clustering

As stated above, the goal of the initial clustering is to split the large set of extracted relations into groups of similar relations with only slight syntactic differences. To implement this kind of similarity between relations, we used the standard *Cosine* similarity on a bag-of-word representation of the *Cmid* part of the relation, which is an interesting option due to its efficiency. Moreover, this calculation was made faster by the application of the *All Pairs Similarity Search* (APSS) algorithm [3], which builds the similarity matrix of relations very efficiently by exploiting a fixed similarity threshold, given as a parameter, to avoid the computation of all pairwise similarities.

All the words in the linguistic expression of a relation do not have the same importance, which we characterize in our framework by a weight. More precisely, we experimented three methods for weighting the words of relations.

Binary. All words appearing in *Cmid* are given the same weight (w=1.0).

tf-idf. Words are weighted by the *tf-idf* score. *idf* corresponds in our case to the inverse relation frequency and measures the specificity of a word among the extracted relations.

POS. Specific weights are given to words according to their part-of-speech (POS) category. An analysis of POS categories led us to divide them into four classes (plus a default class with weight w=0.5) according to the importance of their contribution to the semantic expression of a relation:

Direct (w=1.0)	words directly linked to the meaning of a relation, including verbs, nouns, adjectives, prepositions and particles;
Indirect (w=0.75)	words that are not directly linked to the meaning of the relation but are characteristic of a form of its expression, such as adverbs and pronouns;
Complement (w=0.5)	words that provide complementary information in the relation, such as proper nouns and interjections;
Noise (w=0.0)	words, such as symbols, numbers, determiners, coordinating conjunctions or modal words, that are not considered as relevant to the expression of the relation.

This initial clustering procedure groups most similar expressions into the same cluster in a precise way. Nevertheless, some relation instances are missed because the general weighting schemes do not always give high weights to the most significant words of the linguistic form of a relation. Furthermore, we observed that most of the relation instances are characterized by either a verb (e.g., *founded* for *a group founded by, which is founded by*) or a noun (e.g., *head* for *who is the head of, becomes head of*). In an initial cluster, this characterizing keyword has generally a much higher frequency than the other words. Hence, following [12], we consider the most frequent word (verb or noun) of an initial cluster as its *label* and use it to add a refinement step to the initial clustering in which the initial clusters that share the same label are merged to form bigger initial clusters.

3.2   Semantic Clustering

The second clustering step aims at grouping the initial clusters according to a
more semantic similarity in order to gather equivalent relations expressed differently, such as *based in* and *which is located in*. This clustering relies on cluster-to-
cluster comparisons, which actually implies to consider three levels of semantic
similarity. The first level is the targeted similarity between the initial clusters to
merge. This similarity relies on the similarity between relations, which are the
basic elements of clusters, which itself relies on a semantic similarity between
words because they are the basic elements of the linguistic form of the relations.
We describe how these three levels of semantic similarity are implemented in the
increasing order of their granularity (words, relations, clusters).

Word-Level Similarity. Semantic similarity measures between words are usually separated into two categories: measures based on manually-built resources
such as WordNet and distributional measures, based on corpus-based data. We
compared the two types of measures for our task of semantic clustering.

WordNet-based measures. Numerous types of measures were proposed to compute similarities between the synsets of WordNet by relying on their hierarchy.
We considered two measures that are complementary and representative of different families of measures: on the one hand, the measure from Wu and Palmer [26]
(Sim_{wup}), which takes into account both the depth of the two synsets to compare
in the WordNet hierarchy and the depth of their least common subsumer; on the
other hand, the measure from Lin [15] (Sim_{lin}), which also includes statistical
information about synsets (Information Content) derived from a corpus.

  These similarities are defined between synsets, each of which may contain
several words. In the same way, each word may be included in different synsets.
A simple way of mapping synset similarity to word similarity is to choose the
highest synset similarity among all possible synset pairs [17].

Distributional measures. Distributional measures are based on the distributional
hypothesis that words occurring in the same context tend to have similar meanings. Practically, given a large corpus, a set of co-occurrents, either extracted
from a fixed size window or from syntactic dependency relations, is collected for
each word to form its *context vector* and the semantic similarity of two words
is evaluated by computing a standard bag-of-word similarity measure between
their contexts, such as *Cosine, Jaccard* or *Dice* [13].

Relation-Level Similarity. The evaluation of the semantic similarity of two
relations is related to the problem of paraphrase recognition. More precisely, as
each relation is represented by its *Cmid* part, the considered problem can be seen
as the evaluation of the similarity of two phrases P_a and P_b, represented as bag-
of-words. A simple way of computing the similarity of two phrases or sentences
is to take the average of the word-level similarities between all possible word

pairs. However, all word pairings do not have the same relevance, especially for two words that are not important in the expression of the relation. [17] proposed to match each word in one phrase only with the most similar word in the other phrase and to only take into account these most similar matches, as illustrated by the following example, where *part* is only paired with *stake*:

$$P_a = \{W_i\} \qquad \textbf{ORG} \text{ acquire a part of } \textbf{ORG}$$
$$0.8 \quad 0.55 \quad 0.93$$
$$P_b = \{W_j\} \qquad \textbf{ORG} \text{ buy a minority stake in } \textbf{ORG}$$

The similarity is then given by the following equation:

$$S_1(P_a, P_b) = \frac{1}{\sum_{W_i \in P_a} w_i} \sum_{W_i \in P_a} \max_{W_j \in P_b} \{S_{W_{i,j}}\} \cdot w_i \qquad (1)$$

where w_i is the weight given to the word W_i in P_a and $S_{W_{i,j}}$ is the similarity of words W_i and W_j computed following the various options of word-level similarity.

With this definition, the measure is not symmetric ($S_1(P_a, P_b) \neq S_1(P_b, P_a)$): W_i in P_a being the most similar word with W_j in P_b does not guarantee that W_j is the most similar word in P_b for W_i. Therefore, the average of similarities in both directions is taken to make this measure symmetric, defined by:

$$S(P_a, P_b) = \frac{1}{2}\left(S_1(P_a, P_b) + S_1(P_b, P_a)\right) \qquad (2)$$

Cluster-Level Similarity. Each initial cluster contains two or more relation instances. A complete-linkage or average-linkage between clusters is very costly since it requires to compute the similarities between all relation pairs from the clusters. On the other hand, choosing only one relation instance randomly as a representative of an initial cluster is not a reliable procedure, even with the high precision of each cluster. Moreover, the definition of an average linguistic representation for a cluster is not always obvious and may result in an important loss of information, especially when this information was collected without known expectations.

The proposed solution is to merge the bag-of-word representations of all the relation instances of an initial cluster to form a general bag-of-words for this initial cluster $C = \{W_i : f_i\}$, where each word is associated with its frequency in the cluster. The hypothesis is that the most relevant words with respect to the relation appear more frequently in the cluster and should be given a higher weight. The frequency of words in initial clusters is considered as representative of the information redundancy in these clusters. Therefore, the same similarity as the relation-level similarity (Equation 1) can be adopted. However, this weighting scheme faces a frequency bias problem. Let two clusters $C_a = \{\text{found}:3, \text{actor}:3\}$

(*an actor who found*) and C_b = {study:9, actor:1} (*study at, an actor who study at*), which are not semantically similar. However, the similarity from C_a to C_b is high because the shared word *actor* has a high frequency in the first cluster. Even though the inverse similarity (from C_b to C_a) is low, the average similarity is strongly influenced by the first one and has a relatively high level. To solve this frequency bias problem, the frequencies of matching words in both clusters are taken into account for the computation of similarity in each direction. This leads to replace, in Equation 1, the weight w_i by the weight w_{ij}, defined as: $w_{ij} = f_i \cdot f_j$.

The Choice of Clustering Algorithms. The performance of clustering algorithms largely depends on the nature of the considered data and the specific constraints of the targeted tasks. In unsupervised IE tasks, the clustering algorithms must have the capacity to process large data sets and to deal with the unpredictable number of clusters (due to the heterogeneity of open-domain relations). Hierarchical clustering algorithms are often too costly for large data sets and other standard methods such as K-means require a predefined number of clusters. In this study, we considered Markov Clustering (MCL) [6] and Shared Nearest Neighbors clustering (SNN) [8], which are both efficient and do not require a predefined number of clusters. The MCL algorithm generally requires a pruning threshold to ignore all unnecessary values in the similarity matrix for efficiency and noise filtering (which can also be an advantage from a computational point of view since it can be combined with APSS, presented in Sect. 3.1). The SNN algorithm can be very efficient, even without a pruning threshold on similarity, but it is highly parameterized, and most importantly, the number of nearest neighbors to consider is not obvious to determine in all cases. MCL was chosen for the initial clustering step because specifying a similarity threshold corresponds intuitively to fix the proportion of common words between two phrases, whereas the sizes of clusters can be very diverse in an open-domain context. On the contrary, SNN was chosen for the semantic clustering step because the number of neighbors to consider, which is a central parameter of this method, refers to some extent to the average number of synonymous words or paraphrases, which is more stable than the values of the semantic similarity.[1]

4   Experiments and Evaluations

4.1   Evaluation Measures and Dataset

For the evaluation of all our clustering results, we used measures both at the level of relations and clusters. At the relation level, the *precision* (*prec.*) and *recall* measures were applied on pairs of relation instances, considering that the relations can be positively or negatively grouped into the same cluster or separated in different clusters. At the cluster level, the evaluation was performed

---

[1] These choices have also been verified practically by experiments: SNN results are much worse than MCL results for initial clustering and better for semantic clustering.

with the standard *purity, inverse purity* (*inv. purity*) and *Normalized Mutual Information* (NMI) measures. Our experiments were performed on the 159,400 documents of the *New York Times* part of the AQUAINT-2 corpus but the evaluation focused on a set of 4,420 relations extracted from this corpus and manually grouped into 80 clusters [25]. These relations are divided into the six categories of Table 4 according to the types of the named entities in relations. It is important to note that, unlike other evaluations such as in [18] or [19], these reference clusters were built *a priori* and are not the result of the judgment of automatically built clusters. Hence, they represent a less biased form of reference than the usual ones in the field.

4.2   Initial Clustering Experiments

The similarity threshold used to prune the matrix similarity (using the APSS algorithm) was set to 0.45 for the binary weighting MCL (*i.e.* the association of MCL with a binary weighting of words in relation instances). This threshold was set empirically for covering 3/4 of the similarity values between similar sentences based on observations from the Microsoft Research Paraphrase Corpus [5]. The validity of this threshold was confirmed in practice since it outperforms other tested values, ranging from 0.35 to 0.60. The same threshold was adopted for the tf-idf weighting MCL algorithm whereas, considering the looser constraints in the weighting with POS categorization, a more strict threshold was used (0.60) in this case. The results of the initial clustering using the different weighting strategies are presented in Table 2.

Table 2. Results of the initial clustering with different word weighting strategies

	Prec.	Recall	F-score	Purity	Inv. purity	NMI	#clusters	Size
binary	0.756	0.312	0.442	0.788	0.407	0.671	15,833	7.50
tf-idf	0.203	0.445	0.279	0.646	0.573	0.712	11,911	11.44
POS	0.810	0.402	0.537	0.867	0.513	0.739	13,648	7.56
Refinement	0.812	0.443	0.573	0.857	0.552	0.751	11,726	8.80

The MCL algorithm with the similarity measure based on POS weighting outperforms the other two weighting configurations, with a better precision and a relatively satisfying recall. This is understandable since this weighting strategy takes more knowledge into account to emphasize the importance of verbs, nouns, adjectives and prepositions, which carry the meaning of the relation, and gives less weight to words that contribute mainly to linguistic variations ("who" + verb, "the one that" + verb). This distinction enables the pruning threshold to be increased to improve precision without any big loss of recall.

On the other hand, the tf-idf weighting does not lead to good results. It tends to favor words that are rather rare in the corpus while verbs and nouns that are meaningful to relations are often rather frequent and thus, have a small weight. For instance, the verb "*buy*" is frequent in financial documents, which leads to a low *idf* value, but holds the key role in relation instances for the relation BUY(ORG,ORG). On the contrary, words such as proper nouns or specific numbers, which are not linked with the relation type, are not frequent and often obtain a higher weight. As a result, the td-idf weighting disturbs the clustering of relations by producing irrelevant similarities.

As a consequence, the results of the initial clustering kept for the semantic clustering step were the results obtained with the POS weighting strategy,[2] on which the refinement procedure was applied. Table 2 shows that this refinement step leads to a slight improvement of F-score, especially due to the increase of recall; but it is also important to note that this step succeeds in reducing the number of clusters and increasing their average size, as illustrated in the last two columns of Table 2.

4.3 Semantic Clustering Experiments

We evaluated the different methods for semantic clustering presented in Sect. 3.2 and compared them with a theoretical upper-bound result for this step (that we call "*ideal*" clustering), defined to be the best possible performance for semantic clustering, given the results of the initial clustering: each initial cluster is associated with the reference cluster that shares the largest number of relation instances with it; then, the initial clusters that are associated with the same reference cluster are grouped to form the new *ideal semantic clusters*.[3]

Evaluation of Semantic Similarities. The semantic clustering was applied on the initial clusters resulting from the POS weighting initial clustering with refinement. As stated previously, the important words characterizing a relation are generally verbs and nouns. For our semantic similarity, we have chosen to compare only words with the same part-of-speech, with the objective of grouping relation instances that are either mainly characterized by verbs, such as *found by* or *establish by*, or mainly characterized by nouns, such as *be partner of* or *have cooperation with*.

In practice, for WordNet-based measures, the Sim_{wup} similarity performs well for noun-noun comparisons while the Sim_{lin} similarity performs better for verb-verb comparisons. For nouns, we used the *Wup* similarity as implemented by the NLTK package (`nltk.org`) while for the *Lin* similarity, we exploited the pre-computed similarity pairs from [20]. For the distributional similarities, we used the distributional thesaurus built by [10], obtained using co-occurrences in a window of size 3 (which means only one nearest content word was considered on each side) on the whole AQUAINT-2

---

[2] Several different pruning thresholds and different weighting configuration for POS categories have been tested. The presented version (threshold 0.60 and POS weighting) is the one that gives the best results.

[3] Note that this ideal clustering is therefore performed only on the initial clusters that contain relations present in the 80 reference clusters.

corpus. We also tested a distributional similarity based on co-occurrences obtained from syntactic relations. For both types of distributional similarities, only the three most similar words were taken into account. For the SNN clustering algorithm, we limited the number of considered neighbors to 100. The results of the semantic clustering based on these different similarity measures are presented in Table 3.

Table 3. Results of semantic clustering with different similarity measures, compared to the upper-bound *Ideal* clustering

	Prec.	Recall	F-score	Purity	Inv. purity	NMI	#clusters	Size
WordNet	0.821	0.507	0.627	0.846	0.622	0.763	9,403	10.98
Window-based	0.814	0.540	0.649	0.836	0.634	0.764	10,161	10.16
Syntax-based	**0.831**	**0.549**	**0.661**	**0.853**	**0.645**	**0.770**	10,116	10.20
Ideal	0.861	0.701	0.773	0.867	0.770	0.797	13,468	7.66

The syntax-based distributional similarity achieves the best performance but is comparable to the performance achieved by the window-based distributional similarity. Both distributional similarities outperform WordNet-based similarities, which means that this method can be quite easily adapted to other languages since distributional resources are much easier to obtain than manually built lexical resources such as WordNet. Compared to initial clustering results, all semantic similarities succeed in improving both precision and recall.

Concerning the choice of the words on which the similarity is applied, we observed[4] that the performance using only verbs for semantic clustering is a bit inferior to the results using both nouns and verbs. Taking nouns into account especially improves the recall measure and increases the average cluster size. The integration of adjectives in the similarity computation was also experimented but showed a very limited influence on the final results. Moreover, we tested cross-category distributional similarities between verbs and nouns, with no obvious improvement in recall or precision.

Semantic Clusters Examples. Table 4 gives a qualitative view of the semantic clustering results by presenting some examples of semantic clusters formed using the syntax-based distributional similarity. Each word corresponds to the label of an initial cluster. These examples show that distinct words that are semantically similar, and even distant paraphrase forms such as *grab gold in* and *win the race at*, are grouped together. However, certain errors are still present, for instance the presence of *purchase* and *be purchased by* in the same cluster due to the lack of differentiation between passive and active forms by our current preprocessing.

---

[4] We do not present all the quantitative results due to lack of space.

Table 4. Semantic clustering results

Category	Semantic clusters
ORG – ORG	purchase, buy, acquire, trade, own, be purchased by
ORG – LOC	start in, inaugurate service to, open in, initiate flights to
ORG – PER	sign, hire, employ, interview, rehire, receive, affiliate
PER – ORG	take over, take control of
PER – LOC	grab gold in, win the race at, reign
PER – PER	win over, defeat, beat, topple, defend

4.4   The Effects of Multi-level Clustering

As discussed in Sect. 3.1, computing semantic similarities is much more time-consuming than simple *Cosine* similarities. The total number of relation instances reaches up to 165,708 while the number of initial clusters is only 11,726. Therefore, a first advantage of our multi-level clustering is to avoid the heavy calculation of semantic similarities on a huge set of relation instances.

A second advantage is that it exploits the redundancy of information in initial clusters to identify interesting elements and to improve the quality of the semantic clustering. To validate this hypothesis, we compared, based on our reference, the distribution of similarities between relation instances and the distribution of similarities between initial clusters. First, we examined all the similarities between two relation instances in the same reference cluster (intra-distribution D_{intra}) and all the similarities between two instances in different reference clusters (inter-distribution D_{inter}). Ideally, the two distributions D_{intra} and D_{inter} should be well separated, with a high average similarity for D_{intra} and a low average similarity for D_{inter}. Secondly, we associated each reference cluster with the set of initial clusters it covers[5] and we examined all the similarities between two initial clusters in the same reference cluster, which forms a new intra-distribution D'_{intra}, and all the similarities between two initial clusters in different reference clusters, which forms a new inter-distribution D'_{inter}.

These distributions are presented in Fig. 1 for the syntax-based distributional similarity (but similar results are obtained with all types of semantic similarity), with D_{intra}, D_{inter} on the left, and D'_{intra}, D'_{inter} on the right. It is clear from these graphs that the semantic clustering based on the initial clusters achieves a more stable performance since intra and inter-distributions are better separated and initial clusters in different reference clusters have a rather low similarity on average. This confirms our hypothesis that redundant information in initial clusters can be used to filter out the noise brought by irrelevant words.

---

[5] Since our initial clustering method tends to form small but precise clusters, each reference cluster is split into several small clusters.

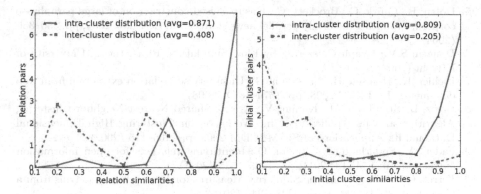

Fig. 1. Distribution of similarities between relations and between initial clusters

5   Conclusion and Perspectives

We have presented in this article a multi-level approach for clustering relation instances extracted in the framework of unsupervised information extraction. This method deals with both problems of scalability and linguistic diversity of relations by using two levels of clustering: a first level builds small clusters in an efficient and precise way while a second level, more semantic, relies on different semantic similarities between words including WordNet-based and distributional similarities. We demonstrate in our experiments the interest of this approach.

In future work, we consider expanding the semantic similarities to take more information into account and using deeper syntactic information about the linguistic expression of the relations in order to spot more precisely the interesting elements of the relations. We also started some experiments to combine the semantic clustering of the relations to a thematic clustering of the contexts in which they appear to give a more precise background for the relation definition.

Acknowledgments. This work was partly funded by European Union Seventh Framework Program FP7/2007–2013 under grant agreement n°FP7-SEC-2012-312651.

References

1. Akbik, A., Broß, J.: Extracting semantic relations from natural language text using dependency grammar patterns. In: SemSearch 2009 Workshop (2009)
2. Banko, M., Cafarella, M.J., Soderland, S., Broadhead, M., Etzioni, O.: Open information extraction from the web. In: IJCAI 2007, pp. 2670–2676 (2007)
3. Bayardo, R.J., Ma, Y., Srikant, R.: Scaling up all pairs similarity search. In: WWW 2007, pp. 131–140 (2007)
4. Bollegala, D.T., Matsuo, Y., Ishizuka, M.: Relational Duality: Unsupervised Extraction of. Semantic Relations between Entities on the Web. In: WWW 2010, pp. 151–160 (2010)

5. Dolan, B., Quirk, C., Brockett, C.: Unsupervised construction of large paraphrase corpora: exploiting massively parallel news sources. In: COLING 2004, pp. 350–356 (2004)
6. Dongen, S.V.: Graph Clustering by Flow Simulation. Ph.D. thesis, University of Utrecht (2000)
7. Eichler, K., Hemsen, H., Neumann, G.: Unsupervised relation extraction from web documents. In: LREC 2008, pp. 1674–1679 (2008)
8. Ertöz, L., Steinbach, M., Kumar, V.: A New Shared Nearest Neighbor Clustering Algorithm and its Applications. In: Workshop on Clustering High Dimensional Data and its Applications of SIAM ICDM 2002, pp. 105–115 (2002)
9. Fader, A., Soderland, S., Etzioni, O.: Identifying relations for open information extraction. In: EMNLP 2011, pp. 1535–1545 (2011)
10. Ferret, O.: Testing Semantic Similarity Measures for Extracting Synonyms from a Corpus. In: LREC 2010, pp. 3338–3343 (2010)
11. Gamallo, P., Garcia, M., Fernández-Lanza, S.: Dependency-Based Open Information Extraction. In: Joint Workshop on Unsupervised and Semi-Supervised Learning in NLP, pp. 10–18 (2012)
12. Hasegawa, T., Sekine, S., Grishman, R.: Discovering relations among named entities from large corpora. In: ACL 2004, pp. 415–422 (2004)
13. Heylen, K., Peirsmany, Y., Geeraerts, D., Speelman, D.: Modelling Word Similarity: An Evaluation of Automatic Synonymy Extraction Algorithms. In: LREC 2008, pp. 3243–3249 (2008)
14. Kok, S., Domingos, P.: Extracting Semantic Networks from Text Via Relational Clustering. In: ECML PKDD 2008, pp. 624–639 (2008)
15. Lin, D.: An Information-Theoretic Definition of Similarity. In: ICML 1998, pp. 296–304 (1998)
16. Mausam, S.M., Soderland, S., Bart, R., Etzioni, O.: Open language learning for information extraction. In: EMNLP-CoNLL 2012, pp. 523–534 (2012)
17. Mihalcea, R., Corley, C., Strapparava, C.: Corpus-based and knowledge-based measures of text semantic similarity. In: AAAI 2006, pp. 775–780 (2006)
18. Min, B., Shi, S., Grishman, R., Lin, C.Y.: Ensemble Semantics for Large-scale Unsupervised Relation Extraction. In: EMNLP 2012, pp. 1027–1037 (2012)
19. Moro, A., Navigli, R.: Integrating syntactic and semantic analysis into the open information extraction paradigm. In: IJCAI 2013, pp. 2148–2154 (2013)
20. Pedersen, T., Patwardhan, S., Michelizzi, J.: WordNet:Similarity Measuring the Relatedness of Concepts. In: HLT-NAACL 2004: Demonstrations, pp. 38–41 (2004)
21. Rink, B., Harabagiu, S.: A generative model for unsupervised discovery of relations and argument classes from clinical texts. In: EMNLP 2011, pp. 519–528 (2011)
22. Rozenfeld, B., Feldman, R.: High-Performance Unsupervised Relation Extraction from Large Corpora. In: ICDM 2006. pp. 1032–1037 (2006)
23. Sekine, S.: On-Demand Information Extraction. In: COLING-ACL 2006, pp. 731–738 (2006)
24. Wang, W., Besançon, R., Ferret, O., Grau, B.: Filtering and Clustering Relations for Unsupervised Information Extraction in Open Domain. In: CIKM 2011, pp. 1405–1414 (2011)
25. Wang, W., Besançon, R., Ferret, O., Grau, B.: Evaluation of unsupervised information extraction. In: LREC 2012, pp. 552–558 (2012)
26. Wu, Z., Palmer, M.: Verbs semantics and lexical selection. In: ACL 1994, pp. 133–138 (1994)
27. Yao, L., Haghighi, A., Riedel, S., McCallum, A.: Structured relation discovery using generative models. In: EMNLP 2011, pp. 1456–1466 (2011)

Automatic Prediction
of Future Business Conditions

Lucia Noce, Alessandro Zamberletti, Ignazio Gallo, Gabriele Piccoli,
and Joaquin Alfredo Rodriguez

University of Insubria,
Department of Theoretical and Applied Science,
Via Mazzini, 5, 21100 Varese, Italy
lucia.noce@uninsubria.it

Abstract. Predicting the future has been an aspiration of humans since
the beginning of time. Today, predicting both macro- and micro-economic
events is an important activity enabling better policy and the potential
for profits. In this work, we present a novel method for automatically
extracting forward-looking statement from a specific type of formal corporate documents called earning call transcripts. Our main objective
is that of improving an analyst's ability to accurately forecast future
events of economic relevance, over and above the predictive contribution of quantitative firm data that companies are required to produce.
By exploiting both Natural Language Processing and Machine Learning
techniques, our approach is stronger and more reliable than the ones
commonly used in literature and it is able to accurately classify forward-
looking statements without requiring any user interaction nor extensive
tuning.

Keywords: Natural Language Processing, Information Retrieval, forward-
looking statement, earning call.

1 Introduction

It is always tough to make predictions, especially about the future; so goes the fa-
mous joke, first appearing as a Danish proverb and brought to mainstream fame
when it was ascribed to Baseball manager Yogi Berra. However, although predictions are indeed difficult, we can gather substantial information about the future
by elaborating increasingly available sources of digital data, such as web pages,
newspapers, blogs, books, and the like. In areas where better forecasting can result in substantial profits, like stock trading, the monitoring and analysis of digital
data streams [1–3] is gaining momentum. Recent works in accounting and finance
demonstrate that both the quantitative data as well as the narrative/qualitative
information enclosed in corporate filings (e.g., 10-K) and earning call transcripts
can be successfully exploited to predict both short term firm-specific performance
and future macroeconomics fluctuations [1, 2, 4–6].

An earning call transcript is the verbatim textual record of a conference call
between the management of a public company, analysts, investors and/or the

A. Przepiórkowski and M. Ogrodniczuk (Eds.): PolTAL 2014, LNAI 8686, pp. 371–383, 2014.

media to discuss the financial results the firm achieved during a specific past reporting period (e.g., quarter, fiscal year). Thus, through these earning calls, firms provide a substantial amount of future-related information to investors, journalists, policy makers and the public at large. Moreover, earning call transcripts are formal and well-formed documents that have few of the typical NLP challenges offered by other forms of communication (e.g., blog posts, tweets, etc.) For these reasons, earning calls, and company filings more generally, have been the subject of inquiry by a rapidly growing literature seeking to extract forward-looking statements. The goal is to help business analysts and policy makers to more accurately forecast firm's future behavior and performance [7–10].

Formally, a *forward-looking statement* is defined as a short sentence that contains information likely to have, or reft to, a direct effect in the foreseeable future (e.g., plans, predictions, forecasting, expectations and intentions). Recent research shows that, by analyzing the information included in these future-looking statements it is possible to improve the analysts' ability to accurately forecast future earnings, over and above the predictive contribution of quantitative firm data and consensus analysts forecasts [11, 12].

Given its relative novelty, the accounting and finance literature approaches extraction of forward-looking statements in a basic manner. For example, the classification of a sentence as a forward-looking statement occurs when it contains at least one term from a human trusted custom dictionary [13]. As a consequence, the Computer Science community can contribute to formulate better and analyze these important documents using the instruments and the knowledge provided by the Information Extraction and Information Retrieval literature. The goal of our work is to provide a simple yet effective method for combining Natural Language Processing and Supervised Machine Learning models in order to automatically extract future-looking statements from generic earning call documents. Even though the proposed method does not require any user interaction nor extensive tuning, it achieves excellent results for a dataset comprising human tagged earning calls from different corporations and it sensibly overcomes other similar approaches. Both the source code and our dataset of sentences extracted from recent earning calls and manually tagged by experts will be available for download.[1]

In Sect. 2, we present related research work. In Sect. 3 and 4, we explain in detail the process pipeline involved in our proposed method and we describe the results of our extensive experimental analysis. Finally, in Sect. 5, we discuss the results underlining possible future points.

2   Related Works

The proposed task is deeply related to several areas, including: Natural Language Processing, Information Extraction, Information Retrieval, Data Mining, Accounting and Finance. Earning calls have always been an important information source for market analysts. However, their automated processing is a new

---

[1] http://artelab.dicom.uninsubria.it/

trend and a growing literature investigates whether conference calls can positively affect analysts' forecasts by helping in defining more precise indices [1, 2]. The outcome of those works prove that the exploitation of the information contained in conference calls can indeed increase analysts' ability to forecast earnings accurately, suggesting that companies release additional and significant business information during those periodical meetings.

Various and wide studies were conducted with the aim of extracting information from earning calls [6] and, recently, it has been proved that the most significant part of information that can be derived from those earning call documents can be obtained by analyzing the tones of small subset of sentences that mainly talk about future business plans: the so-called *forward-looking statements* [4, 5].

Many recent works extract forward-looking statements in a very basic manner, e.g. by classifying a sentence as forward-looking if it contains at least one term from a set of expert selected significant words [4, 5, 13, 14]. Since these kinds of approaches proved to be effective in detecting forward-looking statements from generic documents, we decided to evaluate the feasibility of a similar approach in which, however, the dictionary of terms manually identified by experts is replaced with an automatically learned one.

Surprisingly, if we move from the narrow task of extracting forward-looking statements from business documents to the more generic task of retrieving future related information from generic documents (a task known in literature as *future retrieval*), we observe that the techniques used to approach those two problems are much different, even though the basic goal is practically the same. It is indeed interesting to analyze whether the techniques commonly used in future retrieval can be successfully applied to the forward-looking sentence extraction task; for this reason, in the following paragraphs we will introduce some of the most important future retrieval works that can be found in literature.

Speaking of future retrieval works, the first who mentioned and approached this task was Baeza-Yates [15], in his work he presents the idea of extracting future temporal information from news and considers the time as a formal attribute that can be used for any information retrieval problem. Later, Jatowt *et al.* [16] proposed a query-dependent system that, exploiting future-related information in documents, supports users in detecting, summarizing and forming predictions of recurrent events related to a specific object. In another interesting work, Kawai *et al.* [17] considered automatic ways to extract future-related information from documents using Support Vector Machine (SVM) classifiers [18]. Similarly, Jatowt *et al.* [19] proposed a model-based clustering algorithm that effectively detects future events based on information extracted from a text corpus, and estimate their probabilities over time. Others works arose as extensions of the above: (i) Dias *et al.* [20] used six different learning paradigms to see through effects of temporal features upon clustering and classification of three future-related text classes; they obtained the overall best results using a SVM classifier, (ii) Matthews *et al.* [21] proposed *Time Explorer*, a system that allows users to search information through time. It is important to point out that another slight variant of future retrieval consists in understanding the flow of

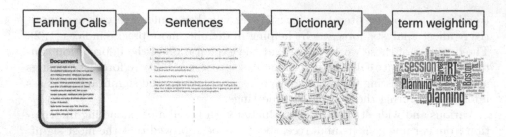

Fig. 1. A visual overview of the sequence of steps performed by the proposed model to extract forward-looking statements from earning call documents. From left to right, the starting earning call documents is split into sentences that are further split into words to build a dictionary, as in classic BOW approaches; each term in the dictionary is given a weight that is computed using 1, only the terms having the highest weights appear in the final dictionary that is used to build the feature vectors provided as input to the supervised classifier model.

people's sentiments along with temporal feature, this task is commonly called *Temporal Sentiment Analysis* [22].

Most of the above cited works adopt machine learning methods to extract future-related information from text documents, suggesting that it is indeed possible to effectively solve the problem of extracting forward-looking statements from earning call transcripts. We build on previous work but we strive to synergistically combine economic and financial studies with methods from Natural Language Processing, Information Extraction, Information Retrieval and Data Mining to provide the first automatic method of information extraction that can be used to inform economic predictions based on aggregate analysis of forward looking statements from the business community as vehicled by the earning call transcripts.

3   Proposed Method

In this work, the task of identifying and extracting forward-looking statements from earning call documents has been approached as a classical text classification problem, in which each sentence is assigned one or more labels from a finite set of predefined categories. More formally, given a finite non-empty set of documents $\mathcal{X} = \{x_0, \ldots, x_n\}$ and a set of categories $\Omega = \{\omega_1, \ldots, \omega_c\}$, the proposed task requires to assign to every pair $(x_i, y_j) \in \mathcal{X} \times \Omega$ a boolean label yes/no. Since it is difficult and ineffective for a supervised classifier to interpret natural text as is, each document $x_i \in \mathcal{X}$ is usually represented as a vector $\phi(x_i)$ in which each element measures the number of times that a given term in x_i, contained into a finite dictionary of terms $\mathcal{D}$, appears in x_i. More in detail, given a document $x_i \in \mathcal{X}$ and a dictionary of terms $\mathcal{D} = \{t_1, \ldots, t_d\}$, the document is represented as $\phi(x_i) = (x_i^1, \ldots, x_i^d)$, where each $x_i^j \in \phi(x_i)$ measures the frequency of occurrence of the term $t_j \in \mathcal{D}$ in the document $x_i \in \mathcal{X}$.

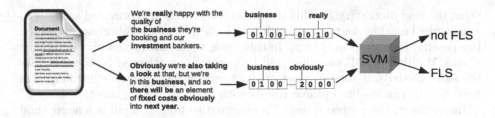

Fig. 2. An example showing how the feature vectors provided as input to the supervised classifier are generated. From left to right, the sentences extracted from the starting earning call documents are given a vectorial representation using the final weighted dictionary $\mathcal{D}$, as in Sect. 3; those feature vectors are finally provided as input to the SVM classifier that assigns them a label denoting wether they represent forward-looking statements or not.

The main objective of this work is to extract forward-looking statements from earning call transcripts, for this reason, each earning call is split into a finite set of sentences that were previously tagged by different experts as either forward-looking (FLS) or not (NFLS), for this reason we have that: $\Omega = \{\text{FLS}, \text{NFLS}\}$.

As summarized in Fig. 1, once the starting set of earning call documents has been split into the set of sentences $\mathcal{X}$, we proceed to further split those sentences into single terms and use them to build a primitive dictionary $\mathcal{D}'$, as in classic Bag of Words (BOW) approaches. Instead of representing each sentence as a vector using the whole dictionary $\mathcal{D}'$, we prune it to obtain the final dictionary $\mathcal{D}$ by assigning a weight to each of its element $t_i \in \mathcal{D}'$, using the weighting formula proposed by Peñas *et al.* [23], with the main goal of detecting a small subset of terms that are characteristics of forward-looking sentences. In detail, the relevance of a term $t_i \in \mathcal{D}'$ is computed as follows:

$$Relevance(t_i, sc, gc) = 1 - \frac{1}{\log_2 \left(2 + \frac{F_{t_i,sc} \cdot D_{t_i,sc}}{F_{t_i,gt}}\right)} \tag{1}$$

where: (i) sc is the *specific corpus*, it corresponds to the subset of sentences from $\mathcal{X}$, extracted from the starting set of earning call documents, that were tagged by experts as FLS, (ii) gc is the *generic corpus*, it is composed by the whole set of sentences $\mathcal{X}$, (iii) $F_{t_i,sc}$ is the relative frequency of the term $t_i \in \mathcal{D}'$ in the specific corpus sc, (iv) $F_{t,gc}$ is the relative frequency of the same term $t_i \in \mathcal{D}'$ in the generic corpus gc and (v) $D_{t,sc}$ is the relative number of documents of sc in which the term $t_i \in \mathcal{D}'$ appears.

Once the relevance of every term $t_i \in \mathcal{D}'$ has been computed using 1, it is possible to obtain the final smaller dictionary $\mathcal{D}$ simply by removing all those terms whose relevance value is lower than a threshold ψ, as follows:

$$\mathcal{D} = \{t_i \in \mathcal{D}' \mid Relevance(t_i, sc, gc) > \psi\} \tag{2}$$

As shown in Fig. 2, given a sentence $x_i \in \mathcal{X}$ extracted from an earning call document, we compute its vectorial representation $\phi(x_i)$ as previously described, using the final dictionary $\mathcal{D}$. This vectorial representation is provided as input to a supervised machine learning algorithm with the goal of assigning one among the two possible categories in Ω to x_i. In this work, we decided to employ a Support Vector Machine (SVM) classifier [18]; this choice is motivated by the fact that the same model was used in most of the previous related works introduced in Sect. 2, as it can lead to optimal results with minimal tuning effort.

In summary, the proposed model is expected to take as input a sentence and to return a label that represents whether that sentence is a forward-looking statement or not. In Sect. 4, we present the results of an extensive experimental analysis that was conducted to identify an efficient and effective way of dealing with the forward-looking sentence extraction task. The obtained results proved that the previously described pipeline is very effective in detecting forward-looking statements; in fact, it can substantially outperform other recent methods that use hand crafted dictionaries for the same set of earning call documents without requiring any user interaction.

4   Experiments

Here we present the results obtained by performing an extensive experimental analysis of the proposed forward-looking statements identification method: (i) in Sect. 4.1 we provide a detailed description of the dataset used in our experiments; (ii) in Sect. 4.2 we present the metrics used to evaluate the effectiveness of the proposed model; (iii) in Sect. 4.3 we compare the results achieved by our model with those obtained by other related approaches while also providing comparisons with other techniques for building the final dictionary of terms.

4.1   Dataset

In order to build a reliable dataset of forward-looking sentences, we downloaded the earning call documents provided by three leading firms: International Business Machines Corporation (IBM), Exxon Mobil Corp (XOM) and J.P. Morgan Chase & Co. (JPM). More in detail, for each of those firms, we picked the earning call documents for the third quarter of the years 2013, 2012 and 2011.

The earning call documents were split into sentences using a classic sentence detector; those sentences were then manually labeled by a team of experts in economic and finance to identify which ones represent possible forward-looking statements. In order to determine the degree of reliability of the dataset, each sentence has been processed multiple times by different experts. By doing so, we determined that the degree of inter-rater reliability among the experts is equal to 89.73%, this high rate of accordance suggests that most of the future-looking statements in our dataset share some basic characteristics that can be easily spot by experts; this proves that, if a significant set of features is identified, the task of identifying forward-looking statements can effectively be automatized using a properly trained supervised machine learning classifier.

We collected a total amount of 3148 tagged sentences. In order to train the supervised classifier we split the dataset into train and test sets following a classic

Table 1. Comparison between different parameter configurations for the classic Bag of Words (BOW) approach. The best results are obtained when considering single words (*ngram* = 1) without exploiting the position of the sentences within the earning call documents to which they belong (*context* = *no*)

Parameters		OA (%)	
ngram	context	prune	no-prune
1	✗	84.93	84.93
1	✓	84.57	84.57
2	✗	83.51	83.51
2	✓	81.38	81.38

$\frac{2}{3}$-train, $\frac{1}{3}$-test split rule; this provided us with a train and test sets containing 2092 and 1046 documents respectively. It is important to point out that the sentences from the three previously cited firms were evenly split among the train and test sets in order to make the dataset as heterogeneous as possible and independent from any firms' specific language style.

4.2 Metrics

In order to compare the results of the classifier with trusted human classification, we measured the performance achieved by the different configurations of the proposed model using the Overall Accuracy (OA) metric, defined as follows:

$$OA = \frac{TP + TN}{TP + TN + FP + FN} \tag{3}$$

where (i) TP and TN are the number of true positives and true negatives, they denote the number of sentences that were correctly classified by the model as FLS or NFLS respectively, (ii) FP and FN are the number of false positives and false negatives, they denote the number of sentences that were wrongly labeled by the model in respect to their ground-truth classes.

For our overall comparison table, in addition to OA, we also measure the performance of the different algorithms using three additional metrics: *Precision* (*p*), *Recall* (*r*) and *F-measure* (*f*).

Precision represents the probability that a randomly chosen sentence $x_i \in \mathcal{X}$ gets correctly tagged by the classifier as either FLS or NFLS. *Recall* represents the probability that a randomly chosen sentence $x_i \in \mathcal{X}$ that has to be tagged as either FLS or NFLS is identified by the classifier. They are defined as follows:

$$p = \frac{TP}{TP + FP} \tag{4}$$

$$r = \frac{TP}{TP + FN} \tag{5}$$

Table 2. Comparison between different parameter configurations for the Part of Speech (PoS) based approach. The best results can be obtained when using bigrams and defining the feature vectors using the positions in which the sentences appear within their respective earning call documents

Parameters		OA (%)	
ngram	context	prune	no-prune
1	✗	80.50	80.50
1	✓	80.14	79.96
2	✗	80.67	80.85
2	✓	81.03	81.03

where TP, FP and FN are defined as in 3. It should be noted that *Precision* and *Recall* are not meaningful taken in isolation. In poor systems, a high p value corresponds to a small r value, and viceversa. For this reason, we also take into account the harmonic average between *Precision* and *Recall*: the *F-measure*. It is defined as follows:

$$f = \frac{2 \cdot p \cdot r}{p + r} \tag{6}$$

4.3   Results

In this section, we prove that the proposed future-looking statements identification method is able to reach and overcome other state-of-the-art approaches for the given classification task. Many experiments were conducted on the dataset described in Sect. 4.1; for each of them we describe the pipeline used to carry out the experiment and provide a table in which we present the obtained results while varying the different possible parameter configurations.

In our first group of experiments we used a classic Bag of Words (BOW) approach; a dictionary composed by all the words from the sentences in the training set was built. The sentences from the train and test sets were then represented as BOW feature vectors and an RBF SVM [18] classifier model was trained using 5-fold cross validation. In order to find the best parameter configuration for the classifier, we ran a grid search in the following parameter space: $c \in [-5, 15]$ and $\gamma \in [-15, 3]$. In Table 1 we present the obtained results, while varying both the number of *ngram* taken into account during the building phase of the dictionary $\mathcal{D}$ and the context of a sentence into the original earning call documents. It is important to note that we did not take into account words whose length was lower than 3 characters. In detail, when using an *ngram* value of 1, we build $\mathcal{D}$ using the single words extracted from the sentences in $\mathcal{X}$. On the other hand, when using an *ngram* value of 2, we not only consider the single words but also all the possible sequences of two consecutives words inside each sentence. For example, given the following sentence: *"We plan more*

Table 3. Accuracies achieved by the proposed method while varying the value of the threshold parameter ψ. A significant drop in performance can be observed when moving the value of ψ from 0.3 to 0.4; the reason behind this behavior is that high values of ψ overprune the dictionary by removing terms that are highly discriminative for forward-looking statements.

threshold ψ	OA (%)
0.2	87.57
0.3	87.57
0.4	83.69
0.5	83.16

investements in the future", the first approach ($ngram = 1$) would lead to a dictionary $\mathcal{D}_{n=1}$ defined as follows: $\mathcal{D}_{n=1} = \{plan, more, investements, future\}$, while the second approach ($ngram = 2$) would generate the following dictionary of terms: $D_{n=2} = D_{n=1} \cup \{planmore, moreinvestements, investements future\}$.

For each experiment, we also measure the obtained results while varying the context considered during the building phase of the feature vectors; meaning that we exploit the position of the sentences within their starting earning call documents. More specifically, given a sentence $x_i \in \mathcal{X}$ in position r within the earning call EC to which it belongs, when *context* is equal to 1, we do not take into account the information related to the position of x_i in EC while building the feature vectors. On the other hand, when *context* is equal to 2, we build the feature vector for x_i by summing the contributes from the sentences in positions $r - 1$, r and $r + 1$ within EC.

Moreover, for the BOW approach, we also evaluate the classification performances of the model while varying the size of the dictionary of terms $\mathcal{D}$; in detail, the dictionary is pruned by removing all its entries having frequency values lower than the median computed for all the frequencies of the terms in $\mathcal{D}$. The obtained results for this first experiment are shown in Table 1, it is possible to observe that the best OA values are achieved when ignoring the context of the sentences while using a dictionary built considering only unigrams. It is interesting to observe that most of the terms in the non pruned dictionary are useless, since equal results are obtained when using its pruned version. The size of the final pruned dictionary D for the best configuration of this BOW approach is equal to 2077, while the non pruned variant contains 3460 terms.

In our second experiment, we pair the previously described BOW approach with a Part of Speech (PoS) tagger algorithm to build a dictionary of tags that is used to build the feature vectors. The number of possible tags assigned to each word by the PoS tagger is equal to 42. Results are shown in Table 2, it is possible to observe that the best accuracies are achieved when building the dictionary using both 2-grams and context information. Similarly to the previous experiment, the pruning procedure does not affect the overall performances but halve the size of the final dictionary (from 2193 to 1110 terms). The results obtained by the PoS based approach are lower that the ones achieved by the

Table 4. Comparison between the results achieved by the proposed method and other approaches for the dataset described in Sect. 4.1, measured using the metrics introduced in Sect. 4.2

method	OA (%)	p (%)	r (%)	f (%)
Bozanic *et al.* [13]	35.56	20.05	78.21	31.91
BOW	84.93	69.37	48.94	57.39
PoS	81.03	69.10	48.07	56.69
Proposed	87.57	74.82	53.23	62.21

previous BOW model; this suggests that the information produced by the POS tagger is not a characteristic feature of forward-looking sentences and this proves that the decisions of the experts of tagging a sentence as either FLS or NFLS are not based on the grammar of those phrases.

In the third experiment, we reproduced and tested the approach proposed by Bozanic *et al.* [13]. In order to do so, we labeled a sentence as FLS iff it contained at least one of the forward-looking terms from Table 5. Results are shown in Table 4; even though this method is quite fast if compared to the ones described in the previous experiments, it achieves poor results since the list of terms it uses is probably not discriminative enough to correctly classify sentences from documents that were never seen by the authors when they defined their dictionary of terms.

In our last experiment we evaluated the performance achieved by the proposed method, described in Sect. 3, while varying the value of the parameter ψ. As previously described, the latter denotes the threshold value at which the dictionary $\mathcal{D}$, weighted using the formula of 1, is pruned. Results are presented in Table 3, it is possible to observe that the OA significantly drops when moving the value of ψ from 0.3 to 0.4. This is not surprising since, by taking a close look to the dictionary $\mathcal{D}$, we observed that the increment of the threshold value removed some terms that were highly discriminative for forward-looking statements, suggesting that the Peñas et al. [23] weighting formula can probably be improved to obtain even better results. Nevertheless, as shown in the overall comparison Table 4, the proposed method can substantially outperform the others. Using all the sentences from our training set, the proposed model requires roughly 7 min to be trained on a CPU i7 with 4gb of RAM; the final dictionary contains 1035 highly discriminative forward-looking terms. This fast training time is motivated by the fact that, in our final pipeline, we only use unigrams and we do not consider the contextual information of each sentence. Experiments using contextual information and different *ngram* values were performed, they are not reported due to lack of space; they will be made available online along with the dataset of Sect. 4.1. The accuracies we obtained in all those experiments were lower that the ones reported in Table 3.

Table 5. Forward-looking terms from Bozanic *et al.* [13]

anticipate	continue	intends	ought
anticipated to be	continues	intent	plan
anticipates	could	intention	planning
are anticipated	estimate	intentions	plans
are anticipating	estimated to be	is anticipated	potential
are estimated	estimates	is anticipating	potentially
are estimating	expect	is estimated	predict
are expected	expectation	is estimating	predicted to be
are expecting	expected to be	is expected	predicts
are forecasted	expects	is expecting	project
are forecasting	forecast	is forecasted	projected to be
are intended	forecasted to be	is forecasting	projection
are intending	forecasts	is intended	projections
are predicted	forward	is intending	projects
are predicting	future	is predicted	schedule
are projected	goal	is predicting	scheduled
are projecting	goals	is projected	scheduled to be
are scheduled	guidance	is projecting	schedules
are scheduling	guide	is scheduled	see
are targeted	guides	is scheduling	sees
are targeting	guiding	is targeted	shall
belief	hope	is targeting	should
beliefs	hopes	may	target
believe	hoping	might	targeted to be
believes	intend	objective	targets
can	intended to be	objectives	will

5 Conclusion

The effectiveness of the proposed future-looking statements identification model has been proved by the results of our extensive experimental analysis. Even though our pipeline is very simple, it can overcome both standard methods based on human made dictionaries and other classic approaches. The choice of using the weighting function proposed by Peñas *et al.* [23] proved to be very effective in building a dictionary that well describes the most significant features of forward-looking sentences. The proposed work will be used to define more precise forecast indices to help business experts in better predicting the future fluctuations of the world economy. Moreover, the highly reliable expert tagged dataset that we have built can be used by other researchers to propose new forward-looking statements identification algorithms; this could significantly boost the interest in this narrow research field and lead to better predictions of the future business economy.

References

1. Bassemir, M., Novotny-Farkas, Z., Pachta, J.: The effect of conference calls on analysts' forecasts. European Accounting Review 22(1), 151–183 (2013)
2. Robert, M., Bowen, A.K.D., Matsumoto, D.A.: Do conference calls affect analysts' forecasts? The Accounting Review 77(2), 285–316 (2002)
3. Piccoli, G., Pigni, F.: Harvesting external data: The potential of digital data streams. MIS Quarterly Executive 12(1), 143–154 (2013)
4. Li, F.: The information content of forward-looking statements in corporate filings – A Naive Bayesian Machine Learning approach. Journal of Accounting Research 48(5), 1049–1102 (2010)
5. Li, F.: The determinants and information content of the forward-looking statements in corporate filings – a naive Bayesian Machine Learning approach. In: Proceedings of Financial Accounting and Reporting Section (2008)
6. Muslu, V., Radhakrishnan, S., Subramanyam, K.R., Lim, D.: Forward-looking disclosures in the MD&A and the financial information environment. Management Science (2014)
7. Beyer, A., Cohen, D.A., Lys, T.Z., Walther, B.R.: The financial reporting environment: Review of the recent literature. Journal of Accounting and Economics 50(2-3), 296–343 (2010)
8. Rogers, J.L., Buskirk, A.V.: Bundled forecasts in empirical accounting research. Journal of Accounting and Economics 55(1), 43–65 (2013)
9. Ball, R., Shivakumar, L.: How much new information is there in earnings? Journal of Accounting Research 46(5), 975–1016 (2008)
10. Baginski, S.P., Kimbrough, J.M.H., The, M.D.: The effect of legal environment on voluntary disclosure: Evidence from management earnings forecasts issued in U.S. and Canadian markets. The Accounting Review 77(1), 25–50 (2002)
11. Frankel, R., Mayew, W., Sun, Y.: Do pennies matter? Investor relations consequences of small negative earnings surprises. Review of Accounting Studies 15(1), 220–242 (2009)
12. Kimbrough, M.: The effect of conference calls on analyst and market underreaction to earnings announcements. The Accounting Review 80(1), 189–219 (2005)
13. Bozanic, Z., Roulstone, D.T., Van Buskirk, A.: Management earnings forecasts and forward-looking statements (2013)
14. Wasley, C.E., Wu, J.S.: Why do managers voluntarily issue cash flow forecasts? Journal of Accounting Research 44(2), 389–429 (2006)
15. Baeza-Yates, R.: Searching the future. In: Proceedings of the ACM SIGIR Workshop of Mathematical/Formal Methods in Information Retrieval. SIGIR (2005)
16. Jatowt, A., Kanazawa, K., Oyama, S., Tanaka, K.: Supporting analysis of future-related information in news archives and the web. In: Proceedings of the ACM/IEEE-CS Joint Conference on Digital Libraries, JCDL (2009)
17. Kawai, H., Jatowt, A., Tanaka, K., Kunieda, K., Yamada, K.: Chronoseeker: search engine for future and past events. In: Proceedings of the International Conference on Uniquitous Information Management and Communication, ICUIMC (2010)
18. Cortes, C., Vapnik, V.: Support-vector networks. Machine Learning 20(3), 273–297 (1995)
19. Jatowt, A., Au Yeung, C.M.: Extracting collective expectations about the future from large text collections. In: Proceedings of the ACM International Conference on Information and Knowledge Management, CIKM (2011)

20. Dias, G., Campos, R., Jorge, A.: Future retrieval: What does the future talk about? In: Proceedings of the ACM SIGIR Workshop on Enriching Information Retrieval, SIGIR (2011)

21. Matthews, M., Tolchinsky, P., Blanco, R., Atserias, J., Mika, P., Zaragoza, H.: Searching through time in the New York Times. In: Proceedings of the Symposium on Human-Computer Interaction and Information Retrieval, HCIR (2010)

22. Fukuhara, T., Nakagawa, H., Nishida, T.: Understanding sentiment of people from news articles: Temporal sentiment analysis of social events. In: Proceedings of the International Conference on Weblogs and Social Media, ICWSM (2007)

23. Penas, A., Verdejo, F., Gonzalo, J.: Corpus-based terminology extraction applied to information access. In: Proceedings of Corpus Linguistics, CL (2001)

Evaluation of IR Strategies for Polish

Mitra Akasereh[1], Piotr Malak[2], and Adam Pawłowski[3]

[1] Computer Sciences Department, University of Neuchatel, Neuchâtel, Switzerland
mitra.akasereh@unine.ch
[2] Institute of Information Sciences and Book Studies, Nicolaus Copernicus University, Poland
piomk@uni.torun.pl
[3] Institute of Information and Library Science, University of Wroclaw, Poland
apawlow@uni.wroc.pl

Abstract. The paper presents results and conclusions of an ad-hoc evaluation lab concerning information retrieval for Polish. A corpus of ca. million document descriptions of Polish Europeana resources was indexed and matched against a set of fifty test queries. Different pre-processing procedures as well as different indexing and term weighting approaches were used and evaluated. Efficiency of different IR models was compared. Finally human-based relevance assessment was provided for retrieved documents.

Keywords: Cultural Heritage IR, Polish IR, searching strategies for Polish, IR strategies evaluation.

1 Introduction

The Cranfield experiment initialized systematic research on Information Retrieval systems. Analyzing and improving weaknesses of the Cranfield experiment, the Text Retrieval Conference (TREC: http://trec.nist.gov/) delivered useful measures and measurement of this kind, as well as appropriate tools and data sets [2, 7, 19]. Those are:

- Large collections of documents on a variety of subjects, necessary to obtain reliable statistical assessment.
- Test topic sets. Used for testing efficiency of information retrieval over the collection. Topics should be relevant to the collection profile.
- Weighting algorithms. Used for delivering lists of matched documents rated by relevance to query.
- Relevance assessment tools. They include objective methods of testing whether matched documents are actually relevant or irrelevant to the topic.
- Evaluation procedures of the IR system efficiency.

Findings of TREC were used also within the Cultural Heritage in CLEF lab (*CHiC*) [3] during the Conference and Labs of the Evaluation Forum (formerly known as Cross-Language Evaluation Forum), *CLEF* (http://www.clef-initiative.eu/).

A. Przepiórkowski and M. Ogrodniczuk (Eds.): PolTAL 2014, LNAI 8686, pp. 384–391, 2014.
© Springer International Publishing Switzerland 2014

CHiC was a typical ad-hoc IR systems evaluation lab. Its goal was to establish systematic and large-scale evaluation of information systems designed to access cultural heritage objects in digital libraries. Using a data set provided by Europeana (*Europeana: think culture:* http://europeana.eu/) CHiC aimed at evaluating the performance of search engines [3, 15]. In 2013 CHiC launched three tasks: a multilingual one, an interactive one, and for the first time, a monolingual task for the Polish resources (referred to as – Polish task) [16]. The main objective of the Polish Task was to gain experimental knowledge of the efficiency of different IR techniques used for languages of relatively complex morphology. This task was a joint project of the University of Neuchâtel (CH), the University of Wroclaw (PL), and the Nicolaus Copernicus University (PL).

This paper is organized as follows. In Sect. 2 we present the text collection and the topic set. Section 3 describes the experiment setup and its execution. The presentation of the results and conclusions are given in Sect. 4 and 5, respectively. The paper is an extension of [10], and [15].

2   The Test Collection

2.1   The Documents

The task in question was carried out on a set of 1,093,705 text documents. Those were descriptions of Polish cultural heritage objects stored by Europeana. The documents in the collection vary in size (counted as number of tokens) but in average they consist of 35 terms. Each document, describing one of Europeana's objects includes meta-data in regard to different schema:

- Dublin Core (tags starting with dc: prefix).
- Qualified Dublin Core (tags starting with dcterms: prefix).
- Europeana Semantic Elements (tags with europeana: prefix).

From the various tags in the documents the following set was chosen for indexing procedures in our experiments [10, 15]: <dc:contributor>, <dc:creator>, <dc:description>, <dc:date>, <dc:language>, <dc:subject>, <dc:title>, <dc:type>, <dcterms:alternative>, <dcterms:created>, <europeana:language>, <europeana:type>, <europeana:uri>, <europeana:year>.

2.2   The Topics

For the evaluation purposes a set of fifty test topics was prepared. Those were search queries to be matched over the Polish collection. The set consisted of a mixture of topical and named-entity queries, based mainly on the original Europeana query logs but also concerning the 150th anniversary of the January Uprising. Topics are short and they tend to reflect the information needs (and its expression) of Europeana users. Additionally, English translation of full topic set was provided.

In general our topics consist of:

1. Chronological topics:
    (a) 8 topics with time frames given explicitly (18th or 19th century),
    (b) 8 topics concerning particular historical period, e.g. *Barok* (Baroque).
2. Named entities:
    (c) 12 topics with personal names, e.g. *genera Józef Bem* (general Josef Bem),
    (d) 6 topics with geographical names, e.g. *Kraków* (Cracow),
    (e) 5 topics concerning historical names, e.g. *Powstanie Styczniowe* (January Uprising).
3. General entities:
    (f) 5 topics concerning religion or beliefs, e.g. *diabe* (Devil),
    (g) 7 topics concerning social groups or functions, e.g. *robotnicy* (workers).

All the topics consist of 141 word tokens (in title part), which gives an average length of 2.82 token per topic.

3   Experiment Setup and Execution

Two types of submissions were accepted, automatic (using only original topics), and a semantically enriched. Each submission was called a run. The run was the result of indexing collection documents, and searching over them with topic set, all with the use of one approach [6].

3.1   Text Pre-processing

There were runs using original, unmodified texts, and some using pre-processing of documents and topics. One of the applied modifications was stop-words removal and another was a light stemming (mostly of nouns).

We have prepared frequency lists of words in collection documents in order to create an adequate stopwords list. As expected, the grammatical vocabulary was highest ranked, but there were also several proper names. For obvious reasons these were not considered as stopwords. Finally, from a set of 304 the most frequent terms we defined the list of 138 stopwords.

For the task a light stemmer, affecting mostly nouns, was prepared and applied. Since earlier experiments had proven high efficiency of light stemming for Czech (a language of similar morphology as Polish) [5, 18], we wanted to examine the influence of stemming on the efficiency of information retrieval in Polish texts. There are stemmers available for Polish (*Stempel*: www.getopt.org/stempel/ or *Morfologik*: https://github.com/morfologik/morfologik-stemming), but due to some implementation problems, we decided to create our own, with the hope of improving stemming efficiency. A purely algorithmic stemming approach was applied, which involved automatic cutting of word endings from the words of given length. During the process of relevance assessment it became evident that stems of 4 or 5 characters of length were very polysemic: such short stems represented numerous classes of words and led to numerous false positive matches during the retrieval process.

3.2   The Relevance Assessment

For each run, the 100 top ranked answers for each topic were submitted to the analysis. Pooling procedure was provided by the DIRECT system (Distributed Information Retrieval Evaluation Campaign Tool, University of Padua: http://direct.dei.unipd.it/) generating a set of 32,144 documents for the relevance assessment. The process was executed by two specialists. A three grade relevance scale was applied: "fully relevant" – documents fully answering the information need expressed in the query; "partially relevant" – documents with some level of relevance but not fully satisfying information needs; "not relevant" – documents not related to the topic. For the final evaluation only items marked as "fully relevant" were accepted, while partially relevant items were ignored. On average there were 170 fully relevant documents per topic (min. 5, 562 max.). Using a lenient approach (accepting also partially relevant documents) there were 265 relevant documents per topic on average (the minimum was 22, the maximum 562).

As the main evaluation measure we used the *Mean Average Precision* (MAP). Another used measure was *Precision at cut-off n*, for *n*=10 (P@10), reflecting one screen of results in Europeana web interface.

3.3   Runs

For a comparison baseline an automatic run without stemming but with stopwords removal was carried out. For this run the *Okapi* was applied, with a classical *tf.idf* (with cosine normalization) [11]. This run was then used as a baseline for both automatic and manual runs.

Table 1. Evaluation of automatic runs for *Polish Task*

N	Run	Parameter Setting	MAP	P@10
1	Torun_Auto	tf.idf, stopwords rem., light stem., Boolean	**0.348**	
2	UniNE_Fusion	data fusion[1] (Okapi: light stem., trunc-5)	0.343	0.614
3	UniNE_DFR	DFR-I(n_e)B2, light stem., stopwords rem.	0.331	0.568
4	UniNE_PRF	data fusion, PRF (Rocchio, 5 docs, 10 terms)	0.258 †	0.494
5	UniNE_Baseline	*tf.idf* (cosine), no stemming, stopwords rem.	0.257 †	0.492
6	UniNEGramPRF	data fusion, 5-gram, PRF	0.220 †	0.472
	Baseline run	Okapi, no stemming, stopwords removing	0.314	0.520

---

[1] For data fusion a Z-score scheme was used.

Automatic Runs. In total there were six automatic runs submitted. Table 1 presents their summary, describing parameter settings, achieved *MAP* and *P@10* value. A paired *t-test* (two-sided, α=5%) was applied in order to detect if there were any significant differences (demoted by "†") [16].

Two groups (four types) of indexing strategies were used for automatic runs. The first one was using the n-gram [12] and trunc-n (truncating a word by keeping its first n characters and removing the remaining letters.with n = 4, 5, and 6) strategies, which tend to be a good alternative to stemming when facing a language without a good available stemmer. As a second indexing strategy, whole words were stored in indexes files, with applying a light stemmer in some of the runs. For most runs we used one of the probabilistic models: the Okapi (BM25) [17] or the DFR-I(ne)B2 based on the Divergence From Randomness (DFR) paradigm [1]. But *tf.idf* was also tested.

Additionally a Boolean conjunction of matched items was applied for one run. This approach did limit the false positives, and led to better relevance of retrieved documents. For example for topic #24: *Fryderyk Szopen* (Frederick Chopin) we achieved the best individual precision of all topics in all runs. An AP for this topic was 0.9959. The Boolean approach gave better results for longer topics (two or more searching terms), while for single word topics the Okapi (BM25) model dealt better.

Manually Enriched. The manual, semantic enrichment of topics tended to emulate a different level of user experience. As Europeana provides specific contents, cultural heritage, Torun enrichments aimed at two groups of users: educated (in terms of education level at least college graduate), and specialists (in terms of knowledge of additional information sources, historical contexts, etc.). Wrocław manually enriched topics reflected the following groups of users: students of information science, professionals experienced in information retrieval, and humanities-oriented educated persons.

For educated users simulation synonyms of topic terms were mostly used. Specialists' enrichment was supported by the use of encyclopedias. Statistically, original topic titles consisted of 141 tokens (2.82 token per topic on average), while educated enrichment resulted in 303 tokens (6.1 per topic av.), and specialists in 489 tokens in total (av. 9.78/topic). As mostly broad terms had been added, there were no improvements in general, in comparison to the baseline run.

For these runs the Okapi (BM25) model was used as the term weighting algorithm and stopwords removal during pre-processing. Table 2 presents achieved relevance measures for manually enriched runs.

All semantically enriched runs performed worse, than the baseline, automatic run. Two reasons of such poor performance of manual enrichment can be defined. The first is too extensive coverage of searching terms over the collection documents. The second one is using specific vocabulary in enrichment. Part of such vocabulary was not present in collection at all, and part indicated documents irrelevant to topic.

Table 2. MAP, and P@10 values for manually enriched Polish Task runs

Name	Parameter Setting	MAP	% of change in MAP	P@10
PLTO1EduLS	Educated, light stemmer	**0.2774**	−11.66%	0.454
PLTO1EduNO	Educated, no stemmer	0.2724	−13.25%	**0.460**
PLTO2HighLS	High, light stemmer	0.2709	−14.33%	0.528
PLTO2HighNO	High, no stemmer	0.2690	−13.73%	0.528
PLWR2Exp	Experts (no stemming)	0.1795 †	−42.83%	0.378
PLWR1Edu	Educated (no stemming)	0.1529 †	−51.31%	0.350
PLWR3Stu	Students (no stemming)	0.1279 †	−59.27%	0.268
Base Line	Basic (no stemming)	**0.3140**	n/a	**0.552**

4   Results and Discussions

Data presented in Table 3 show that the trunc-n indexing strategy tends to produce better retrieval effectiveness than n-gram. In Polish language there are seven grammatical cases, three genders, and two numbers with appropriate functional suffixes. This could be one of the reasons that the truncation performs well for this language. In this case the best value of the parameter n is 6 (larger than for languages such as English or French).

Table 3. MAP of runs based on n-gram or trunc-n approaches

| | DFR-I(n_e)B2 | | Okapi | |
Parameter	n-gram	Trunc-n	n-gram	Trunc-n
$n = 4$	0.2350	0.2268	0.2466	0.2532
$n = 5$	0.2610	0.2968	0.2577	0.3038
$n = 6$	0.2611	**0.3078**	0.2640	**0.3211**

The performance measure shown in Table 4 for different IR models indicates that the Okapi probabilistic model proposes the best performance in comparison to the other two models. Moreover, the use stopwords removal clearly improved the overall effectiveness in comparison to the baseline run (for 2.6% for the Okapi model and 5% for DFR-I(ne)B2 model). The improvement in MAP achieved by applying our stemmer is of 9.3% for the Okapi and 9.2% for the DFR-I(ne)B2. Improvement of MAP value achieved this way was not as good as we expected. For example, for Czech language, applying a stemming stage improves the retrieval effectiveness of about 40% [4]. For a language with a complex morphology such as Polish, a simple algorithmic stemmer does not provide the expected improvement [8]; this is mainly due to numerous exceptions or spelling irregularities.

Table 4. MAP of runs based on word-based indexing (Polish task)

IR Model	Original texts	Stopwords removal no stemming	Stopwords removal with stemming
tf.idf	0.2558	0.2566	0.2579
Okapi	0.3060	0.3140	**0.3433**
DFR-$I(n_e)B2$	0.2883	0.3028	0.3308

However worse than automatic runs, in terms of relevance, manually semantic enriched runs achieved few much better results (in Average Precision) than the baseline run. For example topic #29: *Warszawa w 19 wieku w sztuce* (Warsaw in 19 century in art) for PLTO1EduNo had improved relevance by +34630% (MAP = 0.3463 to MAP = 0.001). For this topic the following enriching terms were added: *architektura* (architecture), *dzielnica* (district).

5   Conclusions

Results lead us to the conclusion that a unigram indexing strategy can be improved (in terms of relevance and precision) by applying Boolean conjunction of searching terms after matching is found. For example, one of the topics concerned personal names *Jaros aw Kaczyński* and *Lech Kaczyński*. However, there is a Polish town called *Jaros aw*, and many municipal documents are available in Europeana. Thus for the topic #031: Jarosław or Lech Kaczyński, there were numerous false positive retrievals, for most of the retrieved documents considered the town, not the person. The matching was made for the first name and the last name separately. There were 3,318 documents pertinent to *Lech* term, 1,049 pertinent to *Kaczyński* name, and 9,253 to both *Jaros aw* a name and a town. Using classical tf.idf or Okapi approaches this topic was of the worst relevance ratio – for the 731 assessed documents only 16 items (2%) were considered relevant or partially relevant.

Another conclusion regarding personal names became evident during the relevance assessment process. For any person the last name should be weighted higher than the first name during the ranking process. Another suggestion is to try to set collocating names for a name given in query.

Neither vector-space, nor probabilistic models can impose relevant retrieval of all keywords from the query. Using a semi-Boolean approach (at least as logical conjunction of query terms) seems to be a better strategy.

With the Polish language, we show that a stemming stage enhances the final retrieval effectiveness. However, we cannot specify whether a more aggressive stemmer (affecting also verbs) or a statistical one may further enhance the performance [9, 13, 14].

Acknowledgements. This research was supported in part by the Sciex-NMS under grant *POL 11.219 Information Retrieval and Text Categorization for Polish*.

References

1. Amati, G., van Rijsbergen, C.J.: Probabilistic Models of Information Retrieval Based on Measuring the Divergence from Randomness. ACM Transactions on Information Systems 20, 357–389 (2002)
2. Buckley, C., Voorhees, E.M.: Retrieval Systems Evaluation. In: Vorhees, H. (ed.) TREC: Experiment and Evaluation in Information Retrieval (Digital Libraries and Electronic Publishing), pp. 53–75. MIT Press, Cambridge (2005)
3. CHiC: Cultural Heritage in CLEF (2013), http://www.promise-noe.eu/chic-2013/home (access date: March 25, 2014) (retrieved)
4. Dolamic, L., Savoy, J.: Indexing and Stemming Approaches for the Czech Language. Information Processing & Management 45, 714–720 (2009)
5. Fautsch, C., Savoy, J.: Algorithmic Stemmers or Morphological Analysis: An Evaluation. JASIST 60, 1616–1624 (2009)
6. Guidelines for participation and submission, http://www.promise-noe.eu/chic-2013/guidelines-for-participation-and-submission (access date: March 25, 2014) (retrieved)
7. Harman, D.K.: The TREC Test Collections. In: Vorhees, H. (ed.) TREC: Experiment and Evaluation in Information Retrieval (Digital Libraries and Electronic Publishing), pp. 21–52. MIT Press, Cambridge (2005)
8. Korenius, T., Laurikkala, J., Järvelin, K., Juhola, M.: Stemming and Lemmatization in the Clustering of Finnish Text Documents. In: Proc. of the ACM-CIKM, pp. 625–633 (2004)
9. Majumder, P., Mitra, M., Parui, S.K., Kole, G.: YASS: Yet Another Suffix Stripper. ACM-Transactions on Information Systems 25, Article #18 (2007)
10. Malak, P.: The Polish Task within Cultural Heritage in CLEF (CHiC) 2013. Torun Runs. In: Forner, P., Navigli, R., Tufis, D. (eds.) CLEF 2013 Evaluation Labs and Workshop Working Notes, Valencia, Spain, September 23-26 (2013), http://www.clef-initiative.eu/documents/71612/b00f7561-fadb-47a8-ab67-74f116ce062a (access date: March 25, 2014) (retrieved)
11. Manning, C.D., Raghavan, P., Schütze, H.: Introduction to Information Retrieval. Cambridge University Press (2008)
12. McNamee, P., Mayfield, J.: Character n-gram Tokenization for European Language Text Retrieval. IR Journal 7, 73–97 (2004)
13. Paik, J.H., Mitra, M., Parui, S.K., Jarvelin, K.: GRAS: An Effective and Efficient Stemming Algorithm for Information Retrieval. ACM-Transactions on Information Systems 29, Article #19 (2011)
14. Paik, J.H., Parui, S.K., Pal, D., Robertson, S.E.: Effective and Robust Query Biased Stemming. ACM-Transactions on Information Systems 31 (2013)
15. Petras, V., et al.: Cultural Heritage in CLEF (CHiC) 2013. In: Forner, P., Müller, H., Paredes, R., Rosso, P., Stein, B., et al. (eds.) CLEF 2013. LNCS, vol. 8138, pp. 192–211. Springer, Heidelberg (2013)
16. Pl Task Unine: Polish Track at CLEF (2013), http://members.unine.ch/jacques.savoy/Polish/ (access date: March 25, 2014) (retrieved)
17. Robertson, S.E., Walker, S., Beaulieu, M.: Experimentation as a Way of Life: Okapi at TREC. Information Processing & Management 36, 95–108 (2000)
18. Savoy, J.: Light Stemming Approaches for the French, Portuguese, German and Hungarian Languages. In: Proceedings ACM-SAC, pp. 1031–1035. The ACM Press (2006)
19. Voorhees, E.M., Harman, D.K.: The Text REtrieval Conference. In: Vorhees, H. (ed.) REC: Experiment and Evaluation in Information Retrieval (Digital Libraries and Electronic Publishing), pp. 3–20. MIT Press, Cambridge (2005)

Word, Syllable and Phoneme Based Metrics Do Not Correlate with Human Performance in ASR-Mediated Tasks

Anne H. Schneider[1], Johannes Hellrich[2], and Saturnino Luz[3]

[1] School of Computing Science, University of Aberdeen, Scotland
a.schneider@abdn.ac.uk
[2] JULIE Lab, Friedrich Schiller University Jena, Germany
johannes.hellrich@uni-jena.de
[3] School of Computer Science and Statistics, Trinity College Dublin, Ireland
luzs@scss.tcd.ie

Abstract. Automatic evaluation metrics should correlate with human judgement. We collected sixteen ASR mediated dialogues using a map task scenario. The material was assessed extrinsically (i.e. in context) through measures like time to task completion and intrinsically (i.e. out of context) using the word error rate and several variants thereof, which are based on smaller units. Extrinsic and intrinsic results did not correlate, neither for word error rate nor for metrics based on characters, syllables or phonemes.

1 Introduction

A good evaluation metric should be able to distinguish between systems of similar performance even if small variations are considered [4]. As language technology improves, however, questions arise as to whether current metrics are fine-grained enough to capture system performance. This can be observed not only in the field of machine translation and the search for alternatives to BLEU [3] but also in automatic speech recognition (ASR), where the word error rate (WER) is usually employed to determine system performance.

We collected a corpus of ASR mediated dialogues using a map task scenario.[1] Based on this material, interaction data such as dialogue turns and time to task completion were logged and the WER was calculated along with several more fine-grained, yet easy to implement error rates, i.e. character error rate (CER) and two new simple evaluation metrics, syllable error rate (SylER) and sound error rate (SER). Assuming that a good ASR performance would be reflected in the logging data (i.e. if the ASR quality is good, exhibiting low error rates, then the number of dialogue turns should be low) we compared the interaction data with the error rate results. We discovered, however, that the logging data is uncorrelated with the results of the four automatic evaluation metrics (WER, CER, SylER and SER).

---

[1] The collected corpus is available from the author on request, for replication of the results presented here or further independent analysis.

A. Przepiórkowski and M. Ogrodniczuk (Eds.): PolTAL 2014, LNAI 8686, pp. 392–399, 2014.
© Springer International Publishing Switzerland 2014

2   Related Work

The word error rate is commonly used to assess performance of ASR systems. WER is the ratio of the Levenshtein distance between the hypothesis into its aligned reference to the total number of words in the reference. While WER is an objective and direct metric, it has been criticised for measuring the cost of recreating the reference given the system's output rather than the proportion of information actually communicated [11]. Alternatives have been proposed which weigh information-carrying words differently [5] or use metrics such as precision, recall and F-scores [8]. However, these proposals have not gained wide acceptance. Research on spoken language understanding (SLU) has shown that an improved accuracy in speech recognition is not always a good indicator for a better SLU accuracy [19]. There have also been various efforts to identify factors that determine a successful performance of spoken dialogue systems (SDS) and to fit them into a useful framework to facilitate SDS evaluation [18,9,10]. One aim of this research, however, is to avoid using metrics which strongly depend on external resources or manual annotations, such as the concept error rate [7].

In a map task experiment sixteen ASR mediated dialogues were collected as basis for our data collection of intrinsic and extrinsic measures. The map task was introduced in the eighties by Anderson et al. [1]. With the emergence of new forms of communication (e.g. online chatting) it was applied widely with varying input and output modalities. In an explorative experiment to elicit human error recovery strategies in computer mediated communication automatic speech recognition was used as input [16], similar to the experiment described in this paper

3   Experimental Setting

In the experiment described in this paper the instructor had to dictate directions to an ASR system and the recognised text was sent to the follower. The sent ASR utterances were protocoled. The directions given by the instructor and the follower's responses were recorded and transcribed.

Participants: Sixteen participants performed the map task in groups of two. Participants, who performed the task together, knew each other before the experiment. Their ages ranged from 25 to 42 (avg 29, stdev 4.5). The twelve male and four female participants were native German speakers and had little or no experience with ASR systems. None of them had taken part in a map task experiment before.

Material: We designed four pairs of instructor and follower maps with eight to twelve landmarks, labelled in German. The instructor maps show a route with a designated starting point and goal. Fig. 1 shows an example of an instructor map. In the follower maps the route and the goal have been left out.

Procedure: First each participant executed a standard initial model training with the DRAGON dictation software [12]. Further training mechanisms for the speech model were switched off to enable comparability between subjects.

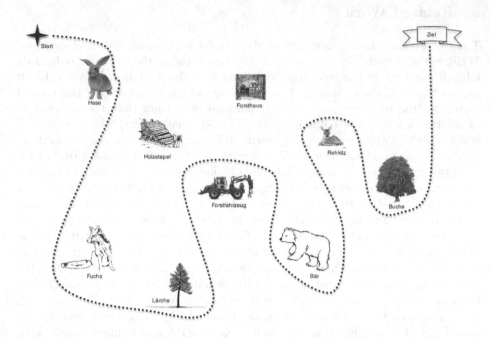

Fig. 1. Example of instructor map used in the study

After a general introduction, each pair of participants was separated in two different rooms with a computer and the respective maps (A4 printouts on paper) in front of them. SKYPE was used to mediate the communication between subjects. Furthermore, the experiment was audio recorded with an audio channel for each participant. DRAGON was used to input the instructor's directions into the SKYPE chat field. The instructor could see what was recognised but could not make any corrections to the recognized text with the keyboard. By hitting the return button utterances could be sent to the follower, who received the messages on screen and was allowed to talk freely. The instructor could hear what the follower said. After the follower reached the goal, roles were swapped and the experiment was repeated with another map. In order to avoid map specific bias and order effects, the arrangement of the maps were equally shuffled.

4   Data Analysis

We retrieved different measures from the logging data such as time to task completion and the number of words uttered. Furthermore, the WER was calculated on basis of the reference transcriptions and the protocoled ASR output for each of the 16 participants acting as instructor. An overview of the WER results and two extrinsic measures (i.e. time to task completion and a measure for task success, that determines the performance on drawing the route in percent) is presented in Table 2. We calculated the correlation between the WER results and the extrinsic measure, hypothesising that a bad word error rate would be reflected in low extrinsic measures. However, we were unable to identify significant correlations between the data sets. For example, the Pearson correlation coefficient [14] between WER and the time to task completion is $r = 0.04$ ($p = 0.88$) and between WER and task success it is $r = -0.15$ ($p = 0.57$).

Table 1. Selection of frequent ASR mistakes

reference word	ASR output
Compound mistakes	
geschlossenen Koffer	Geschlossenenkoffer
Richtung Saloon	Richtungssalon
Richtung Katze	Richtungskatze
Richtung Kaktus	Richtungskaktus
Decomposition mistakes	
Indianerzelt	Indianer Zelt
Linksbogen	links Bogen
Vorderbeinen	vorder Beinen
Other	
Forstfahrzeug	vors Fahrzeug
nordöstlich	nordwestlich
Start	Staat

In order to understand why this is the case a deeper inspection of the transcripts of the reference and the ASR output was performed, focussing on the error types that could be found in the ASR output. A selection of the most frequent mistakes can be found in Table 1. Looking at the most frequent mistakes it can be observed that compounding mistakes occur regularly. Compounds are words that contain more than one stem (e.g. *darkroom or bittersweet*), and are frequent in the German language. In the experiment an instructor told his follower to walk into the direction of the cat (*'Richtung Katze'*), the ASR system, however, compounded *direction* and *cat* to a new word (*Richtungskatze*), which is not a meaningful word. Another very frequent type of error was the decomposition of German compounds into their stems. For example the German word for tepee (*Indianerzelt*), which can be literally translated to *'indian's tent'* was always split into the two words (*Indianer, Zelt* for *indian* and *tent*). Furthermore, misrecognition often resulted in words that only differ by a small number of characters from the original uttered word (e.g. *Nordlicht* and *nördlich*). While these mistakes have a considerable impact on WER, their effect on our other metrics (see Sect. 5 below) seem to be less pronounced. Their influence on human comprehension might be negligible. Hence, we assume that such mistakes are not captured by the extrinsic measures.

Table 2. Overview of intrinsic and extrinsic evaluation results of the ASR study. CER and SER have variants respecting word boundaries (CER, SER) and variants ignoring those (CER2, SER2). Time to task completion (time) in seconds and task success (task) in percent.

| | | | intrinsic | | | | extrinsic | |
	WER	SylER	CER	CER2	SER	SER2	Time	Task
01	0.22	0.16	0.09	0.09	0.20	0.06	417	100.0
02	0.22	0.19	0.12	0.13	0.20	0.11	379	11.1
03	0.25	0.18	0.06	0.05	0.24	0.03	1085	36.8
04	0.36	0.30	0.17	0.18	0.34	0.16	747	66.7
05	0.22	0.21	0.12	0.13	0.21	0.12	318	100.0
06	0.22	0.17	0.10	0.10	0.18	0.08	825	66.7
07	0.39	0.29	0.17	0.18	0.36	0.14	1651	63.2
08	0.16	0.12	0.07	0.07	0.15	0.05	592	55.6
09	0.23	0.17	0.08	0.09	0.20	0.07	1453	100.0
10	0.12	0.08	0.04	0.04	0.10	0.03	1308	61.9
11	0.34	0.21	0.10	0.10	0.33	0.07	550	96.4
12	0.18	0.13	0.06	0.07	0.16	0.05	566	85.7
13	0.20	0.16	0.09	0.10	0.18	0.08	695	92.9
14	0.17	0.15	0.09	0.09	0.16	0.08	413	61.9
15	0.19	0.16	0.08	0.08	0.17	0.07	432	67.9
16	0.48	0.46	0.39	0.39	0.47	0.34	326	38.1

5   Alternatives to WER

These observations suggest that a measure for ASR quality based on words might not be fine-grained enough. We considered further error rates to assess the ASR accuracy intrinsically which are based on smaller units than a word. Therefore, we implemented a variation based on WER that calculates the number of miss-recognised syllables divided by the number of syllables in the reference, the *syllable error rate* (SylER). The implementation segments the reference transcriptions and ASR output texts into syllables, using the HyFo library [6]. We treated the syllables of the segmented words as a continuous stream, i.e. we ignored word segmentation. SylER was intended as a linguistically plausible upgrade to WER, motivated by the human ability to use contextual clues to process and correct mistakes in language processing.

The *character error rate* (CER) is often used for languages like Japanese and Chinese, which do not have word boundaries. CER differs from WER by counting misrecognised characters instead of words. To assess the influence of segmentation errors such as compound or decomposition mistakes, we implemented two variants of CER; one scoring whitespace errors (CER) and one ignoring them (CER2). The latter version results in lower error rates and we believed it to be a better model for human language processing since humans are still able to understand texts that lack segmentation. CER and CER2 are especially resilient

Table 3. Overview of the PEARSON COEFFICIENT for the correlation of the intrinsic evaluation methods (SylER, CER, CER2, SER, and SER2) with time to task completion (time) and task success (task). P-values in brackets.

	SylER		CER		CER2		SER		SER2	
Time	−0.09	(0.72)	−0.18	(0.49)	−0.20	(0.45)	0.02	(0.94)	−0.19	(0.43)
Task	−0.26	(0.33)	−0.30	(0.24)	−0.28	(0.28)	−0.17	(0.53)	−0.27	(0.26)
WER	0.95	(< 0.01)	0.85	(< 0.01)	0.84	(< 0.01)	0.99	(< 0.01)	0.86	(< 0.01)

against miss-recognized compounds, e.g. for a reference text containing only *Indianerzelt* recognised as *Indianer Zelt* they are $\frac{1}{14}$ or 0, respectively, whereas the WER in this case is 2.

Another approach would be to match homophone units, which show minor differences in spelling. We utilized a German version of the SOUNDEX algorithm, which was initially developed to index names by sound, by encoding homophones to the same representation [17], the so called *Kölner Phonetik* [13,2]. Our implementation of the *sound error rate* (SER) is based on an encoding scheme, which maps letter combinations representing similar sounds onto the same numerical code, mostly ignoring vowels and repeated letters. Again we implemented two variants, one comparing the codes for whole words (SER), and one treating the encoded input as a continuous string of numbers, calculating the edit distance for swapped numbers (SER2). Similar to CER2, we believe SER2 better matches human comprehension, since word segmentation is not essential for comprehension. Table 2 shows a summary of CER, SylER, SER and SER2 results.

6 Discussion and Conclusion

From our results we see that WER is not a good indicator of time to task completion or route drawing accuracy. The resulting five sets of new error rate results (i.e. SylER, CER, CER2, SER and SER2) have been used to compare the correlation with the extrinsic measures, however, we could not show a significantly higher correlation than WER had with the extrinsic results. Table 3 shows an overview for Pearson correlation coefficient and p-values between the new intrinsic error rates and WER, time to task completion, and task success. In summary we can say that the extrinsic measures used in this experiment do not correlate well with the intrinsic evaluation and that our new intrinsic error rates have a very high correlation with each other and with WER.

The human ability to process languages is still unmatched by technology. Human conversation is full of errors, ill-formed sentences, false starts and hesitations. These complicate matters for NLP components, however, if humans are faced with them they are well able to filter out the relevant content, sometimes even without recognising the mistakes. The way WER is calculated does not take into account the ability of a human listener to compensate for small mistakes. This possibly explains the lack of correlation between WER and task performance observed in our experiment. However, more forgiving metrics based on phonemes and syllables do not seem to correlate better with human performance in the map task, despite being more robust towards other distortions such as miss-recognised compounds. A finer-grained intrinsic metric thus remains elusive. The low correlation between these intrinsic metrics and time to task completion and accuracy indicates that evaluations that employ such metrics need to be complemented with user trials. This is specially true of evaluation of systems that combine several NLP components, since ASR errors can result in unpredictable errors if used as input to a machine translation module [15].

Acknowledgements. This research is supported by the Science Foundation Ireland (Grant 07/CE/I1142) as part of the Centre for Next Generation Localisation (www.cngl.ie) at Trinity College Dublin.

References

1. Anderson, A., Brown, G., Shillcock, R., Yule, G.: Teaching Talk: Strategies for Production and Assessment. Cambridge University Press (1984)
2. Apache: ColognePhonetic (2013), http://commons.apache.org/codec/apidocs/org/apache/commons/codec/language/ColognePhonetic.html
3. Callison-Burch, C., Osborne, M., Koehn, P.: Re-evaluating the Role of BLEU in Machine Translation Research. In: Proceedings of EACL 2006 (2006)
4. Doddington, G.: Automatic Evaluation of Machine Translation Quality Using N-gram Co-Occurrence Statistics. In: Proceedings of HLT 2002 (2002)
5. Garofolo, J., Voorhees, E., Auzanne, C., Stanford, V., Lund, B.: 1998 TREC-7 Spoken Document Retrieval Track Overview and Results. In: Proceedings of TREC, vol. 7 (1998)
6. HyFo: HyFo library (2013), http://defoe.sourceforge.net/hyfo/hyfo.html (accessed January 2013)
7. Kawahara, T.: New perspectives on spoken language understanding: Does machine need to fully understand speech? In: ASRU 2009. IEEE Workshop on Automatic Speech Recognition & Understanding (2009)
8. McCowan, I., Moore, D., Dines, J., Gatica-Perez, D., Flynn, M., Wellner, P., Bourlard, H.: On the Use of Information Retrieval Measures for Speech Recognition Evaluation- Research Report 04-73. Tech. rep., IDIAP Research Institute (2005)
9. Möller, S.: Parameters for quantifying the interaction with spoken dialogue telephone services. In: Proc. of SIGDial 2005 (2005)
10. Möller, S., Ward, N.: A framework for model-based evaluation of spoken dialog systems. In: Proc. of Workshop on Discourse and Dialogue, SIGDial 2008 (2008)
11. Morris, A., Maier, V., Green, P.: From WER and RIL to MER and WIL: Improved Evaluation Measures for Connected Speech Recognition. In: Proceedings of International Conference on Spoken Language Processing (2004)
12. Nuance: German academic version of Dragon naturally speaking version 11 (2012)
13. Postel, H.: Die Kölner Phonetik. Ein Verfahren zur Identifizierung von Personennamen auf der Grundlage der Gestaltanalyse. IBM-Nachrichten 19, 925–931 (1969)
14. Rodgers, J., Nicewander, W.: Thirteen ways to look at the correlation coefficient. The American Statistician 42(1), 59–66 (1988)
15. Schneider, A., Luz, S.: Speaker Alignment in Synthesised, Machine Translated Communication. In: Proceedings of IWSLT 2011 (2011)

16. Skantze, G.: Exploring Human Error Recovery Strategies: Implications for Spoken Dialogue Systems. Speech Communication 45(3), 325–341 (2005)
17. Stanier, A.: How Accurate is SOUNDEX Matching? Computers in Genealogy 3(7) (1990)
18. Walker, M., Kamm, C., Litman, D.: Towards developing general models of usability with PARADISE. Natural Language Engineering 6(3-4), 363–377 (2000)
19. Wang, Y.Y., Acero, A., Chelba, C.: Is word error rate a good indicator for spoken language understanding accuracy. In: 2003 IEEE Workshop on Automatic Speech Recognition and Understanding, ASRU 2003, pp. 577–582 (November 2003)

Concatenative Hymn Synthesis
from Yared Notations

Girma Zemedu and Yaregal Assabie

Department of Computer Science, Addis Ababa University, Ethiopia
gzemedu@yahoo.com, yaregal.assabie@aau.edu.et

Abstract. Yared musical notation is widely used in Ethiopian liturgical and non-liturgical songs since it was invented about 1500 years ago. This paper presents automatic synthesis of Yared hymn from the notations available in text form in different contexts where a specific note may produce the different utterance depending on the contexts. Concatenative approach with unit selection is used to synthesize the hymns. We also apply contextual pitch shifting and amplitude modification to remove discontinuities caused by concatenation. The system is evaluated using a test data collected from various hymn lyrics. We use Mean Opinion Score to evaluate the performance of the system for both naturalness and intelligibility, which is computed through an assessment made by domain experts. Experimental results are reported.

Keywords: Yared hymn notes, hymn synthesis, unit selection.

1   Introduction

Text-to-speech synthesis systems have been widely used as a technological aid to produce speech from written text. It has a very useful application by assisting people with visual impairments or reading disabilities to listen to written works [6,7]. Similar to text-to-speech systems are singing voice synthesizers that artificially produce a human singing voice from a musical notation [15]. However, singing voice synthesizers differ from the traditional text-to-speech systems as the desired output is a natural-sounding beautiful voice in stead of intelligible speech. Moreover, in singing voice synthesis, the precise control of vowel duration and pitch is very important in achieving a singing voice that closely matches the melody [14]. Despite the differences in the desired output, singing voice synthesis and text-to-speech systems share synthesizer technologies and signal processing components. Several techniques have been proposed to synthesize singing voice [3,4]. Among the most widely used techniques is a concatenative synthesis using unit selection method which is based on concatenation of segments of recorded speech [2]. Historically, the method has been successfully applied to text-to-speech synthesis systems [8]. The method uses a large database of singing voice recordings segmented into units of successive phones. For a given score and lyrics, a sequence of phones corresponding to notes is synthesized, along with the specific characteristics of the performance (pitch, duration, tone, etc.). For

A. Przepiórkowski and M. Ogrodniczuk (Eds.): PolTAL 2014, LNAI 8686, pp. 400–411, 2014.
© Springer International Publishing Switzerland 2014

each successive part of this sequence, the best unit identified as closest to the phone and notes to be produced is then selected from the database [12]. Hidden Markov Models (HMMs) are also used to synthesize singing voice [13,17]. Attempts have been made to synthesize singing voice from speech instead of text [10,14]. The focus is now geared towards increasing the quality of the singing voice. As a result, issues such as pitch shifting, amplitude modification, and other signal processing components are being investigated [2]. Since the singing voices produced are required to look natural, such systems are language dependent as sound utterances associated with each language may differ. Accordingly, attempts have been made to synthesize singing voices for languages such as English [12], Spanish [11], Mandarin [6], Greek [9], etc. However, to the best of our knowledge, there is no any research and development work on the synthesis of singing voice for Ethiopian languages such as Geez, Amharic, Oromiffa, Tigrigna, etc.

This work is aimed at developing hymn synthesis from Yared notation which is an ancient Ethiopian musical notation. The remaining part of this paper is organized as follows. Section 2 presents Yared hymn and musical notations. In Sect. 3, we present the proposed Yared hymn synthesis model. Experimental results are presented in Sect. 4, and conclusion and future works are highlighted in Sect. 5. References are provided at the end.

2   Yared Hymn and Musical Notations

Ethiopia is one of the oldest countries in the history of early civilization with its own script for writing. Christianity was introduced into the country in the 4th century and was made state religion, the second such country in the world after Armenia. The adoption of Christianity as state religion paved the way for the growth of literary and musical works in the country [1,16]. There were many translation works of religious materials from Greek and Aramaic to Geez which is the liturgical language of Ethiopian Orthodox Church [18]. Moreover, many Ethiopians have also contributed literary and musical works in line with the dogma and canon of Ethiopian Orthodox Church. One of such contributions which made a lasting influence on Ethiopian liturgical and non-liturgical songs is Yared hymn notation, named after its inventor. Yared was born in the early 6th century in the city of Axum which was by then the capital of Ethiopia [1]. He received church education from his family members who were known to be reputed scholarly priests. Later on, he studied and specialized on the arts of vocal performance, composition, poetry, versification and improvisation which led him to the invention of hymn notations known as Yared hymn. Since the invention of Yared hymn, the Ethiopian Orthodox Church strictly follows this notation to produce church songs and Yared is regarded as a saint by church. His musical note invention has its own impact even in non-liturgical music of Ethiopia until the present time [1].

Hymn produced by Yared originally had eight musical notations. The notations have their own names which is stated as follows: *yizet, deret, rikrik, difat,* ĉiret, qinat, hidet, and qurt. It was believed that two other notations dirs, and anbir were added later by other scholars. Alphabet symbols are also added to simplify hymn notations [18] which they called it sirey ('root'). Table 1 shows the notations with their descriptions and each note has its own meaning in the religion. Basically, these symbols are the core notes of the Yared hymn which had the capability to describe all his songs when they appear with Geez letters.

Table 1. Yared hymn note symbols with their descriptions

N⁰	Note	Symbol	Description
1	yizet	▪	letters or words are emphasized with louder chant in another wise regular reading form of chant
2	deret	⌣	a form of chant that comes from chest
3	rikrik	⦙ or ⦂	layered and multiple chants conducted to prolong the chant
4	difat	⌐	a method of chanting where the voice is suppressed down in the throat and inhaling air
5	qinat	⌐	highlights the last letter of a chant
6	ĉiret	⌐	highlights with louder notes letter or words in between regular reading of the text. The highlighted chant is conducted for longer period of time.
7	hidet	⌣ or —	it is a chant by stretching one's voice; it is resembled to a major highway or a continuous water flow in a creek
8	qurt	⊢	a break from an extended chant that is achieved by withholding breathing

The note has a context dependent sound and, as a result, it needs additional descriptors for correct interpretation. It is to be noted that the Western music has seven notes represented by 'A' to 'G' and have additional descriptors represented by the digits for respective octaves. Likewise, Yared hymn notes have also seven descriptors as shown in Table 2. This results in 56 (=8x7) possible combination of notes with markers.

The notes are used to decide how to produce the sound in articulatory organs while the descriptors are used to produce the amount of vibration for the produced sound known as fundamental frequency or the pitch of the sound. Thus, the descriptors guide the singer to produce the correct pitch of the song. A sample of Yared notation in Geez language is shown in Fig. 1.

Yared had written five volumes of chants: digua and tsome digua, miraf, zimarie, mewasit and qidasie [1]. All his compositions follow three musical categories (qiñit), each of which are further classified with three musical scales that are reported to contain all the possible musical scales. It is believed that he got the three main hymn scores from three birds and he named them as follows [1]:

- Geez: represents first and straight note. It is described in its musical style as hard and imposing. Scholars often refer to it as dry and devoid of sweet melody.

Table 2. Yared descriptors for his notation

Nº	Note descriptor		Abbreviation used in Geez song scripts	Abbreviation we used in our work
	Name in Geez	Transliteration		
1	ዕለት ብርሃን	*ilete birhan*	ዕለ	I
2	ዓይኑ ዘርግብ	*aynu zergb*	ዓይ	A
3	ቡበይ እርዳበይ	*bubay irdabey*	ቡብ	B
4	በዕንቁ ሰንፔር	*be'inku senpier*	ፔር	P
5	ሠላሳ	*selasa*	ሰ̈	S
6	ቡርከት	*burkt*	ቡር	R
7	ጸጋ ዘተውሂበ	*tsega zetwuhibe*	ጸጋ	T

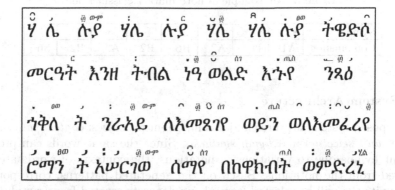

Fig. 1. Sample Yared notation from the song 'halie luya'

- *Izil*: represents melodic, gentle and sweet note, which is often chanted after Geez. It is also described as affective tone suggesting intimation and tenderness.
- *Araray*: represents melodious and melancholic note often chanted on somber moments, such as fasting and funeral mass.

3 The Proposed System for Synthesis of Yared Hymn

3.1 Methodological Approach

The proposed model for Yared hymn synthesis considers two core points: collection of Yared notations and differentiating among the note marker of the given note. Hymns that hold the notes of *araray* and *geez* were collected from different books, especially *digua* and *qidasie*, the purpose of which was to accommodate the seven note markers listed in Table 2. However, this was met with difficulty as songs written in *digua* have not specified all notes needed, rather they use special tags. Tags are a kind of code which had one or two letters and used as an index for other words in a different song having a more described note on it. For example, we may get ⍵ᣟልይᣟእᣟᢞᒐᒐᣟᢛᒫᒐᒐ where " ᣟᢅ ᢅ " stands for " ᣟᒐᣟᢅ " and the notes on them become " ᣟᢅ ᢅ " for " ᣟᒐᒐᣟᒐᒐᣟᒐᒐᢅ ". Once special tags are recognized, the note marker of a given note is identified since an individual note on a single letter has its class of note marker. This is achieved by singing each note in a recursive fashion to classify its marker. Table 3 shows the classification of note marker for subpart of the aforementioned note sequence "ይᣟᒐ " which could simply be identified by note marker " ᣟᢅ ᢅ " on the Ethiopic character "ይ". Letter and number combinations represent notes and markers, respectively as presented in Tables 1 and 2. For example, **B6** represents the combinations of *bubay irdabey* marker combined with *ĉiret* note.

Table 3. An example of note marker classification

Symbol	▪	⌒	ᴗ	ᴗ	ᴗ	⌒	ᴗ	ᴗ
Combination	A1	I5	A2	B6	B2	A5	B2	S6

3.2   System Architecture

The proposed system has three major component processes: *context pattern generation, unit selection* and *hymn synthesis*. Since the same words can produce different hymns due to variations in notations, the contexts of words are first captured from the notations. Based on the generated patterns, corresponding sound units the will be selected from phone datasets created from sample notations by taking the contexts of patterns into consideration. Since it is difficult to cover all the contexts of Yared notations, we also created *character phone dataset* to be used as a reference for unknown contexts during the unit selection process. After appropriate phone units are selected for a given Yared notation, the synthesis of hymn is achieved by concatenating sequences of phone units into a single audio file from which a hymn in waveform is synthesized. Figure 2 shows the system architecture of the proposed model.

3.3   Generation of Context Patterns

To achieve a more natural voice quality of hymns for the notations, the contexts of a word is taken into account. The position of the notes in different contexts usually leads to the variation in pitch. The first step towards this is to perform the transcription of the given word. While handling the transcription, one phoneme context from the left and one phoneme context from the right of the target note are considered. It is empirically seen that the target note does not depend much on contexts beyond the immediate neighborhoods. Table 4 shows the transcription of some words along with their contexts of phone patterns. The first two rows in the table have the same and the target note to describe is also the same (*difat*) but the notes to be transcribed have different context

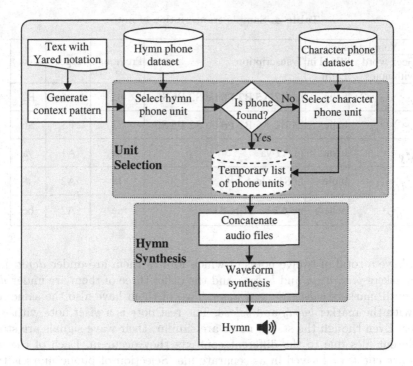

Fig. 2. System architecture

and note markers. The symbol '#' in the table denotes a null value, where we do not have any context parameter point. For instance, the word *hiqle* in the fourth row of the table has two notes **S1** and **A2**. The target note **A2** does not have any succeeding parameter and thus assigned null value for the particular context.

3.4 Unit Selection

After contextual patterns of Yared notations are generated, the next step is to select phone units corresponding to context pattern generated. Phones are selected from two phone datasets: hymn phone dataset and character phone dataset. Hymn phone dataset is created from audio files recorded from sample hymns and contains hymns in various contexts. Character phone dataset contains phone units representing all Geez characters. After recording hymns and phones of characters, phone units are extracted from the audio files. First each word in a lyric is separately labeled as a word in a one labeling row for the specific hymn. On the second iteration, the words are divided into unit labels. Here, when we say unit it stands for notes or phonemes. Figure 3 shows one of the labeling for the word *semaye*. There is a possibility of having two or more similar notes with different markers. In such cases, they will have different signal characteristics depending on the context they are found. For instance, the word *semaye* stated

Table 4. Sample transcription of notes

Yared notation		Transcription	Previous	Target	Next
Geez word with notes	Word in Latin				
ሃ◡⌒◡ ሌ◡⌒	halie	ha B2 **A5** B2 P6 lie S2 P5 S6	B2	A5	B2
ሃ◡⌒◡ ሌ◡⌒	halie	ha B2 A5 B2 P6 lie S2 **P5** S6	S2	P5	S6
ሉ▪ያ	luya	lu **A1** ya	lu	A1	ya
ኂ▪ቅለ◡	hiqle	hi S1 q le **A2**	le	A2	#
ነ◡በ▪ረ▪	nebere	ne **A2** be S1 re S1	ne	A2	be

above have a total of twelve notes of which four of them are under *deret* notes with markers *pier*, *aynu* and *bubay*. And the other three of them are under *difat* notes with markers *bubay*, *ile* and *aynu*. Two of them have also the same note *ĉiret* with the marker *bubay* and *selasa*. The rest note is a *yizet* note with *aynu* marker. Even though the stated notes are similar, their wave signals are stored in different files due to the different contexts they occur in. Each of the unit waves are cutoff and saved in as separate file. Selection of phone units is then

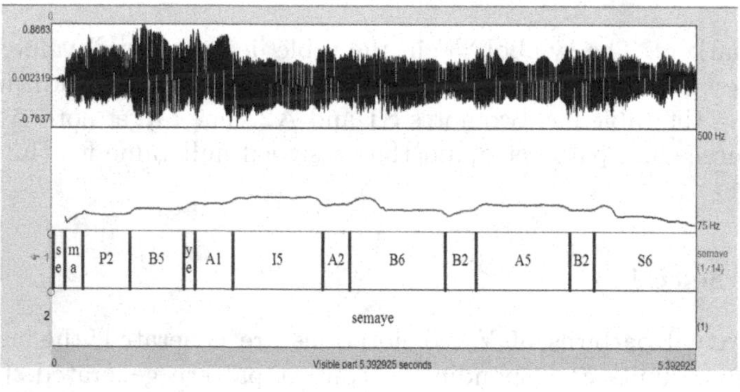

Fig. 3. Unit labeling for the word *semaye* from the song *twiedso*

made based on comparisons of the input notation contextual pattern against the hymn phone dataset. Comparison of the input pattern is made in three steps.

1. *Full matched search*: the system looks for the exact copy of the input context pattern in the hymn phone dataset.
2. *Partial matched search*: the system looks for data that contain the target element with either left or right context element.
3. *Target only search*: it looks for the target element only.

There is a case where a given phoneme is not found in the phone data list since we did not record a song which has all combinations of notes with the whole set of Geez character sets. For the reason described above, we set up a set of audio files for Geez alphabets. This was done by recording sounds for each letter with the voice of the same person and prepare for both audio file and phone data set.

The priority of the candidate selection is given to fully matched result and then for partial matched, and finally for the target only matched results. The second step is to select the most appropriate phone from candidate list by comparing its context behavior. Context is the target note or phoneme plus the left and right strings in a list. The model selects the units to smoothly concatenate the sounds based on the parameters of previous selected unit and target. We use the pitch parameters of both the target and immediate unit before target phone. Since we have pitch information for each unit during sample preparation, which includes the initial and last pitch of the unit. For the comparison of candidate units the absolute pitch difference which was calculated as the square of the difference between candidate pitch and a unit selected before the target pitch is used.

3.5 Hymn Synthesis

In concatenation of the audio files, all audio files which are selected for the given input word for synthesizing are put together in their order of occurrences. Since the audios are collected from different parts of songs, it does not match with the neighboring phone even though special attention is considered to minimize the effect of co-articulation during unit selection. The problem occurs due to various reasons where the amplitude, pitch and duration of the phone are among the very important attributes for the synthesized singing voice. Each note has different frequency when they appear on different phonemes. In other words, the notes are assigned to different pitch according to the hymn. This produce unnecessary sound effect when different units from different words are concatenated to produce hymn for new word. Smoothing of the transition from relatively low frequency to higher or from higher to low frequency at a spot of time is difficult. Thus, in order to alleviate such problems, we employed *contextual pitch shifting* and *amplitude modification*.

The proposed contextual pitch shifting technique has two steps. The first step is to shift the pitch according to the neighboring context. Then, spaces created between the slices are removed. To accomplish this task we first evaluate the cost of each unit signal. The cost is a value given for each unit after unit selection. The cost values given for the units are 1 for full match unit, 2 for partial match unit, and 3 for target only unit. To shift the selected unit, the pitch of the signals at the starting position is taken and compared to the previous unit in the context. If current signal pitch is less than the previous pitch, some parts of the previous unit signal frames will be down shifted. Otherwise, some parts of the previous signal will be shifted up. For our proposed system we adopted the pitch shifter algorithm suggested by Grondin [5]. Figure 4 shows the result of contextual pitch shifting process at concatenation points.

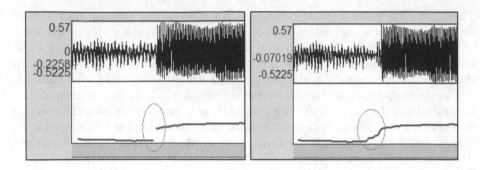

Fig. 4. Result of contextual pitch shifting process

Hearing a sound that has different intensity in a short time in a single word is unnatural and not comfortable to our ear. This phenomenon is also one of the challenges in concatenative synthesis using unit selection as the concatenation of units from different words result in different amplitudes. To overcome this problem, smoothing the amplitude of the synthesized word is recommended at concatenation points by fading-in and fading-out the signal. This can be achieved by amplifying the wave signal of the unit under consideration by a certain amount of multiplying factor. Figure 5 shows the result of amplitude modification.

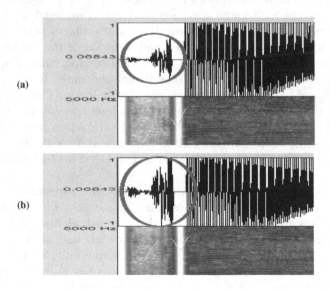

Fig. 5. (a) Concatenation of units, (b) amplitude modification

4   Experiment

4.1   The Corpus

A corpus of Yared notations was collected from various sources. A representative sample lyric for each note along with acoustic details was collected from different scripts having hymn notation. The collected lyrics were transcribed to Yared notations manually by the domain experts. The hymns of transcribed lyrics were also recorded to get the corresponding acoustic data. The contexts of notes within words were also systematically captured. In addition, the expert was recorded while uttering the phones of all Geez characters. From these sets of data, we generated the *hymn* and *character* phone datasets. Hymn phone dataset and character phone dataset would be used to select phone units during the unit selection process.

4.2   Implementation

The proposed Yared hymn synthesizer was implemented using various tools. We used Praat to record the hymns with a sampling frequency of 44100 Hz and 16 bit resolution of data representation. In addition, Praat was also used for pitch and duration extraction. We used Java programming language is to process the input text representing Yared notations. Matlab was used for the synthesis of the wave signal from the concatenated units to give an appropriate pitch and intensity.

4.3   Test Results

To evaluate the performance of the system, we used Mean Opinion Score (MOS) which is among the widely used technique for speech evaluation by perception of human subjects. MOS is an evaluation technique where evaluators indicate their assessments on a scale of bad (1) to excellent (5). Then the average of the opinion will be taken as the performance of the system. Accordingly, eight users who are experts of Yared hymn but nave for synthetic sound were selected. The domain experts were requested to assess the system and provide their evaluations by marking the level of their acceptance for both intelligibility and naturalness. The evaluation results for both naturalness and intelligibility is shown in Tables 5 and 6. The analysis of these evaluation results using MOS yields 3.9 for naturalness and 3.4 for intelligibility.

Table 5. Test result for naturalness of synthesized Yared hymn

Scale	Scale description	Percentage of evaluators' decision
1	Very unnatural	0%
2	Unnatural	0%
3	Good	25.0%
4	Natural	62.5%
5	Very natural	12.5%

Table 6. Test result for intelligibility of synthesized Yared hymn

Scale	Scale description	Percentage of evaluators' decision
1	Very difficult	0%
2	Difficult	25.0%
3	Good	25.0%
4	Easy	37.5%
5	Very easy	12.5%

5 Conclusion

As speech synthesis systems are becoming more applicable, different speech synthesis techniques are being developed and implemented to generate a natural and intelligible speech. However, it is more difficult to directly apply the techniques to generate hymns from music notations as the contexts of notes may significantly vary. We proposed Yared hymn synthesizer model which handles contexts and promising results are obtained. By selecting the best phone unit from the hymn and character phone datasets, the synthesizer produced a hymn in a given note context. After sound units are selected and concatenated we applied contextual pitch shifting and amplitude modification with the aim of smoothing utterances at concatenation points. This has contributed to the improvement in intelligibility and naturalness of synthesized hymns. The performance of proposed system can still be improved in many ways. Since we could not include all the contexts in the hymn phone dataset, further works can be made to consider more contexts of Yared notes. We also could not find duration demarcation for Yared notes. The system can be improved if we incorporate duration demarcation as a parameter in the selection of appropriate phone units. In addition, intelligibility and naturalness of the synthesized hymns can be improved if we work more on pitch shifting and amplitude modification components.

References

1. Bekerie, A.: St. Yared - the great Ethiopian composer. Tadias Magazine (2007)
2. Bonada, J., Serra, X.: Synthesis of the Singing Voice by Performance Sampling and Spectral Models. IEEE Signal Processing Magazine 24(2), 67–79 (2007)
3. Cooper, C., Murphy, D., Howard, D., Tyrrell, A.: Singing Synthesis with an Evolved Physical Model. IEEE Transactions on Audio, Speech and Language Processing 14(4) (2006)
4. Duh, K.: Singing Voice Synthesis, http://ssli.ee.washington.edu/people/duh/projects/singing.html (date accessed; March 03, 2014)
5. Grondin, F.: Guitar Pitch Shifter, http://www.guitarpitchshifter.com/ (date accessed; March 03, 2014)
6. Gu, H., Liao, H.: Mandarin Singing-voice Synthesis Using an HNM Based Scheme. Journal of Information Science and Engineering 27, 303–317 (2011)

7. Jurafsky, D.: An Introduction to Natural language processing, Computational Linguistics and Speech Recognition. Prentice Hall, Englewood Cliffs (1999)
8. Kasie, E.B., Assabie, Y.: Concatenative Speech Synthesis for Amharic Using Unit Selection Method. In: Proc. of International ACM Conference on Management of Emergent Digital EcoSystems (MEDES 2012), Addis Ababa, Ethiopia, pp. 27–31 (2012)
9. Kyritsi, V.: Score-To-Singing voice for Greek language. In: Proceeding of the Inter. Computer Music Conference (ICMC 2007), Copenhagen, Denmark (2007)
10. McNuty, J.: Text to speech to singing voice synthesis. Sonic Arts Research Center, Queens University, Belfast (2009)
11. Ramos, A.: Singing voice synthesis in Spanish by concatenation of syllables based on the TD-PSOLA algorithm. In: Proc. of Latest Advances in Acoustics and Music Conference, Copenhagen, Denmark, pp. 210–215 (2012)
12. Rodet, X.: Synthesis and processing of the singing voice. In: Proc. of the 1st IEEE Benelux Workshop on Model based Processing and Coding of Audio (MPCA 2002), Leuven, Belgium (2002)
13. Saino, K., Zen, H., Nankaku, Y., Lee, A., Tokuda, K.: An HMM-based singing voice synthesis system. In: Proceeding of INTERSPEECH 2006 - ICSLP, Ninth International Conference on Spoken Language Processing, Pittsburgh, PA, USA (2006)
14. Saitou, T., Goto, M., Unoki, M., Akagi, M.: Speech-to-singing synthesis: converting speaking voices to singing voices by controlling acoustic features unique to singing voices. In: Proc. of IEEE Workshop on Applications of Signal Processing to Audio and Acoustics, New Paltz, NY, USA, pp. 215–218 (2007)
15. Sundberg, J.: The KTH Synthesis of Singing. Advances in Cognitive Psychology 2(2-3), 131–143 (2006)
16. Tsegaye, M.: Traditional Education of the Ethiopian Orthodox Church and its Potential for Tourism Development (1975-Present). Addis Ababa University, Addis Ababa (2011)
17. Veaux, C., Astrinaki, M., Oura, K., Clark, R., Yamagishi, J.: Gesture Control of HMM-Based Singing Voice Synthesis. In: Proc. of 8th ISCA Speech Synthesis Workshop, Barcelona, Spain, pp. 247–248 (2013)
18. Workneh, H.: The ancient curriculum of Ethiopia, http://www.ethiopianorthodox.org/ (date accessed; March 03, 2014)

NLP-Oriented Contrastive Study of Linguistic Productions of Alzheimer's and Control People

Maïté Boyé[1], Thi Mai Tran[1,2], and Natalia Grabar[2]

[1] Institut d'Orthophonie, Université Lille 2, Lille, France
[2] CNRS UMR 8163 STL, Université Lille 3,
59653 Villeneuve d'Ascq, France
maite.boye@gmail.com, thimai.tran@univ-lille2.fr,
natalia.grabar@univ-lille3.fr

Abstract. The increase in Alzheimers disease is due to the aging of the population and is the first cause of neurodegenerative disorders. Progressive development of cognitive, emotional and behavior troubles leads to the loss of autonomy and to dependency of people, which corresponds to the dementia phase. Language disorders are among the first clinical cognitive signs of the disease. Our objective is to study verbal communication of people affected by the Alzheimer's disease at early to moderate stages. One particularity of our approach is that we work in ecological conversation situation: people are faced to persons they know. We study verbal productions of five people affected by the Alzheimer's disease and of five control people. The conversations are transcribed and processed with the NLP methods and tools. Over thirty features grouped in four categories are studied. Our results indicate that the Alzheimer's patients present lexical and semantic deficit and that, in several ways, their conversation is notably poorer than the conversation of the control people.

Keywords: conversation, linguistics, Alzheimer's disease, speech and language therapy, Natural Language Processing.

1 Introduction

The Alzheimer's disease (AD) is related to the aging of the population and is the first cause of neurodegenerative dementia. In 2005, 26 million people suffered from the disease and it is estimated that the number can reach over 100 million people worldwide by 2050 [1,2]. The disease is widespread in the modern society and researchers from various areas address its detection and therapy.

Progressive appearance of cognitive, emotional and behavior troubles causes the loss of autonomy and the dependency, which correspond to the demential stage of the disease. Language troubles, together with memory and executive function troubles, are the first clinical cognitive signs of the disease [3]. During the early stage, patients are mainly affected by mnesic troubles and keep their language capacity almost intact. With the evolution of cognitive and language troubles, the communication troubles become irreversible and severely impact

A. Przepiórkowski and M. Ogrodniczuk (Eds.): PolTAL 2014, LNAI 8686, pp. 412–424, 2014.
© Springer International Publishing Switzerland 2014

daily life of the affected persons and of their families. Preservation of communication ability as long as possible is one of the main objectives of the AD therapy. It helps to save social interactions and prevents people from isolation. Hence, it is important to detect the disease as early as possible. In the following of this section, we present the existing studies on assessment of the language capacity in clinical practice and analysis of the conversation AD language.

1.1 Assessment of Language and Communication Troubles

Assessment of language troubles is part of the Speech and Language Therapy used for the diagnosis and therapy of cognitive neurodegenerative pathologies [4,5,6]. Tests of language capacity (spoken and written, production and perception) are usually performed during dedicated interviews. The most frequent tests are related to the assessment of lexical production (fluency, naming of pictures...) and of oral expression (description of pictures, chat on a given topic...). They allow performing quantitative and qualitative assessment of language capacity in controlled situations. More natural and ecological language productions are studied very poorly, mainly because of large number of features to consider. If the controlled tests can reveal language deficiency, they cannot assess the impact of language troubles on the daily communication of patients. Besides, supervised situations may penalize those people who are not used to them. We assume that non-supervised conversation tests correspond to more natural language exchanges and are more suitable for diagnosing the disease and assessing the language ability of AD patients.

1.2 Conversation Language Analysis

Some previous studies addressed the analysis of free or semi-directed conversation of elderly AD and healthy people. The studied groups contain as least five people. It has been observed that linguistic performance of healthy persons is dependent on their education level and age [7]. Inter-individual difference is frequently reported because, contrary to supervised tests, it becomes salient in free conversation. Among the available observations, we can find that elderly people use the same vocabulary than young people but produce sentences that are more complex syntactically [8]; they have more approximate and ambiguous discourse, with more periphrases and redundancies; they often refer to the past events during the conversation, are more egocentric and seldom consider what their interlocutor is saying; the proposed topics are respected but with frequent digressions while the semantic coherence is poor [9]. Concerning the comparison between AD and healthy people, it appears that: length of utterances and lexical variation is not age-dependent [10]; utterances of AD patients are less diversified syntactically [10,11]; AD patients mention less ideas and produce less words [12]; they show redundant and less informative discourse [13,14,15,16]; they use lesser number of modalizers [13]; pronouns miss reference and reveal inconsistency [15,16]. We propose to perform a more complete analysis of the AD conversations and support it by the use of the NLP methods and tools (Sect. 2).

The experiments are described in Sect. 3 and 4. We then present and discuss the results (Sect. 5 and 6) and indicate the directions for future work (Sect. 7).

2   Objectives and Hypotheses

Our objective is to study the language of AD patients produced in conversation context with known interlocutor. These are more ecologic conditions than usual, and we assume they are more relevant for studying the specificity of the AD language. We propose to enrich the existing knowledge on the disease for its better diagnosis and therapy, and perform for this contrastive analysis of AD and healthy people. To make the analysis systematic and to study wide range of features, we use the NLP methods and tools.

3   Linguistic Material Studied

The two groups studied include homogeneous people as for their age (average=90), gender (women), social and cultural level, and place of life (institutions for elderly people in North of France). Conversations of five AD patients (89 to 99 years) and five control people (83 to 102 years) with known interlocutor are recorded. The AD patients are checked up to make sure they do not have previous neurological and psychiatric history. They must have low or moderate stage of the disease, and be communicating. Medical history of control people is also checked up.

To start the conversation, two pictures are presented (departure by train for vacations and bicycle ride in 1950s). The preferred picture is first chosen, but later the discussion is oriented on the other picture. The interviews are not limited in time: they last from 20 to 40 minutes. The interviews are then transcribed with Transcriber [17], which allow marking additional information. Transcription is done with the standard spelling [18] and some adaptations:

- transcription is orthographic (and not phonetic);
- transcription is lexically and syntactically correct to make possible the use of the NLP tools;
- transcription can contain marking of some features relevant for the study (disfluencies, hesitations, pauses...).

The marking up the features like disfluencies, hesitations and pauses, that are very frequent in spoken language productions, makes the transcriptions close to the original productions and no information is losed at this step. Still, as indicated, the transcription with lexically and syntactically correct sentences makes possible the use of the NLP tools. These two aspects are important for our study. If necessary, it is possible to go back to the original record and to re-analyze the conversation.

Twenty minutes of conversation are transcribed for each speaker, which required over 100 hours of work. Corpora are anonymized.

4 Methodology for Studying the Transcribed Conversations

The analysis proposed addresses two aspects: verbal interaction (Sect. 4.1) and content of conversations (Sect. 4.2). Corpora are POS-tagged with TreeTagger [19] and then corrected with Flemm [20]. Lemmas are morphologically analyzed by Dérif [21]. Data encoded during transcription (time of speech, shift of speakers, disfluencies, hesitations, pauses...) are also exploited. Some features are proposed in the literature, while others correspond to the original contribution of our work. The feature values (average, minimal or maximal) are computed at the level of persons and then at the level of the (AD and control) groups. We also use a set of interjections, such as *oh, ah, bon (well), là (here), donc (hence)*.

4.1 Analysis of Verbal Interaction

Two verbal interaction features studied are provided by Transcriber:

- *Turns of speech.* Turns of speech allow studying the dynamics of exchanges;
- *Time of speech and overlapping.* Time of speech and overlappings are indicative of the participation of people in conversations and of their interactions.

4.2 Analysis of Conversation Content

We distinguish spoken, lexical and syntactic features. We give examples in English, although work is done on the French data.

Spoken Features. Among the spoken features, we study the following:

- *Breath groups.* Interruptions of utterances are considered as breath groups, and can be linked with people physiology and with their mental processes;
- *Empty pauses.* Empty pauses are integral part of conversations. They ensure fluency of discourse and give time for understanding and thinking. Empty pauses last at least one second and contain no words;
- *Non-empty pauses.* Non-empty pauses contain only disfluencies (*euh, hum...*) and lengthened syllables;
- *Primes.* Primes cover different situations [22,23]: the corrected words may keep the same syntactic function (e.g. *Wha- what*), or change it (e.g. *He will be retired in t- when he is seventy-two*);
- *Repetitions and stutters.* We study disfluency repetitions [23], like in *And then you arrive in in beautiful gardens.* Stylistic repetitions (e.g. *It was raining it was raining* meaning *it was raining a lot*) are not considered. Stutters are repetitions usually caused by articulatory problems;
- *Self-corrections.* Self-corrections can be related to grammatical corrections (*well port-work have has been better paid*), paradigmatic corrections using the same syntactic structure (*we all meet in a at one person home*), or different syntactic structure (*the buildings belong to she is not the owner*);

- *Interruptions.* Interruptions occur when speaker stops the utterance without completing, repeating or correcting it (e.g. *Well, there is...*) [24]. The speaker can then start another utterance or stop speaking at all;
- *Verbal output.* Verbal output is the ratio between the total number of words and the total time of speech (without empty pauses). It is estimated that normal speakers produce 200 words/minute on average.

Lexical Features. Lexical features address lexical variability and complexity:

- *Number of words.* This measure gives global quantitative indication on verbal productions. We compute total number of words without disfluencies;
- *Informativity.* The lack of informativity can be revealed with lexical units such as interjections (*oh, ah, well, here*);
- *Yes/No utterances.* Some utterances may contain only *yes* and *no* words. They also indicate the lack of informativity;
- *Lexical diversity.* Lexical diversity is the number of types of lemmas (nouns, verbs and adjectives);
- *Ratio lemmas/total words.* Ratio lemmas/total words is representative of semantic content of conversations and gives another assessment on lexical diversity of speakers;
- *Morphological complexity.* Morphological complexity of words, computed with the Dérif analyzer, corresponds to number of bases and affixes they contain;
- *Lexical frequency.* Lexical frequency of each lemma is computed in the corpora processed and on the web (in August 2013).

Syntactic Features. Syntactic features address complexity of sentences:

- *Average length of utterances.* Utterance is defined as linguistic unit with common semantics and showing decrease in voice frequency at the end [25]. Average length of utterances, computed without disfluencies, corresponds to the ratio between total number of words and number of utterances;
- *Interpolated clauses and reported speech.* Interpolated clauses and reported speech indicate the distance the speaker has with his own conversation [26]:
  - interpolated clauses are often used to add new information necessary for the understanding (*Then I ask the driver he is very kind the driver to stop near the church.*)
  - reported speech means that the speaker is able to use statements produced by other people and to keep the right references (*He says: "Mary you are not right..."*)
- *Personal pronouns.* The ratio of personal pronouns is computed within the whole number of lemmas;
- *Verbs.* The number of verb lemmas is computed;
- *Distribution within syntactic categories.* We compute the distribution of lemmas among verbs, nouns and adjectives.

Table 1. Results observed for various features studied with AD and control people

Features	AD	Control	Diff.
Verbal interaction			
Turns of speech (avg.)	194.80	181.80	-7
Time of speech (min.)	*7'56"*	*11'08"*	-46
Time of speech (max.)	*13'11"*	*17'24"*	-31
Time of speech (avg.)	11'16"	14'12"	-26
Overlapping (avg.)	*0.59*	*1.01*	-71
Spoken features			
Breath groups (avg. words)	3.90	4.31	-10
Empty pauses (avg.)	*65.00*	*31.00*	52
Non-empty pauses (avg.)	*13.00*	*34.60*	-166
Disfluencies (avg.)	*30.00*	*37.00*	-24
Primes of words (avg.)	*2.8*	*8.4*	-200
Interrupted sentences (avg.)	18.8	19.40	-3
Speech output (max. words/min.)	184.23	> 190	-3

4.3 Comparison of Average Values

The average values for each feature are compared at the level of groups to detect those features that show important deviation from normal (control) values. Given two average values, A_1 (normal) et A_2 (AD), we compute: $\frac{A_1 - A_2}{A_1} * 100$.

5 Results

In Tables 1 and 2, we present the results obtained for the comparative analysis between AD and control groups. Features are grouped as previously in four sets (verbal interaction, spoken, lexical and syntactic features). Difference between the values is considered as notable if it is higher than 20% in either way. Notable features are marked in italics. We assume these features can be used for the differentiation between AD and control people, and for the diagnosis of the Alzheimer's disease.

Almost all features related to verbal interaction show notable difference: minimal, maximal and average time of speech (usually higher in control group), and overlappings (also higher in control group). The number of turns of speech is similar in the two groups. Also, it should be noticed that we observe important intra-group and personal difference. Indeed, independently on the disease, some people are used to speak more easily and rapidly, while other people may remain reticent even when speaking with know interlocutor.

Several features from the spoken feature set show notable difference. The AD patients produce more empty pauses and less non-empty pauses; less repetitions, self-corrections and primes of words than control people, which is contrary to what is usually observed in the literature [13,14,15], although their distribution is balanced. Among the features that do not show notable difference, we can find length of breath groups, number of interrupted sentences, and speech output.

Table 2. Results observed for various features studied with AD and control people (continued)

Features	AD	Control	Diff.
Lexical features			
Number of words (avg.)	*1,778.4*	*2,407.4*	-35
Informativity (avg.)	9	8	11
Yes/No utterances (avg.)	*88.80*	*52.40*	40
Lexical diversity (avg.)	*168.60*	*302.40*	-79
Lemmas/total words ratio (avg. %)	10	*14*	-32
Morphological complexity	1	1.07	-3
Lexical frequency in corpora (avg.)	11	8	25
Lexical frequency on the web (avg.)	*304,000,000*	*227,000,000*	25
Syntactic features			
Length of utterances (avg.)	*4.79*	*7.26*	-51
Interpolated clauses (avg.)	*1.20*	*4.80*	-300
Reported speech (avg.)	*1.60*	*14.20*	-787
Personal pronouns (avg. %)	*36*	*23*	37
Pronoun/noun ratio	1	0.65	6
Verbs (avg.)	*377*	*553*	-46
Ratio of verbs	32	32	0
Ratio of nouns	51	55	-5
Ratio of adjectives	16	14	14

Here again, we observe great inter-personal variability. For instance, 2 out of 5 patients in the AD group show high speech output, although people with higher speech output always belong to the control group.

Almost all lexical features show notable difference: number of words (higher in control group), number of *Yes/No* utterances (higher in AD group), lexical diversity (higher in control group), lemmas/total occurrence ratio (higher in control group), average frequencies in the corpora studied (higher in AD group because the lexicon is more redundant) and on the web (higher in AD group because the AD patients use more common and frequent words). Informativity and morphological complexity features are comparable in both groups. Concerning morphological complexity, several common affixes (e.g. *-ment, -tion, dé-, re-*) are not processed currently by Dérif. We assume their treatment may change the impact of this feature.

Syntactic features that show notable difference are: average length of utterances (higher in control group), interpolated clauses and reported speech (higher in control group), personal pronouns (higher in AD group), number of verbs (higher in control group). The two remaining features (pronoun/noun ratio and distribution within syntactic categories) show no notable difference.

6   Discussion

The results we propose contain diversified analysis of the AD patients free conversation. Some features provide salient information while others do not appear to be specific to a given group. With the use of the NLP tools we are able to propose systematic and reproducible analyses of corpora. We propose now the discussion of the results: their interest for the task and current limitations.

6.1   Specificities of the AD Group

Our study points out several linguistic specificities of the AD patients through their comparison with the control group:

- they speak less for the equal speaking time,
- they show higher number of turns of speech, they segment more frequently their discourse and produce shorter utterances,
- their speech output is slower,
- they do not speak at the same time as their interlocutor (overlapping is very low) and do not interrupt sentences,
- they often produce minimal utterances with *Yes/No* words only,
- their discourse contains small number of disfluencies,
- they have more empty pauses and less non-empty pauses,
- they seldom use interpolated clauses and reported speech,
- they use more pronouns, especially at the first person in singular,
- the lexical diversity is poor, whatever the part of speech,
- they use more common and frequent words with higher frequency.

Our hypothesis concerning the possible identification of lexical and semantic troubles seems to be confirmed. Indeed, verbal production of AD patients can be quantitatively and qualitatively differentiated from the verbal production of control people: AD patients speek less, produce shorter and less elaborated utterances. They show tendency to use frequent words, to produce redundant vocabulary and more pronouns without reference. They do not take the initiative for verbal exchanges, remain centered on their own opinions and show low distance from what they are saying. These facts can explain why their discourse does not contain disfluencies, non-empty pauses, interpolated clauses and reported speech, and why they use a lot of *yes* and *no* statements. These are the main findings of our work.

Besides, consideration of a large set of conversation features open new perspectives on the pathological language. If the existing studies of lexical and semantic troubles of the AD patients concentrate on surface features (paraphasias, pauses, self-corrections, repetitions and interrupted utterances), our study indicates that non-observation of these features does not imply the absence of troubles but indicates that the troubles can be revealed differently, such as less elaborated and poorer utterances or lesser participation in verbal interactions. We assume that our study provides the clinicians with additional features to explore for the correct and more sensitive diagnosis of the disease and for its therapy.

In a recent study on the analysis of semi-directed interviews with aphasic patients, the researchers have also proposed to explore features issued from the NLP methods (e.g. semantic categorization of answers, their coherence and semantic distance [27]). This study revealed that classical clinical features (number of correct answers, time needed for answers...) are more efficient than those issued from the NLP. Our work points out that the application of NLP to this kind of corpora can bring new light and reveal additional information useful for the identification of lexical and semantic troubles and for a better understanding of mechanisms involved in language production by AD and control people. The findings also propose new solutions for helping the patients to save their verbal communication and social interactions. The positive impact of the NLP methods and tools for the study of pathological AD language is another positive finding of our work.

6.2   Other Results

As we have observed, some features are not specific for a given group and do not show notable difference between AD and control groups:

- number of turns of speech and of breath groups, as well as the average speech output,
- distribution of disfluencies,
- distribution of lemmas within syntactic categories and ratio between pronouns and nouns,
- informativity and morphological complexity of words.

We assume that, at early stage of the disease, some aspects of conversation language are preserved by the AD patients and remain similar to those observed with healthy people. The speech therapy should rely on these features to maintain language performance of the AD patients.

As we have noticed, the results must be looked at in the context of their intra-group and inter-individual variability. The main reason of this finding is that people are interviewed in natural situation without constraints, which increases the inter-individual variation: personal factor is more real than in artificial directed test conditions. Independently on pathology, the personality of every people has impact on their performance in real conditions. Similarly, it is necessary to consider how the people are used to be, because some of them may have naturally high or low output, or tendency to interrupt other people. In such conditions, it is more difficult to define the normal values and to compare the performance of speakers. On contrary, larger set of features may partially inhibit this variable. Still, it is necessary to study larger group of people in order to take into account larger set of population and also to take into account previous known performance of people. Another possibility is to consider the combination of features and to study the potential relations these features have among them.

6.3   Limitations of the Current Study

The limitations of current study are related to the exploitation of features, to the data processed and to the NLP and statistical methods.

The Features Exploited. The impact and relevance of features are studied individually, thanks to the comparison of their average values from the two groups (AD and control people). This gives first indications on their importance. We plan to combine these features within common models built with machine learning algorithms for instance. Besides, this perspective can help to test the features during the automatic categorization of new people conversations in order to define whether these people are affected by the disease or not. Concerning the word frequency collected on the web, this feature is not fully reliable *per se*. In the experiments proposed, we wanted to test it, because it was available, and to see whether this feature is interesting to be explored individually or in combination with other features. Our current intuition is that this feature should be combined with other features. The word frequency can also be computed in reference corpora.

The Corpus Studied. Currently, we exploit language production of small number of people: five AD and five healthy people. First of all, the number of people is comparable to the one usually used in this kind of studies and containing at least five people. Indeed, it remains complicated to recruit larger number of homogeneous groups of participants with health specificities such as those aimed in our study. The transcription of the conversations is also a long and tedious task: for the currently studied corpora, totiling ten interviews, we needed up to 100 hours to transcribe the conversations. Nevertheless, we plan to interview larger number of participants in future. One particular reason is that we would like to inhibit the inter-individual variable and to be able to generalize the observations on a better basis. The main difficulty is then to collect the necessary data and to perform their transcription. Despite this limitation, we feel that the currently studied corpora and features provide strong indicators on the difference between the conversations collected with healthy people and the AD patients.

The NLP Methods Used. Concerning the limitations related to the NLP tools, the morphological analyzer Dérif does not fully cover the affixes of the French language. For instance, affixes such as *-ment, -tion, dé-* or *re-*, that are part of the common competence of speakers, are not analyzed in the current version of the tool. We think that their treatment may allow obtaining more relevant results concerning the morphological complexity of words. Besides, other NLP tools can be used. For instance, syntactic parsers can be used to assess the syntactic complexity of sentences, which can complete the set of features we currently study for the analysis of the AD conversation language.

The Contrastive Measure Used. For the contrastive analysis of the data, we use the numerical difference in percentage between the average values from the two datasets. If this gives some indications, we plan to improve this aspect and to use more sophisticated statistical significance measures, such as *Student's t-test*.

7   Conclusion and Future Work

We have proposed a comparative analysis of spoken transcribed corpora collected with five healthy and five Alzheimer's disease patients, all of which are over 80-year old. These data provided us the possibility to study the conversation language of AD patients and to compare it with productions of healthy people. The analysis of the conversation appears to be complex because of the variety of features involved. In the existing studies on the Alzheimer's disease language, researchers usually analyze small number of features on language productions collected in semi-directed context with non-known interviewers. As we indicated, language productions collected during directed standard language tests are not always good indicators to understand the communication troubles specific to AD patients. We work in the context of free conversation and observe that lexical and semantic deficit is real with AD patients in such natural conditions. It is particularly observable through the reduced conversation competence of speakers. Thus, even if people are communicating, the decrease of the utterance length, the reduced quantity and diversity of vocabulary, and of speech output indicate that AD patients do not have the same linguistic capacity than the one of healthy people. Our analysis also indicates that AD patients find it difficult to communicate in natural situations: they can hardly take the initiative in verbal exchanges, produce complex utterances, and maintain the conversation. Often, AD patients are helped for this by the existing language automatisms elaborated lifelong and preserved despite the loss of informativity, spontaneity and dynamics of language exchanges.

Conclusions of this first study propose a better understanding of the language troubles observed with the AD patients and their possible impact on daily communication. These findings encourage to continue studying conversation productions to propose better protocols for the diagnosis of Alzheimer's disease and its therapy, and to help people maintaining social interaction and autonomy as long as possible. New data to be collected will enrich the corpora studied. Their exploitation will diversify (age, gender of participants for instance) and complete the current study. As we have mentioned, it is necessary to improve the analysis obtained with the NLP tools and to use additional tools for performing syntactic and semantic analysis. For instance, we observed that the use of verbs shows notable difference in the two studied groups. We can make this analysis more detailed and take into account the semantic categorization of verbs [28] and to explore whether the two groups use the same types of verbs. The morphological analyzer is not sufficient and should be enriched with the analysis of more affixes. The features concerned with the verbal interaction have been studied poorly and should be addressed better in future studies because they are particularly important for the communication situations. The use of the NLP methods and tools provided the possibility to perform quantitative analysis of a large set of features issued from pathological language, and shows the advantage of this kind of approach. Complementarity between language and gesture exchanges can also provide directions for future work. A stronger statistical significance measures will complete the analysis proposed. Finally, the automatic categorization of persons on the basis of their linguistic productions in order to make the diagnosis of the AD is another perspective of our work.

Acknowledgements. This work is partially funded by the COMAZ project, under the CPER-MESHS framework.

References

1. Ferri, C.P., Prince, M., Brayne, C., Brodaty, H., Fratiglioni, L., Ganguli, M., Hall, K., Hasegawa, K., Hendrie, H., Huang, Y., Jorm, A., Mathers, C., Menezes, P.R., Rimmer, E., Scazufca, M.: Global prevalence of dementia: a Delphi consensus study. Alzheimer's Disease International 366(9503), 2112–2117 (2006)
2. Brookmeyer, R., Johnson, E., Ziegler-Graham, K., Arrighi, M.: Forecasting the global burden of Alzheimer's disease. Alzheimer's and Dementia 3(3), 186–191 (2007)
3. Derouesné, C.: Maladie d'Alzheimer: données épidémiologiques, neuropathologiques et cliniques. In: Belin, C., Ergis, S., Moreaud, O. (eds.) Actualités Sur Les démences: Aspects Cliniques et Neuropsychologiques, pp. 25–34. Solal, Marseille (2006)
4. McKhann, G., Drachman, D., Folstein, M., Katzman, R., Price, D., Stadlan, E.: Clinical diagnosis of Alzheimer's disease: report of the NINCDSADRDA work group under the auspices of department of health and human services task force on Alzheimer's disease. Neurology 34, 939–944 (1984)
5. Dubois, B., Feldman, H., Jacova, C., Cummings, J., Dekosky, S., Barberger-Gateau, P., Delacourte, A., Frisoni, G., Fox, N., Galasko, D., Gauthier, S., Hampel, H., Jicha, G., Meguro, K., O'Brien, J., Pasquier, F., Robert, P., Rossor, M., Salloway, S., Sarazin, M., De Souza, L., Stern, Y., Visser, P., Scheltens, P.: Revising the definition of Alzheimer's disease: a new lexicon. Lancet Neurology 9(11), 1118–1127 (2010)
6. McKhann, G., Knopman, D., Chertkow, H., Hyman, B., Jack, C., Kawash, C., Klunk, W., Koroshetz, W., Manly, J., Mayeux, R., Mohs, R., Morris, J., Rossor, M., Scheltens, P., Carrillo, M., Thies, B., Weintraub, A., Phelps, C.: The diagnosis of dementia due to Alzheimer's disease: Recommendations from the national institute on Aging-Alzheimer's association workgroups on diagnostic guidelines for alzheimer's disease. Alzheimer's & Dementia 7(3), 263–269 (2011)
7. Mackenzie, C.: Adult spoken discourse: the influences of age and education. International Journal of Language and Communication Disorders 35(2), 269–285 (2000)
8. Nef, F., Hupet, M.: Les manifestations du vieillissement normal dans le langage spontané oral et écrit. L'année Psychologique 9(3), 393–419 (1992)
9. Rousseau, T., De Saint-André, A., Gatignol, P.: Évaluation pragmatique de la communication des personnes âgées saines. Neurologie Psychiatrie Gériatrie 9, 271–280 (2009)
10. Kynette, D., Kemper, S.: Aging and the loss of grammatical forms: A cross-sectional study of language performance. Language and Communication 6, 65–72 (1986)
11. Kemper, S., Rash, S., Kynette, D., Norman, S.: Telling stories: The structure of adults' narratives. European Journal of Cognitive Psychology 2, 205–228 (1990)
12. Croisile, B., Ska, B., Brabant, M., Duchene, A., Lepage, Y., Aimard, G., Trillet, M.: Comparative study of oral and written picture description in patients with Alzheimer's disease. Brain and Language 53, 1–19 (1996)

13. Duong, A., Tardif, A., Ska, B.: Discourse about discourse: What is it and how does it progress in Alzheimer's disease? Brain and cognition 53, 177–180 (2003)
14. Berrewaerts, J., Hupet, M., Feyereisen, P.: Langage et démence: examen des capacités pragmatiques dans la maladie d'Alzheimer. Revue de Neuropsychologie 13(2), 165–207 (2003)
15. Ska, B., Duong, A.: Communication, discours et démence. Psychologie et NeuroPsychiatrie du Vieillissement 3(2), 125–133 (2005)
16. Lee, H.: Vieillissement normal et maladie d'Alzheimer: analyse comparative de la narration semi-dirigée au niveau lexical. In: Méthodes et Analyses Comparatives en Sciences du Langage (2011)
17. Barras, C., Geoffrois, E., Wu, Z., Liberman, M.: Transcriber: a free tool for segmenting, labeling and transcribing speech. In: Conference on Language Resources and Evaluation (LREC), pp. 1373–1376 (1998)
18. Blanche-Benveniste, C., Bilger, M., Rouget, C., Van Den Eynde, K.: Le français parlé. Études grammaticales. CNRS Éditions, Paris (1991)
19. Schmid, H.: Probabilistic part-of-speech tagging using decision trees. In: ICNMLP, Manchester, UK, pp. 44–49 (1994)
20. Namer, F.: FLEMM: un analyseur flexionnel du français à base de règles. Traitement Automatique des Langues (TAL) 41(2), 523–547 (2000)
21. Namer, F.: Morphologie, Lexique et TAL: l'analyseur DériF. TIC et Sciences cognitives. Hermes Sciences Publishing, London (2009)
22. Pallaud, B.: Les amorces de mots comme faits antonymiques en langage oral. Recherches Sur le Français Parlé 17, 79–102 (2002)
23. Henry, S., Pallaud, B.: Amorces de mots et répétitions dans les énoncés oraux. Recherches Sur le Français Parlé 18, 201–229 (2004)
24. Bove, R.: Analyse syntaxique automatique de l'oral: étude des disfluences. Thèse de doctorat, Université d'Aix-Marseille I, Marseille (2008)
25. Dewaele, J.: Saisir l'insaisissable? Les mesures de longueur d'énoncés en linguistique appliquée. International Review of Applied Linguistics in Language Teaching 38, 17–34 (2009)
26. Blanche-Benveniste, C.: Approches de la langue parlée en français. Ophrys, Paris (1997)
27. Gaspers, J., Thiele, K., Cimiano, P., Foltz, A., Stenneken, P., Tscherepanow, M.: An evaluation of measures to dissociate language and communication disorders from healthy controls using machine learning techniques. In: IHI 2012, pp. 209–218 (2012)
28. Levin, B.: English Verb Classes and Alternations: A Preliminary Investigation. University of Chicago Press (1993)

Paraphrastic Reformulations in Spoken Corpora

Iris Eshkol-Taravella[1] and Natalia Grabar[2]

[1] CNRS UMR 7270 LLL, Université d'Orléans, 45100 Orléans, France
iris.eshkol@univ-orleans.fr
[2] CNRS UMR 8163 STL, Université Lille 3,
59653 Villeneuve d'Ascq, France
natalia.grabar@univ-lille3.fr

Abstract. Our work addresses the automatic detection of paraphrastic reformulation in French spoken corpora. The proposed approach is syntagmatic. It is based on specific markers and the specificities of the spoken language. Manual multi-dimensional annotation performed by two annotators provides fine-grained reference data. An automatic method is proposed in order to decide whether sentences contain or not paraphrastic relations. The obtained results show up to 66.4% precision. Analysis of the manual annotations indicates that few paraphrastic segments show morphological modifications (inflection, derivation or compounding) and that the syntactic equivalence between the segments is seldom respected, as these segments usually belong to different syntactic categories.

Keywords: reformulation, paraphrase, spoken corpora.

1 Introduction

The acquisition of paraphrases is an important research topic in the NLP area, as the paraphrase plays several intra and inter-speaker functions in language: it guarantees the natural character and the beauty of language, as it avoids repetitions and redundancies for instance; in language learning, in indicates the capacity to master the language; it helps the understanding and communication [1, 2], as it is widely used for the interpretation of religious, philosophical and literary texts; but it can also prevent the clarity of the communication [3], such as happens in the specialized fields with the technical paraphrases that cannot be easily understood by laymen. In any case, the speakers must share common knowledge and background to be able to detect and understand the paraphrases, and to appreciate them.

For the NLP applications, the paraphrase remains a real challenge for the same reasons: it involves a great variety of linguistic and referential mechanisms in both detection and production of paraphrases. Several NLP applications are concerned with the use of paraphrases. For information retrieval and extraction, the paraphrases allow accessing more complete and exhaustive information in documents despite the surface dissimilarity of the linguistic expressions. For question-answering and language generation systems, the paraphrases allow producing less redundant sentences that show more natural character. For

A. Przepiórkowski and M. Ogrodniczuk (Eds.): PolTAL 2014, LNAI 8686, pp. 425–437, 2014.
© Springer International Publishing Switzerland 2014

textual entailment, the paraphrases allow making inferences on the semantic relations between two statements despite the formal and lexical difference of these statements. For semantic interoperability between terminologies, ontologies and textual documents, the paraphrases allow creating links between terms from different semantic resources processed and also between these semantic resources and information contained in documents. For these different applications, the discovery and acquisition of paraphrases play very important role.

In what follows, we present work on paraphrase (Sect. 1.1) and on paraphrastic reformulation (Sect. 1.2) in linguistics and in NLP (Sect. 1.3). We then specify our objectives (Sect. 1.4).

1.1   Linguistic Description of Paraphrases

The paraphrase can be described from different points of view. For instance, it can refer to the utterance situation [4–6] and receive contextual values, such as in *two year ago* and *in 2012*. It is then opposed to the linguistic paraphrase, that involves linguistic transformations. Several typologies of linguistic transformations are proposed [7, 8, 2, 9] (the paraphrased elements are underlined):

- morphological paraphrase involves morphological processes (i.e., inflection, affixation and compounding), such as in *We need an improvement of recycling system* and *We need an improved recycling system*;
- lexical paraphrase involves changes at the lexical level with synonyms, hyperonyms, antonyms, etc., such as in *There's a risk of receiving a severe wound* and *There's a possibility of receiving serious injure*;
- semantic paraphrase often covers segments larger than words, such as in *Emma burst into tears* and *Emma cried*;
- syntactic paraphrase reorganizes sentences with the shifting of components or diathesis, such as in *The riddle is solved by him* and *He solved the riddle*, or *Bill sold a car to Tom* and *Bill sold Tom a car*;
- mixed paraphrase may involve various combination of these modifications.

Paraphrase can also be described according to the size of linguistic units involved [10, 11, 2], that distinguishes lexical, sub-phrastic and sentence paraphrases.

The existing classifications of paraphrase are often oriented on one dimension, described with more or less detail (e.g. up to 67 lexical functions [7] or 25 categories [9]). As far as we know, the only multidimensional classification considers five dimensions [12]: type of the knowledge required for the production of paraphrases; involved meaning modifications; types of linguistic modifications, which is close to the classifications above; accuracy of the paraphrastic relation; and mode of production. Besides, paraphrase may also cover two additional dimensions: register of language (e.g. specialized vs. non-specialized, spoken vs. literary [13]); and language (equivalences that correspond to translations [6, 12]). Our acceptance of paraphrase is large but reserved to one language only. Hence, we consider that paraphrase can also be used for description, precision or explanation of ideas previously expressed by a speaker.

1.2 Paraphrastic Reformulation

Reformulations occur in written and spoken language, although they show differences [10, 14]. Thus, in oral speech we can observe the elaboration of ideas, that often contains hesitations, false starts, repetitions [15], while in written documents, we find rather its final result [16]. It is usually considered that reformulation is the activity of speakers built on their own linguistic production or on the one of their interlocutor, with or without specific markers. The objective is then to modify some aspects (lexical, syntactic, semantic, pragmatic) but to keep the semantic content constant [17, 18]. Not every reformulation corresponds to paraphrase, and two categories of markers can be thus distinguished (we give examples in French): markers of non-paraphrastic reformulation (e.g. *en somme, en tout cas, de toute façon, enfin*, etc.) and markers of paraphrastic reformulation (called MPRs), like *c'est-à-dire, je m'explique, ça veut dire, en d'autres termes* [19]. With the paraphrastic reformulation, we can distinguish source and target (or paraphrased) entities, usually linked by an MPR. The following criteria are typically used for the distinction of paraphrastic reformulations [19]:

- three phonetic criteria: the repetition of the intonation contour of the sentence; the decrease of the output speed; and a very clear articulation of last syllables at the end of the paraphrase;
- syntactic parallelism of the source and paraphrased entities;
- occurrence of the MPRs, although it is possible to find paraphrases without markers. Among the MPRs, the authors distinguish markers which main task is to establish paraphrastic relations (e.g. *c'est-à-dire*), markers that can establish this relation, and markers that seldom play this role.

The MPRs provide formal mark-up of paraphrastic relations. Notice that the semantic properties of the MPRs allow creating the paraphrase relation among entities that show no semantic equivalence or similarity otherwise [19].

1.3 Natural Language Processing

The two recent literature reviews of methods proposed for the automatic detection of paraphrases [20, 21] state about the increasing importance of this topic for the NLP research. The methods used usually depend on the type of material that is exploited. Often, these methods are based on the paradigmatic properties of linguistic entities and on their capacity to be replaced by each other:

1. *Monolingual corpora.* In monolingual corpora, the string edition similarity [22] and distributional methods are mainly used. In this last case, the linguistic entities (words, phrases, etc.) have to share similar vectors to be considered as good candidates for the paraphrase [23, 24];
2. *Monolingual parallel corpora.* When a given text is translated more than once in another language, these translations allow building the monolingual parallel corpus. One of the most used is built with the English translations of *20,000 lieux sous la mer* by Jules Verne. Exploitation of such corpora becomes possible thanks to the methods for word alignment. Various approached are proposed for processing this kind of corpora [25–27];

3. *Monolingual comparable corpora.* Monolingual comparable corpora contain texts on the same event but created independently, like news articles on a given political or social event. The thematic consistency of such texts, the distributional methods and the alignment of comparable sentences allow inducing paraphrastic relations between linguistic entities [28, 29];

4. *Bilingual parallel corpora.* Bilingual parallel corpora, that typically contain translations of a given text in another language, can also be used for the acquisition of paraphrases. In this case, different translations of a given linguistic entity can provide paraphrases [30–33].

1.4 Objectives

Our objective is to work on the detection of paraphrastic reformulations. The originality of our work is related to the following points: (1) The work is done on spoken corpora, that have been very little exploited up to now for the detection of paraphrases [2]; (2) The method for the detection of paraphrastic reformulations is syntagmatic, but not paradigmatic, that is usually dedicated to monolingual corpora [20]; (3) Manual multidimensional annotation of paraphrases is done and provides the reference data; (4) Method for the automatic distinction between paraphrastic and non-paraphrastic reformulations is proposed and tested.

In the following of the paper, we describe the data exploited (Sect. 2) and the method proposed (Sect. 3). We then present and discuss the results (Sect. 4), and outline the directions for future work (Sect. 5).

2 Linguistic Data

2.1 Corpora

We use the *ESLO* (Enquêtes SocioLinguistiques à Orléans) corpora [34]: *ESLO1* and *ESLO2*. *ESLO1*, the first sociolinguistic survey in Orléans, France, has been done between 1968 and 1971 by the French department staff from the Essex University, UK in collaboration with the B.E.L.C. (Bureau pour l'étude de l'enseignement de la langue et de la civilisation françaises de Paris) lab. The corpus contains 300 hours of speech, with over 4,500,000 occurrences. The building of the corpus *ESLO2* started in 2008. The objective is to collect over 350 hours of speech with 10 M occurrences. The two corpora are available online.[1] The transcriptions apply two principles: use of the standard spelling and non-use of the written language punctuation. The segmentation is done on *breath groups* detected by the transcribers and on *turns of speech* detectable with the shift of speakers. The corpora provide different genres, such as meetings, interviews, shop and school discussions. In order to study comparable data, we use 260 interviews from *ESLO1* (2,349,829 occurrences) and 308 interviews from *ESLO2* (1,412,891 occurrences).

---

[1] http://eslo.tge-adonis.fr/

2.2 Markers of Paraphrastic Reformulation (MPR)

We exploit three MPRs: *c'est-à-dire*, *je veux dire* and *disons*. These MPRs can be translated as *in other words*, *that is to say* and *let's say* respectively. Their common feature is that they are coined on the verb *dire (to say)*. *c'est-à-dire* is the most lexicalized and the most studied [35, 36]: (1) it is used in monologues and dialogues, both spoken and written; (2) the linguistic entities in relation of paraphrase cannot be interchanged because they are not semantically equal; (3) *c'est-à-dire* can shift for instance with *autrement dit* and *en d'autres termes*; (4) it creates paraphrase relation between entities without semantic equivalence; (5) in addition to the three prototypical functions (correction, reformulation and argumentation) it can also mark conclusion, justification and hesitation. Concerning *disons*, its known characteristics are [37]: (1) it is semantically close to *je veux dire*; (2) *disons* and *eh bien* present analogy because they mark the break between two utterances; (3) *disons* and *enfin* present analogy in correction contexts. It is impossible to remove *disons* because the target entity conveys different semantics [38]. Finally, *je veux dire* is known to have several meanings and can be replaced by *autrement dit, c'est-à-dire* [39]. As indicated by the studies cited, these markers can have several functions. We use them for their paraphrastic reformulation function.

3 Methodology for the Detection of Paraphrases

Utterances that contain one of the MPRs studied are extracted from the corpora and pre-processed (Sect. 3.1). The method relies on manual (Sect. 3.2) and automatic (Sect. 3.3) processing of corpora. The analysis and evaluation of the results is done (Sect. 3.4).

3.1 Pre-processing of Corpora

In order to rebuild the utterances, the transcription files are segmented in turns of speech: new utterance begins with the shift of speakers. In case of speaker overlapping, the overlapped segments are associated with all the involved speakers. If the speaker continues the speech after the overlapping, his turn of speech continues as well. The corpora are then POS-tagged and analysed with the SEM chunker [40] adapted to spoken language. SEM detects minimal chunks.

3.2 Manual Annotation of Paraphrastic Reformulations

The manual annotation aims at thwo steps: distinguishing between paraphrastic and non-paraphrastic reformulations, and fine-grained annotation of the paraphrastic reformulations. The annotations applie to the source and target entities related by the MPRs, and to the paraphrastic relation. The annotation is done along several dimensions, some of which are inspired by the existing classifications (Sect. 1):

1. *Syntactic tag*: each entity is annotated with its POS-tag (e.g. N, A, V, Prep) or syntactic constituent (e.g. NP, VP, AP, PP). Size of entities is defined according to the semantics of the paraphrase, but not on the basis of chunks;
2. Each relation is annotated with:
    - *rel-lex*: type of lexical relation among the two paraphrased entities (e.g. hyperonym, synonym, antonym, instance, meronym);
    - *modif-lex*: type of lexical modification (i.e. remplacement, deletion, insertion);
    - *modif-morph*: type of morphological modification (i.e. inflection, derivation or compounding);
    - *modif-synt*: type of syntactic modification (e.g. active/passive);
    - *rel-pragm*: type of pragmatic relation, linked to the function of paraphrase and reformulation, inspired by the existing typologies [17, 18]. We distinguish: definition, explanation, exemplification, precision, denomination, result, linguistic or referential correction, and equivalence.

Annotation examples can be found in (1) and (2): annotation is in gray, the file reference between brackets. We can see for instance that entities {*Saint Jean de la Ruelle, Orléans*} in (1) and {*démocratiser l'enseignement (democratize the education), permettre à tout le monde de rentrer en faculté (allow everybody to enter the university)*} in (2) have the paraphrase relation.

(1)   *pendant nous avons fait grève à la Régie Renault euh de <NP1> Saint Jean de la Ruelle</NP1> <MPR>c'est-à-dire</MPR> <NP2 rel-lex="mer (Saint Jean de la Ruelle/Orléans)" rel-pragm="cor-ref">Orléans </NP2> parce que c'est ça fait partie d'Orléans* [ESLO1_ENT_149_C]

(2)   *<VP1>démocratiser l'enseignement</VP1> <MPR>c'est-à-dire </MPR> <VP2 rel-lex="syn(démocratiser/permettre à tout le monde) syn (enseignement/ faculté)" modif-lex="ajout(rentrer à)" rel-pragm="explic"> permettre à tout le monde de rentrer en faculté</VP2>* [ESLO1_ENT_121_C]

3.3   Automatic Detection of Paraphrastic Reformulations

The automatic processing addresses the first step: its objective is to decide whether a given occurrence of MPR creates the paraphrastic reformulation relation or not. On the whole set of utterances with MPRs, we apply several filters:

- if the MPR is at the beginning or end of utterance, the consider that the context is not sufficient to establish paraphrastic relation;
- if the MPR is found in specific lexical contexts, such as occurrence of *nous* with *disons (we say)*, we consider that such contexts are not paraphrastic;
- if the MPR occurs with other repeated discursive markers (*donc, enfin, quoi*), hesitation euh, interjections (*ben hm ouais*), primes (*s-*), etc., we consider that the MPR is part of oral disfluencies [15] and is not paraphrastic;
- if the MPR occurs within expression or phrase, like *indépendamment de (independently of)* in example (3), we consider that the context is not paraphrastic. This test is done with the syntactically chunked output. In order

to verify whether the expression or phrase exist, we query an online search engine and analyse the frequencies attensted on the web. We assume that the web frequencies provide with information that is more exhaustive than frequencies found in reference corpora. Thus, each segment is tested in three ways: with one, two or three chunks on the right and on the left of the MPR, excepting the disfluence markers. Size of the tested segments is empirically set to seven words at most. Then, we compute the average frequency for the three kinds of segments (one, two or three chunks on the right and on the left of the MPR). The average frequency of the segments must not be lower than the threshold tested, that is between 10 and 6,000. If the average frequency is higher than the threshold, the test indicates that the expression or phrase exist in the language and that the MPR represents the disfluency.

3.4   Analysis and Evaluation

The annotation protocol has been fixed on a subset of *ESLO1*, while the evaluation is done on the remaining *ESLO1* subset and on the *interviews* from *ESLO2*. Two kinds of evaluation are performed: (1) manual annotation is checked for the inter-annotator agreement at the level of the paraphrastic relation. With two sets of annotations, we apply the Cohen kappa [41] measure; (2) precision of automatic detection of the paraphrastic relation is evaluated against the manual annotation. The analysis of results addresses the frequency of relations and their attributes. We are particularly interested in the existence of paraphrastic relations, in the syntactic equivalence between the entities and the existence of morphological modifications (inflection, derivation or compounding), that can give formal indications on the paraphrastic relation between two segments.

4   Results and Discussion

4.1   Building and Pre-processing of Corpora

In Table 1, we indicate the size of corpora in number of words, the number and average size of turns of speech, the number of utterances with the MPRs studied, and the size of these utterances. The average size of turns of speech is between 14 and 19 words, with many minimal utterances (one or two words). *c'est-à-dire* is the most frequent, as it provides over half of utterances. *disons* is particularly frequent in *ESLO1*, but the less frequent in *ESLO2*. The difference may be due to the diachronic evolution, as other words may have taken the corresponding discursive function. Concerning the average size of utterances with MPRs, it is quite high (62.88 in *ESLO1* and 86.34 in *ESLO2*). We assume that these utterances can contain paraphrases and show the genesis of the speaker ideas [16, 15]. This can also explain the fact that the average size of these utterances is higher than the global average observed in corpora. We can observe that the maximal size of utterances is very high and can reach up to 1,050 in *ESLO2* and 6,382 in *ESLO1*.

Table 1. Description of corpora: size, number and average size of turns of speech, number of utterances with three MPRs studied, size of utterances with the MPRs

	ESLO1	ESLO2
number of transcription files	260	308
size of corpora (occ of words)	2,349,829	1,412,891
average size of transcription files	9,037,80	4,587,31
number of turns of speech	166,602	70,707
average size of turns of speech	14.10	19.98
c'est-à-dire	1,849	594
je veux dire	285	291
disons	1,068	183
total number of utterances with MPRs	3,202	1,068
size of utterances with MPRs (minimal)	1	1
size of utterances with MPRs (maximal)	6,382	1,050
size of utterances with MPRs (average)	62.88	86.34

Table 2. Jugment on the paraphrastic relation for the two annotators A1 and A2

	ESLO1			ESLO2		
	A1	A2	agr.	A1	A2	agr.
	yes no	yes no		yes no	yes no	
c'est-à-dire (number)	96 193	66 223	249	74 124	65 137	162
je veux dire (number)	16 49	8 57	57	47 91	27 110	107
disons (number)	18 104	8 115	106	10 45	9 46	46
total utterances with MPRs (number)	130 346	82 395	412	131 260	101 293	315
total utterances with MPRs (%)	27 73	17 83		33 67	26 74	

4.2   Manual Annotation of Paraphrases

We annotated 476 utterances in *ESLO1* and 394 utterances in *ESLO2* (54 and 30 interviews respectively) that contain the MPRs. These annotations correspond to our reference data. Table 2 indicates the annotation results provided by the two annotators. The annotators state that between 17 and 27% of utterances are paraphrastic in *ESLO1*, and between 26 et 33% in *ESLO2*. Annotator *A1* accepts more contexts as paraphrastic. The inter-annotator agreement is substantial (0.617) in *ESLO1* and moderate (0.526) in *ESLO2*. This is acceptable agreement rate given the inherent subjectivity induced by these data. As observed in the literature, these MPRs can occur in paraphrastic and nonparaphrastic contexts. In example (3), that does not contain paraphrases, the MPR is to be associated with discursive markers and disfluencies.

(3)   *différence sensible entre vos différents clients dans leur façon de choisir la viande dans ce qu'ils achètent et caetera indépendamment <MPR>disons </MPR> de leurs oui origines de classe* [ESLO1_ENT_001_C]

Table 3 indicates the percentage of paraphrastic and non-paraphrastic constructions with MPRs. The two annotators consider that *c'est-à-dire* is the most

Table 3. Percentage of paraphrastic and non-paraphrastic constructions with MPRs

| | ESLO1 | | | | ESLO2 | | | |
| | A1 | | A2 | | A1 | | A2 | |
	yes	no	yes	no	yes	no	yes	no
c'est-à-dire (%)	33	67	22	78	37	63	32	68
je veux dire (%)	25	75	12	88	34	66	20	80
disons (%)	15	85	7	93	18	82	6	94

grammaticalized in this function because it introduces the largest number of paraphrases, while *disons* is the less grammaticalized. Concerning *disons*, we assume that it is ambiguous: in addition to the paraphrase, it can also mean *dire (to say)* and show discursive [17] or disfluency function (example (3)).

In over 70% of contexts, there is no syntactic equivalence between the entities in paraphrastic relations (examples (4) and (5)). This aspect depends also on the annotator choice: for instance, in (4), various segments can be selected (*les gens me semblent plus plus affables*, *plus affables* or *affables*). Another interesting fact is related to the morphological modifications: only ten such modifications are observed in each corpus (e.g. {*achat (purchase)*, *achète (buy)*}, {*connais (know)*, *connu (known)*}, {*pourrait (could)*, *pouviez (can)*}, {*client (client)*, *clientèle (clientele)*}, {*manoeuvres (manoeuvres)*, *manuel (manual)*}). This means that very few formal cues are available for the detection of paraphrases in this kind of corpora. Besides, we find only one occurrence with syntactic modifications (active/passive). Concerning lexical modifications, we mainly observe replacements. As noticed in the literature [19], we can find several paraphrastic reformulations in which entities have no strong semantic relation except the one marked by the MPR, such as {*conférences (conferences)*, *causeries (chat)*} in (6).

(4)   *je préfère mieux le le nord de la France franchement le département du Nord et le département du Pas-de-Calais où <P1>les gens me semblent plus plus affables</P1> <MPR>disons</MPR> euh <PP2 rel-lex="syn" rel-pragm="explic">avec qui j'ai on a plus facilement des des rapports agréables</PP2>* [ESLO1_ENT_003_C]

(5)   *y a le euh le le plus grand goup- groupe et puis euh ce qu'on appelle <NP1>toujours les mêmes</NP1> <MPR>c'est-à-dire</MPR> euh <P2 rel-lex="syn" rel-pragm="equiv">tous ceux qu'on connait</P2> quoi* [ESLO2_ENT_1004_C]

(6)   *des conférences y en a assez souvent sur France culture enfin <MPR>disons</MPR> des causeries* [ESLO1_ENT_121_C]

Among the lexical relations, synonymy and hyperonymy are the most frequent, followed by instances with named entities, equivalence and result. According to the pragmatic relations, we can distinguish three functions of MPRs:

– possibility to add new information with explanation, precision, exemplification and definition. This function can be associated with the known functions

Table 4. Precision of the automatic detection of paraphrastic reformulations

	ESLO1		ESLO2	
	A1	A2	A1	A2
lexical and discursive filters	40.5	40.5	37.7	37.8
lexical and discursive filters + frequency (>6000)	25.8	25.9	18.7	18.9
lexical and discursive filters + priority frequency (>6000)	63.0	63.0	66.4	66.3

(correction, reformulation and argumentation [35]). In these situations, the target entity is richer and clairer, like in examples (4) and (2);

— possibility to tell the same thing, but using other linguistic means with the equivalence relation, such as in (5) and (6). We consider that, contrary to what has been noticed in the literature [38], with these relations the source and target entities are exchangeable and we can remove the MPR;

— with the relation *result*, we can observe inverse situation to the explanation: the target entity can be shorter than the source entity (example (7)).

(7)   *voilà <P1>le côté très betonné voilà c'est pas ils ont pas développé les les logements étudiants suffisamment ils ont pas développé l'off- l'offre culturelle euh en même temps</P1> donc enfin <MPR>je veux dire</MPR> voilà <P2 rel-pragm="res">c'est mort</P2>* [ESLO2_ENT_1012_C]

4.3 Automatic Detection of Paraphrastic Reformulations

Precision of the automatic detection of paraphrastic reformulations is indicated in Table 4. The results are coherent between the two annotators in the two corpora processed, although it is more complicated to correctly process the *ESLO2* corpus. The lexical and discursive filters reach 40% and 38% precision in *ESLO1* and *ESLO2* respectively. The additional use of the frequency filters decreases the results to 26% and 19%. But, when the frequency filters have priority on the lexical and discursive filters, we improve precision to up to 63% and 66%: in this case, we consider that frequency is indicative of the paraphrases even if the utterance contains oral disfluencies. Notice that precision is improved with with the increasing of the threshold. The highest threshold tested is 6,000, while the improvement of precision is observed with the average frequency between 10 and 4,500. Above that threshold, we observe no evolution of the precision values.

The precision we obtain can be considered as acceptable. It is higher than the inter-annotator agreement. It is comparable or even superior to the precision obtained in previous work [2]. By comparison with paraphrase recognition results obtained on another written corpus, that is annotated mainly with lexical, syntactic and contextual paraphrases [9], our results are similar to those provided by the baselines and some of the systems reported [21]. We expect to improve ou current results in the next future.

5   Conclusion and Future Work

We have proposed a method for the detection of paraphrastic reformulations in monolingual spoken corpora in French (*ESLO1* and *ESLO2*). One original-ity is that we take into account the specificity of the spoken data through the building of utterances, the consideration of oral disfluencies, and the use of the NLP tool adapted to spoken corpora [40]. Another originality is that we address the detection of paraphrastic reformulations with syntagmatic approach, while usually paradigmatic approaches are used with this kind of data. We accept a large acceptance of paraphrase [7, 9], that also covers clarification, explanation, or synthesis of ideas uttered previously by a speaker. We perform manual an-notation and automatic detection of paraphrases. The manual multidimensional annotation allows producing the reference data and observations on the para-phrase relations in spoken corpora. These data allow evaluating the automatic method. The inter-annotator agreement is 0.617 and 0.526 in *ESLO1* and *ESLO2* respectively. The automatic recognition of paraphrases relies on a set of filters (lexical, discursive and frequency) and reaches up to 66.4%. The comparison with the existing work confirms some previous observations [19]: (1) reformula-tions are not always paraphrastic, and can perform other functions; (2) MPRs can create paraphrastic relations between entities that do not show semantic equivalence otherwise. On contrary, we seldom observe syntactic equivalence be-tween source and target entities, we assume that it is possible to exchange the places of entities with the equivalence relation, and that it is possible to remove the MPRs.

We have several directions for future work. We plan to involve additional an-notators and organize conciliation meetings to obtain more consensual reference data. Other MPRs can be studied and compared among them. For the auto-matic detection of paraphrastic reformulations, we can improve the current per-formance thanks to a better recognition of repetitions and to machine learning. The automatic detection of boundaries of source and target entities in another perspective. We plan also to compare the paraphrastic reformulations in spoken and written corpora: we assume the process is similar, as it allows making ideas clearer, and dissimilar from the cognitive point of view [16, 15]. As indicated, the two corpora exploited have been built with similar principles but with 40 year difference. This offers the possibility to perform the diachronic study of MPRs. Besides, a similar study can be done on corpora from other languages. We can also combine this study with the social data on speakers.

Acknowledgements. We are grateful to Yoann Dupont for this help in adap-tation of the SEM chunker to oral corpora and for making it available.

References

1. François, F.: La communication inégale. Heurs et malheurs de l'interaction verbale. In: Actualités pédagogiques et psychologiques. Delachaux & Niestlé, Neuchâtel-Paris (1990)

2. Bouamor, H., Max, A., Vilnat, A.: Étude bilingue de l'acquisition et de la validation automatiques de paraphrases sous-phrastiques. TAL 53(1), 11–37 (2012)
3. Boucheron, S.: La langue de l'un, et celle de l'autre: l'entre parenthèses comme aire de reformulation. In: Répétition, Altération, Reformulation, pp. 113–118. Presses Universitaires Franc-Comtoises, Besançon (2000)
4. Martin, R.: Inférence, antonymie et paraphrase. Klincksieck, Paris (1976)
5. Vezin, L.: Les paraphrases: étude sémantique, leur rôle dans l'apprentissage. L'année Psychologique 76(1), 177–197 (1976)
6. Fuchs, C.: Paraphrase et énonciation. Orphys, Paris (1994)
7. Melčuk, I.: Paraphrase et lexique dans la théorie linguistique sens-texte. Lexique et paraphrase Lexique 6, 13–54 (1988)
8. Vila, M., Antònia Mart, M., Rodríguez, H.: Paraphrase concept and typology. a linguistically based and computationally oriented approach. Procesamiento del Lenguaje Natural 46, 83–90 (2011)
9. Bhagat, R., Hovy, E.: What is a paraphrase? Computational Linguistics 39(3), 463–472 (2013)
10. Flottum, K.: Dire et redire. La reformulation introduite par "c'est-à-dire". Thèse de doctorat, Hogskolen i Stavanger, Stavanger (1995)
11. Fujita, A.: Typology of paraphrases and approaches to compute them. In: CBA to Paraphrasing & Nominalization, Barcelona, Spain (2010) (Invited talk)
12. Milicevic, J.: La paraphrase: Modélisation de la paraphrase langagière. Peter Lang (2007)
13. Elhadad, N., Sutaria, K.: Mining a lexicon of technical terms and lay equivalents. In: BioNLP, pp. 49–56 (2007)
14. Rossari, C.: De l'exploitation de quelques connecteurs reformulatifs dans la gestion des articulations discursives. Pratiques 75, 111–124 (1992)
15. Blanche-Benveniste, C., Bilger, M., Rouget, C., Van Den Eynde, K.: Le français parlé. Études grammaticales. CNRS Éditions, Paris (1991)
16. Hagège, C.: L'homme de paroles. Contribution linguistique aux sciences humaines. Fayard, Paris (1985)
17. Gülich, E., Kotschi, T.: Les actes de reformulation dans la consultation La dame de Caluire. In: Bange, P. (ed.) L'analyse des Interactions Verbales. La Dame de Caluire: une Consultation, pp. 15–81. P Lang, Berne (1987)
18. Kanaan, L.: Reformulations, contacts de langues et compétence de communication: analyse linguistique et interactionnelle dans des discussions entre jeunes Libanais francophones. Thèse de doctorat, Université d'Orléans, Orléans (2011)
19. Rossari, C.: Les opérations de reformulation. Analyse du processus et des marques dans une perspective contrastive français-italien (1993)
20. Madnani, N., Dorr, B.J.: Generating phrasal and sentential paraphrases: A survey of data-driven methods. Computational Linguistics 36, 341–387 (2010)
21. Androutsopoulos, I., Malakasiotis, P.: A survey of paraphrasing and textual entailment methods. Journal of Artificial Intelligence Research 38, 135–187 (2010)
22. Malakasiotis, P., Androutsopoulos, I.: Learning textual entailment using SVMs and string similarity measures. In: ACL-PASCAL Workshop on Textual Entailment and Paraphrasing, pp. 42–47 (2007)
23. Lin, D., Pantel, L.: Dirt - discovery of inference rules from text. In: ACM SIGKDD Conference on Knowledge Discovery and Data Mining, pp. 323–328 (2001)
24. Paşca, M., Dienes, P.: Aligning needles in a haystack: Paraphrase acquisition across the Web. In: Dale, R., Wong, K.-F., Su, J., Kwong, O.Y. (eds.) IJCNLP 2005. LNCS (LNAI), vol. 3651, pp. 119–130. Springer, Heidelberg (2005)

25. Barzilay, R., McKeown, L.: Extracting paraphrases from a parallel corpus. In: ACL, pp. 50–57 (2001)
26. Ibrahim, A., Katz, B., Lin, J.: Extracting structural paraphrases from aligned monolingual corpora. In: International Workshop on Paraphrasing, pp. 57–64 (2003)
27. Quirk, C., Brockett, C., Dolan, W.: Monolingual machine translation for paraphrase generation. In: EMNLP, pp. 142–149 (2004)
28. Shinyama, Y., Sekine, S., Sudo, K., Grishman, R.: Automatic paraphrase acquisition from news articles. In: Proceedings of HLT, pp. 313–318 (2002)
29. Sekine, S.: Automatic paraphrase discovery based on context and keywords between NE pairs. In: International Workshop on Paraphrasing, pp. 80–87 (2005)
30. Bannard, C., Callison-Burch, C.: Paraphrasing with bilingual parallel corpora. In: ACL, pp. 597–604 (2005)
31. Madnani, N., Resnik, P., Dorr, B., Schwartz, R.: Applying automatically generated semantic knowledge: A case study in machine translation. In: NSF Symposium on Semantic Knowledge Discovery, Organization and Use, pp. 60–61 (2008)
32. Callison-Burch, C., Cohn, T., Lapata, M.: Parametric: An automatic evaluation metric for paraphrasing. In: COLING, pp. 97–104 (2008)
33. Kok, S., Brockett, C.: Hitting the right paraphrases in good time. In: NAACL, pp. 145–153 (2010)
34. Eshkol-Taravella, I., Baude, O., Maurel, D., Hriba, L., Dugua, C., Tellier, I.: Un grand corpus oral disponible: le corpus d'Orléans 1968-2012. Traitement Automatique de Langues 52(3), 17–46 (2012)
35. Hölker, K.: Zur Analyse von Markern. Franz Steiner, Stuttgart (1988)
36. Beeching, K.: La co-variation des marqueurs discursifs bon, c'est-à-dire, enfin, hein, quand même, quoi et si vous voulez: une question d'identité? Langue Française 154(2), 78–93 (2007)
37. Hwang, Y.: Eh bien, alors, enfin et disons en français parlé contemporain. L'Information Grammaticale 57, 46–48 (1993)
38. Petit, M.: Discrimination prosodique et représentation du lexique: application aux emplois des connecteurs discursifs. Thèse de doctorat, Université d'Orléans, Orléans (2009)
39. Teston-Bonnard, S.: Je veux dire est-il toujours une marque de reformulation? In: Bot, M.L., Schuwer, M., Richard, E. (eds.) Rivages linguistiques. La Reformulation. Marqueurs linguistiques. Stratégies énonciatives, pp. 51–69. PUR, Rennes (2008)
40. Dupont, Y., Tellier, I., Courmet, A.: Un segmenteur-étiqueteur et un chunker pour le français. Technical report, LIFO, Université d'Orléans, demo (2012)
41. Cohen, J.: A coefficient of agreement for nominal scales. Educational and Psychological Measurement 20(1), 37–46 (1960)

Between Sound and Spelling: Combining Phonetics and Clustering Algorithms to Improve Target Word Recovery

Marcos Zampieri[1] and Renato Cordeiro de Amorim[2]

[1] Saarland University, Germany
marcos.zampieri@uni-saarland.de
[2] Birkbeck University of London and Glyndŵr University, United Kingdom
r.amorim@glyndwr.ac.uk

Abstract. In this paper we revisit the task of spell checking focusing on target word recovery. We propose a new approach that relies on phonetic information to improve the accuracy of clustering algorithms in identifying misspellings and generating accurate suggestions. The use of phonetic information is not new to the task of spell checking and it was used successfully in previous approaches. The combination of phonetics and cluster-based methods for spell checking was to our knowledge not yet explored and it is the new contribution of our work. We report an improvement of 8.16% accuracy when compared to a previously proposed spell checking approach.

Keywords: spell checking, clustering, phonetic algorithm.

1 Introduction

Spelling English words correctly can be a rather daunting task to children and adults alike. English spelling is particularly difficult being very much unlike other languages, particularly those with a recent written form, in which the spelling of a word is directly related to the way it is pronounced. Mitton [17] discusses the pros and cons of the current English spelling system through a historical perspective as the English language changed in different directions: foreign words were incorporated, pronunciation patterns changed and new words were created. The orthography may either change accordingly to reflect those changes and make spelling easier and often simpler or it may conserve the old forms which inevitably creates difficulties to writers. Over the centuries, several attempts have been made to reform the English orthography and bring it closer to the current pronunciation of words, but there are still a great number of exceptions and difficulties in the English spelling system that writers must cope with.

Various surveys have shown that correct spelling is a difficult task even for those that are native speakers [17,9]. Gorman [8] reported that seventy-two percent of adults, participating in a literacy scheme, seemed to have particular

A. Przepiórkowski and M. Ogrodniczuk (Eds.): PolTAL 2014, LNAI 8686, pp. 438–449, 2014.

difficulty with spelling . In such scenario, one can safely assume that matters are even worst among non-native speakers, particularly those exposed to the English language for the first time during their adult life.

In order to provide help to such large population, word processors and even other types of applications such as browsers, tend to have some form of spel checking built in. The task of the latter is well-known in computational linguistics, its origins can be traced back to the work of Blair [4] and Damerau [6]. These spell checkers have two main functions, explained below.

First, to identify possible misspellings committed by users. Probably the easiest, and the most common, way to establish whether a given word is misspelled is by checking if it belongs to a list of correctly spelled words, a dictionary. Of course this presents difficulties, dictionaries tend to be rather large making this task computationally demanding. Another problem is that the misspelling may match a correct word that was not intended by the user, this is sometimes referred as *real-world* errors.

Second, to suggest the users' target word or a list of candidates containing the target word. In the case of the latter, a spell checker should aim for a short list, preferably ordered with the most likely intended spelling on the top. This list of possible target words produced by a spell checker can be seen as a ranked cluster of words with a degree of similarity to the misspelled word.

Clustering algorithms aim to group a set of N entities $y_i \in Y$, each described over V features, into K homogeneous groups $S = \{S_1, S_2, ..., S_K\}$. There are indeed a number of such algorithms, following partitional and hierarchical approaches. Hierarchical algorithms form a clustering tree in which a given entity y_i may belong to different clusters, as long as these are found at different levels in the hierarchy. Although useful these algorithms tend to have a high time complexity, in most cases of $\mathcal{O}(N^3)$. Partitional algorithms follow a different approach. They produce disjoint clusters in which an entity $y_i \in Y$ is assigned exclusively to a single cluster S_k. Such algorithms tend to have a much more favourable time complexity.

Clustering algorithms were recently applied to spell checking [2], proving to be an effective alternative to other spell checking methods, particularly those that rely on finite state automata (FSA) -based dictionaries. However, the success of the former strongly depends on the distance measure used to form clusters. The chosen distance has to accurately represent the similarity between the misspelling and the target word.

We find it very difficult to design a single distance measure that can handle all the different sources of misspellings, such as *typos* coming from neighbour letters in a keyboard, and those based on the difference between the sound and the spelling of a word. The former may be represented well by the Levenshtein metric [12], or one of its many variants. By using this metric, the distance between two words is given by the number of single character edits necessary to make them equal. Phonetic information, such as that produced by the Soundex algorithm [27], may provide us some clues regarding the target word as it may have a similar sound to that of the misspelling. Such information has been used in spell checking in different occasions such as the work of [30,28] and [29]. Selecting an inappropriate measure may result in a large measured distance between a given misspelling and its respective target word, making it difficult for a clustering algorithm to suggest the correct target.

In this paper we aim to address the above issue, as well as the fact that measuring the distances between a given misspelling and each word in a dictionary can be quite computationally demanding. We do so by introducing a clustering-based method that allows the combination of different distance metrics. Our method improves the performance of spell checkers in terms of both, processing time and target word recovery. We believe this to be the first time that a clustering based spell checker is set to use phonetic as well as a string metric to form clusters of target words, given a misspelling.

2   Related Work

Spell checking has been substantially studied over the years. The aforementioned studies published by Blair [4] and Damerau [6] were, according to Mitton [19], the pioneers for this task. The majority of spell checking methods (including the one we present) use dictionaries as a list of correct spellings (target words). A few attempts, however, try to address this problem without the use of dictionaries. Morris and Cherry [20] was the first to use the frequency of character trigrams to calculate an 'index of peculiarity'. This coefficient estimates the probability of a given trigram occurring in English words. If a word contains a rare trigram the algorithm considers this word a good candidate for misspelling.

Some years later, McIlroy [15] introduced a technique called affix-stripping. This algorithm stores a stem word, e.g. *walk*, instead of all its possible derivations: *walking, walks, walked* and apply a set of rules to handle affixes. The method was effective in identifying misspellings but a shortcoming of this technique is the generation of non-existing words.

As previously mentioned, most state of the art spell checkers use dictionaries and similarity measures to calculate the distance between the target word and the possible misspellings [1]. This approach, however, tends to be computationally demanding when working with large dictionaries (the bigger the dictionary, the greater the number of distances to be calculated). There are two alternatives that have been propose to cope with this limitation, the first of them is the use of dictionaries organized as Finite State Automata (FSA) such as in Pirinen and Linden [24] and Pirinen [25] and the second is partitioning the dictionary is smaller clusters [2]. The latter is the strategy that will be used in our experiments, which is discussed in more details in Sect. 3.

A difficulty that most dictionary-based systems face is recognizing the so-called *real-word errors*, which are misspellings that generate words that exist in the dictionary. To identify *real-world errors* spell checkers must take context into account. For example, the word *haze* exists in English and will certainly be part of every dictionary and therefore not generate any alert when it is used in a text. However, when the sentence written is *I haze you* the distribution can tell the spell checkers that there is something wrong with the sentence, turning *haze* in this particular context in a good candidate of a misspelled word.

Confusion sets [7,5,21] are an interesting technique to cope with *real-world errors*. They are small groups of words that are often confused with one another, e.g. *(there, their, they're)* or *(we're, were)* or *(than, then, them)*. There are a number of methods developed to address the question of *real-word errors* such as Verberne [31], which proposed a probabilistic context-sensitive word trigram-based method. The method works under the assumption that the misspelling of a word often results in an unlikely sequence of three words.

Among the most recent studies, Islam and Inkpen [10] uses the Google Web IT 3-gram dataset to improve recall, Xue et al. [32] using syntactic and distributional information and more recently Lin and Chu [13] using N-gram models, similar character replacement, and filtering rules for Chinese.

3   Methods

There are indeed many clustering algorithm, generally divided into partitional and hierarchical. The latter have a considerably higher time complexity than the former, hence we focus on partitional clustering algorithms.

K-Means [14] is probably the most well-known partitional clustering algorithm there is. It partitions a data set $Y = \{y_1, y_2, ..., y_N\}$ into K disjoint partitions $S = \{S_1, S_2, ..., S_K\}$, each represented by a single centroid in $C = \{c_1, c_2, ..., c_K\}$. The centroid c_k of a given cluster S_k is the center of $y_i \in S_k$. In most cases K-Means is used in conjunction with the squared Euclidean distance, defined for the V-dimensional x and z as $d(x, z) = \sum_{v \in V}(x_v - z_v)^2$. In this case the definition of centroid is rather straight forward $c_{kv} = \frac{1}{|S_k|}\sum_{y_i \in S_k} y_{iv}$. Unfortunately finding the minimization of other distance measures, such as the Levenshtein distance, is not always possible. One should note centroids were designed to be synthetic representations of clusters, in other words, in most cases $c_k \notin Y$. This means that even if we were to somehow minimize the Levenshtein distance, we would be potentially representing clusters of words by a non-existing words.

The Partition Around Medoids (PAM) [11] addresses the above problem by replacing each centroid by a medoid. The medoid m_k of a given cluster S_k is equivalent to the entity $y_i \in S_k$ with the minimum sum of distances to all other $y_j \in S_k$. PAM minimises the criterion below.

$$W(S, M) = \sum_{k=1}^{K} \sum_{i \in S_k} \sum_{v \in V}(y_{iv} - m_{kv})^2, \tag{1}$$

where M represents a set of medoids $m_1, m_2, ..., m_K$. The criterion above represents the sum of distances between each medoid $m_k \in M$ and each entity $y_i \in S_k$, for $k = 1, 2, ..., K$. This Criterion is based on the squared Euclidean distance, but clearly we can substitute $(y_{iv} - m_{kv})^2$ for $d(y_{iv}, m_{kv})$, a function returning the distance between y_{iv} and m_{kv}. PAM allow us to use a distance measure whose center is unknown. We can minimize (1) with three simple steps.

1. Randomly select K entities from Y as initial medoids.
2. Assign each entity $y_i \in Y$ to the cluster formed by its closest medoid.
3. Update each medoid $m_k \in M$ to be equivalent to the entity $y_i \in S_k$ with the minimum sum of distances to all other $y_j \in S_k$, for $k = 1, 2, ..., K$.

The partitions produced by PAM are heavily influenced by the initial medoids found ramdonly in step one. There are of course many solutions for this problem. One could run PAM a number of times and choose as final partition the one that produces the smallest $W(S, M)$ (1), but this would have an impact on the amount of time taken to produce S. Another common solution is to apply a initialization algorithm so that the medoids are not simply found at random, but instead represent the final medoids, to some level. Regarding the second solution there are a various algorithms one could you, here we choose a variation of the original intelligent K-Means (iK-Means) [16], mainly because of our previous success in clustering dictionaries with it.[2].

The iK-Means algorithm was originally designed to find the number of clusters in a data set Y, as well as representative initial centroids for K-Means. This initialization does so by extracting one anomalous cluster from Y at a time, and initializes the K-Means centroids with those of the anomalous clusters. We have adapted iK-Means to work with medoids, rather than centroids, so it can initialize the PAM algorithm. We refer to our modification as the intelligent PAM (iPAM) algorithm, it alternatingly minimizes the below.

$$W(S_A, m_t) = \sum_{y_i \in S_A} d(y_i, m_t) + \sum_{y_i \in S_A} d(y_i, m_c), \qquad (2)$$

where S_A is an anomalous cluster in Y, m_c is the entity $y_i \in Y$ with the smallest sum of distances to all other entities $y_j \in Y$, and m_t is the entity $y_i \in S_A$ with the smallest sum of distances to all other entities $y_j \in S_A$, we can see m_t as a tentative medoid based on S_A, and m_c as the medoid of the data set Y. Below, the algorithm for this minimization.

1. Set m_c as the entity with the smallest sum of distances to all other entities in the dataset Y.
2. Set m_t to the entity farthest away from m_c.
3. Apply PAM to Y using m_c and m_t as initial medoids, m_c should remain unchanged during the clustering.
4. Add m_t to M.
5. Remove m_t, and each $y_i \in S_A$ from Y. If there are still entities to be clustered go to Step 2.
6. Apply PAM to the original dataset Y, initialized by the medoids in M and $K = |M|$.

In order to complete (1) and (2) we need to define d, the function returning the distance between two strings. Using single distance measure would probably bias the clustering towards a single type of error. For instance, if we were to simply use the Levenshtein distance each cluster would contain words that are close to each other in terms of character insertions, deletions and substitutions. However, these clusters would not take into account phonetic similarities between words. We have decided to create a framework allowing the use of more than a single distance.

Our framework follows the idea that multiple distances can be combined, so that the distance between two strings x and z is given by

$$d(x, z) = \sum_{t=1}^{T} w_t d_t(x, z) \qquad (3)$$

where w_t is the weight of distance d_t, allowing different distances to have different degrees of relevance for our particular task, and T is the total number of distances used. Equation (3) is subject to $\sum_{t=1}^{T} w_t = 1$. In the experiments for this paper we have decided to use only two distances, as per below.

$$d(x, z) = w_1 * Levenshtein(x, z) + w_2 * Levenshtein(Soundex(x), Soundex(z)), \qquad (4)$$

where *Levenshtein* is a function returning the Levenshtein distance between two strings, and *Soundex* is a function returning the four digits Soundex code of a single string. This Soundex code is the same for homophones, allowing the matching of two strings that only have minor spelling differences. In our experiments we have set w_1 and w_2 to 0.5. Using (1), (2), and (4) we have developed a method used to find the target words of a misspellings. Our method is open to the use of virtually any quantity of distance measure, as long as they are valid for strings.

1. Apply the iPAM initialization to the dictionary using (4) to find the number of clusters K and a set of initial medoids M_{init}
2. Using the medoids in M_{init}, apply PAM with (4) to the dictionary to find K clusters. This should output a final set of medoids $M = \{m_1, m_2, ..., m_K\}$.
3. Given a misspelling z, calculate its distance (4) to each medoid $m_k \in M$. Save in M_* the medoids that have the distance to z equal to the minimum found plus a constant c.
4. Calculate the distance between z and each word in the clusters represented by the medoids in M_*, outputting the words whose distance is the minimum possible to z.
5. Should there be any more misspellings, go back to Step 3.

We have added a constant c to avoid our method getting trapped in local minima, increasing the chances of the algorithm finding the target word. Clearly a large c will mean more distance calculations. In our experiments we have used $c = 1$. By calculating distances initially between the misspelling z and each of the medoids $m_k \in M$ in Step 3 we reduce considerably the number of distance calculations needed to output our cluster of words. Instead of calculating 57,046 distances (the size of our dictionary, as described in Sect. 4), we reduce this number to $|M| + |S_{M_*}|$. In Sect. 5 we empirically show the latter to be much smaller than the former.

4   Experimental Setting

In order to perform our experiments we first acquired an English dictionary containing 57,046 words. This dictionary is used as our list of correctly spelled words. We have also downloaded the Birkbeck spelling error corpus[1], consisting of a list of 36,133 misspellings of 6,136 target words. This corpus has been used by various publications, including [18,2].

The corpus includes misspellings from young children and extremely poor spellers who took part in spelling tests way beyond their ability. For this reason, some of the misspellings are completely different from their target words. As stated in the corpus description, the misspellings compiled were often very distant from the target words, examples include the misspellings *o*, *a*, *cart* and *sutl* for the targets *accordingly*, *above*, *sure* and *suitable*, respectively. As a second step, we removed from the corpus all misspellings whose targets were not present in the dictionary. These words would be impossible to be handled in a dictionary-based approach such as ours, thus there was no reason to keep them in the dictionary. This reduced the corpus to 34,956 misspellings, just under 97% of the original dataset.

Our method requires the clustering of our dictionary. This clustering has to be done only once, afterwards one can use our method as many times as required. However, this can be a time consuming experience as our dictionary is rather large. In order to reduce this processing time we divided the dictionary into 26 sub-datasets, based on the first letter of each word. Subsequently, we have then applied the first and second steps of our method to each of these 26 sub-datasets. This segmentation, however, does not mean that our method will not find the target word when the misspelling happens in the first letter.

5   Results

We are interested in evaluating our method in terms of processing time and target word recovery. Calculating the processing time of a particular algorithm can be a delicate matter. Computers using different hardware, and even different operating systems, may output different times for the same algorithm. An important point not to be overlooked is that one needs to be careful to measure the algorithm, and not the skills of a particular programmer. With these points in mind we have decided that a better processing time will be given by an algorithm that requires less distance calculations. Since our dictionary has 57,046 words, this is the maximum number of calculations needed to find a set of words in a dictionary that is similar to a particular misspelling.

Regarding target word recovery, there are two measures that we need to take into account. First, given a misspelling z, our method returns a cluster S_k containing candidate words. Our method should include the target word of z in S_k. Second, we aim to have a rather small $|S_k|$, preferably $|S_k| = 1$, since S_k would be meaningless to most users if for example $|S_k| = 100$.

---

[1] http://www.dcs.bbk.ac.uk/roger/corpora.html

Table 1 presents the results of our main experiments. We take as baseline the previous research carried out by de Amorim and Zampieri (2013) [2], which used a similar clustering algorithm but did not include the use of phonetic information. It is important to point out that in [2] researchers reported results in terms of success rate. They considered as success: 1) cases in which the method returned the correct target word 2) cases in which the method returned the word with the smallest Levenshtein distance as the target word. In this paper we consider accuracy to be a more interesting measure than the above defined success rate. This allows us to have an estimation of how suitable is our method to real-world applications and for this reasons we take only the correct target words into account.

We can clearly see that by including the phonetic information in the distance calculation (4) our method has improved considerably. The new version we present in this paper is able to recover the correct target word of 2,146 more misspellings than if phonetic information was not used. We can also see that $|S_k|$, the cluster of candidate words, has decreased considerably in terms of mean size, as well as median. Given a misspelling z our method presents, in average, a cluster of potential targets S_k whose cardinality is of 2.17. In fact, in 59.72% of cases, our method suggests a single candidate word.

The number of distance calculations has also decreased considerably. For a given misspelling z, our method calculates its distance, in average, to 3,175.3 words in the dictionary. This number represents 5.57% of the total number of words in our dictionary.

Table 1. Results: The use phonetic information in comparison to those presented in de Amorim and Zampieri (2013)

Method	Clustering	Clustering + Phonetics
Accuracy (%)	41.71%	47.85%
Accuracy (Nominal)	14,579 words	16,725
Total Number of Clusters	1,570 clusters	1,215
Cluster Size (Mean)	3.78 words	2.17 words
Cluster Size (Median)	2 words	1 word
Average Distance Calculations	3,251.4	3,175.3

We find that our method compares favourably to others in the literature. Mitton[18] presents various experiments using the Birkbeck spelling error corpus. In one set of experiments variations were generated of a given misspelling z, if this variation belonged to the dictionary it would be kept as a potential target word. This method meant that for each variation of z one would need to verify if it was equivalent to each of the words in the dictionary Clearly this generates a considerable computational effort, by generating these variations and checking each one of them against a dictionary. The success rate of this method was of 33%, however, it did not produce a single candidate word to 44% of the misspellings.

Mitton [18] also experimented by assembling a collection of potential target words by representing words and misspellings with a phonetic key originated from the SPEEDCOP project [26]. Under this approach the success rate improved radically to 74.2%, however, each cluster had an average of 600 candidate words, hardly realistic. The number of distance calculations used in this approach would be equal to the size of the dictionary for any given misspelling.

The cluster of candidate words generated by the above method can be ranked. Mitton [18] experimented a ranking based on letter by letter matching between the misspelling and the candidate words. The correct target word was found among the top three ranked candidates in 47.4% of cases, however, the number of distance calculations increases considerably. With an average cluster of 600 words this means a total number of calculations, just for the ranking part of this method, equivalent to the number of misspellings times 600. This is over 20 million distance calculations for our data set. Instead of matching letters one could use the edit distance, or even other methods suggested by Mitton [18], which do indeed increase the success rate, however, these do not seem to drastically decrease the required number of distance calculations to the level of our method.

5.1   Parameter Tuning: The Constant c

Our method uses a constant c in order to reduce the chances of our method getting trapped in local optima (see Sect. 3 for details). Originally we experimented with $c = 1$, by increasing this constant one can expect to increase the accuracy of our method, at the cost of increasing the number of distance calculations as well. Table 2 presents our experiments with $c = 2$ and $c = 3$.

Table 2. Results: Tuning the constant c

Constant c Value	$c = 2$	$c = 3$
Accuracy (%)	50.44%	50.59%
Accuracy (Nominal)	17,631 words	17,270 words
Total Number of Clusters	1,570 clusters	1,215 clusters
Cluster Size (Mean)	2.23 words	2.24 words
Cluster Size (Median)	1 word	1 word
Average Distance Calculations	13,573	35,450

We obtained the best performance for our method when using $c = 3$. Along with the performance improvement we can see that the number of distance calculations increases rather rapidly, which makes the method slower than when using the constant $c = 1$. Therefore, for this setting, we do not believe this variable should be fine tuned.

6   Conclusion and Future Work

In this paper we have introduced the use of phonetic information to clustering based spell checking methods. We do so by proposing a method that allows the combination of the distance bias of different distance measures. Here, we have experimented with the popular Levenshtein distance to find the distance between two strings in terms of character insertion, deletion and substitution, as well as the Levenshtein distance between the Soundex code of the the strings in order to detect the phonetic similarity between them.

The success of our method can be measured in different ways. Given a misspelling, our method produces a cluster of potential target words whose cardinality is smaller than that of previous research. That means that the list of suggestions given to a user has in average only 2.17 words with a median of 1 – for most misspellings our method suggested a single target word. We can also see a considerable improvement in terms of having the correct target word in the cluster, or one with a distance to the misspelling that is smaller than that of the target word. Finally, we have reduced the number of distance calculations from 57,046 (if no clustering is used), which is the cardinality of our dictionary, to an average of 3,175.3, about 5.57% of the original number.

We find it important to point out that even if another method is used to reduce the number of distance calculations, for instance by assuming that the first letter of the misspelling is correct, one can still reduce calculations further by implementing our method as a second stage to that.

Nevertheless, we still see considerable room for improvement. At the present we treat each distance equally by setting each w in (4) to 0.5. In future research we intend to investigate the possibility of designing an adaptive weighting system that would optimize each value of w to the best values for a particular user. To carry out such training there is unfortunately a lack of language resources available. Research in spell checking depends heavily on the availability of spelling error corpora and to our knowledge very few suitable resources are available.

In future work we would like to investigate whether performance can be increased by using other phonetic algorithms such as the Metaphone [22] or Double Metaphone [23] as well as by using other string distance functions [6]. Another direction that our work will take is the application of this clustering method to similar problems, particularly those in which phonetic information plays an important role. The method can be applied, for example, to carry out spelling normalization in internet texts [33] or historical data [3].

References

1. de Amorim, R.: An adaptive spell checker based on ps3m: Improving the clusters of replacement words. Computer Recognition Systems 3, 519–526 (2009)
2. de Amorim, R., Zampieri, M.: Effective spell checking methods using clustering algorithms. In: Proceedings of Recent Advances in Natural Language Processing (RANLP 2013), Hissar, Bulgaria, pp. 172–178 (2013)

3. Baron, A., Rayson, P.: Vard2: A tool for dealing with spelling variation in historical corpora. In: Postgraduate Conference in Corpus Linguistics (2008)
4. Blair, C.: A program for correcting spelling errors. Information and Control 3, 60–67 (1960)
5. Carlson, A., Rosen, J., Roth, D.: Scaling up context-sensitive text correction. In: Proceedings of the 13th Innovative Applications of Artificial Intelligence Conference, pp. 45–50. AAAI Press (2001)
6. Damerau, F.: A technique for computer detection and correction of spelling errors. Communications of the ACM 7, 171–176 (1964)
7. Golding, A., Roth, D.: A winnow-based approach to context-sensitive spelling correction. Machine Learning 34, 107–130 (1999)
8. Gorman, T.P.: A survey of attainment and progress of learners in adult literacy schemes. Educational Research 23(3), 190–198 (1981)
9. Hamilton, M., Stasinopoulos, M.: Literacy, numeracy and adults: Evidence from the national child development study (1987)
10. Islam, A., Inkpen, D.: Real-word spelling correction using googleweb 1t 3-grams. In: Proceedings of Empirical Methods in Natural Language Processing (EMNLP 2009), Singapore, pp. 1241–1249 (2009)
11. Kaufman, L., Rousseeuw, P.: Finding groups in data: an introduction to cluster analysis, vol. 39. Wiley Online Library (1990)
12. Levenshtein, V.I.: Binary codes capable of correcting deletions, insertions and reversals. Soviet Physics Doklady 10, 707 (1966)
13. Lin, C., Chu, W.: Ntou chinese spelling check system in sighan bake-off 2013. In: Proceedings of the Seventh SIGHAN Workshop on Chinese Language Processing (SIGHAN-7), Nagoya, Japan, pp. 102–107 (2013)
14. MacQueen, J.: Some methods for classification and analysis of multivariate observations. In: Proceedings of the Fifth Berkeley Symposium on Mathematical Statistics and Probability, California, USA, vol. 1, p. 14 (1967)
15. McIlroy, M.: Development of a spelling list. IEEE Transactions on Communications 1, 91–99 (1982)
16. Mirkin, B.: Clustering for data mining: a data recovery approach, vol. 3. CRC Press (2005)
17. Mitton, R.: English spelling and the computer. Longman (1996)
18. Mitton, R.: Ordering the suggestions of a spellchecker without using context. Natural Language Engineering 15, 173–192 (2009)
19. Mitton, R.: Fifty years of spellchecking. Writing Systems Research 2, 1–7 (2010)
20. Morris, R., Cherry, L.: Computer detection of typographical errors. IEEE Transactions on Professional Communication 18, 54–64 (1975)
21. Pedler, J., Mitton, R.: A large list of confusion sets for spellchecking assessed against a corpus of real-word errors. In: Proceedings of LREC 2010, Malta (2010)
22. Philips, L.: Hanging on the metaphone. Computer Language 7 (December 12, 1990)
23. Philips, L.: The double metaphone search algorithm. C/C++ Users Journal 18(6), 38–43 (2000)
24. Pirinen, T., Linden, K.: Finite-state spell-checking with weighted language and error models. In: Proceedings of the Seventh SaLTMiL Workshop on Creation and Use of Basic Lexical Resources for Less-resourced Languages, Malta (2010)
25. Pirinen, T.: Weighted Finite-State Methods for Spell-Checking and Correction. Ph.D. thesis, University of Helsinki (2014)
26. Pollock, J.J., Zamora, A.: Automatic spelling correction in scientific and scholarly text. Communications of the ACM 27(4), 358–368 (1984)

27. Russel, R.: Soundex (1981)
28. Stüker, S., Fay, J., Berkling, K.: Towards context-dependent phonetic spelling error correction in children's freely composed text for diagnostic and pedagogical purposes. In: INTERSPEECH, pp. 1601–1604 (2011)
29. Toutanova, K., Moore, R.C.: Pronunciation modeling for improved spelling correction. In: Proceedings of the 40th Annual Meeting on Association for Computational Linguistics, pp. 144–151. Association for Computational Linguistics (2002)
30. Uzzaman, N., Khan, M.: A bengali phonetic encoding for better spelling suggestions. In: Proceeding of the 7th International Conference on Computer and Information Technology (ICCIT), Dhaka, Bangladsh (2004)
31. Verberne, S.: Context-sensitive spell checking based on word trigram probabilities. Master's thesis, University of Nijmegen (2002)
32. Xu, W., Tetreault, J., Chodorow, M., Grishman, R., Zhao, L.: Exploiting syntactic and distributional information for spelling correction withweb-scale n-gram models. In: Proceedings of Empirical Methods in Natural Language Processing (EMNLP 2011), Edinburgh, Scotland, pp. 1291–1300 (2011)
33. Zampieri, M., Hermes, J., Schwiebert, S.: Identification of patterns and document ranking of internet texts: A frequency-based approach. In: ZSM Studien, Special Volume on Non-Standard Data Sources in Corpus-Based Research, vol. 5. Shaker (2013)

Experiments with Language Models
for Word Completion and Prediction in Hebrew

Yaakov HaCohen-Kerner, Asaf Applebaum, and Jacob Bitterman

Dept. of Computer Science, Jerusalem College of Technology, Lev Academic Center,
21 Havaad Haleumi St., P.O.B. 16031, 9116001 Jerusalem, Israel
kerner@jct.ac.il, {asaf.app,Jacobitterman}@gmail.com

Abstract. In this paper, we describe various language models (LMs) and combinations created to support word prediction and completion in Hebrew. We define and apply 5 general types of LMs: (1) Basic LMs (unigrams, bigrams, trigrams, and quadgrams), (2) Backoff LMs, (3) LMs Integrated with tagged LMs, (4) Interpolated LMs, and (5) Interpolated LMs Integrated with tagged LMs. 16 specific implementations of these LMs were compared using 3 types of Israeli web newspaper corpora. The foremost keystroke saving results were achieved with LMs of the most complex variety, the Interpolated LMs Integrated with tagged LMs. Therefore, we conclude that combining all strengths by creating a synthesis of all four basic LMs and the tagged LMs leads to the best results.

Keywords: Hebrew, Keystroke savings, Language models, Word completion, Word prediction.

1 Introduction

Word Prediction is the process in which words that are likely to follow in a given text are suggested by displaying a list of most relevant words. Word Prediction enables a user to select a word from a given list and so to reduce the number of keystrokes necessary for typing, saving physical effort, time and spelling errors. The process of Word Prediction is as follows: As the writer types a word(s), the software produces one word or a list of word choices. Each time a word is added, the list is updated. When the target word appears in the list, it can be selected and inserted into the ongoing text with a single keystroke. Typically, word lists are numbered and words can be selected by typing the corresponding number. If the word the user seeks is not predicted, the writer must enter the word.

Word Completion is the process in which words that include the entered letter sequence are suggested by displaying a list of the most relevant words. The process of Word Completion is as follows: As the writer types the first letter or letters of a word, the software produces one word or a list of words beginning with the entered letter sequence. Each time a letter is added, the list is updated. When the target word appears in the list, it can be selected and inserted into the ongoing text with a single keystroke. Typically word lists are numbered and words can be selected by typing the corresponding number. If the word that the user seeks is not predicted, the writer must enter the next letter of the word.

A. Przepiórkowski and M. Ogrodniczuk (Eds.): PolTAL 2014, LNAI 8686, pp. 450–462, 2014.
© Springer International Publishing Switzerland 2014

The original purpose of word prediction and completion was to assist people with physical disabilities to increase their typing speed [1] and to decrease the number of keystrokes required to either write or complete a word [2]. The main aims of word prediction are to speed up typing and to reduce writing errors (particularly for dyslexic people). However, word prediction and completion have also been shown to be helpful for everyone who types text, especially those who use text editors, mobile phones search engines and short messages services. In addition, Word Prediction software often allows users to enter their own words into the word prediction dictionaries either manually, or by "learning" inputted words [3,4].

Augmentative and alternative communication (AAC) is used by individuals to compensate for their impairments in the expression or comprehension of written or spoken language [5,6]. AAC is utilized by individuals with a variety of congenital conditions such as autism, cerebral palsy, intellectual disability, and acquired conditions such as amyotrophic lateral sclerosis, aphasia, and traumatic brain injury [7]. AAC's users are able to input (letter by letter) over 10–15 words per minutes (WPM), despite the fact that many users suffer from motor impairment resulting in a typing speed of 1 WPM or less [8].

Word prediction and completion programs commonly utilize LMs that attempt to capture the properties of a language. These LMs provide the user with one or more suggestions for completing the current or, occasionally, the following word [8-10].

The main evaluation measure for word prediction is keystroke savings (KS) (Newell et al., [11]; Carlberger et al. [12]; Li and Hirst [13]; Trnka and McCoy [14]). KS measures the saving percentage in keys pressed compared to letter-by-letter text entry. KS is computed using the following formula: (chars – keystrokes) / chars × 100, where chars represents the number of characters in the text, including spaces and newlines, and keystrokes is the minimum number of key presses required to enter the text using word prediction, including the keystroke, to select a prediction from the list and a key press at the end of each utterance.

This paper is organized as follows. Section 2 presents a number of previous prediction systems as well as several previous prediction systems for Hebrew. Section 3 describes LMs in general and LM combinations. Section 4 presents the proposed combinations of LMs for word completion and prediction in Hebrew. Section 5 describes the examined corpora and the experimental results and analyzes them. Section 6 includes a summary and proposes future suggestions for research.

2   Previous Word Completion and Prediction Systems

PAL [15], of the first word prediction systems, suggests most frequent words that match the partially typed word, ignoring the context of the typed word. Profet (for Swedish) [12,16], and WordQ (for English) [17] use not only word unigrams but also bigrams. Concentrating solely on n-grams' distribution statistics for prediction, occasionally results in the suggestion of syntactically inappropriate words. Whereas, utilizing syntactic knowledge, such as PoS tags of a language, can filter such suggestions. Systems such as Syntax PAL [18] (for English), and Prophet [16] (for Swedish) base their predictions on syntactic knowledge of a language. Syntax PAL and Prophet are improvements of their previous systems PAL and Profet, respectively. Syntax

PAL has decreased the errors caused by PAL and has made it possible for users to write longer and more complicated sentences. Prophet required 33% less keystrokes than Profet, its predecessor.

2.1   Previous Hebrew Word Prediction Systems

Netzer et al. [19] were the first to present the results of a Hebrew word prediction system. Their system applied three methods: (1) Statistical knowledge – Markov LMs for unigrams, bigrams, and trigrams, (2) Syntactic knowledge – PoS tags, and phrase structures, and (3) Semantic knowledge – Associating words into categories and finding a group of rules that constrain the prospective following word.

They tested their system on three corpuses of 1M words, 10M words, and 27M words. The best results were obtained after training a hidden Markov LM on the largest corpus. In contrast to expectation, the use of morpho-syntactic information such as PoS tags lowered the prediction results. The best results were achieved using statistical data, resulting in keystroke saving of 29% with 9 word proposals, 34% for 7 proposals, and 54% for a single proposal.

HaCohen-Kerner and Greenfield [20] present another Hebrew word prediction system containing the following six components: (1) sorted lists of words, frequent nouns, and frequent verbs, (2) 6 corpuses containing around 177M words, (3) 3 LMs (trigram, bigram, and unigram) that were generated using the Microsoft Research Scalable Language-Model-Building Tool[1] and the aforementioned corpora, (4) Results of queries sent to the Google Search Engine,[2] (5) A morphological analyzer[3] generated by MILA,[4] and (6) a cache containing the 20 most recently typed words.

The KS rates (between 54% to 72%) reported in HaCohen-Kerner and Greenfield [20] are higher than those reported in Netzer et al. [19]. However, these two systems were trained and tested on different corpora. In any event, it seems that larger corpora enable improvement in the prediction results.

In this research, we concentrate on KS using various combinations of LMs, unlike the two mentioned systems that use basic LMs. In addition, we are working on 3 corpuses that are dissimilar and smaller than the corpora examined in the two previous systems. Thus, no comparisons were made between our system and these systems.

3   Language Models and Their Combinations

The most commonly used LMs are statistical N-gram LMs. An n-gram is a contiguous sequence of n items (e.g., letters, words, phonemes, and syllables) from a given sequence of text or speech. These LMs try to capture the syntactic and semantic properties of a language by estimating the probability of an item in a sentence given the

---

[1] http://research.microsoft.com/en-us/downloads/
78e26f9c-fc9a-44bb-80a7-69324c62df8c/
[2] http://www.google.com/
[3] http://www.mila.cs.technion.ac.il/eng/tools_analysis.html
[4] http://www.mila.cs.technion.ac.il/eng/about.html

preceding n–1 items. An N-gram (N = 1, 2, 3, 4, ...) word LM for a specific corpus, gives a probability to each sequence of N consecutive words in the corpus, according to its distribution in the discussed corpus. LMs with N = 1, 2, 3, and 4 are referred to as unigram, bigram, trigram, and quadgram (also known as fourgram), respectively.

Various combinations of LMs have been performed by several systems. For instance, Kimelfeld et al. [21] use combinations of LMs and HITS algorithm for XML retrieval. They show that combined LMs generally yield better results in identifying large collections of relevant elements. Kirchhoff et al. [22] applied various combinations of LMs for large-vocabulary conversational Arabic speech recognition including an LM, which uses a backoff procedure where both words and/or additional morphological features can be used in tandem. They report that combinations of more than one LM usually have a more significant effect; the greatest reduction (0.5% absolute) is obtained by combining stream models with a class-based model, involving all three morphological components (roots, stems, and morph classes). McMahon [23] in the 2nd chapter of his Ph.D. dissertation supplies an overview of word based LMs in general and combinations of LMs in particular. McMahon describes a few combination variations using several LMs, such as weighted combinations in order to produce an interpolated LM and a backoff use of LMs that either uses one LM or, alternatively utilizes an additional LM, contingent on a strict set of conditions. Beyerlein [24] has found that integrating bigram, trigram and quadgram LMs (with two additional acoustic models) into one LM leads to fewer error rates than searching for the 'best' combination of a specific LM and a particular acoustic model.

4   LMs for Word Completion and Prediction in Hebrew

As mentioned in sub-section 2.1, Netzer et al. [19] applied Markov LMs for unigrams, bigrams, and trigrams. HaCohen-Kerner and Greenfield [20] also utilized three LMs (unigrams, bigrams, and trigrams) that were automatically generated using the Scalable Language-Model-Building Tool of Microsoft Research. Similarly to many other LM systems, none of these systems applied a quadgram LM. A quadgram LM has been proven to be superior to a trigram LM in certain (e.g.,[25]). Furthermore, to the best of our knowledge, of these systems none define and apply LM combinations.

Thus, we decided to explore whether at least one combination of LMs exists that significantly improves the prediction and completion abilities of all unique LMs. To achieve this, we defined, applied, tested and compared a variety of four unique LMs (unigrams, bigrams, trigrams, and quadgrams) and combinations of these LMs as follows: (1) Basic LMs (unigram, bigram, trigram and quadgram), (2) Backoff LMs, (3) Backoff LMs integrated with tagged LMs, (4) Interpolated LMs, and (5) Interpolated, integrated and tagged LMs. A description for each specific type of LM will be provided later. In order to build the tagged LMs we used a tagger built by Adler Meni (see [26–28]). This tagger achieved 93% accuracy for word segmentation and POS tagging when tested on a corpus of 90K tokens.

(1) **Basic LMs** – The most elementary LMs. These LMs are word-based: unigrams, bigrams, trigrams, and quadgrams and solely utilize a single LM.

(2) **Backoff LMs** – Our implemented backoff LM is based on the exclusive use of the highest n-gram basic LM. If this fails to yield results, we then attempt to use the (n–1)-gram LM, etc. There are three variants of the backoff LMs. The most complex variant is the backoff quadgram LM, which initially activates the quadgram LM; if it provides at least one proposal (for prediction or completion) it is selected; if this does not occur, we then activate the trigram LM; if this yields at least one proposal it is selected; if this is also unsuccessful, we move on to the bigram LM; if this produces at least one proposal it is selected; otherwise we use the unigram LM. In the event that there several proposals have the same highest result, a proposal is selected those results using a random function. Two additional variants of the backoff LMs are the backoff trigram and the backoff bigram.

(3) **Backoff integrated LMs and Tagged LMs** – Backoff Integrated LMs with Tagged LMs (the tagged LMs are also referred to as POS-based LMs) is the first variant of this method. This variant uses the back-off LM mentioned in the previous paragraph. This variant is called conservative since only in a case that there are at least two proposals with the same highest result proposed by the n-gram LM we attempt to choose between them based on the compatible tagged n-gram LM. If no selection was made, we attempt the (n–1)-gram LM and so on. If several proposals with the same highest result are proposed by the unigram LM, one is selected using a random function. The second variant of this method is Backoff Integrated Tagged LMs with Basic LMs. In contrast to the previous model, this model first activates the tagged n-gram LM. According to the likely POS-tag, it retrieves in context the word that is most likely to fit this POS-tag using the compatible n-gram LM.

(4) **Interpolated LMs** – This LM is considered to be a general LM as it synthesizes all 4 types of n-gram LMs (unigram, bigram, trigram, and quadgram). There are various possible techniques to determine how much weight to give to each n-gram LM. We have implemented 4 specific variants of this type of LM as follows: (1) Fixed Equal Weights (0.25) for each n-gram LM. Equal weight is given to all n-gram LMs due to the assumption that each n-gram LM carries equal strength, unless proven otherwise. We refer to this specific model as Synthesized LM with Equal Fixed Weights; (2) Fixed Unequal Weights distributes the weights as follows: 0.4, 0.3, 0.2, and 0.1 for the quadgram, trigram, bigram and unigram LMs, respectively. The main reason for giving higher weights to higher n-gram LMs is that they contain larger context environments and therefore are supposed to be more successful than lower n-gram LMs. We call this model interpolated LM with fixed unequal weights; (3) a different weight is given to each n-gram LM according to its relative rate of successful predictions and completions of words. We call this model interpolated LM with relative weights; and (4) similar to the previous variant (i.e., weight is given to each n-gram LM in accordance with its relative rate of successful word predictions and completions) with a special statistical treatment of the first word in each sentence. This treatment is based on the distribution of the first word in all sentences. When we wish to predict or to complete the first word in any sentence, a special treatment is added, based in the assumption that the distributions of the first word in all sentences differ from the distributions of words in other parts of the sentences. This treatment means that we concentrate on distributions of the first word in all sentences rather than the distributions of regular words.

(5) Interpolated and integrated LMs and tagged LMs – This LM includes 3 variants. These variants correspond to variants 2–4 of the previously mentioned LMs (Interpolated LMs). However, in addition to those methods, this LM also takes into consideration the tagged LMs in a similar way to that presented in the third kind of LM (Backoff Integrated LMs and Tagged LMs). The 3 variants are as follows: (1) Fixed unequal weights of 0.4, 0.3, 0.2, and 0.1 for the quadgram, trigram, bigram and unigram LMs, respectively; (2) weights are given to each n-gram LM according to its relative rate of successful word predictions and completions; and (3) weights are given to each n-gram LM according to its relative rate of successful word predictions and completions with a special statistical treatment for the first word in each sentence based on the distribution of the first word in all sentences.

5   Experimental Results

5.1   Examined Corpora

The examined corpora are three Israeli news web corpuses: *NRG,*[5] *TheMarker (TM),*[6] and *Arutz Sheva (A7)*.[7] These corpora contain 2,130,477 tokens. In this research, we work with a simplified definition of a token in Hebrew as follows: A token is a string of contiguous characters between two spaces, or between a space and punctuation marks. A token can also be an integer, real, or a number with a colon (time, for example: 2:00). All other symbols are tokens themselves except apostrophes and quotation marks in a word (with no space), which in many cases symbolize acronyms or citations. A token can present a single Hebrew word or a group of words such as the following token "ולאחי" (VeLeAhi) that includes 4 words "And to my brother".

These corpora represent three different types of Israeli web newspapers: Israel's political left-center, economic, and national religious. The diversity of three different corpora allows for testing of the performance of all LMs on various news genres. Table 1 presents general information regarding the examined corpora.

5.2   Experimental Set

We have tested 16 specific LMs belonging to 5 kinds of LMs, which were detailed in the previous section. Each LM was tested on the 3 corpora mentioned in Table 1. We simulated a process of user interaction in the following manner: we went over each word in each sentence in the test corpus. Firstly, we attempted to predict the next word in its entirety. If we failed to do so, we tried to predict a single character at a time until a space or a dot (or any other punctuation mark) was reached or until the next word was correctly proposed.

---

[5] NRG is the online edition of the Israeli newspaper Maariv. While most of the content on the website comes from the print edition, some of the material is written exclusively for the web edition.

[6] TheMarker is an economic news website that offers ongoing coverage of the capital markets in Israel and globally, high-tech, advertising and media industries, real estate, labor market, consumer markets, the Israeli legal world, communication, vehicles and transportation.

[7] Arutz Sheva is an Israeli national religious media network. It offers free podcasts, live streaming radio, a daily email news update, streaming video and 24 hour updated text news.

Table 1. General information regarding the examined corpora

News-paper	Period	# of files	# of sen-tences	# of tokens	Avg. # of to-kens per file	Avg. # of to-kens per sen-tence	Avg. # of cha-racters per token
NRG	MAY-OCT/04	2,500	26,313	551,518	220.61	20.96	3.82
The-Marker	MAY-OCT/02	835	32,462	698,577	836.62	21.52	3.99
A7	FEB-AUG/01	8,724	52,497	880,382	100.92	16.77	4.04

The KS results are reported when only one suggestion (with the highest result) is proposed. For each specific model we present a unique table. The leftmost column presents the tested corpus. The second-to-left column shows the KS result for prediction of a word in its entirety. The third-to-left column shows the KS result for the completion of a word after having the first letter of the discussed word. The fourth-eighth left columns present the KS result for the completion of a word after having at least the first 2–6 letters of the discussed word, respectively. The rightmost column shows the KS result for the completion of a word after having at least the first 7 or more letters of the discussed word.

Table 2. KS results of the basic LM

LM	Cor-pus	Pre-dic-tion	1	2	3	4	5	6	7+
1-gram	NRG	0.34	5.47	11.69	20.66	26.44	28.02	28.37	**28.65**
	TM	1.25	4.54	10.14	18.72	23.92	25.97	26.60	**27.20**
	A7	0.81	3.81	10.51	19.58	24.63	26.17	26.61	**27.11**
2-gram	NRG	13.86	23.62	28.96	31.42	31.85	31.93	31.95	**32.76**
	TM	11.22	19.41	26.02	28.82	29.47	29.62	29.66	**30.68**
	A7	15.28	27.35	36.47	40.09	40.84	40.97	41.02	**41.64**
3-gram	NRG	13.75	15.99	17.00	17.22	17.26	17.26	17.26	**18.24**
	TM	11.21	13.65	14.72	14.87	14.88	14.90	14.91	**16.23**
	A7	26.34	31.99	34.28	34.68	34.71	34.72	34.73	**35.64**
4-gram	NRG	12.35	13.12	13.38	13.43	13.44	13.44	13.44	**14.46**
	TM	6.36	6.80	6.88	6.89	6.89	6.90	6.90	**8.37**
	A7	22.79	24.1	24.5	24.55	24.55	24.55	24.56	**25.55**

Table 2 presents the KS results for the basic LMs: unigram, bigram, trigram and quadgram. The results of the Backoff LMs are presented in Table 3. Table 4 presents the KS results for the two variants of the Conservative Integrated LMs. Table 5 presents the KS results for the 3 variants of the Interpolated LMs. Table 6 presents the results of the 3 variants of the Interpolated LMs Integrated with Tagged LMs.

Main conclusions drawn from Table 2:

- The n-gram models can be ranked according to their KS results (especially the results from 3 known letters or more) in the following descending order: bigram (the best model), unigram, trigram, and quadgram. The limited success of the trigram and quadgram LMs is probably due to the fact that in many cases the discussed word is not presented at all or it lacks the highest frequency as the last word in n-gram strings (for n ≥ 3). Furthermore, there are many more potential 3 or 4 gram strings than potential 2 gram strings; therefore, it is less likely to predict or complete the correct word. Due to these, the results of the quadgram produced poorer results than the trigram.
- The KS results of the unigram model are relatively low for fewer than 3 known letters (especially for 0 or 1 known letters). This finding is probably due to the fact that the unigram LM for fewer than 3 known letters lacks the necessary context to successfully predict or complete a word or its beginning; this is in contrast to higher n-gram models that usually possess at least one previous word.
- In most cases, all LMs produced higher KS results the more known letters they had.
- However, even with at least 5 letters, KS improvement rates were less than 1% for all corpora.

Table 3. KS results of the Backoff LMs

LM	Corpus	Prediction	1	2	3	4	5	6	7+
Backoff 2-gram	NRG		19.13	34.75	43.20	48.56	51.76	52.75	52.96	**53.13**
	TM		11.66	20.56	28.63	35.17	39.06	40.65	41.14	**41.59**
	A7		14.96	27.46	37.46	43.61	46.65	47.68	47.99	**48.23**
Backoff 3-gram	NRG		19.91	27.82	33.38	39.16	43.28	44.74	45.11	**45.41**
	TM		16.13	23.92	31.22	37.54	41.27	42.81	43.33	**43.77**
	A7		28.35	38.11	45.21	50.13	52.98	54.00	54.32	**54.56**
Backoff 4-gram	NRG		20.74	28.81	34.27	40.01	44.03	45.54	45.88	**46.17**
	TM		15.78	22.72	29.62	35.79	39.81	41.41	41.93	**42.40**
	A7		32.55	40.67	47.27	52.27	54.99	55.97	56.24	**56.47**

Main conclusions drawn from Table 3:

- Concerning completion of words for the NRG corpus, the Backoff Bigram LM was the superior backoff LM. The second best Backoff LM was the one which begins from the Quadgram and the poorest performance was from the Trigram. This finding suggests that the Backoff Bigram LM had a better context environment than all other higher n-gram LMs for the word completion task for this corpus. However, concerning completion of words for the other corpora, the Backoff Bigram LM produced the poorest results. Thus, as of writing this paper, no superior Backoff Integrated LM was found. These LMs seem to be corpus-dependent.
- Concerning prediction of words, the Backoff Quadgram LM was found to be the best Backoff LM (for two corpora). The second best Backoff LM was the Bbackoff Trigram LM and least successful one was the Backoff Bigram LM. This finding is in contrast to the previous conclusion. This suggests that for the task of word prediction task the Backoff Quadgram LM possesses a superior context environment than all other Backoff LMs due to its larger word context.
- All Backoff LMs are better than all basic n-gram models including the bigram LM (the best basic LM). This finding means that the simple combination of LMS using a backoff LM is much better than using only one basic LM.

Table 4. KS results for the LMs integrated with tagged LMs

LM	Corpus	Prediction	1	2	3	4	5	6	7+
Backoff Integrated LMs with Tagged LMs	NRG	33.66	41.49	46.56	51.06	54.18	55.24	55.49	**55.71**
	TM	18.34	25.46	32.21	38.41	42.05	43.74	44.24	**44.71**
	A7	32.85	40.85	47.28	52.07	54.91	55.94	56.27	**56.55**
Backoff Integrated Tagged LMs with LMs	NRG	21.58	28.22	33.60	38.60	42.21	43.69	44.06	**44.34**
	TM	17.76	24.42	30.71	36.41	40.05	41.58	42.09	**42.54**
	A7	30.96	38.74	45.02	49.67	52.43	53.51	53.79	**54.09**

Main conclusions drawn from Table 4:

- The KS results of the Conservative LM integrated with Tagged LMs proved superior to those of the Backoff Tagged LMs integrated with LMs. This finding suggests that activating the regular LM before the Tagged LM is preferred to the opposite method (Tagged LM and then the LM).

- The KS results of the LM integrated with Tagged LMs proved superior to those of the corresponding Ba ckoff LMs. This finding suggests that integration of an n-gram LM and a Tagged n-gram LM before using the (n–1)-gram LM leads to better results.

Table 5. KS results for the Interpolated LMs

Inter-polated LM	Cor-pus	Pre-dic-tion	1	2	3	4	5	6	7+
Fixed equal weights	NRG	22.72	30.13	35.97	41.45	45.12	46.57	46.93	**47.19**
(0.25 for each LM)	TM	16.02	23.03	30.00	36.27	40.35	42.04	42.58	**43.05**
	A7	31.77	39.62	46.57	51.85	54.74	55.86	56.17	**56.52**
Fixed unequal weights	NRG	36.02	44.18	49.26	53.79	56.8	57.81	58.07	**58.28**
(0.4, 0.3, 0.2, 0.1)	TM	16.97	23.88	30.83	37.15	41.07	42.64	43.14	**43.57**
	A7	31.71	39.75	46.41	51.34	54.19	55.57	55.81	**56.31**
	NRG	21.16	28.63	34.12	39.7	43.66	45.13	45.51	**45.77**
Relative weights	TM	17.18	24.09	30.91	36.99	40.89	42.43	42.95	**43.41**
	A7	31.73	39.94	46.59	51.34	54.19	55.54	55.82	**56.28**
Relative weights with treatment for the first word in each sentence	NRG	39.82	48.38	53.33	57.66	60.51	61.44	61.67	**61.83**
	TM	16.91	23.98	31.14	37.39	41.33	42.88	43.39	**43.93**
	A7	31.88	40.02	46.67	51.65	54.49	55.54	56.37	**56.62**

Main conclusions drawn from Table 5:

- The leading KS results were achieved by the most complex type of Interpolated LM, the LM with relative weights in which the first word in each sentence were treated.
- The results of the best Interpolated LM are also slightly higher than those of the best integrated LM and far superior to those of the Backoff LMs and the Basic LMs. The main reason for this might be that a real synthesis of all four basic LMs leads to better results than all other LMs, especially in comparison to using only one LM according to the Backoff LMs or a unique Basic LM.

Table 6. KS results for the Interpolated LMs Integrated with Tagged LMs

Interpo-lated LM	Cor-pus	Pre-dic-tion	1	2	3	4	5	6	7+
Fixed unequal weights (0.4, 0.3, 0.2, 0.1)	NRG	42.14	50.39	54.95	59.09	61.74	62.64	62.85	63.01
	TM	18.00	24.82	31.77	37.92	41.76	43.33	43.85	**44.39**
	A7	33.12	40.91	47.19	52.03	54.87	55.85	56.34	**56.78**
Relative weights	NRG	40.96	49.14	53.93	58.22	61.02	61.98	62.19	**62.35**
	TM	17.84	24.85	31.76	37.75	37.74	43.50	43.90	**44.44**
	A7	33.91	41.85	48.34	52.93	55.75	56.72	57.03	**57.29**
Relative weights with treatment for the first word in each sentence	NRG	42.06	50.46	55.11	59.28	62.06	62.98	63.19	**63.37**
	TM	18.00	25.03	31.67	37.88	41.80	43.34	43.87	**44.33**
	A7	32.77	40.66	46.94	51.72	54.51	55.56	55.88	**56.15**

Main conclusions that may be drawn from Table 6:

- It is unclear which variant is ideal. The results of all 3 variants are rather similar. The second variant presents the best results for two corpora, while the third variant presents the best results for the third corpus. Thus, the first variant (the most basic variant) is least successful for all corpora.
- The KS results achieved by all 3 variants of this kind of LM are superior to the results achieved by the best variant so far, the most complex type of Interpolated LM. This finding suggests that this LM is superior to others.

6   Summary, Conclusions and Future Work

We have implemented five types of LMs (including 16 variants) to support word prediction and completion in Hebrew. The best KS results were achieved by the two most complex variants of the most complex kind of LM, the Interpolated and Integrated LMs and Tagged LMs. Sharing all the strengths in the form of a real synthesis of all 4 basic LMs and the Tagged LMs leads to the best results.

The improvements of the KS rates for completion of a word after having at least 5 letters was less than 1% for all corpora. That is to say, the contribution of the various

LMs and the combinations of LMs is primarily expressed for either prediction or completion of words for less than 5 letters. The KS rates for the *TheMarker* corpus (the economic corpus) are significantly lower than those achieved for the other corpora. The reason for this may be that this corpus has relatively larger diversity and contains fewer repetitions of the same n-grams. Integration of other components to the LMs (see next paragraph) might lead to improvements in KS for all corpora in general and for the economic corpus in particular.

Future directions for research would be to define and apply other combinations of LMs for this task as well as for other domains, applications and languages. Examples of possible new LMs are LMs acquired by sampling of n-grams from documents or sampling of documents in order to speed up the construction of LMs. Another general potential direction is integrating LMs and combinations of LMS with other software components (e.g., POS-tagger, morphological analyzer, and cache model) in order to improve word prediction and completion in general and for Hebrew in particular.

Acknowledgments. The authors would like to thank Meni Adler for providing his tagger, MILA and the Natural Language Processing Project in the computer science department at Ben-Gurion University of the Negev for providing their Hebrew resources, and Maayan Sarig for her edits and suggestions.

References

1. Tam, C., Wells, D.: Evaluating the Benefits of Displaying Word Prediction Lists on a Personal Digital Assistant at the Keyboard Level. Assistive Technology 21, 105–114 (2009)
2. Anson, D., Moist, P., Przywara, M., Wells, H., Saylor, H., Maxime, H.: The Effects of Word Completion and Word Prediction on Typing Rates Using On-Screen Keyboards. Assistive Technology 18, 146–154 (2006)
3. Beukelman, D., Mirenda, P.: Augmentative and Alternative Communication: Supporting Children and Adults with Complex Communication Needs, 3rd edn., p. 77. Brookes Publishing, Baltimore, MD (2008)
4. Darragh, J.J., Witten, I.H., James, M.L.: The Reactive Keyboard: A Predictive Typing Aid. Computer 23(11), 41–49 (1990)
5. Calculator, S., et al.: Roles and Responsibilities of Speech-Language Pathologists With Respect to Augmentative and Alternative Communication: Position Statement. Technical report, American Speech-Language-Hearing Association (2004), http://www.asha.org/docs/html/PS2005-00113.html
6. Fossett, B., Mirenda, P.: Augmentative and Alternative Communication. In: Odom, S.L., Horner, R.H., Snell, M.E. (eds.) Handbook of Developmental Disabilities, pp. 330–366. Guilford Press (2009) ISBN 978-1-60623-248-4
7. Beukelman, D., Mirenda, P.: Augmentative & Alternative Communication: Supporting Children & Adults with Complex Communication Needs, 3rd edn. Paul H. Brookes Publishing Company (2005) ISBN 978-1-55766-684-0
8. Trnka, K., McCaw, J., Yarrington, D., McCoy, K.F.: User Interaction with Word Prediction: The Effects of Prediction Quality. Special Issue of ACM Trans. on Accessible Computing (TACCESS) on Augmentative and Alternative Communication 1(3), 1–34 (2009)
9. Darragh, J.J., Witten, I.H.: Adaptive Predictive Text Generation and the Reactive Keyboard. Interacting with Computers 3(1), 27–50 (1991)

10. Wandmacher, T., Antoine, J.-Y.: Methods to Integrate a Language Model with Semantic Information for a Word Prediction Component. In: Proc. ACL SIGDAT Joint Conference EMNLP-CoLLN 2007, Prague, Tchéquie, pp. 503–513 (2007)
11. Newell, A., Langer, S., Hickey, M.: The Rôle of Natural Language Processing in Alternative and Augmentative Communication. Natural Language Engineering 4(1), 1–16 (1998)
12. Carlberger, A., Carlberger, J., Magnuson, T., Hunnicutt, M.S., Palazuelos-Cagigas, S.E., Navarro, S.A.: Profet, a New Generation of Word Prediction: An Evaluation Study. In: Copestake, A., Langer, S., Palazuelos-Cagigas, S. (eds.) Natural Lang. Processing for Communication Aids, Madrid. Proc. of a Workshop Sponsored by acl, pp. 23–28 (1997)
13. Li, J., Hirst, G.: Semantic Knowledge in Word Completion. In: Proceedings of the 7th Int. ACM SIGACCESS Conf. on Computers and Accessibility (ASSETS), pp. 121–128 (2005)
14. Trnka, K., McCoy, K.F.: Evaluating Word Prediction: Framing Keystroke Savings. In: ACL (Short Papers) 2008, pp. 261–264 (2008)
15. Swiffin, A.L., Pickering, J.A., Arnott, J.L., Newell, A.F.: PAL: An Effort Efficient Portable Communication Aid and Keyboard Emulator. In: Proceedings of the 8th Annual Conference on Rehabilitation Technology, pp. 197–199 (1985)
16. Carlberger, J.: Word Prediction: Design and Implementation of a Probabilistic Word Prediction Program. Master dissertation. Royal Institute of Technology, Stockholm (1997)
17. Shein, F., Nantais, T., Nishiyama, R., Tam, C., Marshall, P.: Word Cueing for Persons with Writing Difficulties: WordQ. In: Technology and Persons with Disabilities Conference, Los Angeles, CA (2001)
18. Morris, C., Newell, A., Booth, L., Ricketts, I., Arnott, J.: Syntax Pal: A System to Improve the Written Syntax of Language-Impaired Users. Assistive Techn. 4(2), 51–59 (1992)
19. Netzer, Y., Adler, M., Elhadad, M.: Word Prediction in Hebrew: Preliminary and Surprising Results. In: ISAAC (2008)
20. HaCohen-Kerner, Y., Greenfield, I.: Basic Word Completion and Prediction for Hebrew. In: Calderón-Benavides, L., González-Caro, C., Chávez, E., Ziviani, N. (eds.) SPIRE 2012. LNCS, vol. 7608, pp. 237–244. Springer, Heidelberg (2012)
21. Kimelfeld, B., Kovacs, E., Sagiv, Y., Yahav, D.: Using Language Models and the HITS Algorithm for XML Retrieval. In: Fuhr, N., Lalmas, M., Trotman, A. (eds.) INEX 2006. LNCS, vol. 4518, pp. 253–260. Springer, Heidelberg (2007)
22. Kirchhoff, K., Vergyri, D., Bilmes, J., Duh, K., Stolcke, A.: Morphology-based Language Modeling for Conversational Arabic Speech Recognition. Computer Speech & Language 20(4), 589–608 (2006)
23. McMahon, J.G.G.: Statistical Language Processing Based on Self-organising Word Classification. Doctoral dissertation, Queen's University of Belfast (1994)
24. Beyerlein, P.: Discriminative Model Combination. In: Proceedings of the 1998 IEEE International Conference on Acoustics, Speech and Signal, vol. 1, pp. 481–484 (1998)
25. Badaskar, S., Agarwal, S., Arora, S.: Identifying Real or Fake Articles: Towards better Language Modeling. In: IJCNLP, pp. 817–822 (2008)
26. Adler, M.: Hebrew Morphological Disambiguation: An Unsupervised Stochastic Word-based Approach. Ph.D. Dissertation, Ben Gurion University, Israel (2007)
27. Adler, M., Netzer, Y., Gabay, D., Goldberg, Y., Elhadad, M.: Tagging a Hebrew Corpus: The Case of Participles. In: LREC 2008, European Language Resources Association, Marrakech, Morocco (2008)
28. Goldberg, Y., Adler, M., Elhadad, M.: EM Can Find Pretty Good HMM POS-Taggers (When Given a Good Start). In: Proceedings of the ACL 2008 Conference, pp. 746–754 (2008)

Slovak Web Discussion Corpus

Daniel Hládek, Ján Staš, and Jozef Juhár

Department of Electronics and Multimedia Communications, Technical University
of Košice, Park Komenského 32, 04001 Košice, Slovak Republic
{daniel.hladek,jan.stas,jozef.juhar}@tuke.sk

Abstract. This contribution aims to provide a representative sample
of Slovak colloquial language in an organized corpus. The corpus makes
it possible to study spontaneous, interactive communication that often
includes various incorrect or unusual words. The corpus includes a complete set of web discussions about various topics from a single site. Each
discussion is marked with a topic and talking person and is assigned to
a specific section. The corpus includes an index for easy searching using
regular expressions. Text of the discussions is processed with our tools for
word tokenization, sentence boundary detection and morphological analysis. Token annotations include a correct word, proposed by a statistical
correction system.

Keywords: web discussions, Slovak language, spelling correction.

1 Introduction

Most of the research corpora for various languages contain grammatically correct, well-formed language, as it can be found in newspaper, fiction or science
works. People usually use phrases and word forms in a different way, sentences are
usually shorter. This type of communication in contemporary research corpora
is partially covered only in fiction literature. Because this type of communication
is very rare, especially in case of Slovak, this effort brings a collection of web
discussions processed and prepared for studying.

In the case of Slovak language only a small amount of language resources is
available in form of a corpus, ready for natural language processing or statistical
language modeling. The only existing formal research corpora are maintained by
the Slovak National Corpus organization,[1] but the corpora are publicly available
only as a limited full-text search interface. On the other hand, the results of our
research will be fully available on web site.[2]

A set of informal corpora and processing tools have been prepared in our
previous research. As a part of the Slovak Judicial Domain Dictation System
applied research task [8], a web-crawling, gathering and processing agent has
been proposed [3]. A similar attempt was performed for Czech language [10].

---

[1] http://korpus.juls.savba.sk/index_en.html
[2] http://nlp.web.tuke.sk

A. Przepiórkowski and M. Ogrodniczuk (Eds.): PolTAL 2014, LNAI 8686, pp. 463–469, 2014.
© Springer International Publishing Switzerland 2014

2 Contents of the Corpus

The first step in corpus creation was selection of the source. It should be accessible and easy to parse. For this purpose, a public discussion web site[3] has been selected. The site is not thematically oriented, but rather provides space for talks about common subjects. Because of its focus on broad audience, the used language can mostly be considered as representative for all web discussions.

A specialized web agent [3] is used to explore the discussion web site. After HTML code is downloaded, content of each web page is analyzed and saved in a database. A custom parser is designed to extract interesting meta-information (described below) about each discussion. Discussions are sorted into sections according to their theme, as it was found on the web site. A summary of domains and number of discussions can be found in the Table 1.

Table 1. Contents of the corpus

section	# discussions	# sentences	# tokens
philosophy	41	11 127	94 057
culture	29	4 700	42 214
relationships	881	175 796	1 504 175
religion	834	239 079	2 546 434
computers	210	16 488	139 275
politics	552	79 195	808 724
miscellaneous	3 320	511 600	4 646 054
sport	10	1 257	11 614
health	17	4 463	38 727
together	5 894	1 043 705	9 831 274

The contents of the corpus are stored in a form that should be easy to process using any common programming language or tool. The data processing step should preserve as much information as possible. Only those parts that are not interesting for the intended use of the corpus are removed.

There are many possible ways to do so and none of them can work for all use-cases. Probably the most popular form of organization of structured data is XML as it is in [1]. Its biggest advantage is a precise description of the stored information and relatively easy parsing. On the other hand, it is also very verbose, requires working with external parsing library and it is unable to read improperly formatted data.

For this reason, our own way of data organization has been designed. It has a more informal structure that can be processed using only very basic tools and can be easily viewed and queried using only shell commands or common text editor. It does not require working with an additional library or any knowledge of the XML format.

---

[3] http://diskusneforum.sk

The index is a separate file where meta-information about documents is stored. Each line in the index file contains the following information, separated by a tabulator:

- name of the file, including section name;
- topic of the discussion;
- date of publication;
- author starting the discussion.

After identification of words and sentences in the text, each identified token has annotations assigned. Token annotations are stored together with tokens in the document files. Each line of the document file contains a single sentence with tokens separated by a space. In addition, each token has appended annotations, separated by a vertical line character. Token annotations do not contain any spaces (if space is a part of a proposed correction, it is replaced by underscore). Thanks to this simple format of annotations it is possible to parse a document file easily and utilize only meta-information that is useful for the given task.

3 Word and Sentence Boundary Detection

Classical regular expressions are very hard to read and complicated to find potential errors. For the task of the tokenization of the natural written language it is necessary to perform a separate matching operation for each character and each tokenization rule. If the proposed system contains more regular expressions, the system becomes too complex.

In order to overcome these problems, each proposed rule of the formal grammar is compiled into a single state-machine, using a specialized parser generator Ragel [11]. Rules in the Ragel language are more readable than rules in classical regular expression engines and the whole system is later translated to a single state machine in the C language.

The result is a word and sentence boundary detector, written in the C language, that has no external dependencies and is usable as a library in other systems. A similar rule-based system for lexical and morphological analysis has been developed for Croatian language [2].

The main goal is to distinguish between types of tokens that are interesting for further processing by adding and removing spaces and unnecessary characters. The following types of tokens are recognized:

- words and acronyms;
- abbreviations;
- various number representations;
- emotional marks;
- URLs and e-mails;
- punctuation.

The regular expressions in the proposed tokenizer try to identify tokens that should be preserved, and tokens that should be joined by removing spaces from the input stream of characters. After these tokens are correctly found in the input text, it is possible to identify the end of the sentence by the punctuation used.

Special attention should be paid to the tokens that contain a dot. The dot is probably the most ambiguous character in the Slovak language – it can be a part of abbreviation, a part of an ordinal numeral or it can mean the end of a sentence. Fortunately, the Slovak convention is to use comma as a floating point mark.

4 Part of Speech Tagging

An important part of the annotation process is the morphological tagger Dagger [4]. This classifier uses second-order hidden Markov model (HMM) and Viterbi algorithm and can utilize grammatical features for smoothing of the observation and transition matrix for improvement classification accuracy. The whole classification process is briefly depicted in Fig. 1.

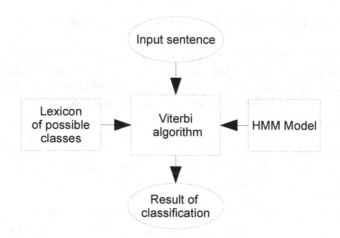

Fig. 1. Tokens Classification

The model has been trained on trigram counts from the Slovak National Corpus [6] and uses their tag set containing $3,500$ distinct tags. The search space of the classifier is restricted by a lexicon that contains a list of possible tags for each known word. Observation probabilities are smoothed using custom algorithm that takes morphological features of words into account. The classifier is 86% correct [4]. The reason for the relatively low accuracy of the morphological analysis is a high number of possible classes and the low quality of the used training data. Detailed documentation for all possible tags and flags is available on the web site of the Slovak National Corpus.[4]

---

[4] http://korpus.juls.savba.sk/attachments/morpho/tagset-www.pdf

5   Automatic Correction

Spontaneously written text usually contains a large amount of misspelled words, unusual forms, words without diacritical markings or inappropriate words. The biggest problem with processing of this type of data is a high occurrence of typographical errors, misspelling, omission of diacritical marks, slang and dialect words, omission or overuse of punctuation, random capitalization of words and emotion markings.

The most common error is incorrect typing of words, where certain letters are replaced with incorrect equivalents with diacritical markings removed. In the case of Slovak language, it is possible that one incorrect form can have more possible correct forms. The incorrect words are still readable and are recognizable to a human reader, but automated processing requires some kind of disambiguation that can distinguish the correct form of a word meant by the author. This feature is given special attention. Statistical correction system [5] is used and each possibly incorrect token has assigned its most probable correct form as annotation.

The first part of the system is a correction lexicon that identifies possible erroneous forms and proposes a set of candidate corrections. The lexicon of possible corrections is created by taking a large vocabulary of correct words and generating a set of possible incorrect forms for each word.

This approach allows to find the most common incorrect forms and possible corrections.

The list of possibly incorrect words is generated by recursive aplication of formal grammar rules. The most common errors in the written Slovak are omission of diacritical markings (such as ň, or á) and incorrect usage of letters y and i. For each word containing one or more letters that can be typed incorrectly, a list of all possible incorrect forms is generated. This approach does not solve upper/lower case normalization, this problem is solved separately during tokenization.

The second part of the correction system is a statistical language model trained on a corpus of texts that are considered to be well written and correct. It can express probability of occurrence of any given word sequence.

A HMM-based classifier takes the lexicon and the language model into account and Viterbi algorithm is used to find the most probable sequence of correct words for the given sentence that can contain misspellings. The classification algorithm is very similar to the one used for morphological analysis, as it is described above (Fig. 1).

The language model is used as a state-probability matrix, the lexicon provides a list of possible corrections (states) for each seen token (observation). Experiments in [5] show that classification accuracy of the proposed system is very high, over 99 %. Weak part of this approach is that it is unable to correct extra spaces in the middle of word.

Example of the correction result is in Fig. 2; corrected form is appended after each token.

```
rychla|rýchla hneda|hnedá liska|líška skace|skáče cez|cez
maleho|malého psa|psa
```

Fig. 2. Example of sentence correction

6 Conclusion

This corpus is one of a few available language resources for the Slovak language. It is focused on a type of speech that is very close to the one used in a common conversation. The corpus contains common types of spelling errors, jargon and dialect expressions. The proposed form and annotations should enable further classical and computational linguistic research of a contemporary way of communication – web discussions. Its size should be sufficient for statistical analysis of word connotations, language modeling or document classification, clustering or information retrieval tasks such as [9,7]. Future effort will be focused on processing data from social networks.

A major problem of all content-providing web sources that provide discussions about presented topics, such as newspaper or blog sites, is a high occurrence of offenses and swear words. If the discussion is not moderated and has not been reviewed, due to the anonymity, people are tempted to be more verbally aggressive and personal. Hopefully, this corpus can serve as a basis for research in this field and, after careful analysis, it will be possible to propose a tool for automatic detection of improper talk.

Acknowledgements. The research presented in this paper was partially supported by the Research and Development Operational Program funded by the ERDF under the projects ITMS-26220220141 (50%) and ITMS-26220220182 (50%).

References

1. Böhmová, A., Hajič, J., Hajičová, E., Hladká, B.: The Prague dependency treebank. In: Treebanks, pp. 103–127. Springer (2003)
2. Ćavar, D., Jazbec, I.P., Stojanov, T.: Cromo-morphological analysis for standard Croatian and its synchronic and diachronic dialects and variants. In: Finite-State Methods and Natural Language Processing. Frontiers in Artificial Intelligence and Applications, vol. 19, pp. 183–190 (2009)
3. Hládek, D., Staš, J.: Text gathering and processing agent for language modeling corpus. In: Proceedings of the 12th International Conference on Research in Telecommunication Technologies, RTT, pp. 200–203 (2010)
4. Hládek, D., Staš, J., Juhár, J.: Dagger: The Slovak morphological classifier. In: ELMAR, 2012 Proceedings, pp. 195–198. IEEE (2012)
5. Hládek, D., Staš, J., Juhár, J.: Unsupervised spelling correction for Slovak. Advances in Electrical and Electronic Engineering 11(5), 392–397 (2013)

6. Horák, A., Gianitsová, L., Šimková, M., Šmotlák, M., Garabík, R.: Slovak national corpus. In: Sojka, P., Kopeček, I., Pala, K. (eds.) TSD 2004. LNCS (LNAI), vol. 3206, pp. 89–93. Springer, Heidelberg (2004)
7. Rosenthal, S., McKeown, K.: Detecting opinionated claims in online discussions. In: Proceedings - IEEE 6th International Conference on Semantic Computing, ICSC 2012, pp. 30–37 (2012)
8. Rusko, M., Juhár, J., Trnka, M., Staš, J., Darjaa, S., Hládek, D., Cerňak, M., Papco, M., Sabo, R., Pleva, M., et al.: Slovak automatic transcription and dictation system for the judicial domain. In: Human Language Technologies as a Challenge for Computer Science and Linguistics: 5th Language & Technology Conference, pp. 365–369 (2011)
9. Saxe, J., Mentis, D., Greamo, C.: Mining web technical discussions to identify malware capabilities. In: Proceedings - International Conference on Distributed Computing Systems, pp. 1–5 (2013)
10. Spoustová, J., Spousta, M.: A high-quality web corpus of Czech. In: LREC, pp. 311–315 (2012)
11. Thurston, A.D.: Parsing computer languages with an automaton compiled from a single regular expression. In: Ibarra, O.H., Yen, H.-C. (eds.) CIAA 2006. LNCS, vol. 4094, pp. 285–286. Springer, Heidelberg (2006)

Towards the Development of the Multilingual Multimodal Virtual Agent

Inese Vīra, Jānis Teseļskis, and Inguna Skadiņa

Tilde, Vienības gatve 75a, Riga, Latvia
{inese.vira,janis.teselskis,inguna.skadina}@tilde.lv

Abstract. The mobile virtual agent (assistant) is one of today's most intriguing new technologies. Development of such an agent is multidisciplinary work, and natural language processing is an indispensable part of this work. Our goal is to develop a multilingual multimodal virtual agent. In this paper, we describe the first steps towards this goal – the design, development, and evaluation of the intelligent translation agent. The agent provides speech to speech translation of words, phrases, and sentences from English into Spanish, French, or Russian. The initial evaluation performed for natural language components, as well as for the agent in general, indicated that there is user interest and that such an application is useful.

Keywords: intelligent virtual agent, multimodal systems, speech interfaces, machine translation, question answering, multilingual information systems.

1   Introduction

Nowadays, the entire world is in our pocket – mobile, fast, and smart. Mobile devices have become powerful, and communication through virtual agents seems to be the most appropriate way to interact. Multimodal systems (including dialog systems), such as smart home or Artificial Intelligence robots, are often used in real life situations to help people with daily tasks. According to Ian Urbina's article in The New York Times [1]: "this new breed of bots is being designed not just with greater sophistication but also with grander goals: to sway elections, to influence the stock market, to attack governments, even to flirt with people and one another."

Commercial applications, such as *Apple Siri*[1] and *Google Now,*[2] are important steps towards natural language communication between a user and a computer. Information extraction and question answering systems, such as *Ask.com, WolframAlpha,*[3] provide a wide range of information and a significant knowledge base for virtual agents. A wide range of multimodal chatbots and virtual agents designed to maintain a conversation are available (e.g., *Virtual Assistant Denise*[4] for *Personal Computer*). Most of the mobile chatbots with 2D or even 3D avatars include text

---

[1] http://www.apple.com/ios/siri/
[2] http://www.google.com/landing/now/
[3] https://www.wolframalpha.com/
[4] http://guile3d.com/

A. Przepiórkowski and M. Ogrodniczuk (Eds.): PolTAL 2014, LNAI 8686, pp. 470–477, 2014.
© Springer International Publishing Switzerland 2014

message templates for responses. For instance, *Talking Pocket Bot Lucy*[5] and *Pocket Blonde Smart Assistant*[6] on *Android* help to find information, make calls, and send messages.

Globalisation requires people to have more and more tools that facilitate multilingual communication and are available everywhere. A simple and widely used solution is the mobile dictionary (*Oxford Dictionary of English*,[7] *Dictionary.com*,[8] etc.). There are also some machine translation solutions for mobile devices (e.g., *Google Translate*,[9] *Tilde Translator*,[10] *ABBY TextGrabber+Translator*[11]).

However, a more attractive and user-friendly way to communicate would be through a multilingual multimodal agent. In this paper, we present ongoing work towards the development of a prototype for a new 3D multimodal virtual translation agent – *Laura*[12] for *Android* mobile devices. The visual interface of our prototype is a 3D head which responds with natural mimics, emotions, and synchronous lip movements. Our agent provides speech to speech translation of words, phrases, and sentences from English into Spanish, French, or Russian. This agent can also answer questions and handle simple dialog in English.

2 Related Work

Human computer interaction has been actively researched for many decades, starting with the first chatbot ELIZA [2]. Creating a multimodal virtual agent is a multidisciplinary effort, but its core components are natural language processing tools that can handle dialog, understand user input, and generate meaningful output. Different aspects are studied, for example, dialog management, interactivity, reactive behaviour, and others. Based on these studies, different applications are proposed – assistants, tutors, simple chatbots, etc.

However, to our knowledge, although various virtual agents have been developed, the translation task is not their primary focus. Some mobile phone applications provide multilingual translation services in a particular domain [3], but do not provide

---

[5] https://play.google.com/store/apps/details?id=com.kauf.botv3.talkingpocketbotlucy&hl=lv
[6] https://play.google.com/store/apps/details?id=com.brainyfriends.voice.game.pocket.blonde
[7] http://www.oxforddictionaries.com/words/oxford-dictionary-apps
[8] https://play.google.com/store/apps/details?id=com.dictionary&hl=en
[9] https://play.google.com/store/apps/details?id=com.google.android.apps.translate
[10] https://play.google.com/store/apps/details?id=com.tilde.translator
[11] https://play.google.com/store/apps/details?id=com.abbyy.mobile.textgrabber.full&hl=en
[12] First version of *Laura* is available from
https://play.google.com/store/apps/details?id=com.tilde.laura&hl=en

the multimodal component. Others are developed as language teachers [4]. Applications such as *Voice Translator Free*,[13] *Translator With Speech*,[14] *Jibbigo*[15] and *Spectra* [5] provide speech translations, but do not have personality.

3   Concept, Design and Architecture

Mobile devices are powerful and serve as small computers for many of us, but their intelligence and possibilities to communicate through natural language are limited. Therefore, our work aims to investigate new ways of how natural language processing tools can be used in applications for mobile devices. As mobile devices are used in different environments and situations, speech input becomes an indispensable and convenient way of communication with mobile applications. Moreover, some authors think that speech technology becomes critical for the mobile industry [6]. On the other hand, while speech is one possible way of communication, multimodal agents allow a wider range of interaction, including speech and emotions. Multimodality is important not only for conversational agents, but also for translation applications, e.g., according to Delmonte [7], translation systems that can take advantage of both – speech and visual input – seem more promising.

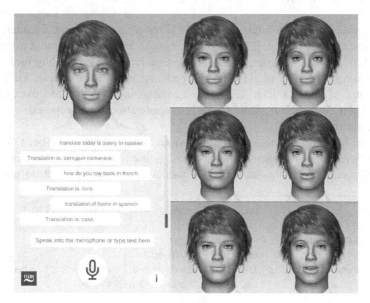

Fig. 1. Overall design and different facial expressions of the virtual agent

---

[13] https://play.google.com/store/apps/details?id=
   com.smartmobilesoftware.voicetranslatorfree
[14] https://play.google.com/store/apps/details?id=
   com.proxy.translator
[15] http://jibbigo.com/

Thus, our work aims to develop a multilingual multimodal 3D virtual agent prototype that allows to communicate through text and speech interfaces and has some emotions, mimics, gaze, and head nods. Taking into account the limited size of a screen, the 3D head is designed to take up about 40% of the screen, while the rest of the screen is used for displaying dialog between the user and the application. The overall design of the virtual agent is shown on Fig. 1.

At the moment, we have incorporated a few basic facial expressions and emotions, including, smile, wink, sadness, excitement, surprise, scepticism, and grumpiness while the avatar is in the idle state.

We used *Autodesk Maya*[16] and *Unity Game Engine*[17] for fast and powerful modelling, animation, and flexible application development. For proper lip synchronisation, we used the *Unity Speecher* plugin[18] that analyses the incoming audio file spectrum data and generates lip and head movements accordingly.

The architecture of the translation agent is shown in Fig. 2. It consists of five main constituents: mobile application, virtual agent web service, automatic speech recognition (ASR) service, Text Translation Service, and speech synthesis service (TTS).

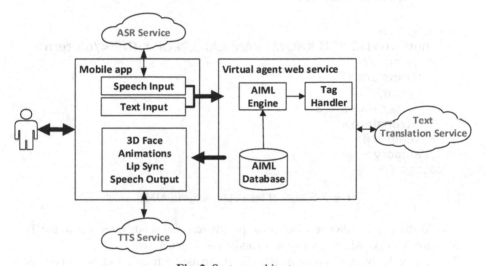

Fig. 2. System architecture

The virtual agent web service is responsible for intelligent conversation between the agent and the user. Dialogue management is based on *Artificial Intelligence Markup Language* – AIML [8]. ASR and TTS components support voice interaction. Text translation Service supports translation of the user's queries. The application can be easily extended with more services, supporting multilingual and multimodal communication.

---

[16] http://www.autodesk.com/products/autodesk-maya/overview
[17] https://unity3d.com/unity
[18] http://u3d.as/content/achest-software-design/speecher-2-1/2Lm

4   Translation and Dialogue Management

Voice input from a user is transformed into text by the ASR service and then sent to the Virtual agent web service that is responsible for dialogue management and finding an adequate response in the AIML database. The basic AIML file describes patterns of a user's input and corresponding answers. A more advanced AIML database can include tags for defining the avatar`s emotions, tasks for AIML connectivity with a mobile device (writing messages and notes, opening e-mails or web pages, etc.), and properties to connect with external services, such as text translation.

For text translation, we have designed a new AIML tag <translate> with two input fields for the word or sentence and the translation language. Different patterns of how a user can ask for translation are defined through the <pattern> tag (see Fig. 3). The sentence/word that needs to be translated and the language of translation are then passed to the <translate> tag through the "star" tags that correspond to asterisks in the pattern. The first asterisk represents language, but the second represents text – asterisks are counted from right to left.

```
<category>
  <pattern>DO YOU KNOW TRANSLATION OF * IN *</pattern>
  <template>
   <translate>
    <star/>
    <star index="2"/>
   </translate>
  </template>
 </category>
<category>
```

Fig. 3. Sample of translation script in AIML

The AIML Tag Handler receives these parameters and sends a request to the Text Translation Service. After translation, an audio file is generated.

To identify the ways in which people usually request a translation, we conducted a small user study. It showed that one of the most popular phrases is "translate WORD to LANGUAGE", as well as "WORD translate in LANGUAGE", and "WORD in LANGUAGE". The translation queries were shorter for text input, while translation patterns were longer and similar to real life dialogue for voice input, for instance, "How to translate WORD into LANGUAGE", "How do I say WORD in LANGUAGE", "Do you know the translation for WORD in LANGUAGE", or even "Hi, LAURA, could you, please, translate WORD in LANGUAGE".

The Translation Service allows to translate not only simple words or phrases, but can also translate sentences or paragraphs. Examples of different translation patterns are shown in Fig. 4.

5   Evaluation

Two evaluations were performed to assess the overall concept of the virtual agent, the translation scenario, and the importance of different modalities. The goal of the first evaluation was to obtain an overall assessment of the virtual agent, while the second evaluation aimed to evaluate different modalities and get a deeper insight into the translation scenario and the natural language processing tools used in this scenario.

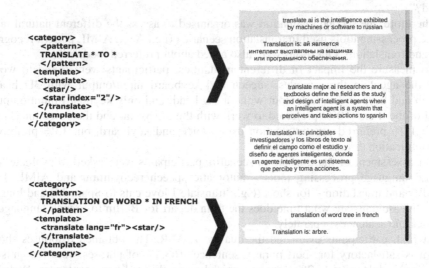

Fig. 4. A fragment of AIML translation script and its corresponding output on a user's screen

Both evaluations were performed through on-line questionnaires. 18 people participated in the first evaluation, with a 50:50 gender breakdown and 84% in the 20–40 age group. There were 23 participants in the second evaluation, with mostly the same characteristics as for the first evaluation. In both evaluations, participants preferred voice communication (83% in the first evaluation and 77% in the second evaluation). However, time spent on the evaluation was shorter in the first evaluation task (56% spent less than 10 minutes), while we encouraged longer communication in the second evaluation – only 13% of participants spent less than 10 minutes and 18 respondents (78%) spent 10–30 minutes.

Questions were divided into three groups for the first evaluation: questions related to the user's experience, questions about the visuals and general features, and questions related to the translation task. Most of the respondents were satisfied with the interface of the application – the application was rated as 'simple and easy to understand' by 50% and 'simple, but needs improvements' by the remaining 50%. Also, most of the participants (83%) thought that the virtual agent is friendly and suggested to add an option for changing the hair style or eye colour.

Regarding the dialogue and answers by the agent, 76% of respondents were satisfied with the replies provided. 83% of users replied with a "yes" to the question of whether they found the interaction to be simple.

Questions related to the translation task involved two important aspects: interaction with the user and quality of the given translation answers. To the question "Was the answer correct?" – 11% responded with "yes, always", 72% with "yes, mostly", 11% with "no, most of the time", and 6% with "answer was incorrect". However, only 33% responded positively to the question "Did Laura understand the question asked?".

Therefore, the second evaluation was organised to assess the different natural language processing tools used in translation scenarios (i.e., ASR, AIML pattern recognition and translation). Questions were also asked about preferred modalities.

To measure the impact of different modalities, participants were asked to work with the agent in three modes: speech and keyboard input/output with 3D head; speech and keyboard input/output without 3D head; and only keyboard input/output. Most of the users (70%) preferred to work with the 3D avatar and the voice interface, while 17% preferred communication using voice and keyboard, but 13% preferred communication through text only.

For assessment of the translation scenario, participants were asked to evaluate the natural language processing tools – automatic speech recognition and AIML Tag Handler and translation – for short (e.g., 'translate I love cats to Spanish') and longer (e.g., 'please translate what time does the train depart for Berlin to French language') phrases and sentences separately.

At first, participants evaluated the quality of ASR. The obtained responses show that it is satisfactory for short phrases/sentences (63.4% of phrases were recognised correctly), while only 45.2% were recognised correctly for longer sentences. Perhaps the rather low recognition rate was influenced by the fact that participants were not native speakers of English.

To evaluate the coverage of our translation patterns, participants were asked to tell "How often did *Laura* correctly understand the question, phrase, or sentence as the one that needs to be translated?". According to the answers, most (72.2%) short translation requests were correctly understood, but only 53.6% were correctly understood in the case of longer requests. This indicates that we have received much better results after the update: the rate of understanding increased from 33% to 53.6%–72.2% for different sentences. Finally, participants evaluated translation quality – 61.8% short sentence translations and 42.8% longer sentences translations were correct.

To identify the ways in which translations were requested, we asked participants to write down the specific patterns and questions. The sentence structures were: "Say in LANGUAGE: SENTENCE", "Translate in LANGUAGE: SENTENCE", and "translate please SENTENCE to LANGUAGE".

6   Conclusion

In this paper, we presented the ongoing work towards the development of the intelligent translation agent. Evaluation results show that the possibility to translate phrases and short utterances is not only an interesting research direction, but also a useful and interesting feature for the user. Following user suggestions, we plan to add new languages for the translation task in the near future.

The next step will be to achieve fully multilingual conversation with our agent. Currently, *Laura* can answer questions and handle simple dialog in English. As the next step, we plan to extend her functionality to allow dialog and question answering in several languages.

While our virtual agent has simple multimodal communication skills, our current work also involves emotion recognition and emotional response when speaking about certain topics. The next tasks will be to work on fusion and fission.

Acknowledgments. The research leading to these results has received funding from the research project "Information and Communication Technology Competence Center" of EU Structural funds, contract nr. L-KC-11-0003 signed between the ICT Competence Centre and the Investment and Development Agency of Latvia, research No. 2.1 "Natural language processing in mobile devices". We would also like to thank all evaluation participants for their time and suggestions for further work.

References

1. Urbina, J.: I Flirt and Tweet. Follow Me at #Socialbot. In: The New York Times (August 10, 2013), http://www.nytimes.com/2013/08/11/sunday-review/i-flirt-and-tweet-follow-me-at-socialbot.html?_r=0
2. Weizenbaum, J.: ELIZA – A Computer Program for the Study of Natural Language Communication between Man and Machine. In: Communications of the Association for Computing Machinery, vol. 9, pp. 36–45. ACM, New York (1966)
3. Paul, M., Okuma, H., Yamamoto, H., Sumita, E., Matsuda, S., Shimizu, T., Nakamura, S.: Multilingual Mobile-Phone Translation Services for World Travelers. In: COLING 2008 22nd International Conference on Computational Linguistics: Demonstration Papers, pp. 165–168. COLING Demos (2008)
4. Hazel, M., Mervyn, A.J.: Scenario-Based Spoken Interaction with Virtual Agents. In: Computer Assisted Language Learning, vol. 18, pp. 171–191. Routledge, part of the Taylor & Francis Group (2005)
5. Rangarajan, V., Bangalore, S., Jimenez, A., Golipour, L., Kolan, P.: SPECTRA: A Speech-to-Speech Translation System in the Cloud. In: IEEE International Conference on Emerging Signal Processing Applications (2013)
6. Mozer, T.: Speech's Evolving Role in Consumer Electronics ...From Toys to Mobile. In: Neustein, A., Markowitz, J.A. (eds.) Mobile Speech and Advanced Natural Language Solutions, pp. 23–34. Springer (2013)
7. Delmonte, R.: Getting Past the Language Gap: Innovations in Machine Translation. In: Neustein, A., Markowitz, J.A. (eds.) Mobile Speech and Advanced Natural Language Solutions, pp. 103–181. Springer (2013)
8. Marietto, M.D.G.B., Varago, R.A., Oliveira, G.B., Botelho, W.T., Pimentel, E., Robson, S.F., Silva, V.L.: Artificial Intelligence Markup Language: A Brief Tutorial. International Journal of Computer Science & Engineering Survey 4(3) (2013)

The WikEd Error Corpus:
A Corpus of Corrective Wikipedia Edits
and Its Application
to Grammatical Error Correction

Roman Grundkiewicz and Marcin Junczys-Dowmunt

Faculty of Mathematics and Computer Science,
Adam Mickiewicz University in Poznań,
ul. Umultowska 87, 61-614 Poznań, Poland
{romang,junczys}@amu.edu.pl

Abstract. This paper introduces the freely available WikEd Error Corpus. We describe the data mining process from Wikipedia revision histories, corpus content and format. The corpus consists of more than 12 million sentences with a total of 14 million edits of various types. As one possible application, we show that WikEd can be successfully adapted to improve a strong baseline in a task of grammatical error correction for English-as-a-Second-Language (ESL) learners' writings by 2.63%. Used together with an ESL error corpus, a composed system gains 1.64% when compared to the ESL-trained system.

Keywords: error corpus, Wikipedia revision histories, grammatical error correction.

1   Introduction

Machine learning approaches in the field of natural language processing are data-hungry. In the ideal case, large and diversified data sets are available that can be used directly or easily adapted to the investigated problem. An example where large amounts of data can be beneficial is automated grammatical error correction for English-as-a-second-language (ESL) learners.

Although some types of errors, for instance subject-verb mistakes can be corrected using heuristic rules, others, like preposition errors, are difficult to correct without substantial amounts of corpus-based information [10]. The above is especially true when Statistical Machine Translation (SMT) toolkits are applied as error correction systems [15]. Compared to multilingual translation corpora which today are plentiful or can be easily collected, genuine error corpora are not easy to come by. If copyright and licensing issues are taken into account as well, the number of resources becomes very scarce.

In this paper, we introduce the — to our knowledge — largest free corpus of corrective edits available for the English language: the WikEd Error Corpus, version 0.9. This corpus consists of edited sentences extracted from Wikipedia

A. Przepiórkowski and M. Ogrodniczuk (Eds.): PolTAL 2014, LNAI 8686, pp. 478–490, 2014.
© Springer International Publishing Switzerland 2014

revisions, and as such inherits the user-friendly CC BY-SA 3.0 license of the original resource.

In contrast to other works that use Wikipedia to build various NLP resources [13,21,3], we processed the entire English Wikipedia revision history[1] and gathered ca. 12 million sentences with annotated edits. Possible application include, but are not limited to, sentence paraphrasing, spelling correction, grammar correction, etc. Both, the WikEd Error Corpus and the tools used to produce it have been made available for unrestricted download.[2]

In the next section we describe related work in the domain of error corpora collection. Section 3 presents our language-independent method of edit operation mining from Wikipedia's revision histories and contains descriptions of the collected data, error types, and formats. In Sect. 4, we demonstrate the usefulness of WikEd to automated ESL grammatical error correction: an SMT-based system is adapted for ESL error correction. Unlike in numerous previous works, we do not restrict ourselves to only a few chosen error types, but attempt a full correction as it has been introduced in this year's CoNLL Shared Task [17]. Finally, we conclude in Sect. 5 with comments on planned improvements of the WikEd Error Corpus.

2   Related Work

While reviewing related work, we restrict ourselves to approaches to error corpora gathering. For a review of the field of grammatical error correction, we refer the reader to Leacock et al. [10], for the current state-of-the-art, we recommend the proceedings of the 2013 and 2014 CoNLL Shared Tasks [18,17].

Three main approaches to gathering error corpora are present in literature: manual annotation of students' writings, artificial errors generation within well-formed sentences, and the extraction of errors and their corrections from edit histories. A fourth possibility are social networks for language learners.

2.1   Learner's Corpora

As noted by Leacock et. al [10], even if large quantities of students' writings are produced and corrected every day, only a small number of them is archived in electronic form. Most of the available error-annotated corpora has been created from ESL learners' writings. Examples are the NUS Corpus of Learner English [5] (NUCLE), the dataset of FCE scripts[3] extracted from the Cambridge Learner Corpus, and the International Corpus Network of Asian Learners of English.[4]

They are usually small, a few hundreds sentences. NUCLE is a notable exception, but for machine learning approaches even the ca. 50,000 sentences from

---

[1] Wikipedia database dump from January 2nd, 2014:
   http://dumps.wikimedia.org/enwiki/20140102/
[2] http://romang.home.amu.edu.pl/wiked/wiked.html
[3] http://ilexir.co.uk/applications/clc-fce-dataset/
[4] http://language.sakura.ne.jp/icnale/

NUCLE are a rather small resource. It is also worth noting that errors made by learners differ from errors made by native-speakers, therefore, the use of ESL corpora for the correction of native speaker errors may be limited, and vice versa.[5]

2.2 Artificial Errors

One proposed solution to overcome data sparseness is the creation of artificial data. In the case of artificial error corpora, grammatical errors are introduced by random substitutions, insertions, or deletions according to the frequency distribution observed in seed corpora.

Brocket et al. [1] introduce mass/count noun errors with hand-constructed rules. Wagner et al. [19] produce ungrammatical sentences based on an error analysis carried out on a corpus formed by roughly 1,000 error-annotated sentences. Foster and Andersen [6] introduce *GenERRate*, a tool for the production of artificial errors that imitate genuine errors from two data sets: a grammatical corpus and a list of naturally-occuring errors. Yuan and Felice [20] extracted lexical and part-of-speech patterns for five types of errors from NUCLE and applied them to well-formed sentences.

Admittedly, artificial error generation is an efficient and economic way to increase the size of training datasets, but there are drawbacks. The diversification of errors in such corpora can be lower due to small set of real seed data. For specific error types it may be difficult to create descriptive patterns that can be applied to well-formed sentences. Furthermore, it has been reported that artificial data can be less suited for evaluation purposes [21].

2.3 Text Revision Histories

An alternative solution consists in the extraction of errors from text revision histories. The most frequently used are Wikipedia revisions.

Miłkowski [14] proposes the construction of error corpora from text revision histories based on the hypothesis that the majority of frequent minor edits are error corrections. A Polish corpus of errors automatically extracted from Wikipedia revisions has been created by Grundkiewicz [7]. To distinguish error corrections from unwanted edits and to determine error categories the author used handwritten rules.

Wikipedia revisions have been used for the creation of sentence paraphrase corpora by Max and Wisniewski [13], real-word spelling error correction by Zesch [21] and preposition error correction by Cahill et al. [3]. Cahill et al. confirm that data from Wikipedia is useful for both, training a correction system and creating artificial data. This research is the closest to our work, but focuses only on prepositions, whereas we perform experiments on a much larger scale and cover all error types.

---

[5] We show that this is not necessarily true.

The main advantage of Wikipedia-extracted data sets is their size, but there are also disadvantages, for instance Wikipedia's encyclopedic style and an abundance of vandalism.

2.4 Social Networks for Language Learners

Probably the best resource for language errors has made a very recent appearance in the form of social networks for language learners, an example being Lang-8.com. Learners with different native languages correct each others texts based on their own native-language skills. See Sect. 4.3 for more information. However, this resources are not free for all purposes, special license agreements are required.

3 The WikEd Error Corpus

In this section we describe our method of edit extraction from Wikipedia revisions which leads to the creation of the WikEd Error Corpus, version 0.9.

3.1 Extracting Edits from Wikipedia

Wikipedia dumps with complete edit histories are provided in XML format.[6] Similarly to Max and Wisniewski [13] and Grundkiewicz [7], we iterate over each two adjacent revisions of every Wikipedia page, including articles, user pages, discussions, and help pages. To minimize the number of unwanted vandalism, we skip revisions and preceding revisions if comments contain suggestions of reversions, e.g. *reverting after (...)*, *remove vandalism*, *undo vandal's edits*, *delete stupid joke*, etc. This is done by a few hand-written rules involving regular expressions.

Next, we remove markup[7] from each article version and split texts into sentences with the NLTK toolkit.[8] Pairs of edited sentences are identified with the Longest Common Subsequence algorithm (LCS) [12]. Edits consisting of additions or deletions of full paragraphs are disregarded.

Two edited sentences[9] s_i and s_j are collected if they meet several surface conditions

- the sentence length is between 2 and 120 tokens,
- the length difference is less than 5 tokens,
- the relative token-based edit distance $ed(s_i, s_j)$ with respect to the shorter sentence is smaller than 0.3.

---

[6] http://dumps.wikimedia.org/
[7] http://medialab.di.unipi.it/wiki/Wikipedia_Extractor
[8] http://nltk.org/
[9] In the remainder of this paper we will refer to two corresponding edited fragments as sentences, even if they are not well-formed.

The threshold values in the above restrictions were chosen experientially. The relative token-based edit distance is defined as:

$$\mathrm{ed}(s_i, s_j) = \frac{\mathrm{dist}(s_i, s_j)\min(|s_i|, |s_j|)}{\log_b \min(|s_i|, |s_j|)},$$

where $\mathrm{dist}(s_i, s_j)$ is the token-based Levenshtein edit distance [11], $|s|$ is the length of the sentence s in tokens, and the logarithm base b is empirically set to 20. This formula implies that the longer the sentence is, the more edits are allowed, but it prevents the acceptance of too many edits for long sentences.

3.2 Collected Corrective Edits

At this stage, 12,130,508 pairs of edited sentences from the English version of Wikipedia have been collected. The most useful edits include:

- spelling error corrections:
  `You can use rsync to [-donload-] {+download+} the database .,`
- grammatical error corrections:
  `There [-is-] {+are+} also [-a-] two computer games based on the movie .,`
- stylistic changes:
  `[-Predictably , the-] {+The+} game ended [-predictably-] when she crashed her Escalade...,`
- sentence rewordings and paraphrases:
  `These anarchists [-argue against-] {+oppose the+} regulation of corporations .,`
- encyclopaedic style adjustments:
  `A [-local education authority-] {+Local Education Authority+} ( LEA ) is the part of a council in England or Wales.`

The WikEd corpus contains also less useful edits for grammatical error correction task, e.g.:

- time reference changes:
  `The Kiwi Party [-is-] {+was+} a New Zealand political party formed in 2007 .,`
- information supplements:
  `Aphrodite is the Greek goddess of love {+, sex+} and beauty .,`
- numeric information updates:
  `In [-May 2003-] {+August 2004+} this percentage increased to [-62-] {+67+}% .,`
- item additions/deletions to/from bulleted lists:
  `Famous Bronxites include {+Regis Philbin ,+} Carl Reiner , Danny Aiello...,`
- amendments of broken MediaWiki's markups:
  `The bipyramids are the [-[ [ dual polyhedron |-] dual polyhedra [-[ [-] of the prisms.,`

Table 1. 30 most frequent edits in the WikEd 0.9 corpus

Edits	Freq.	Edits	Freq.	Edits	Freq.
ins(")	667,098	ins(a)	45,870	ins(and)	28,518
ins(,)	348,341	ins(')	41,473	del(of)	26,257
del(")	226,854	del(.)	41,161	sub(a,an)	24,626
del(,)	158,324	sub(is,was)	40,062	ins(was)	23,670
ins(.)	138,322	sub(',")	37,236	del(])	22,443
del('s)	80,669	del(')	36,051	sub(was,is)	21,372
ins(the)	79,708	del())	34,401	ins(()	20,079
del(the)	61,999	del(persons)	33,773	del(a)	19,615
ins())	60,852	ins(The)	32,819	ins(in)	18,651
ins(< br >)	51,802	sub(it,its)	31,171	ins(is)	18,647

– changes made by vandals:
David Zuckerman is a writer and [-producer-] {+poopface+} for
television shows.

The total number of edits is 16,013,830 among which 3,273,862 (20,44%) are
deletions and 4,829,019 (30.16%) insertions. The most frequently occurring edits
are presented in Table 1.

3.3 Filtering

As shown by Grundkiewicz [7], sentences with potentially unwanted edits, e.g.
updates of bulleted list, amendments of MediaWiki markup, and vandalism can
be effectively filtered out using heuristic rules. For example, all pairs of sentences
s_i and s_j that satisfy the following conditions can be disregarded:

– Either the sentence s_i or s_j consists of a vulgar word (determined by the
  list of vulgarisms) or a very long sequence of character with no spaces (e.g.
  produced by random keystrokes).
– Any of the sentences s_i or s_j contains fragments of markup, e.g. <ref>,
  or [http:.
– All edits concern only changes in dates or numerical values.
– The only edit made consists of removing a full stop or semicolon at the end
  of the sentence s_i.
– The ratio of non-words tokens in s_j to word tokens is higher than a given
  threshold (we used 0.5).

In the end, 1,775,880 (14.63%) pairs of sentences are marked as potentially
harmful, but not removed. For instance, vandalized entries may be useful for
various tasks by themselves.

3.4   Corpus Format

It is our intention to release the WikEd Error Corpus in a machine-friendly format. We chose a representation based on GNU `wdiff` output[10] extended by comments including meta-data. For example, for a sentence *This page lists links about ancient philosophy.* with the following two edits: insertion of *some* at third position and substitution of *about* with *to*, the WikEd entry corresponds to:

> This page lists {+some+} links [-about-] {+to+} ancient philosophy.

Meta-data consists of:

- the revision id, accompanying comment, and revision timestamp,
- the title and id of the edited Wikipedia page,
- the name of the contributor or IP address if it is an anonymous edition.

All sentences preserve the chronological order of the original revisions.

4   Application to ESL Error Correction

In the second part of the paper we examine the usefulness of the WikEd Error Corpus in an automated ESL error correction scenario. Despite the fact, that WikEd is not an English-as-a-Second-Language (ESL) learners' error corpus — although it may contain a substantial number of errors contributed by non-native English users — we demonstrate that it is possible to select mistakes in such a way that an ESL error correction system can benefit from WikEd.

4.1   Task Description

We take advantage of the training data published during the CoNLL-2014 Shared Task on Grammatical Error Correction [17]. The aim of the shared task was to automatically correct essays written by Singaporean ESL learners. Training data has been made available in form of the previously mentioned NUCLE corpus [5]. NUCLE consists of 1,414 essays (57,151 sentences) which cover a wide range of topics, such as environmental pollution and health care. The sentences have been corrected by professional English instructors and annotated with 44,385 corrections in 28 error categories, such as article or determiner errors, wrong collocation or idiom, noun number errors, etc.

System performance is measured by the MaxMatch (M^2) metric [4] which computes the F-score for the proposed corrections against a gold standard that has been similarly annotated as NUCLE. It is not necessary to correctly classify error types, only the text of the correction is compared. In this paper, we use the test set (ST-2013) from the previous edition of the CoNLL shared task for evaluation. It has been made available as training data for the current shared task and contains annotation for all 28 error types. Apart from testing on ST-2013, we also report results for 4-fold cross validation on NUCLE.

---

[10] https://www.gnu.org/software/wdiff/manual/wdiff.html#wdiff

4.2   System Description

Due to our background in statistical machine translation and the rising popularity of grammatical error correction by SMT, we decided to use the Moses [9] toolkit to built our demonstration system. Our baseline is a re-implementation of an intermediate system from Junczys-Dowmunt and Grundkiewicz [8] which is labeled by the authors as NUCLE+ CCLM. This system uses NUCLE as the sole parallel training data. It also adds a web-scale language model estimated from English CommonCrawl data made available by Buck et al. [2]. For the 28 error categories, the baseline achieves $F_{0.5}$=27.43%.

We can assume that this is a strong baseline. For the previous 5 error-type task from the CoNLL-2013 Shared Task the same system achieves F_1=29.84% (CoNLL-2013-ST used F_1, CoNLL-2014-ST changed to $F_{0.5}$). Had it taken part in the task, it would have ranked on second place among 17 teams, only 1.36% below the winning system and over 4% higher than the next best submission.

During training, 4-fold cross validation has to be adjusted to accommodate parameter tuning as is common practice in SMT. This results in a testing/tuning scheme labeled 4×2-fold cross validation (4×2-CV). The original test sets from 4-fold cross validation are divided into two halves and both are used for cross tuning and testing. This results in four training steps with two testing/tuning steps each. The eight tuned parameter weight vectors are averaged and the centroid vector is used to translate ST-2013 with a translation model estimated from the complete NUCLE data.

In the grammatical error correction scenario where source and target phrases are often identical or similar, it might be useful to inform the decoder about the differences in a phrase pair. Similarly to Junczys-Dowmunt and Grundkiewicz [8] we extend translation models with a word-based Levenshtein distance feature [11] that captures the number of edit operations required to turn the source phrase into the target phrase. Each phrase pair in the phrase table is scored with $e^{d(s,t)}$ where d is the word-based distance function, s is the source phrase, t is the target phrase. The exponential function e^x is used because Moses scores translations e of string f by a log-linear model

$$\log p(e|f) = \sum_i \lambda_i \log(h_i(e, f)),$$

where h_i are feature functions and λ_i are feature weights. That way the model score include the total numer of edits in a sentence counted by the Levenshtein distance feature for individual phrases pairs. This feature should also be helpful for reducing noise in the translation output. During evaluation, we refer to this component as "LD".

4.3   True ESL Error Data

We also compare our data to a true ESL corpus. Mizumoto et al. [16] published[11] a list of learners' corpora that were scraped from the social language learning site

---

[11] http://cl.naist.jp/nldata/lang-8

Table 2. The comparison of the WikEd 0.9 and Lang-8 NAIST corpora

Statistics	WikEd 0.9	+Select	L8-NAIST	+Select
sentences	12,130,508	—	2,567,964	—
tokens (source side)	292,570,716	294,965,241	28,506,516	34,351,819
edits	16,013,830	5,327,293	3,408,834	1,066,690
sentences with ≥1 edits	91.79%	32.62%	53.86%	28.15%
edits per sentence	1.32	0.44	1.33	0.42

Lang-8 (`http://lang-8.com`). Version 1.0 is available for academic purposes, commercial applications require special licenses from the copyright owner. Newer versions (2.0) require special license agreements for any usage.

We collect all entries from "Lang-8 Learner Corpora v1.0" with English as the learned language, we do not care about the native language of the user. Only entries for which at least one sentence has been corrected are taken into account. Sentences without corrections from such entries are treated as error-free and mirrored on the target side of the corpus. Eventually, we obtain a corpus of 2,567,969 sentence pairs with 28,506,540 tokens on the uncorrected source side. We call this resource "L8-NAIST". The comparison with the WikEd 0.9 corpus is presented in Table 2.

4.4   Error Selection

As mentioned before, the WikEd Error Corpus is not an ESL error corpus and may contain a very different type of errors from those made by language learners. We try to mitigate this by selecting errors that resemble mistakes from NUCLE, other errors are replaced by their corrections.

For each pair of uncorrected and corrected sentences from NUCLE, we compute a sequence of deletions and insertions with the LCS algorithm that transform the source sentence into the target sentence. Adjacent deleted words are concatenated to form a phrase deletion, adjacent inserted words result in a phrase insertion. A deleted phrase followed directly by a phrase insertion is interpreted as a phrase substitution. Substitutions are generalized if they consist of common substrings, again determined by the LCS algorithm, that are equal to or longer than three characters. We encode generalizations by the regular expression (`\w{3,}`) and a back-reference, e.g. `\1`.

Patterns can contain multi-word strings, e.g. `sub((\w{3,}) is,\1s are)` models a case of subject-verb agreement. Sometimes, more than one generalization is possible, e.g. `sub((\w{3,})-(\w{3,}),\1\2)`. Table 3 contains the some of the most frequent patterns extracted from NUCLE for all 28 error types. The table includes also the most frequent error categories matching the pattern. A frequency threshold is defined at 5, patterns that occur less often are discarded, in the end 666 patterns remain.

Table 3. 14 most frequent patterns extracted from NUCLE 3.0

Pattern	Freq.	Categories with Freq.
sub((\w{3,}),\1s)	2864	Nn(2188) SVA(395) Wform(146)
ins(the)	2494	ArtOrDet(2424)
del(the)	1772	ArtOrDet(1696)
sub((\w{3,})s,\1)	1317	Nn(651) SVA(263) Wform(141) Rloc-(92)
ins(,)	971	Mec(733) Srun(196)
ins(a)	679	ArtOrDet(646)
sub((\w{3,}),\1d)	300	Vt(112) Vform(105) Wform(62)
del(,)	266	Mec(175) Rloc-(83)
sub((\w{3,}),\1ed)	252	Vt(138) Vform(75) Wform(29)
ins(an)	246	ArtOrDet(234)
del(of)	222	Prep(202)
sub(is,are)	219	SVA(198)
del(.)	205	Rloc-(135) Mec(60)
sub((\w{3,})d,\1)	202	Vt(109) Wform(46) Vform(28) Rloc-(11)

Next, we perform the same computation for sentence pairs from WikEd. Edits that result in patterns from our list are not modified and remain in the data, for all other edits, the selected correction is applied to the source sentence. Error types not covered by the patterns thus disappear. Noise like vandalism is either removed or reduced to identical sentences on both sides for the training corpus. In both cases this cannot harm our systems. Eventually, 3,957,547 (32,62%) sentences remain that still contain edit pattern. We keep all sentences with surviving errors and randomly select sentences without edits to be kept as well. The final parallel corpus consists of 4,703,353 sentence pairs. Two versions are used in our experiments: the error selected corpus which is labeled "WikEd+Select" and a second version consisting of the same sentences but with all errors present (a proper subset of the unprocessed WikEd), this version is denoted as "WikEd".

Error selection is also applied to L8-NAIST, resulting in L8-NAIST+Select, all sentences remain in this resource.

4.5   Results

Table 4 contains results for our experiments with WikEd and L8-NAIST. Unadapted WikEd used as parallel training data lowers the results drastically, which is not surprising, many edits may be different in style and scope and considered harmful for the ESL-based NUCLE. Adding the LD feature makes it even worse.

Error selection changes the picture. Only errors similar to NUCLE data remain in the training corpus. Results improve for both, NUCLE cross-validation (4×2-CV) and the unseen test set ST-2013. The latter excludes the possibility of overfitting to NUCLE due to error selection as might be postulated based on cross-validation results alone. Adding LD to the error-selected version of WikEd

Table 4. Evaluation for grammatical error correction task

System	4×2-CV	ST-2013
NUCLE+CCLM	22.19	27.43
+WikEd	18.96	26.12
+LD	18.21	23.63
+Select	23.80	29.49
+LD	**24.33**	**30.06**

(a) WikEd Error Corpus 0.9

System	4×2-CV	ST-2013
NUCLE+CCLM	22.19	27.43
+L8-NAIST	23.34	31.20
+LD	24.44	34.06
+Select	25.43	33.89
+LD	**26.40**	**34.15**

(b) Lang-8 Error Corpora 1.0

System	4×2-CV	ST-2013
Joint-Translation	26.01	32.33
Composition	**26.63**	**35.79**

(c) System combination results

leads to further gains in both cases. Eventually, improvements of 2.24% and 2.63% $F_{0.5}$-score over the baseline can be observed.

We perform the same experiments with the ESL corpus L8-NAIST. It is, of course, not surprising that the in-domain L8-NAIST performs much better than WikEd. It should however be noted that the significant performance improvements stem from our error selection procedure and the Levenshtein distance feature. Final results achieve 4.21% and 6.72% over the baseline.

For both error corpora, the improvements are due the combined effects of additional parallel data, error selection, and the task-specific LD feature. In order to verify that this is not alone the beneficial effect of the LD feature on data present in NUCLE, we also test NUCLE+CCLM+LD with does not improve the baseline (22.10% and 27.62%).

We also evaluated two simple system combinations for which results are presented in Table 4c:

- "Joint-Translation", which is a single Moses system configured to use both separately tuned phrase tables from the best two systems — NUCLE+CCLM +WikEd+Select+LD and NUCLE+CCLM+L8-NAIST+Select+LD;
- "Composition", which is a chain of NUCLE+CCLM+WikEd+Select+LD and NUCLE+CCLM+L8-NAIST+Select+LD. The output of the first system is corrected a second time by the latter.

In the case of Joint-Translation we see worse results than for the best L8-NAIST-trained system. It seems that the parameter tuning process could not take advantage of the two phrase tables. However, results for "Composition" are significantly better (+1.64%) than for the single systems, both systems tend to correct slightly different errors which results in a more correct composed output. Experiments with other system combination techniques from SMT should be performed in the future.

5   Conclusions and Future Work

With this paper, we introduced the WikEd Error Corpus — a publicly available large corpus of corrective Wikipedia edits. It consists of more than 12 million sentences with a total of 14 million edits of various types.

A certain portion of noisy edits is included, but as was shown in this work it can be adapted to specific tasks when seed data is available. There is nothing that prevents other researchers from tailoring the corpus to their own purposes. Advantages of WikEd are its size and friendly license, and we believe the collected data to be more reliable than artificially created errors.

As demonstrated, despite not being an ESL corpus, WikEd can be successfully adapted to improve a strong baseline in an ESL grammatical error correction task by 2.63%. When used together with an ESL error corpus, a composed system gains 1.64% when compared to the ESL-trained system alone.

Future work should concentrate on better cleaning methods and additional meta-data. Version 1.0 should also see automatically added linguistic knowledge, for instance part-of-speech tagging. An obvious and planned extension is the expansion to other languages and the addition of new sources in the form of other wiki-sites with publicly available revision histories.

References

1. Brockett, C., Dolan, W.B., Gamon, M.: Correcting ESL Errors Using Phrasal SMT Techniques. In: Proceedings of ACL, pp. 249–256. ACL (2006)
2. Buck, C., Heafield, K., van Ooyen, B.: N-gram Counts and Language Models from the Common Crawl. In: Proceedings of LREC (2014)
3. Cahill, A., Madnani, N., Tetreault, J.R., Napolitano, D.: Robust Systems for Preposition Error Correction Using Wikipedia Revisions. In: Proceedings of NAACL: HLT, pp. 507–517. ACL (2013)
4. Dahlmeier, D., Ng, H.T.: Better Evaluation for Grammatical Error Correction. In: Proceedings of NAACL: HLT, pp. 568–572. ACL (2012)
5. Dahlmeier, D., Ng, H.T., Wu, S.M.: Building a Large Annotated Corpus of Learner English: The NUS Corpus of Learner English. In: Proceedings of the Eighth Workshop on Innovative Use of NLP for Building Educational Applications, pp. 22–31. ACL (2013)
6. Foster, J., Andersen, O.E.: GenERRate: Generating Errors for Use in Grammatical Error Detection. In: Proceedings of the Fourth Workshop on Innovative Use of NLP for Building Educational Applications, pp. 82–90. ACL (2009)
7. Grundkiewicz, R.: Automatic Extraction of Polish Language Errors from Text Edition History. In: Habernal, I., Matoušek, V. (eds.) TSD 2013. LNCS (LNAI), vol. 8082, pp. 129–136. Springer, Heidelberg (2013)
8. Junczys-Dowmunt, M., Grundkiewicz, R.: The AMU System in the CoNLL-2014 Shared Task: Grammatical Error Correction by Data-Intensive and Feature-Rich Statistical Machine Translation. In: Proceedings of CoNLL: Shared Task. ACL (2014)

9. Koehn, P., Hoang, H., Birch, A., Callison-Burch, C., Federico, M., Bertoldi, N., Cowan, B., Shen, W., Moran, C., Zens, R., Dyer, C., Bojar, O., Constantin, A., Herbst, E.: Moses: Open source toolkit for statistical machine translation. In: Proceedings of the 45th Annual Meeting of the ACL on Interactive Poster and Demonstration Sessions, ACL 2007, pp. 177–180. ACL (2007)
10. Leacock, C., Chodorow, M., Gamon, M., Tetreault, J.: Automated Grammatical Error Detection for Language Learners. Morgan and Claypool Publishers (2010)
11. Levenshtein, V.I.: Binary Codes Capable of Correcting Deletions, Insertions and Reversals. Soviet Physics Doklady 10 (1966)
12. Maier, D.: The Complexity of Some Problems on Subsequences and Supersequences. J. ACM 25(2), 322–336 (1978)
13. Max, A., Wisniewski, G.: Mining Naturally-occurring Corrections and Paraphrases from Wikipedia's Revision History. In: Proceedings of LREC (2010)
14. Miłkowski, M.: Automated Building of Error Corpora of Polish. In: Corpus Linguistics, Computer Tools, and Applications — State of the Art, pp. 631–639. Peter Lang (2008)
15. Mizumoto, T., Hayashibe, Y., Komachi, M., Nagata, M., Matsumoto, Y.: The Effect of Learner Corpus Size in Grammatical Error Correction of ESL Writings. In: Proceedings of COLING 2012: Posters, pp. 863–872 (2012)
16. Mizumoto, T., Komachi, M., Nagata, M., Matsumoto, Y.: Mining Revision Log of Language Learning SNS for Automated Japanese Error Correction of Second Language Learners. In: IJCNLP, pp. 147–155 (2011)
17. Ng, H.T., Wu, S.M., Briscoe, T., Hadiwinoto, C., Susanto, R.H., Bryant, C.: The CoNLL-2014 Shared Task on Grammatical Error Correction. In: Proceedings of CoNLL: Shared Task. ACL (2014)
18. Ng, H.T., Wu, S.M., Wu, Y., Hadiwinoto, C., Tetreault, J.: The CoNLL-2013 Shared Task on Grammatical Error Correction. In: Proceedings of CoNLL: Shared Task, pp. 1–12. ACL (2013)
19. Wagner, J., Foster, J., van Genabith, J.: A Comparative Evaluation of Deep and Shallow Approaches to the Automatic Detection of Common Grammatical Errors. In: Proceedings of EMNLP-CoNLL, pp. 112–121. ACL (2007)
20. Yuan, Z., Felice, M.: Constrained Grammatical Error Correction using Statistical Machine Translation. In: Proceedings of CoNLL: Shared Task, pp. 52–61. ACL (2013)
21. Zesch, T.: Measuring Contextual Fitness Using Error Contexts Extracted from the Wikipedia Revision History. In: Proceedings of EACL, pp. 529–538 (2012)

Author Index